Nonnegative Matrix Factorization

Data Science Book Series

Daniela Calvetti and Erkki Somersalo, *Mathematics of Data Science: A Computational Approach to Clustering and Classification*
Nicolas Gillis, *Nonnegative Matrix Factorization*

Nonnegative Matrix Factorization

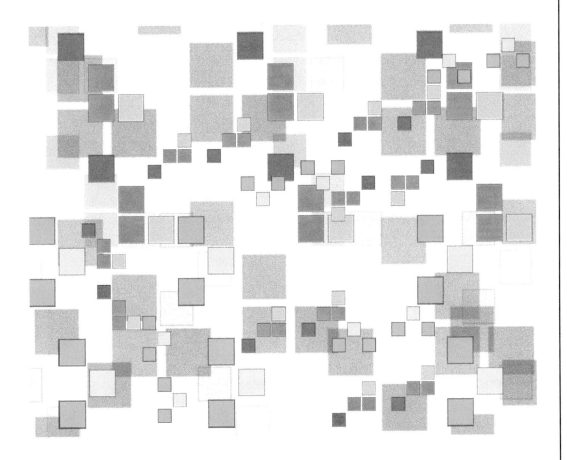

Nicolas Gillis

University of Mons
Mons, Belgium

Society for Industrial and Applied Mathematics
Philadelphia

Publications Director	Kivmars H. Bowling
Executive Editor	Elizabeth Greenspan
Developmental Editor	Mellisa Pascale
Managing Editor	Kelly Thomas
Production Editor	Lisa Briggeman
Copy Editor	Susan Fleshman
Production Manager	Donna Witzleben
Production Coordinator	Cally A. Shrader
Compositor	Cheryl Hufnagle
Graphic Designer	Doug Smock

Library of Congress Cataloging-in-Publication Data
Names: Gillis, Nicolas, author.
Title: Nonnegative matrix factorization / Nicolas Gillis, University of
   Mons, Mons, Belgium.
Description: Philadelphia : Society for Industrial and Applied Mathematics,
   [2021] | Series: Data science ; 2 | Includes bibliographical references
   and index. | Summary: "This book provides a comprehensive and up-to-date
   account of the NMF problem and its most significant features"-- Provided
   by publisher.
Identifiers: LCCN 2020042037 (print) | LCCN 2020042038 (ebook) | ISBN
   9781611976403 (paperback) | ISBN 9781611976410 (ebook)
Subjects: LCSH: Non-negative matrices. | Factorization (Mathematics) |
   Computer algorithms. | Data mining.
Classification: LCC QA188 .G566 2021 (print) | LCC QA188 (ebook) | DDC
   512.9/434--dc23
LC record available at *https://lccn.loc.gov/2020042037*
LC ebook record available at *https://lccn.loc.gov/2020042038*

Pour Aline, Elinor et Rose

Contents

Preface

Identifying the underlying structure of a data set and extracting meaningful information is a key problem in data analysis. Simple and powerful methods to achieve this goal are *linear dimensionality reduction* (LDR) techniques, which are equivalent to low-rank matrix approximations (LRMA). Examples of LDR techniques are principal component analysis (PCA), independent component analysis, sparse PCA, robust PCA, low-rank matrix completion, and sparse component analysis. The reason for the success of this type of methods is that, although simple, they are applicable in a wide range of applications such as recommender systems, model-order reduction and system identification, clustering, image analysis, and blind source separation, to cite a few.

Among LRMA techniques, nonnegative matrix factorization (NMF) requires the factors of the low-rank approximation to be componentwise nonnegative. This makes it possible to interpret them meaningfully, for example when they correspond to nonnegative physical quantities. Applications of NMF include extracting parts of faces (such as eyes, noses, and lips) in a set of facial images, identifying topics in a set of documents, learning hidden Markov models, extracting materials and their abundances in hyperspectral images, separating audio sources from their mixture, detecting communities in large networks, analyzing medical images, and decomposing gene expression microarrays.

Aim of the book   The aim of this book is to provide a comprehensive account of the most important aspects of the NMF problem:

- Theoretical aspects: the nonnegative rank, the nonuniqueness/identifiability of NMF solutions, the geometric interpretation of NMF, and computational complexity issues.

- Models: choice of the objective function and regularizations, link with well-known techniques such as k-means, and use of additional constraints such as orthogonality or symmetry.

- Algorithms: heuristic algorithms using standard nonlinear optimization schemes such as block coordinate descent methods, and provably correct algorithms under appropriate assumptions.

- Applications: they include image analysis, document classification, hyperspectral unmixing, audio source separation, topic modeling, and community detection.

This book is accessible to a wide audience. In particular it is intended for people who want to know about the workings of NMF. It also aims to give more insights to practitioners so that they can use NMF meaningfully. To read this book, basic knowledge of linear algebra and optimization is needed.

Why is this book important?   Although NMF has been studied extensively for the last 20 years, there is currently only one book on the topic, by Cichocki et al. [98] (2009) which is already more than 10 years old. It focuses on iterative algorithms and applications, and many aspects of NMF are not covered in that book—especially since many important results have been obtained in the last 10 years.[1]

The aim of this book is to fill in this gap by providing more insights into the theoretical aspects of NMF. These are key to be able to use NMF effectively and meaningfully in practice. This will allow the reader to make better use of NMF as a computational tool. This book is aimed at researchers who want to understand the NMF problem; for example,

- You do not know (much) NMF and want to discover this problem, why and how it works, and what it can be used for. This book would be ideal for example for a master's or Ph.D. student starting to work on NMF.

- You are using NMF for applications but you would like to understand better its subtly difficult aspects such as its computational complexity, its geometric interpretation, or its nonuniqueness/identifiability issues. Also, you would like to know about the state-of-the-art algorithms.

- You are already rather familiar with NMF but have not yet studied all of its aspects (for example you would like to know more about the nonnegative rank, or the nonuniqueness of NMF solutions). This book will allow you to delve into different aspects of the NMF problem and will give you useful references.

Moreover, this book contains a few new results not present in the literature (as far as I know): bounds on the nonnegative rank under rank-one perturbations (Theorem 3.3), the study of the generic value of the nonnegative rank (Section 3.3.2), the identifiability of orthogonal NMF (Section 4.3.2), and a necessary condition for the sufficiently scattered condition, a crucial notion when studying the uniqueness of NMF solutions (see Theorem 4.28).

MATLAB code   All algorithms and numerical experiments presented in this book are available from bookstore.siam.org/di02/bonus. When we discuss an algorithm, or display results from a numerical experiment, the corresponding MATLAB file will be indicated using

        [Matlab file: Name of file]

It can be found in the folder of the corresponding chapter. Hence the interested reader can easily find the corresponding MATLAB file. To provide a better view of all the NMF algorithms available with this book, there is an *exception* for NMF algorithms: they can all be found in the folder [algorithms]. For example, the separable NMF algorithms presented in Chapter 7 can be found in the folder [algorithms/separable NMF], although the numerical experiments presented in Chapter 7 can be found in the folder [Chapter 7 - Separable NMF].

All tests in this book are performed using MATLAB R2019b on a laptop Intel Core i7-7500U CPU @2.9GHz, 24GB RAM.

How to use this book   The book is organized so that it is possible to read only subsets of the chapters depending on the reader's interests. The book was written so that each chapter is

---

[1]Such as the identifiability results based on the sufficiently scattered condition (Chapter 4), and the polynomial-time algorithms for separable NMF (Chapter 7).

as self-contained as possible. Each chapter can be thought of as a survey on the corresponding topic. Moreover, each chapter ends with take-home messages that summarize and highlight the important results covered in the chapter. The book is organized as follows.

Chapter 1 serves as an introduction and the problem definition. It contains the description of four key applications of NMF that will be used throughout the book as illustrations. It also contain a historical account of the problem explaining when, why, and how the NMF problem came about. Then the book is divided into two parts.

Part I. Exact factorizations The first part considers the Exact NMF problem: given the nonnegative input matrix X and a factorization rank r, find nonnegative factors W with r columns and H with r rows such that $X = WH$. Chapter 2 discusses theoretical aspects of Exact NMF and in particular its geometric interpretation; this is crucial to design algorithms but also to use NMF meaningfully in practice. Section 2.1 on the geometric interpretation of NMF is useful to understand Chapters 4 and 7. The second part of this chapter discusses a more constrained NMF problem where the first factor W is required to have the same rank as the input matrix, which is referred to as restricted Exact NMF, and its link with a geometric problem, namely, the nested polytope problem. The third part of the chapter discusses the computational complexity of these problems. Chapter 3 digs into theoretical aspects of the nonnegative rank (the smallest r such that an Exact NMF exists): its properties, its lower and upper bounds, and its link with extended formulations of polytopes and with communication complexity. Chapter 4 discusses the identifiability issues when using NMF in practice. In fact, NMF decompositions are in general not unique, while most applications are looking for the unique ground truth underlying factors. This chapter explains how to recover the true factors, and under which conditions it is possible. Section 4.2 focuses on uniqueness conditions for the plain NMF model, while Section 4.3 discusses regularized NMF models, namely, orthogonal, separable, minimum-volume, and sparse NMF, that lead to unique decompositions under milder conditions.

Part II. Approximate factorizations The second part of the book considers approximate NMF decompositions, where $WH \approx X$, which we refer to as NMF for short as it is the standard in the literature. Chapter 5 discusses several important variants of the NMF model that use additional constraints, regularizations, and different objective functions; for example symmetric NMF which requires $W = H^\top$. As discussed in Chapter 4, considering such variants in practice is key to obtain unique solutions. This chapter also discusses different models that are closely related to NMF such as k-means and probabilistic latent semantic analysis/indexing (PLSA/PLSI). Chapter 6 discusses the computational complexity of NMF, which is NP-hard[2] in general. Chapter 7 considers NMF under the separability assumption, referred to as separable NMF, where the columns of the first factor W can be found among the columns of X. Although it is a strong assumption, it makes sense in several applications. Moreover, it allows us to provably solve NMF efficiently, that is, in polynomial time, and renders the solution unique, hence resolving two key issues of NMF (NP-hardness and identifiability). This chapter presents the main algorithms for separable NMF, discusses their robustness to noise, and compares them on several synthetic data sets. Chapter 8 focuses on iterative heuristic algorithms to compute NMF solutions. The state-of-the-art algorithms for NMF are presented; they are based on standard optimization techniques. We also discuss convergence guarantees and provide some numerical comparisons. Chapter 9 presents three more applications of NMF, as well as pointers to more applications.

---

[2]NP-hardness of a problem implies that unless P=NP, there exists no algorithm running in a number of operations polynomial in the size of the input that solves the problem.

Disclaimer The book presents the NMF problem from my own perspective and is clearly biased toward my own work and research interests. I apologize for not discussing or referring to many relevant works (that I am either unfamiliar with or unaware of). This book is a summary of my current knowledge about NMF and is by no means comprehensive. Any feedback on the book is more than welcome and is highly encouraged.

Tensors The NMF model can be directly generalized to tensors, in which case it is referred to as nonnegative tensor factorization (NTF) or nonnegative canonical polyadic decomposition (nonnegative CPD). It is out of the scope of this book to discuss this important extension, and we stick to the matrix case. However, one should keep this connection in mind as results from the matrix case can be used in the tensor case. For example, NP-hardness results for NMF (see Chapter 6) directly apply to NTF since NTF is a generalization of NMF, and NMF algorithms based on the block coordinate descent framework described in Chapter 8 directly extend to NTF. We refer the interested reader to [283, 98, 418] and the references therein for more details on tensor factorizations.

Acknowledgments Writing this book would not have been possible without the many fruitful collaborations I have been lucky to have over the years. In particular, I am grateful to my two mentors, François Glineur, who introduced me to NMF (the topic of my master's thesis in 2007), and Steve Vavasis, who welcomed me as a postdoc at the University of Waterloo. I have also been lucky that many enthusiastic researchers joined me in Mons: Arnaud, Punit, Andersen, Jérémy, Valentin, Hien, Junjun, François, Nicolas[2], Tim, Christophe, Pierre, and Maryam.

I am thankful to the persons who gave me feedback on early drafts of this book: Tim Marrinan, Arnaud Vandaele, Ken Ma, Jérémy Cohen, Christophe Kervazo, Pierre De Hanschutter, Le Hien, Andersen Ang, Nicolas Nadisic, Hamza Fawzi, Yaroslav Shitov, Anthony Degleris, Marc Pirlot, Junjun Pan, Fabian Lecron, and the anonymous reviewers. I also thank Kaie Kubjas, Xiao Fu, and Kejun Huang for insightful discussions.

Writing a book on NMF has been in the back of my mind for several years, but I thank Andersen Ang and Jennifer Pestana, who motivated me and pushed me to finally undertake this endeavor.

I would like to thank the SIAM editorial team, who have done a wonderful job improving the book through their reviewing and copyediting. It was a pleasure collaborating with them on this project.

I am grateful for the support from the European Research Council (ERC starting grant 679515, COLORAMAP project), the Fonds de la Recherche Scientifique-FNRS through several grants (Incentive Grant for Scientific Research F.4501.16, and Excellence of Science grant O005318F-RG47 which is also supported by the Fonds Wetenschappelijk Onderzoek-Vlaanderen), and the Francqui Foundation (Francqui research professor).

I thank my family and my friends.

Notation

Sets of scalars, vectors, matrices

$\mathbb{R}$	set of real numbers
$\mathbb{R}_+$	set of nonnegative real numbers
$\mathbb{R}_{++}$	set of positive real numbers
$\mathbb{R}^n$	set of real column vectors of dimension n
$\mathbb{R}^{m \times n}$	set of real m-by-n matrices
$\mathbb{R}^n_+$	set of nonnegative real column vectors of dimension n
$\mathbb{R}^{m \times n}_+$	set of m-by-n nonnegative real matrices
$\mathbb{S}^n$	set of n-by-n symmetric matrices
$\mathbb{S}^n_+$	set of n-by-n positive semidefinite matrices
$\mathbb{S}^n_{++}$	set of n-by-n positive definite matrices
$\mathfrak{C}^n$	set of n-by-n copositive matrices
$\mathfrak{C}^n_+$	set of n-by-n completely positive matrices
$\mathcal{C}$	second-order cone $\left\{ x \in \mathbb{R}^r \mid e^\top x \geq \sqrt{r-1}\|x\|_2 \right\}$
$\mathcal{C}^*$	second-order cone $\left\{ x \in \mathbb{R}^r \mid e^\top x \geq \|x\|_2 \right\}$, the dual of $\mathcal{C}$ (p. 113)
Δ^n	unit simplex of dimension n, that is, $\Delta^n = \{x \in \mathbb{R}^n \mid x \geq 0, \sum_{i=1}^n x_i = 1\}$
$\mathcal{S}^n$	convex hull of the unit simplex and the origin, that is, $\mathcal{S}^n = \{x \in \mathbb{R}^n \mid x \geq 0, \sum_{i=1}^n x_i \leq 1\}$

Submatrices, transpose and inverse

x_i of $x(i)$	ith entry of the vector x
$A_{i:}$ or $A(i,:)$	ith row of A
$A_{:j}$ or $A(:,j)$	jth column of A
A_{ij} or $A(i,j)$	entry at position (i,j) of A
$A(I,J)$	submatrix of A with row (resp. column) indices in I (resp. J)
$[A\ B; C\ D]$	We use Matlab notation: $[A\ B; C\ D] = \begin{pmatrix} A & B \\ C & D \end{pmatrix}$
$A^\top$	transpose of the matrix A, $(A^\top)_{ij} = A_{ji}$
A^{-1}	inverse of the square matrix A, $A^{-1}A = AA^{-1} = I$
$A^{-\top}$	inverse of the transpose of the square matrix A, that is, $A^{-\top}A^\top = A^\top A^{-\top} = I$

Special vectors and matrices

0	matrix of zeros of appropriate dimension
$0_{m \times n}$	m-by-n matrix of zeros
I_n	identity matrix of dimension n
I	identity matrix of appropriate dimension
e	vector of all ones of appropriate dimension
e_k	kth unit vector with $e_k(k) = 1$ and $e_k(i) = 0$ for all $i \neq k$, that is, $e_k = I(:, k)$

Norms

$\|.\|_1$	ℓ_1 norm, $\|x\|_1 = \sum_{i=1}^{n} \|x_i\|$, $x \in \mathbb{R}^n$
	componentwise matrix ℓ_1 norm, $\|A\|_1 = \sum_{i,j} \|A_{i,j}\|$, $A \in \mathbb{R}^{m \times n}$
$\|.\|_2$	vector ℓ_2 norm, $\|x\|_2 = \sqrt{\sum_{i=1}^{n} x_i^2}$, $x \in \mathbb{R}^n$
	matrix ℓ_2 norm, $\|A\|_2 = \max_{x \in \mathbb{R}^n, \|x\|_2 = 1} \|Ax\|_2$, $A \in \mathbb{R}^{m \times n}$
$\|.\|_F$	Frobenius norm, $\|A\|_F = \sqrt{\sum_{i=1}^{m} \sum_{j=1}^{n} A_{ij}^2}$, $A \in \mathbb{R}^{m \times n}$
$\|.\|_\infty$	vector ℓ_∞ norm, $\|x\|_\infty = \max_{1 \leq i \leq n} \|x_i\|$, $x \in \mathbb{R}^n$
	componentwise matrix ℓ_∞ norm, $\|A\|_\infty = \max_{i,j} \|A_{i,j}\|$, $A \in \mathbb{R}^{m \times n}$
$\|.\|_0$	ℓ_0 "norm," $\|x\|_0 = \left\|\{i \| x_i \neq 0\}\right\|$, $x \in \mathbb{R}^m$
$\|.\|_{1,q}$	matrix $\ell_{1,q}$ norm, $\|A\|_{1,q} = \sum_{i=1}^{m} \|A(i,:)\|_q$, $A \in \mathbb{R}^{m \times n}$

Inequalities

$A \geq 0$	A is a nonnegative matrix, that is, $A(i,j) \geq 0$ for all i, j
$A \geq B$	This means $A - B \geq 0$
$A \succeq 0$	A is a PSD matrix
$A \succeq B$	$A - B$ is a PSD matrix

Functions and sets on matrices

$\langle ., . \rangle$	Euclidean scalar product, that is, for $A, B \in \mathbb{R}^{m \times n}$, $\langle A, B \rangle = \sum_{i=1}^{m} \sum_{j=1}^{n} A_{ij} B_{ij}$,
$\sigma_i(A)$	ith singular values of matrix A, in nondecreasing order
$\sigma_{\max}(A)$	largest singular value of A, that is, $\sigma_1(A)$
$\sigma_{\min}(A)$	smallest singular value of $A \in \mathbb{R}^{m \times n}$, that is, $\sigma_{\min(m,n)}(A)$
$\kappa(A)$	condition number of A, $\kappa(A) = \frac{\sigma_{\max}(A)}{\sigma_{\min}(A)}$
$\det(A)$	determinant of A
$\operatorname{tr}(A)$	trace of A, that is, sum of its diagonal entries
$\operatorname{diag}(.)$	For $a \in \mathbb{R}^n$, $A = \operatorname{diag}(a) \in \mathbb{R}^{n \times n}$ is a diagonal matrix such that $A_{ii} = a_i$ for all i
	For $A \in \mathbb{R}^{n \times n}$, $a = \operatorname{diag}(A) \in \mathbb{R}^n$ is the vector containing the diagonal entries of A
$\overline{A}$	$\overline{A}$ is the matrix A without the last row
$\operatorname{rank}(A)$	rank of A
$\operatorname{rank}_+(A)$	nonnegative rank of A (p. 55)
$\operatorname{rank}_+^*(A)$	restricted nonnegative rank of A (p. 35)
$\operatorname{cp-rank}(A)$	completely positive rank of A (p. 79)
$\operatorname{k-rank}(A)$	Kruskal rank of A (p. 151)
$\operatorname{cone}(A)$	conical hull of the columns of A (p. 20)
$\operatorname{conv}(A)$	convex hull of the columns of A (p. 21)
$\operatorname{col}(A)$	column space of A (p. 21)
$\operatorname{aff}(A)$	affine hull of the columns of A (p. 21)
$A \circ B$	componentwise multiplication between A and B, that is, $(A \circ B)_{ij} = A_{ij} B_{ij}$
$\frac{[A]}{[B]}$	componentwise division between A and B, $\left(\frac{[A]}{[B]} \right)_{ij} = \frac{A_{ij}}{B_{ij}}$
$A^{\circ a}$	componentwise exponent of matrix A by the scalar a, that is, $A^{\circ a}(i,j) = A(i,j)^a$ for all i, j
$\operatorname{supp}(A)$	support (index set of nonzero entries) of matrix A, that is, $\operatorname{supp}(A) = \{(i,j) \mid A(i,j) \neq 0\}$
$\omega(A)$	fooling set bound for A (p. 64)
$\operatorname{bin}(A)$	$\operatorname{bin}(A)$ is the binarization of A, that is, the nonzero entries of A are set to 1 (p. 70)
$\operatorname{rank}_{01}(A)$	this is the Boolean rank of $\operatorname{bin}(A)$ (p. 70)
$\operatorname{rc}(A)$	rectangle covering bound for A (p. 72)
$\operatorname{rrc}(A)$	refined rectangle covering bound for A (p. 73)

Miscellaneous

$a\!:\!b$	set $\{a, a+1, \ldots, b-1, b\}$ (for a and b integers with $a \leq b$)
$[a, b]$	closed interval for reals $a \leq b$
(a, b)	open interval for reals $a \leq b$
∇f	gradient of the function f
$\nabla^2 f$	Hessian of the function f
$\lceil \cdot \rceil$	$\lceil x \rceil$ is the smallest integer greater or equal to $x \in \mathbb{R}$
$\lfloor \cdot \rfloor$	$\lfloor x \rfloor$ is the largest integer smaller or equal to $x \in \mathbb{R}$
$\backslash$	subtraction of two sets, that is,
	$R \backslash S$ is the set of elements that are in R but not in S
$|\cdot|$	cardinality of a set, $|S|$ is the number of elements in S
k-sparse	the vector x is k-sparse if it has k nonzero entries, that is,
	$|\operatorname{supp}(x)| = k$
$\mathbb{P}(x)$	probability of the event $X = x$
$\mathbb{E}(X)$	expected value of a random variable X
$\mathcal{N}(\mu, \sigma)$	normal distribution of mean μ and standard deviation σ
$\mathcal{U}(a, b)$	uniform distribution in the interval $[a, b]$
$f(x) = \mathcal{O}(g(x))$	Big $\mathcal{O}$ notation: there exists K and x_0 such that $f(x) \leq Kg(x)$ for all
	$x \geq x_0$
$f(x) = o(g(x))$	small o notation: $\lim_{x \to +\infty} \frac{f(x)}{g(x)} = 0$
$f(x) = \Omega(g(x))$	Big Omega notation, equivalent to $g(x) = \mathcal{O}(f(x))$
$f(x) = \Theta(g(x))$	Big Theta notation, equivalent to $f(x) = \mathcal{O}(g(x))$ and $f(x) = \Omega(g(x))$
$\min_{x \in \mathcal{X}} f(x)$	minimum value of $f(x)$ over the feasible set $\mathcal{X}$
$\operatorname{argmin}_{x \in \mathcal{X}} f(x)$	set of minimizers of $f(x)$ over the feasible set $\mathcal{X}$

Abbreviations

w.l.o.g.	without loss of generality
i.i.d.	independently and identically distributed
w.r.t.	with respect to

Acronyms

Page number indicates where the acronym is first defined.

2-BCD	two-block coordinate descent (p. 261)
ADMM	alternating direction method of multipliers (p. 289)
ANLS	alternating nonnegative least squares (p. 281)
BCD	block coordinate descent (p. 266)
BSUM	block successive upper-bound minimization (p. 268)
EDM	Euclidean distance matrix (p. 64)
Exact NMF	exact nonnegative matrix factorization (p. 19)
FAW	fast anchor words (p. 230)
FPGM	fast projected gradient method (p. 288)
HALS	hierarchical alternating least squares (p. 283)

Acronyms (continued)

HU	hyperspectral unmixing (p. 7)
IS	Itakura–Saito (p. 162)
KKT	Karush–Kuhn–Tucker (p. 264)
KL	Kullback–Leibler (p. 161)
k-sparse MF	k-sparse matrix factorization (p. 151)
LDR	linear dimensionality reduction (p. 1)
LP	linear programming (p. 249)
LRMA	low-rank matrix approximation (p. 1)
min-vol	minimum-volume (p. 138)
MLP	multiple linear programs (p. 249)
MM	majorization-minimization (p. 265)
MVE	minimum-volume ellipsoid (p. 237)
MU	multiplicative updates (p. 270)
NMF	nonnegative matrix factorization (p. 4)
NMU	nonnegative matrix underapproximation (p. 176)
NNLS	nonnegative least squares (p. 280)
NPP	nested polytope problem (p. 36)
ONMF	orthogonal NMF (p. 136)
PCA	principal component analysis (p. 2)
PGM	projected gradient method (p. 287)
PLSA	probabilistic latent semantic analysis (p. 189)
PLSI	probabilistic latent semantic indexing (p. 189)
PMF	positive matrix factorization (p. 4)
PPI	pure-pixel index (p. 247)
PSD	positive semidefinite (p. 79)
RE-NMF	restricted Exact NMF (p. 35)
SD-LP	self-dictionary via linear programming (p. 254)
SMCR	self-modeling curve resolution (p. 12)
SNPA	successive nonnegative projection algorithm (p. 232)
SPA	successive projection algorithm (p. 223)
SSC	sufficiently scattered condition (p. 104)
SSM	stochastic sequential machine (p. 13)
SVD	singular value decomposition (pp. 2, 196)
symNMF	symmetric NMF (p. 178)
tri-NMF	nonnegative matrix trifactorization (p. 181)
tri-symNMF	symmetric nonnegative matrix trifactorization (p. 181)
TSP	traveling salesman problem (p. 15)
VCA	vertex component analysis (p. 232)
WLRA	weighted low-rank matrix approximation (p. 3)

List of Figures

List of Tables

Chapter 1

Introduction

In this chapter, we elaborate on the main reason why nonnegative matrix factorization (NMF) became a popular and standard tool in data analysis, namely because NMF is an easily interpretable linear dimensionality reduction technique for nonnegative data (Section 1.1). After providing a formal definition of NMF, the notation, and the terminology (Section 1.2), we illustrate the ability of NMF to extract meaningful components in nonnegative data sets with four applications: feature extraction in images, hyperspectral unmixing, text mining, and audio source separation[3] (Section 1.3). The last part of the chapter provides a historical overview of how NMF came about (Section 1.4). In each chapter of this book, we conclude with some take-home messages (Section 1.5).

1.1 ▪ Linear dimensionality reduction techniques for data analysis

Most works on NMF are motivated by its applicability in data analysis, more precisely, by the capability of NMF to automatically extract meaningful information in a data set. Extracting the underlying structure within data sets is one of the central problems in data science, and numerous techniques exist to perform this task. One of the oldest approaches is linear dimensionality reduction (LDR). LDR represents each data point as a linear combination of a small number of basis elements. Mathematically, given a data set of n data points $x_j \in \mathbb{R}^m$ ($1 \leq j \leq n$), LDR looks for a small number r of basis vectors $w_k \in \mathbb{R}^m$ ($1 \leq k \leq r$) such that each data point is well-approximated by a linear combination of these basis vectors, that is,

$$x_j \approx \sum_{k=1}^{r} w_k h_{kj} \ \text{ for all } j,$$

where the h_{kj}'s are scalars; see Figure 1.1 for an illustration in three dimensions with $m = 3$, $r = 2$, and $n = 50$.

LDR is equivalent to low-rank matrix approximation (LRMA):

$$\underbrace{[x_1, x_2, \ldots, x_n]}_{X \in \mathbb{R}^{m \times n}} \approx \underbrace{[w_1, w_2, \ldots, w_r]}_{W \in \mathbb{R}^{m \times r}} \underbrace{[h_1, h_2, \ldots, h_n]}_{H \in \mathbb{R}^{r \times n}},$$

---

[3]Sections 1.1 and 1.3 follow closely the introductions from [189, 190, 191].

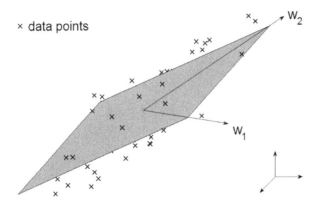

Figure 1.1. *Illustration of LDR: approximation of three-dimensional data points with a two-dimensional subspace generated by w_1 and w_2.*

where

- each column of the matrix $X \in \mathbb{R}^{m \times n}$ is a data point, that is, $X(:, j) = x_j$ for $1 \le j \le n$;

- each column of the matrix $W \in \mathbb{R}^{m \times r}$ is a basis element, that is, $W(:, k) = w_k$ for $1 \le k \le r$; and

- each column of $H \in \mathbb{R}^{r \times n}$ contains the coordinates of a data point $X(:, j)$ in the basis W, that is, $H(:, j) = h_j$ for $1 \le j \le n$.

Hence LDR provides a rank-r approximation WH of X, and each data point is mapped into the basis W using the corresponding column of H:

$$x_j \approx W h_j \quad \text{for all } j.$$

Typically, the number of basis vectors is much smaller than the ambient dimension m and the number of data points n, that is, $r \ll \min(m, n)$. This allows LDR and LRMA to compress the data, which is achieved for $r < \frac{mn}{m+n}$ since X contains mn entries, while W and H contain only $mr + nr$ entries.[4] Note that if the input matrix is sparse, with $\operatorname{nnz}(X)$ nonzero entries, compression requires $r < \frac{\operatorname{nnz}(X)}{m+n}$.

In order to compute W and H given X and r, one needs to define an error measure. For example, when the solution (W, H) minimizes the sum of the squares of the entries of the residual $X - WH$, that is, the squared Frobenius norm of the residual $\|X - WH\|_F^2 = \sum_{i,j} (X - WH)_{ij}^2$, LRMA is equivalent to principal component analysis (PCA) [266] which can be computed via the singular value decomposition (SVD) [216]. LRMA is a workhorse in numerical linear algebra, with the SVD being a central technique, and is closely related to the eigenvalue decomposition and factorizations such as Cholesky, QR, and LU, to cite a few. LRMA is at the heart of many fields in applied mathematics and computer science, for example in

- statistics, data analysis, and machine learning to perform regression, prediction, clustering, classification, and noise filtering [266, 144];

- numerical linear algebra to solve linear systems of equations [216];

---

[4]The factorization WH only has $(m + n - 1)r$ degrees of freedom due to the scaling degree of freedom of the columns of W and rows of H, that is, $W(:, k)H(k, :) = (\alpha W(:, k))(H(k, :)/\alpha)$ for any $\alpha \ne 0$ and any k; see Chapter 4.

- signal and image processing to perform blind source separation [104, 87];

- graph theory to cluster the vertices of a graph [94];

- optimization to gain computational efficiency [68]; and

- systems theory and control to perform model-order reduction and system identification [341].

Although PCA has been around for more than 100 years, LRMA models have gained momentum in the last 20 years. The reason is mostly twofold: (i) data analysis has become more and more important due to the recent deluge of data in many different areas (the big data era), and (ii) despite a simple model, LRMA is very powerful since many high-dimensional data sets are well-approximated by low-rank matrices [458]. LRMA models are used to compress the data, filter the noise, reduce the computational effort for further manipulation of the data, or to directly identify hidden structure in a data set; see for example the survey [457]. Many variants of LRMA have been developed over the last few years [422]. They differ in two key aspects:

1. The error measure can vary and should be chosen depending on the noise statistic assumed on the data. For example, PCA uses least squares, that is, it minimizes $\|X - WH\|_F^2$, which implicitly assumes independently and identically distributed (i.i.d.) Gaussian noise.

   If data is missing or if weights are assigned to the entries of X (for example because the noise is not identically distributed over the entries of X), the problem can be cast as a weighted low-rank matrix approximation (WLRA) problem that minimizes $\sum_{i,j} P_{i,j}(X - WH)_{i,j}^2$ for some nonnegative weight matrix P, where $P_{i,j} = 0$ when the entry at position (i, j) is missing [434]. Note that if P contains entries only in $\{0, 1\}$, then the problem is also referred to as PCA with missing data or low-rank matrix completion with noise. WLRA is widely used in recommender systems [31, 287] for predicting the preferences of users for a given product based on the product's attributes and user preferences, such as in the Netflix prize competition; see [34] and Section 9.5.

   If the sum of the absolute values of the entries of the error $\sum_{i,j} |X - WH|_{i,j}$ is used as the error measure, we obtain yet another variant more tolerant to outliers; this is closely related to robust PCA [73, 212]. It can be used, for example, for background subtraction in video sequences where the noise (the moving objects) is assumed to be sparse while the background has low rank (the background does not change much between consecutive images in video sequences).

2. Different constraints can be imposed on the factors W and H. These constraints depend on the application at hand and allow, for example, for meaningful interpretation of the factors.

   For example, the k-means problem, which is the problem of finding a set of r centroids w_k ($1 \leq k \leq r$) such that the sum of the distances between each data point and the closest centroid is minimized, is equivalent to LRMA where the factor H is required to have a single nonzero entry in each column that is equal to one, so that the columns of W are the cluster centroids.

   If instead one wants to explain each data point using as few basis vectors as possible, each column of the matrix H should contain as many zero entries as possible. This LRMA variant is referred to as sparse component analysis [221] and is closely related to dictionary learning [4, 455] and sparse PCA [115]. It yields a more compact and easily interpretable decomposition.

NMF, an LDR technique for nonnegative data   Among LRMA models, nonnegative matrix factorization (NMF) requires the factor matrices W and H to be componentwise nonnegative, which we denote $W \geq 0$ and $H \geq 0$. These nonnegativity constraints play an instrumental role in various applications as they allow one to extract meaningful and interpretable components in nonnegative data sets. For example, in some applications, the entries in W and H can be interpreted as physical quantities. Before presenting four such applications in Section 1.3, let us first define the NMF problem rigorously.

1.2 ▪ Problem definition

Let us formally define the NMF problem and discuss some of its aspects.

Problem 1.1 (Nonnegative matrix factorization).   *Given a nonnegative matrix $X \in \mathbb{R}_+^{m \times n}$, a factorization rank r, and a distance measure $D(.,.)$ between two matrices, compute two nonnegative matrices $W \in \mathbb{R}_+^{m \times r}$ and $H \in \mathbb{R}_+^{r \times n}$ such that $D(X, WH)$ is minimized, that is, solve*

$$\min_{W \in \mathbb{R}_+^{m \times r}, H \in \mathbb{R}_+^{r \times n}} D(X, WH). \tag{1.1}$$

Terminology   NMF in its modern form was first referred to as positive matrix factorization (PMF) by Paatero and Tapper [371] in 1994, but this name has not been used much, most likely because "positive" means strictly larger than zero, while NMF usually generates a factor with many zeros entries; see the discussion below in the paragraph Sparsity. The name NMF became standard after the paper of Lee and Seung published in 1999 [303]; see Section 1.4 for the historical overview of NMF. In most data analysis applications (see Section 1.3 and Chapter 9), the solution (W, H) is only an approximation of the data matrix X; hence $X \neq WH$ as with most applications using LRMA mentioned in Section 1.1. This is due to the presence of noise and the linear model being, in most cases, only an approximate model ("All models are wrong but some are useful" as mentioned by George Box[5] in 1976). Therefore, the use of the term "factorization" might be misleading since it usually refers to an exact decomposition $X = WH$. Hence some authors have argued that it would, for example, make more sense to refer to (1.1) as nonnegative matrix approximation [433]. However, in this book we subscribe to the widely accepted standard that the name NMF refers to the associated approximation problem, and we will further specify when we are considering the exact factorization, $WH = X$, by calling it Exact NMF. Exact NMF is important in linear algebra as it allows us to compute the nonnegative rank of a matrix X, denoted $\mathrm{rank}_+(X)$, which is the smallest r such that an Exact NMF of X exists; see Chapters 2 and 3.

Objective function   The objective function of the NMF problem is defined as

$$D \colon \mathbb{R}_+^{m \times n} \times \mathbb{R}_+^{m \times n} \mapsto \mathbb{R}_+ \ \text{ given by } \ (A, B) \mapsto D(A, B)$$

and is also referred to as the error measure. The choice of this function is crucial when designing LRMA models and often depends on the assumptions made about the noise statistics. It may greatly influence the solution (W, H) and leads to rather different optimization problems; see

---

[5]https://en.wikipedia.org/wiki/All_models_are_wrong (consulted May 27, 2020).

Chapter 8. Most error measures give the same importance to each entry of the data matrix X and hence have the form

$$D(A, B) = \sum_{i,j} d(A_{ij}, B_{ij})$$

for some function $d \colon \mathbb{R}_+ \times \mathbb{R}_+ \mapsto \mathbb{R}_+$ such that $d(x, y) = 0$ if and only if $x = y$. A standard choice is $d(x, y) = (x - y)^2$ which leads to $D(X, WH) = \|X - WH\|_F^2$. There are two main reasons for this choice: (1) it corresponds to the assumption of i.i.d. Gaussian noise which is reasonable for many data sets, and (2) it leads to a smooth optimization, which is easier to handle (see Chapter 8 for a discussion). We refer the reader to Section 5.1 for a discussion on the choice of D in the context of NMF.

Choice of the symbols In the linear algebra community, authors consistently use the symbols $A = U\Sigma V^\top$ to represent the SVD of matrix A. However, in the NMF literature, there is no consensus on the choice of the symbols used for the data matrix X and the factor matrices (W, H), and many combinations of symbols exist; examples include $V \approx WH$ [303], $X \approx CS^\top$ [263], $A \approx BC$ [433], $X \approx UV$ [478], $Y \approx AX$ [98], $X \approx WH$ [189], or $X \approx WH^\top$ [170]. In this book we choose the notation $X \approx WH$.

Transpose: WH vs. $WH^\top$ In the numerical linear algebra community, most authors would likely prefer the use of $X \approx WH^\top$, similarly as for the SVD that uses $U\Sigma V^\top$ (see Section 6.1.1), as it preserves the symmetry by transposition, that is, $X \approx WH^\top$ if and only if $X^\top \approx HW^\top$. However, we choose $X \approx WH$ for the following reason. When interpreting NMF as an LDR technique, which is the main motivation behind NMF, the columns of H play the role of the coefficients of the columns of X in the subspace spanned by the columns of W since $X(:, j) \approx WH(:, j)$ for all j. In other words, there is a one-to-one correspondence between the high-dimensional columns of X and their low-dimensional representations as the columns of H. We believe this is the reason why it is the most common choice in the NMF literature.

NMF variants The problem (1.1) is the formulation of the standard NMF problem. However, it is important to keep in mind that there exist many variants of this problem. Moreover, as we will argue in Chapter 4, it is in general crucial to consider such variants in practice to obtain unique decompositions and be able to identify the true underlying factors that generated the data. Some variants use regularization in order to obtain solutions with some structure, such as sparsity; see Sections 4.3.4 and 5.3. Other variants use additional constraints; for example symmetric NMF (symNMF) requires $H = W^\top$, and orthogonal NMF (ONMF) requires H to have orthogonal rows (see Section 5.4).

Sparsity Because of the nonnegativity constraints, NMF solutions (W, H) are expected to contain zero entries and hence to naturally have some degree of sparsity; see for example Figures 1.2, 1.4, 1.6, and 1.8 in the next section. Mathematically, this is explained by the first-order optimality conditions of a smooth optimization problem with nonnegativity constraints

$$\min_{x \in \mathbb{R}^n} f(x) \quad \text{such that} \quad x \geq 0,$$

which are given by

$$x \geq 0, \quad \nabla f(x) \geq 0, \quad \text{and} \quad x_i (\nabla f(x))_i = 0 \text{ for all } i.$$

This enforces $x_i = 0$ whenever $(\nabla f(x))_i > 0$; see Section 8.1.2 for more details. In some applications, one might need to obtain even sparser solutions, which requires the use of additional constraints or regularization; see Chapters 4 and 5.

Applications   In the next section, we describe four important applications of NMF in data analysis (see Chapter 9 for other NMF applications). However, it is important to stress that *NMF is not motivated only by its use as an LDR technique for data analysis*; see in particular the historical overview in Section 1.4. Another important motivation is exact factorizations. In particular, the nonnegative rank of a nonnegative matrix X, denoted $\text{rank}_+(X)$, is the smallest r such that there exists an Exact NMF $X = WH$ where W has r columns and H has r rows; see Chapters 2 and 3 for more details. An application where the nonnegative rank has had tremendous impact is in the study of the extension complexity of polytopes; see Section 3.6.

1.3 ▪ Four applications of NMF in data analysis

In this section, we describe four important applications of NMF that will be used throughout the book as illustrative examples. They show that NMF is a particularly meaningful LRMA model as it leads to interpretable decompositions. In Chapter 9, we review other applications of NMF.

1.3.1 ▪ Feature extraction in a set of images

Given a set of gray-scale images of the same dimensions, let us construct the matrix X such that each column of X corresponds to a vectorized gray-level image. Vectorization means that the two-dimensional images are transformed into a long one-dimensional vector, for example, by stacking the columns of the image on top of each other.[6] This means that each row of X corresponds to the same pixel location among the images. The entry of X at position (i, j), that is, $X(i, j)$, is equal to the intensity of the ith pixel within the jth image, which is nonnegative. As explained in Section 1.1, factorizing X with NMF as $X \approx WH$ where $W \geq 0$ and $H \geq 0$ provides an LDR where the columns of W form a basis for the columns of X. Because W is nonnegative, its columns can be interpreted as basis images: the columns of W are vectors of pixel intensities whose linear combinations allow us to approximate each input image. Moreover, because of the nonnegativity constraints on the weight matrix H, no cancellation is possible between the basis images to reconstruct all the input images. Hence these basis images typically correspond to localized features that are shared among the input images, and the entries of H indicate which input image contains which feature. For example, if the columns of X are facial images, the columns of W correspond to facial features such as eyes, noses, mustaches, and lips. Figure 1.2 illustrates such a decomposition and shows that NMF is able to extract a part-based representation of a set of facial images. Note that to obtain such decompositions, the images need to be well-registered/aligned (for example, pixels corresponding to noses should be located at the same position in the input images).

Another example is the synthetic swimmer data set [138], where the columns of X are vectorized images of a swimmer whose four limbs take four different positions for a total of $n = 256$ images; see Figure 1.3. For an NMF of rank 17, the columns of W correspond to the body and the limbs in the 16 different positions; see Figure 1.4.

Remark 1.1 (Uniqueness, minimality, and sparsity). *The decomposition shown in Figure 1.4 is not minimal, that is, there exists an Exact NMF with fewer basis elements. In particular, there exist several Exact NMFs with factorization rank 16: for example the body can be put together with the limbs in all positions with an intensity of 1/4 (since each image in the data set contains four limbs), or the body can be put together with one limb in its four positions with an intensity of 1 (since each image in the data set contains each limb in one position). Hence the Exact NMF of rank 16 of this data set is not unique; see Chapter 4 for a discussion*

---

[6]In MATLAB, this is achieved via the function vec.

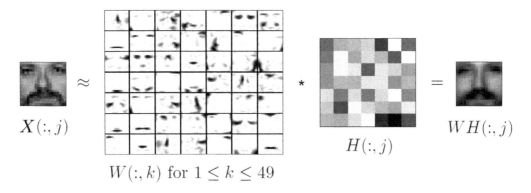

$X(:,j) \approx \quad W(:,k)$ for $1 \leq k \leq 49 \quad * \quad H(:,j) \quad = \quad WH(:,j)$

Figure 1.2. *NMF applied on the CBCL face data set with $r = 49$ (2429 images with 19×19 pixels each), as in the seminal paper of Lee and Seung [303]. On the left is a column of X reshaped as an image. In the middle are the 49 columns of the basis W reshaped as images and displayed in a 7×7 grid and the reshaped column of H in the same 7×7 grid that shows which features are present in that particular face. On the right is the reshaped approximation $WH(:,j)$ of $X(:,j)$ as an image.* [Matlab file: CBCL.m].

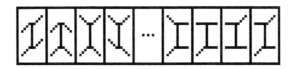

Figure 1.3. *Sample images of the swimmer data set.*

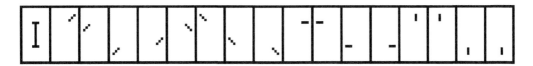

Figure 1.4. *NMF basis images for the swimmer data set.* [Matlab file: Swimmer.m].

on this important issue. However, additional constraints on the decomposition may make it unique. For example, the decomposition of rank 17 from Figure 1.4 can be obtained as the unique decomposition using sparse NMF; see Section 4.3.4. It can also be obtained with separable NMF (Chapter 7), minimum-volume NMF (Section 4.3.3), ONMF (Section 4.3.2), or nonnegative matrix underapproximation (NMU; Section 5.4.5).

1.3.2 ▪ Blind hyperspectral unmixing

A hyperspectral image measures the intensity of the light within a scene for many different wavelengths, typically between 100 and 200 wavelengths; see for example [427] for a gentle introduction to hyperspectral imaging. Hence, for each pixel, a vector of intensities is recorded that is equal to the fraction of light reflected by that pixel depending on the wavelength; this is referred to as the spectral signature of the pixel. Given a hyperspectral image, the goal of blind hyperspectral unmixing (blind HU) is to recover the materials present in an image, referred to as the endmembers, and their proportions in each pixel, referred to as abundances. Under the

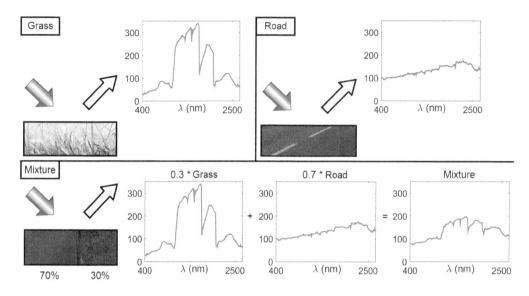

Figure 1.5. *Linear mixing model for hyperspectral imaging.*

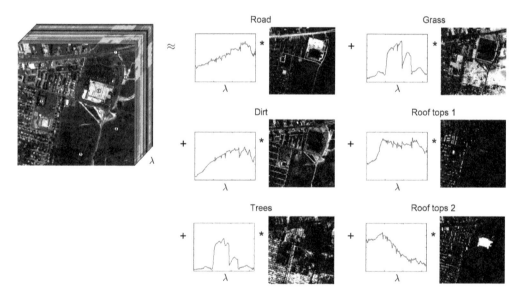

Figure 1.6. *Blind HU of an urban image taken above the Walmart in Copperas Cove, Texas, using NMF with $r = 6$ (162 spectral bands, 307×307 pixels). This is the so-called Urban data set. Each factor corresponds to the spectral signature of an endmember (a column of W) with its abundance map (a row of H, reshaped as an image on the figure), where light tones represent high abundances. The Urban hyperspectral image is mostly made up of six materials, namely road, grass, trees, dirt, and roof tops 1 and 2. In particular, the rank-6 NMF above explains more than 95% of the data, that is, $\|X - WH\|_F \leq 0.05\|X\|_F$. Image adapted from [191]. [*Matlab file:* Urban.m].

linear mixing model, the spectral signature of a pixel is equal to the linear combination of the spectral signatures of the materials it contains. For example, if a pixel contains 30% grass and 70% road surface, its spectral signature is equal to 0.3 times the spectral signature of the grass plus 0.7 times the spectral signature of the road surface; see Figure 1.5 for an illustration. Hence letting each column of the matrix X be the spectral signature of a pixel, blind HU boils down to the NMF of matrix X. In fact, NMF decomposes X as

$$X(:,j) \approx \sum_{k=1}^{r} W(:,k)H(k,j),$$

where the columns of W are the spectral signatures of the endmembers, while the entries of H give the abundance of each endmember in each pixel; see Figure 1.6 for an illustration of NMF on the widely used Urban data set.

1.3.3 ▪ Text mining: topic recovery and document classification

Let each column of the matrix X correspond to a document, that is, a nonnegative vector of word counts. For example, the entry of X at position (i, j) can be the number of times word i appears in document j. The matrix X can also be constructed in different, more sophisticated ways, for example, with the term frequency-inverse document frequency (tf-idf) [389]. This is the so-called bag of words model where the positions of the words in a document are not taken into account. The NMF of X provides the model

$$X(:,j) \approx \sum_{k=1}^{r} W(:,k)H(k,j),$$

where the nonnegative columns of W can also be interpreted as bags of words, that is, as vectors of word count. Since the number of columns of W is much smaller than the number of columns of X ($r \ll n$), each column of W must be used to reconstruct many documents. Moreover, because of the nonnegativity of H, no cancellation is possible, and hence each column of W must contain words that appear simultaneously in these documents. In practice, it has been observed that the columns of W correspond to different topics; see Figure 1.7 for an illustration. Moreover, the columns of the factor H indicate the importance of the topics discussed in the corresponding documents.

Remark 1.2. *We will provide more details on topic modeling and its link with NMF later in the book. In Section 5.5.4, we will show the equivalence between NMF and probabilistic latent semantic analysis/indexing (PLSA/PLSI) which is a simple probabilistic topic model. In Section 5.4.9.1, we will discuss the limitations of NMF for topic modeling and describe a better suited NMF model for this task.*

1.3.4 ▪ Audio source separation

Given an audio signal recorded from a single microphone, its magnitude spectrogram can be constructed as follows. The signal is split into small time frames with some overlap (usually 50%), and each frame is multiplied by some window function (such as the Hamming window) to avoid artifacts due to the truncation of the signal. Each column of X is obtained by taking the short-time Fourier transform of each time frame. More precisely, the entry of X at position (i, j) is the magnitude of the Fourier coefficient for the jth time frame at the ith frequency. Given such

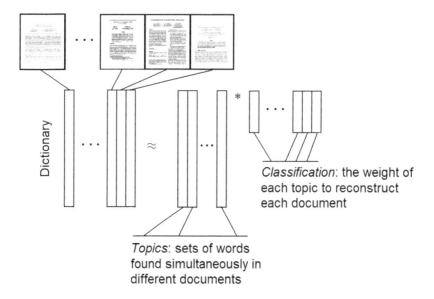

Figure 1.7. *Illustration of NMF for text mining: extraction of topics, and classification of each document with respect to these topics. The columns of X are sets of words, with $X(i, j)$ being the importance of word i in document j. Because H is nonnegative, these sets are approximated as the union of a smaller number of sets defined by the columns of W that correspond to topics.*

a signal, the goal is to blindly separate the sources that compose the signal, for example, separate the voice and the instruments in a song. Under the two following assumptions, this separation problem is yet another NMF problem:

1. The spectrogram of the mixture is a nonnegative linear combination of the spectrogram of the sources. Nonnegativity means that sound cancellation is neglected. Linearity is a natural model; nonlinear effects, such as the saturation of the microphone or reverberations of the sound, are not taken into account.

2. The spectrograms of the sources have low rank. This has been validated on many experiments and makes sense physically. For example, the spectrogram of an instrument is composed of rank-one factors made of the signature of each note in the frequency domain along with its activation over time.

We refer the interested reader to [158] and the references therein for more information on this model.

Let us use a simple monophonic signal for illustrative purposes, namely a piano recording of "Mary Had a Little Lamb," whose musical score is shown below.

The sequence is composed of three notes, C_4, D_4, and E_4, that activate as follows: E_4, D_4, C_4, D_4, E_4, E_4, E_4. For this data set, the three main sources are the piano notes whose spectrograms

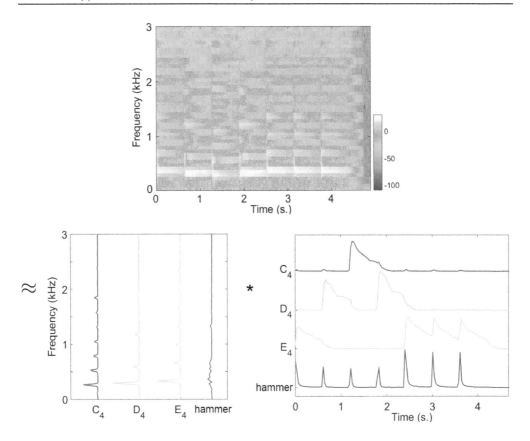

Figure 1.8. *Decomposition of the piano recording "Mary Had a Little Lamb" using NMF: (top) amplitude spectrogram X in dB; (bottom left) basis matrix W corresponding to the three notes C_4, D_4, and E_4, and the hammer noise; (bottom right) activation matrix H that indicates when each note is active. Image adapted from [191]. [Matlab file: Mary_piano.m].*

have rank one: each column of W is the signature of each note in the frequency domain, while the entries of H indicate when a note is active. However, there is a fourth source in this data set: it captures the common mechanism that triggers a note; in particular the hammer within the piano. Figure 1.8 shows the NMF decomposition of the magnitude spectrogram using $r = 4$. As expected, the three notes and the hammer noise are extracted as the columns of W, while the rows of H provide the activation of each note in the time domain. NMF is able to blindly separate the different sources and identify which source is active at which moment in time. NMF has been used, for example, for automatic music transcription; see the survey [33].

NMF shows its full potential for polyphonic music analysis when several notes, and even several instruments, are played at once. However, for such complicated scenarios, refined NMF models should be used, such as sparse NMF [471] or convolutive NMF [424]; see Chapter 5. Moreover sources usually have rank higher than one (for example, the voice of a person), and a postprocessing step is necessary to assign each rank-one factor to its corresponding source. Note that the phase information is missing when reconstructing the sound signal from the spectrogram of the sources. This is an important issue, and in practice one often uses the phase of the input signal; see [337, 468, 338] for more information on this topic.

1.4 ▪ History

The first use of the NMF model can be traced back to the fields of analytical chemistry and of geoscience and remote sensing. In both these fields, NMF corresponds to a meaningful physical model. In both cases,

- the columns of the matrix X correspond to nonnegative spectra of samples (also known as the spectral signature),

- the columns of W correspond to the pure component spectra, and

- the columns of matrix H provide the concentration of the pure components within each sample.

This linear mixing model makes sense physically as the spectrum of a sample is, in ideal conditions, proportional to the spectra of the pure components it contains; this is the so-called Lambert–Beer law [297, 29]; see Figure 1.5 (page 8). This model, which is equivalent to NMF, was discovered and used independently in the early 1960s in these two fields.

1.4.1 ▪ Analytical chemistry

The columns of $X \in \mathbb{R}_+^{m \times n}$ are the spectra of a chemical reaction measured over time, so that m is the number of measured spectral bands and n is the number of time steps for which the reaction is measured. These spectra can be obtained in different ways, for example using Raman spectroscopy; see Section 9.3 for a numerical example. Given these nonnegative spectra, the goal is to recover the pure spectra of the chemical compounds present in the reaction (the columns of W) along with their proportions in each sample (the rows of H). This is an NMF problem which is referred to as self-modeling curve resolution (SMCR); see Section 9.3 for more details.

To the best of our knowledge, Wallace in 1960 [473] was the first to describe the model. He then discussed a way to estimate the number of sources (that is, r). Later in 1971, Lawton and Sylvestre [301] focused on the case $r = 2$ which can be solved easily; see Sections 2.1.3 and 4.1. In 1985, Borgen and Kowalski focused on the case $r = 3$, and, in 1986, Borgen et al. studied the general case [53]. The literature on the topic has grown rapidly since then, and we refer the reader to [263, 390, 366] and the references therein for more information on SMCR. Note that the problem of multivariate curve resolution (MCR) is more general than SMCR as it does not necessarily assume nonnegativity [119, 403].

Interestingly, in the SMCR literature, most works have analyzed the noiseless case, that is, $X = WH$, with the additional assumption that $\text{rank}(X) = \text{rank}(W) = r$. Under these assumptions, Exact NMF is equivalent to finding a transformation of an unconstrained factorization (such as the SVD) to make it nonnegative; see Section 5.5.1. A unique feature within the SMCR literature is that, in most works, the goal is to recover all possible factorizations, that is, all possible nonnegative (W, H) such that $X = WH$, referred to as the set of feasible solutions. The task of choosing the "right" factorization is left to the expert analyzing such chemical reactions. This is a particularly challenging problem, which may explain why an important part of the literature has focused on the exact case, as pointed out in [366].

1.4.2 ▪ Geoscience and remote sensing

The columns of X are the spectra of pixels within a hyperspectral image. The spectrum of a pixel records the fraction of light reflected by that pixel for many wavelengths. The goal is to recover the pure materials present in the image (the columns of W) and the abundance of the materials in each pixel (the columns of H). This is referred to as blind HU; see Section 1.3.2. This model was first used in 1964 by Imbrie and Van Andel for analyzing the mixture of heavy mineral data [254]. Early findings in this field include the works by Craig in the early 1990s [108, 109], by Boardman [48] (1993), and by Winter [482] (1999). The main contribution of these authors is they have shed light on the geometric interpretation of NMF (Section 2.1) and devised algorithms based on this intuition; see Chapters 4 and 7.

The literature on blind HU has grown considerably since the 1990s, motivated by the development of hyperspectral cameras that are becoming higher performing and more affordable as the years go by. We refer the reader to [276, 377, 45, 334] and the references therein for more information on blind HU.

1.4.3 ▪ Stochastic sequential machines

Another very early account of the NMF model appears in the study of stochastic finite state systems, also known as stochastic sequential machines (SSMs). An important problem in this context is to obtain a minimal state representation of the given system. The SSM can be represented with a nonnegative matrix that contains the transition probabilities between the different states. To the best of our knowledge, Ott [369] (1966) was the first to show that computing an Exact NMF of this matrix is equivalent to finding the minimal state representation of the SSM. Moreover, Ott discovered the connection between Exact NMF and the nested polytope problem (NPP), an important problem in computational geometry. Recall that a polytope is a bounded set defined via affine inequalities, that is, a bounded polyhedron. In two dimensions, polytopes are convex polygons. The NPP requires finding a polytope with the minimum number of vertices nested between two given nested polytopes. As we will discuss in length in Chapter 2, the Exact NMF problem with $r = \text{rank}(X)$ is equivalent to the NPP where the sought solution is required to have r vertices. Follow-up works investigating this connection and providing new insights include the papers of Paz [378] (1968) and Bancilhon [21] (1974); see also the first chapter of the book of Paz [379] (1971). The problem of minimal state representation of an SSM is equivalent to the minimal covering of a labeled Markov chain. We refer the reader to [88] for more details and for the proof of the equivalence between Exact NMF and minimal covering of a labeled Markov chain.

1.4.4 ▪ Computational geometry

As explained in the previous section, Exact NMF is equivalent to the NPP. In two dimensions, the NPP requires finding a convex polygon with a minimum number of vertices nested between two given convex polygons. Motivated by its applicability for stochastic sequential machines, Silio [420] (1979) solved this problem by proposing a practical polynomial-time algorithm. However, his paper seems to have been overlooked in the computational geometry literature. The problem was later solved independently in a very similar way by Aggarwal et al. [3] (1989). However, the algorithm of Aggarwal et al. has a lower computational cost; see Section 2.3.1.1. In dimension higher than two, the NPP was shown to be NP-hard by Das and Joseph [114] (1990).

1.4.5 ▪ Linear algebra

The Exact NMF problem, with $X = WH$, is closely related to the notion of nonnegative rank. It started to draw attention after a question of Berman and Plemmons [37] in *SIAM Review* in 1973 (Problem 73-14):

> It is well known that an $m \times n$ matrix A of rank r can be factored (in a variety of ways) in the form $A = BG$ where B is of order $m \times r$ and G is of order $r \times n$. Show by counterexample, that when A is nonnegative there need not exist such a "rank factorization" with both B and G nonnegative. If possible, find a simple characterization of the class of nonnegative matrices A for which a nonnegative rank factorization exists.

In other words, this question asks us to characterize when the rank and the nonnegative rank of a matrix coincide, that is, when $\mathrm{rank}(X) = \mathrm{rank}_+(X)$.

Thomas [450] answered the first part of the question in *SIAM Review* in 1974: he showed that $\mathrm{rank}(X) = \mathrm{rank}_+(X)$ always holds when $\mathrm{rank}(X) \leq 2$ and gave a counterexample to show that this is not necessarily true when $\mathrm{rank}(X) \geq 3$:

$$X = \begin{pmatrix} 1 & 1 & 0 & 0 \\ 1 & 0 & 1 & 0 \\ 0 & 1 & 0 & 1 \\ 0 & 0 & 1 & 1 \end{pmatrix}, \tag{1.2}$$

for which $\mathrm{rank}_+(X) = 4$ while $\mathrm{rank}(X) = 3$; see Section 2.1 for the proofs and more details. It turns out that it is NP-hard to check whether $\mathrm{rank}(X) = \mathrm{rank}_+(X)$ [465]; see Section 2.3, which discusses the computational complexity of Exact NMF.

Early works on the nonnegative rank include the papers of Wall [472] (1979), Jeter and Pye [261] (1981), Campbell and Poole [72] (1981), and Chen [84] (1984). The first comprehensive account of the properties of the nonnegative rank was written by Cohen and Rothblum [100] in 1993; see Section 3.1 for more details.

1.4.6 ▪ Probability

NMF can be used to unravel a particular probabilistic model. Let $Y^{(k)} \in \{1, \ldots, m\}$ and $Z^{(k)} \in \{1, \ldots, n\}$ be two independent random variables for $1 \leq k \leq r$, and let $P^{(k)}$ be the joint distribution with

$$P_{ij}^{(k)} = \mathbb{P}\left(Y^{(k)} = i, Z^{(k)} = j\right) = \mathbb{P}\left(Y^{(k)} = i\right)\mathbb{P}\left(Z^{(k)} = j\right)$$

for $1 \leq i \leq m$ and $1 \leq j \leq n$. Each distribution $P^{(k)}$ corresponds to a nonnegative rank-one matrix. Let us define the joint distribution of two random variables Y and Z as follows:

- Choose the distribution $P^{(k)}$ with probability α_k, where $\sum_{k=1}^{r} \alpha_k = 1$.

- Draw Y and Z from the distribution $P^{(k)}$.

Equivalently, (Y, Z) has the following probability distribution: for $1 \leq i \leq m$ and $1 \leq j \leq n$,

$$\mathbb{P}(Y = i, Z = j) = \sum_{k=1}^{r} \alpha_k P_{i,j}^{(k)} = P_{i,j},$$

where the matrix P is the sum of r rank-one nonnegative matrices. In other words, P is the mixture of r independent distributions. In practice, only P is observed and is referred to as a contingency table, and computing its nonnegative rank and a corresponding factorization amounts to explaining the distribution P with as few independent variables as possible. Early work using this connection includes De Leeuw and Van der Heijden [120] (1991) and Ritov and Gilula [398] (1993); see Cohen and Rothblum [100] (1993) and Kubjas, Robeva, and Sturmfels [292] (2015) for more details. However, the first to discuss the above decomposition were Suppes and Zanotti [443] (1981), although without linking it explicitly with the nonnegative rank. In their terms, the nonnegative rank of P is the smallest support of a hidden random variable, which explains the correlation of the two-valued random variable whose joint distribution is represented by P.

Motivated by this application, Mond, Smith, and Van Straten [353] (2003) also discovered the link between Exact NMF and the NPP (which they call the sandwiched simplices problem) and studied the set of feasible solutions of Exact NMF.

1.4.7 ▪ Extended formulations

A standard approach in combinatorial optimization is to model the convex hull of the set of feasible solutions using affine inequalities, and then solve the problem using linear optimization. These linear formulations are referred to as extended formulations, and their size is defined as the number of inequalities; see Section 3.6 for more details. In the 1980s, some researchers were trying to prove P=NP by constructing such formulations of polynomial size for NP-complete combinatorial optimization problems, such as the traveling salesman problem (TSP).[7] If such formulations have polynomial size, then linear optimization can be used to solve them in polynomial time, which would imply P=NP.

In 1988, Yannakakis proved that this was not possible for the matching and TSP polytopes using extended formulations that are symmetric (for the TSP, this means that they are invariant under permutations of the cities in the input) [492, 493]; see also the discussion by Yannakakis in [494]. To do this, Yannakakis unraveled a key result: the minimum-size extended formulation of a polytope, referred to as its extension complexity, is equal to the nonnegative rank of its slack matrix. Given a polytope

$$\mathcal{P} \;=\; \big\{x \in \mathbb{R}^d \mid a_i^\top x \le b_i \text{ for } i = 1, 2, \dots, m\big\},$$

whose vertices are v_j for $j = 1, 2, \dots, n$, its slack matrix $S \in \mathbb{R}_+^{m \times n}$ is an inequality-by-vertex matrix where

$$S(i,j) = b_i - a_i^\top v_j \ge 0$$

is the slack of the jth vertex for the ith inequality.

Many results were obtained 20 years later (starting around 2010) to provide bounds on the extended formulations of combinatorial problems using bounds for the nonnegative rank. Several long-standing open questions were addressed via the nonnegative rank. For example, Fiorini et al. [163, 164] proved that the extension complexity of the TSP polytope is exponential in the number of cities (the difference with Yannakakis' result is that here asymmetric extended formulations are allowed, and asymmetry may reduce the size of extended formulations [270]). This is not surprising if you believe that $P \neq NP$. Rothvoss proved that the perfect matching polytope[8] has exponential extension complexity so that any extended formulation for the perfect matching polytope must have exponential size [400, 401]. This is somewhat surprising since optimizing a linear function over the perfect matching polytope can be performed in polynomial time [142]. Moreover, Braun and Pokutta [60] later established that for all fixed $0 < \epsilon < 1$, even every linear program approximating the matching polytope by a factor $(1 + \epsilon/n)$ must have exponential size, where n is the number of nodes in the graph. We refer the reader to Section 3.6 for more details and examples.

1.4.8 ▪ The first appearance of NMF in its modern form

As far as we know, the first time the NMF problem was explicitly stated as in (1.1) is in the paper by Paatero and Tapper in 1994 [371]. As discussed above, the models previously studied in the literature either considered exact factorizations, assumed $\mathrm{rank}(X) = \mathrm{rank}(W) = r$, or were only based on geometric representations. In their paper, Paatero and Tapper referred to this problem as positive matrix factorization (PMF), proposed an algorithm based on alternatively

---

[7]Given a set of cities, the TSP requires finding the shortest possible route that visits each city and returns to the origin city.

[8]The perfect matching polytope is the convex hull of the set of all perfect matchings of a complete graph (a perfect matching decomposes the set of vertices into pairs of vertices).

updating W and H (see Section 8.3.1), and explained how it can be used to analyze environmental data, for example, for air emission control. However, until 1999, PMF remained a specialized physical/chemical model confined within the field of chemometrics.

1.4.9 ▪ The "big bang" of NMF: data analysis and machine learning

NMF gained momentum with the seminal paper "Learning the Parts of Objects by Non-negative Matrix Factorization" by Lee and Seung published in *Nature* in 1999 [303]. It explained why NMF is a powerful tool for the analysis of nonnegative data sets. They illustrate their findings on two examples: the extraction of facial features in a set of facial images (see Figure 1.2, page 7) and the identification of topics within a set of documents (see Figure 1.7, page 10). Note that Lee and Seung also proposed algorithms based on multiplicative updates (MU) that became the workhorse in the NMF literature [304]; see Section 8.2.

Let us point out the difference in spirit of Lee and Seung compared to the previous literature by quoting Pentti Paatero:[9]

> The original concept of NMF, as presented in the *Nature* paper, was essentially different from our PMF: they did not search for the (hopefully unique) model of a data set that describes a physical/chemical situation. Instead, their goal could be described as "data compression." Such compressed version of data is normally not unique, it is not "THE solution."

In fact, prior to the paper of Lee and Seung, most works using NMF for data analysis focused on the physical model behind NMF and were hoping that NMF would produce the true underlying sources (such as pure spectra; see Sections 1.4.1 and 1.4.2). In their paper, Lee and Seung did not discuss this issue, and their goal was rather to compute one possible NMF decomposition that compresses the data and can be meaningfully interpreted. We will discuss this key identifiability question for NMF in Chapter 4.

1.5 ▪ Take-home messages

NMF is a linear dimensionality reduction technique with nonnegativity constraints on the factors and is able to extract meaningful components in various applications. From our account of the early literature on the NMF model, we observe that, before the paper of Lee and Seung, the study of NMF was motivated by either specific applications or theoretical questions. Lee and Seung were able to popularize NMF for the analysis of nonnegative data sets via the extraction of sparse and part-based components. They showed that NMF is very versatile and can be used in many different settings. This has been confirmed since then as NMF has been used successfully in many applications; see Chapter 9. In summary, the paper of Lee and Seung set a spark on NMF that has ignited a fire that has been growing steadily since then.

---

[9]Private communication.

Part I

Exact factorizations

The next three chapters of the book make up *Part I: Exact factorizations*, where the theoretical foundations behind NMF are presented.

Chapter 2 provides the geometric interpretation of NMF, introduces the restricted Exact NMF problem which requires $\mathrm{rank}(W) = \mathrm{rank}(X)$, describes its connection with the NPP, and discusses the computational complexity of Exact NMF and restricted Exact NMF.

Chapter 3 discusses the nonnegative rank (the smallest r such that an Exact NMF exists), its properties, lower and upper bounds and its applications, with a focus on the extension complexity of polytopes.

Chapter 4 focuses on the identifiability of Exact NMF and studies the conditions under which an Exact NMF $X = WH$ is essentially unique, that is, unique up to permutation and scaling of the columns of W and rows of H. It also considers regularized NMF models, namely orthogonal, separable, minimum-volume, and sparse NMF, that require weaker conditions for identifiability.

Following Chapter 4, *Part II: Approximate factorizations* will shift the focus from Exact NMF to NMF, that is, to the approximation problem as defined in Problem 1.1 (page 4).

Chapter 2

Exact NMF

In this section, we consider the Exact NMF problem defined as follows.

Problem 2.1 (Exact NMF). *Given a nonnegative matrix $X \in \mathbb{R}_+^{m \times n}$ and a factorization rank r, compute, if possible, two nonnegative matrices $W \in \mathbb{R}_+^{m \times r}$ and $H \in \mathbb{R}_+^{r \times n}$ such that*

$$X = WH.$$

We refer to WH as an Exact NMF of X of size r.

Exact NMF is closely related to the quantity referred to as the nonnegative rank of X which is the smallest r such that X admits an Exact NMF size r, and it is denoted $\mathrm{rank}_+(X)$. The nonnegative rank is the topic of Chapter 3.

Organization of the chapter In Section 2.1, we discuss the geometric interpretation of Exact NMF. In Section 2.2, we study Exact NMF with the additional constraint that $\mathrm{rank}(W) = \mathrm{rank}(X)$, which is referred to as restricted Exact NMF (RE-NMF). The main result of Section 2.2 and, in fact, of this chapter, is the equivalence between RE-NMF and the NPP (Theorem 2.11). In Section 2.3, we elaborate on the computational complexity of solving RE-NMF and Exact NMF. The chapter is concluded with take-home messages (Section 2.4).

   The two main objectives of this chapter are the following:

1. Introduce the geometric interpretation of Exact NMF. This gives crucial insight into the NMF problem, and it is key to understanding the identifiability/nonuniqueness issues of NMF (Chapter 4). It is also key to designing algorithms for NMF, in particular for separable NMF (Section 4.3.1 and Chapter 7) and for minimum-volume NMF (Section 4.3.3).

2. Discuss the computational complexity of Exact NMF, which has direct implications for NMF. The computational complexity of NMF will be discussed further in Chapter 6.

2.1 ▪ Geometric interpretation

The geometric interpretation is a key aspect of NMF. It provides insight into this numerical problem and is, for example, extremely useful in designing models and algorithms; see Chapters 4, 5, and 7. This interpretation dates from the early works on NMF in the fields of analytical chemistry and of geoscience and remote sensing; see Section 1.4.

This section is organized as follows. We describe the geometric interpretation of Exact NMF in terms of nested convex cones in Section 2.1.1 and in terms of nested convex hulls in Section 2.1.2. Based on these geometric interpretations, we prove in Section 2.1.3 the two results of Thomas, namely that $\text{rank}_+(X) = \text{rank}(X)$ for any nonnegative matrix X such that $\text{rank}(X) \leq 2$ (Theorem 2.6) and that the 4-by-4 matrix X from (1.2) satisfies $\text{rank}(X) = 3$ while $\text{rank}_+(X) = 4$ (Theorem 2.10). Finally, we provide in Section 2.1.4 an important example that shows that in an Exact NMF $X = WH$ of size $\text{rank}_+(X)$, we may need the matrix W to satisfy $\text{rank}(W) > \text{rank}(X)$, which is never the case for the usual rank. This example is also analyzed using the geometric interpretation of Exact NMF.

2.1.1 ▪ Interpretation with nested convex cones

Given a matrix $A \in \mathbb{R}^{m \times n}$, $\text{cone}(A)$ is the convex cone generated by the columns of A, that is,

$$\text{cone}(A) = \{x \mid x = Ay \text{ for some } y \in \mathbb{R}^n, y \geq 0\}.$$

The set $\text{cone}(A)$ is referred to as the conical hull of the columns of A, or the conical hull of A for short. All elements of $\text{cone}(A)$ are conic combinations of the columns of A, that is, linear combinations with nonnegative weights. The dimension of $\text{cone}(A)$ is the dimension of the subspace spanned by A and is equal to $\text{rank}(A)$. Let us consider the Exact NMF of matrix $X = WH$. Since $X(:,j) = WH(:,j)$, $W \geq 0$, and $H \geq 0$,

$$X(:,j) \in \text{cone}(W) \subseteq \mathbb{R}_+^m$$

for all j. Equivalently,

$$\text{cone}(X) \subseteq \text{cone}(W) \subseteq \mathbb{R}_+^m,$$

which is a nested cone problem: given two cones nested in each other, namely $\text{cone}(X) \subseteq \mathbb{R}_+^m$, find a cone nested between them, namely $\text{cone}(W)$. Hence, Exact NMF can be formulated as the problem of finding, if possible, r vectors $W(:,k)$ $(1 \leq k \leq r)$ within the nonnegative orthant whose conical hull, $\text{cone}(W)$, contains a given cone generated by the columns of X, namely $\text{cone}(X)$. Figure 2.1 (left) provides such a geometric interpretation for $m = r = 3$ and $n = 25$.

2.1.2 ▪ Interpretation with nested convex hulls

The geometric interpretation of Exact NMF in terms of nested cones can be reformulated into an equivalent interpretation using nested convex hulls, that is, nested polytopes. This nested convex hull interpretation will be the one used in this chapter, rather than the, perhaps more natural, nested convex cone interpretation, for the following reasons:

1. *Intuition.* It is easier to visualize nested convex hulls than nested cones.

2. *Related literature.* The problem of nested convex hulls has a long history in computational geometry, where it is referred to as the NPP [114, 112, 135]. We will discuss this problem in detail in Section 2.2.

3. *Illustration.* Using convex hulls allows us to represent the geometric problems in one dimension lower. We will see that, after normalization, the convex hull of the columns of X has dimension $\text{rank}(X) - 1$ (Lemma 2.5), while the dimension of $\text{cone}(X)$ is $\text{rank}(X)$. For example, looking at Figure 2.1 (right), we observe that after normalization the columns of X and W all belong to the same two-dimensional plane.

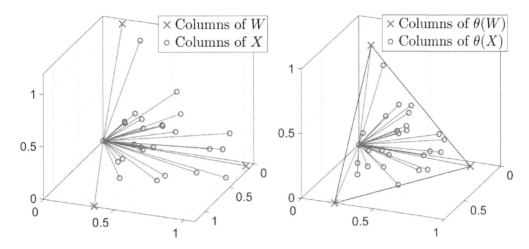

Figure 2.1. *Geometric illustration of Exact NMF for $m = r = 3$ and $n = 25$. Both figures represent the same data set, up to scaling. On the left, we observe that $\mathrm{cone}(X) \subseteq \mathrm{cone}(W) \subseteq \mathbb{R}^3_+$. The figure on the right is the normalization to unit ℓ_1 norm of the columns of X and W from the figure on the left, and we observe $\mathrm{conv}(X) \subseteq \mathrm{conv}(W) \subseteq \Delta^3$.*

In Chapter 4, however, we will return to the nested cone interpretation to characterize the uniqueness of NMF solutions.

Before going further, let us define a few useful notions about convex hulls and polytopes; see [517] for more on this topic. Let us define the convex hull of the columns of a matrix $A \in \mathbb{R}^{m \times n}$ as

$$\mathrm{conv}(A) = \left\{ x \mid x = Ay \text{ for } y \in \mathbb{R}^n, y \geq 0, \text{ and } e^\top y = 1 \right\},$$

where e is the vector of all ones of appropriate dimension, so that $e^\top y = \sum_{i=1}^n y_i$. All elements of $\mathrm{conv}(A)$ are convex combinations of the columns of A, that is, linear combinations with nonnegative weights summing to one. The vertices of $\mathrm{conv}(A)$ are the columns of A that are not contained in the convex hull of the other columns of A. That is, $A(:, j)$ is a vertex of $\mathrm{conv}(A)$ if and only if $A(:, j) \notin \mathrm{conv}(A(:, \mathcal{J}))$ where $\mathcal{J} = \{1, 2, \dots, n\} \setminus \{j\}$. Note that all the vertices of $\mathrm{conv}(A)$ are contained in the set of the columns of A since, by definition, any other point in $\mathrm{conv}(A)$ is a nonnegative linear combination of these columns. Let us also define the unit simplex in dimension r as

$$\Delta^r = \left\{ x \in \mathbb{R}^r \mid x \geq 0, e^\top x = 1 \right\} = \mathrm{conv}(I_r), \tag{2.1}$$

where I_r is the identity matrix of dimension r. In fact, the vertices of Δ^r are the unit vectors, and $I_r = (e_1, e_2, \dots, e_r)$.

A polytope is a bounded polyhedron which is the intersection of half spaces. Hence a polytope can be characterized via a set of equalities and inequalities (half-space representation), but it can also be represented via its vertices (vertex representation). For example, Δ^r is a polytope, and (2.1) provides the half-space and vertex representations.

Given a matrix $A \in \mathbb{R}^{m \times n}$, let us define its column space as

$$\mathrm{col}(A) = \{ x \mid x = Ay \text{ for } y \in \mathbb{R}^n \}$$

and its affine hull as

$$\mathrm{aff}(A) = \{ x \mid x = Ay \text{ for } y \in \mathbb{R}^n \text{ and } e^\top y = 1 \}. \tag{2.2}$$

Note that $\text{conv}(A) \subseteq \text{aff}(A) \subseteq \text{col}(A)$. The dimension of the set $\text{col}(A)$ is the number of linearly independent columns of A, that is, the dimension of $\text{col}(A)$ is equal to $\text{rank}(A)$. The set $\text{aff}(A)$ has dimension d if and only if it contains $d + 1$ columns which are affinely independent, that is, none of these $d + 1$ columns is contained in the affine hull of the other d columns. In other words, the dimension of an affine hull is the dimension of the linear space obtained after any translation of the affine hull onto the origin, for example, translating $\text{aff}(A)$ using $-A(:, 1)$. Note that if $\text{aff}(A)$ contains the origin, then $\text{aff}(A) = \text{col}(A)$.

The dimension of a polytope is the dimension of the affine hull of its vertices (which is the smallest affine set containing it). For example, the dimension of Δ^r is the dimension of the affine hull of Δ^r which is given by $\text{aff}(I_r)$. The dimension of $\text{aff}(I_r)$ is $r - 1$, since all columns of I_r are affinely independent, and $\text{aff}(I_r) = \{x \in \mathbb{R}^r | e^\top x = 1\}$. Another important example is the dimension of $\text{conv}(A)$ which is equal to the dimension of $\text{aff}(A)$ since the set of the columns of A contains all the vertices of $\text{conv}(A)$.

The face of a polytope $\mathcal{P}$ is a subset $\{x \in \mathcal{P} \mid w^\top x = \delta\}$ for some (w, δ) such that $w^\top y \leq \delta$ for all $y \in \mathcal{P}$. A face is also a polytope (and can be the empty set). A k-face is a face of dimension k. A facet of $\mathcal{P}$ is a face whose dimension is one less than that of $\mathcal{P}$. For example, Δ^r has r facets, corresponding to each nonnegativity constraint $x_i \geq 0$ $(1 \leq i \leq r)$: the ith facet of Δ^r is $\{x \in \Delta^r \mid x_i = 0\}$. The 0-faces of a polytope are its vertices. For example, Δ^r has r vertices, namely the unit vectors, as mentioned above. A class of polytopes we will encounter throughout this book are n-gons, that is, two-dimensional polytopes, which have n vertices and n facets (the segments of the n-gons).

Let us describe the geometric interpretation of Exact NMF in terms of nested convex hulls. Given an Exact NMF of $X = WH$, the following two assumptions can be made without loss of generality (w.l.o.g.):

A1 The matrices X and W do not contain columns equal to the zero vector. If this is the case, they can simply be removed as they do not play any role in the factorization. If a column of W is equal to the zero vector, it can be removed along with the corresponding row of H, and r is reduced by one. If a column of X is equal to the zero vector, it can be removed along with the corresponding column of H. Note that this assumption implies that H also does not contain any column equal to the zero vector since $X(:, j) = WH(:, j)$ for all j.

A2 The columns of X and W have ℓ_1 norm equal to one, that is,

$$\|X(:, j)\|_1 = \|W(:, k)\|_1 = 1 \text{ for all } j, k.$$

Given a matrix A with nonzero columns, let D_A be the diagonal matrix whose entries are specified by

$$D_A(i, j) = \begin{cases} \frac{1}{\|A(:, i)\|_1} & \text{if } i = j, \\ 0 & \text{if } i \neq j. \end{cases}$$

Let us also denote $\theta(A) = AD_A$ as the matrix A whose columns have been normalized to have ℓ_1 norm equal to one. (This was referred to as the pullback map in [93].) For any Exact NMF $X = WH$ where the columns of X and W are nonzero, we obtain

$$X = WH \iff \underbrace{XD_X}_{\theta(X)} = \underbrace{(WD_W)}_{\theta(W)} \underbrace{(D_W^{-1} H D_X)}_{H'}$$

$$\iff \theta(X) = \theta(W)H'. \tag{2.3}$$

If the columns of X and W have unit ℓ_1 norm and $X = WH$, then the columns of H also have unit ℓ_1 norm. Let us prove this well-known result; see Figure 2.1 (right) for an illustration.

Lemma 2.1. *[100, Theorem 3.2] Let (W, H) be any factorization of X. If the entries in each column of X and W sum to one, then the entries in each column of H sum to one.*

Proof. The entries in each column of W and X sum to one if and only if $e^\top X = e^\top$ and $e^\top W = e^\top$. (Note that the e's on both sides of these equalities do not have the same size.) This implies

$$e^\top = e^\top X = e^\top W H = e^\top H,$$

hence $e^\top = e^\top H$, that is, the entries in each column of H sum to one.   □

Lemma 2.1 implies that if the columns of X and W are different from the zero vector, then

$$\operatorname{cone}(X) \subseteq \operatorname{cone}(W) \subseteq \mathbb{R}^m_+ \quad \Longleftrightarrow \quad \operatorname{conv}(\theta(X)) \subseteq \operatorname{conv}(\theta(W)) \subseteq \Delta^m,$$

which leads to the following theorem.

Theorem 2.2. *Computing an Exact NMF of X of size r is equivalent to finding r vertices (the columns of $\theta(W)$) within the unit simplex Δ^m whose convex hull contains the columns of $\theta(X)$ (discarding the zero columns of X).*

Proof. This follows directly from (2.3) and Lemma 2.1, which shows that the columns of H' in (2.3) have unit ℓ_1 norm.   □

Figure 2.2 illustrates Theorem 2.2 on the data set from Figure 2.1.

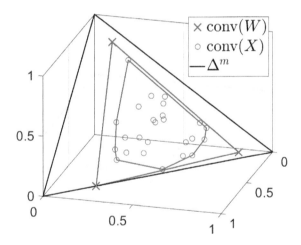

Figure 2.2. *Geometric illustration of Theorem 2.2 with* $\operatorname{conv}(X) \subseteq \operatorname{conv}(W) \subseteq \Delta^m$ *where the columns of X and W have unit ℓ_1 norm. This is the same data set as the one displayed in Figure 2.1 (right).*

Remark 2.1. *In the simple illustrative examples of Figures 2.1 and 2.2, the dimension of* $\operatorname{conv}(\theta(X))$ *is two (= $\operatorname{rank}(X) - 1$; see Lemma 2.5 below) and hence is equal to the dimension of Δ^3. However, it is crucial to note that this is usually not the case: the dimension of* $\operatorname{conv}(\theta(X))$ *is typically (much) smaller than $m - 1$, that is, $\operatorname{rank}(X)$ is typically (much) smaller than m. When $\operatorname{rank}(X) = m$, preforming NMF does not make much sense because the trivial factorization $X = I_m X$ provides an Exact NMF of minimum size, and the factorization does not reduce the dimensionality, which is one of the main purposes of NMF (see Chapter 1). We will encounter several more interesting examples later in this chapter.*

2.1.2.1 ▪ Reducing the dimension by one

A point within the unit simplex contains redundant information, namely, any entry can be deduced from the others: for $x \in \Delta^m$, $x_i = 1 - \sum_{j \neq i} x_j$ for all i. Hence, when considering the interpretation of Exact NMF in terms of nested convex hulls, it is possible to reduce the dimension of the problem by one and represent it in a lower dimensional subspace (in dimension $m - 1$). Let us define

$$\mathcal{S}^r = \left\{ x \in \mathbb{R}^r \mid x \geq 0, e^\top x \leq 1 \right\},$$

which is the convex hull of the unit simplex and the origin, that is, $\mathcal{S}^r = \mathrm{conv}([I_r, 0])$. Given a matrix A such that $A(:,j) \in \Delta^m$ for all j, let us denote $\overline{A} = A(1 : m - 1, :)$ such that $\overline{A}(:,j) \in \mathcal{S}^{m-1}$ for all j. Note that we arbitrarily get rid of the last coordinate, w.l.o.g. We have the following lemma.

Lemma 2.3. *For $x \in \Delta^m$, $W \in \mathbb{R}^{m \times r}$ such that $W(:,j) \in \Delta^m$ for all j, and $h \in \Delta^r$,*

$$x = Wh \iff \overline{x} = \overline{W}h.$$

Proof. Let $x = \binom{\overline{x}}{x_m}$ and $W = \left(\begin{smallmatrix} \overline{W} \\ w_m^\top \end{smallmatrix} \right)$. Since $x \in \Delta^m$ and $W(:,j) \in \Delta^m$ for all j, $x_m = 1 - e^\top \overline{x}$ and $w_m = e - \overline{W}^\top e$. The direction $\Rightarrow$ is straightforward since $x = Wh$ is equivalent to $\binom{\overline{x}}{x_m} = \binom{\overline{W}h}{w_m^\top h}$. For the direction $\Leftarrow$, we need to show that $x_m = w_m^\top h$ given that $\overline{x} = \overline{W}h$,

$$x_m = 1 - e^\top \overline{x} = 1 - e^\top \overline{W}h = 1 - (\overline{W}^\top e)^\top h = 1 - (e - w_m)^\top h = w_m^\top h,$$

since $e^\top h = 1$. □

Lemma 2.3 implies that for $X \geq 0$ and $W \geq 0$ whose columns have unit ℓ_1 norm, we have

$$\mathrm{conv}(X) \subseteq \mathrm{conv}(W) \subseteq \Delta^m \iff \mathrm{conv}\left(\overline{X}\right) \subseteq \mathrm{conv}\left(\overline{W}\right) \subseteq \mathcal{S}^{m-1}.$$

This will be particularly useful to represent graphically problems with $m = 4$ in three dimensions; see the next subsection for an example, and in particular Figure 2.4. Figure 2.3 illustrates Lemma 2.3 on the data set from Figure 2.2.

2.1.3 ▪ Deriving the result of Thomas

Section 1.4.5 introduced the early results of Thomas [450] (1974) regarding the relationship between the rank and the nonnegative rank of a matrix. The geometric interpretation of Exact NMF in terms of convex hulls leads to illuminating proofs of these results, but first we need some additional tools in the form of the following two lemmas.

Lemma 2.4. *Let $X \in \mathbb{R}_+^{m \times n}$ be a matrix whose columns have unit ℓ_1 norm, that is, $X^\top e = e$. We have*

$$\mathrm{aff}(X) = \mathrm{col}(X) \cap \{x \mid e^\top x = 1\}$$

and

$$\mathrm{conv}(X) \subseteq \mathrm{col}(X) \cap \Delta^m.$$

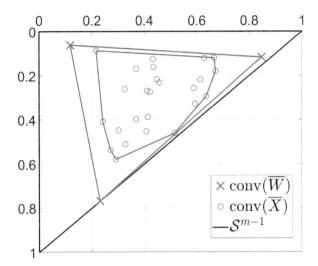

Figure 2.3. *Geometric illustration of Lemma 2.3 on the data set from Figure 2.2. This figure is obtained by looking at Figure 2.2 from the top, so that the vertical axis corresponds to the first coordinate and the horizontal axis to the second coordinate.*

Proof. We have

$$\mathrm{aff}(X) = \left\{ x \mid x = X\alpha \text{ for } \alpha \in \mathbb{R}^n \text{ and } e^\top \alpha = 1 \right\}$$
$$= \left\{ x \mid x = X\alpha \text{ for } \alpha \in \mathbb{R}^n \text{ and } e^\top x = 1 \right\}$$
$$= \mathrm{col}(X) \cap \left\{ x \mid e^\top x = 1 \right\},$$

where

- the first equality follows by definition of the affine hull (2.2),

- the second equality follows from the equality $e^\top x = e^\top X \alpha = e^\top \alpha$ since $e^\top X = e^\top$ (by assumption), and

- the third equality follows from the definition of the column space.

Since $X \geq 0$ and $\mathrm{conv}(X) \subseteq \mathrm{aff}(X)$,

$$\mathrm{conv}(X) \;\subseteq\; \mathrm{aff}(X) \cap \mathbb{R}_+^m \;=\; \mathrm{col}(X) \cap \left\{ x \mid e^\top x = 1 \right\} \cap \mathbb{R}_+^m \;=\; \mathrm{col}(X) \cap \Delta^m,$$

where we used $\Delta^m = \mathbb{R}_+^m \cap \left\{ x \mid e^\top x = 1 \right\}$. $\qquad\square$

Lemma 2.5. *Let $X \in \mathbb{R}^{m \times n}$ be a nonnegative matrix with no column equal to the zero vector. Then, $\mathrm{conv}\left(\theta(X)\right)$ and $\mathrm{col}(X) \cap \Delta^m$ are polytopes of dimension* $\mathrm{rank}(X) - 1$.

Proof. The dimension of $\mathrm{conv}\left(\theta(X)\right)$ is equal to the dimension of $\mathrm{aff}(\theta(X))$; see Section 2.1.2.

Multiplication by a positive diagonal matrix does not change the column space of X, therefore $\mathrm{col}(X) = \mathrm{col}(X D_X) = \mathrm{col}(\theta(X))$, and Lemma 2.4 implies that

$$\mathrm{aff}(\theta(X)) = \mathrm{col}(X) \cap \left\{ x \mid e^\top x = 1 \right\}.$$

Moreover,

$$\mathrm{col}(X) \cap \Delta^m = \mathrm{col}(X) \cap \mathbb{R}_+^m \cap \left\{ x \mid e^\top x = 1 \right\}.$$

Thus $\operatorname{col}(X) \cap \Delta^m$ has the same affine hull as $\operatorname{conv}(\theta(X))$, namely $\operatorname{col}(X) \cap \{x \mid e^\top x = 1\}$, and hence the dimension of the polytopes $\operatorname{col}(X) \cap \Delta^m$ and $\operatorname{conv}(\theta(X))$ coincide by definition.

Geometrically, the affine hull of $\theta(X)$ can be characterized as the intersection of the linear subspace defined by $\operatorname{col}(X)$ and the affine subspace $\{x \mid e^\top x = 1\}$. The dimension of $\operatorname{col}(X)$ is equal to $\operatorname{rank}(X)$, and intersecting it with $\{x \mid e^\top x = 1\}$ removes one degree of freedom, so we can see that the affine hull of $\theta(X)$ has dimension $\operatorname{rank}(X) - 1$.

Let us show this rigorously using algebraic arguments. Let $Y = \theta(X)$. We have

$$
\operatorname{aff}(Y) = \left\{ \sum_{i=1}^{n} \alpha_i Y(:,i) \mid \alpha \in \mathbb{R}^n, \sum_{i=1}^{n} \alpha_i = 1 \right\}
$$

$$
= \left\{ Y(:,n) + \sum_{i=1}^{n-1} \alpha_i \big(Y(:,i) - Y(:,n) \big) \mid \alpha_i \in \mathbb{R} \text{ for } i = 1, 2, \ldots, n-1 \right\}
$$

$$
= Y(:,n) + \operatorname{col}(Y'),
$$

where we use $\alpha_n = 1 - \sum_{i=1}^{n-1} \alpha_i$, and denote $Y'(:,i) = Y(:,i) - Y(:,n)$ for $i = 1, 2, \ldots, n-1$. The dimension of the affine hull of Y is equal to the dimension of $\operatorname{col}(Y')$ which is equal to the rank of Y'. We have to show that $\operatorname{rank}(Y') = r-1$. As $\operatorname{rank}(X) = r$ and $Y = X D_X$ where D_X is a diagonal matrix with positive diagonal elements, $\operatorname{rank}(Y) = r$. W.l.o.g. assume the vectors $Y(:,1), Y(:,2), \ldots, Y(:,r-1), Y(:,n)$ are maximally linearly independent (that is, all other columns of Y are linear combinations of these r columns). This implies that the $r-1$ vectors $\{Y(:,i) - Y(:,n)\}_{i=1}^{r-1} = \{Y'(:,i)\}_{i=1}^{r-1}$ are linearly independent; hence $\operatorname{rank}(Y') \geq r-1$. It remains to show that $\operatorname{rank}(Y') \leq r-1$ or, equivalently, that any subset of r columns of Y' are linearly dependent. For this, let $\mathcal{I}$ be the indices of any subset of r columns of Y', and let us denote $\mathcal{I}_i$ the ith element of $\mathcal{I}$. Since $\operatorname{rank}(Y) = r$, there exists $\beta \in \mathbb{R}^{r+1}$ with $\beta \neq 0$ such that

$$
\beta_{r+1} Y(:,n) + \sum_{i=1}^{r} \beta_i Y(:,\mathcal{I}_i) = 0. \tag{2.4}
$$

Let us premultiply this equality by $e^\top$ to obtain

$$
\beta_{r+1} e^\top Y(:,n) + \sum_{i=1}^{r} \beta_i e^\top Y(:,\mathcal{I}_i) = \beta_{r+1} + \sum_{i=1}^{r} \beta_i = 0,
$$

since $e^\top Y(:,i) = 1$ for all i as $Y = \theta(X)$. This gives $\beta_{r+1} = -\sum_{i=1}^{r} \beta_i$ and implies that $\beta_{1:r} \neq 0$; otherwise $\beta = 0$. Substituting this equality in (2.4) gives

$$
0 = \sum_{i=1}^{r} \beta_i Y(:,\mathcal{I}_i) - \sum_{i=1}^{r} \beta_i Y(:,n) = \sum_{i=1}^{r} \beta_i \big(Y(:,\mathcal{I}_i) - Y(:,n) \big) = \sum_{i=1}^{r} \beta_i Y'(:,\mathcal{I}_i),
$$

where $\beta_{1:r} \neq 0$. This implies that any subset of r columns of Y' are linearly dependent and completes the proof. $\qquad\square$

For example, when $\operatorname{rank}(X) = 1$, $\operatorname{conv}(\theta(X))$ is a zero-dimensional polytope, that is, a single point. In fact, the columns of X are multiples of one another, and hence the columns of $\theta(X)$ are equal to one another.

The first result of Thomas [450] follows directly from Lemma 2.5.

Theorem 2.6. *[450] If X is a nonnegative matrix with $\operatorname{rank}(X) \leq 2$, then $\operatorname{rank}(X) = \operatorname{rank}_+(X)$.*

Proof. This is trivial for $\text{rank}(X) \leq 1$ since either X is the zero matrix or its columns are multiples of one another. For $\text{rank}(X) = 2$, Lemma 2.5 shows that after removing the zero columns of X, $\text{conv}(\theta(X))$ is a one-dimensional polytope, that is, a segment. Since the columns of X must be collinear to lie on a segment, the vertices of this polytope must be two of the nonzero columns of X. Thus there exists an Exact NMF of size $r = 2$ where the columns of W are these two columns of X, and we have $2 = \text{rank}(X) \leq \text{rank}_+(X) \leq 2$, and $\text{rank}_+(X) = 2$ as desired. In fact, it is easy to prove that $\text{rank}_+(X) \geq \text{rank}(X)$ for any nonnegative matrix X; see Theorem 3.1(i). □

Remark 2.2 (Link with separable NMF). *When* $\text{rank}(X) = 2$, *the two columns of W in an Exact NMF of X of size $r = 2$ can be picked from the columns of X; see the proof of Theorem 2.6. We will see how to pick these columns efficiently in Algorithm 4.1. Unfortunately, this fact does not hold for* $\text{rank}(X) \geq 3$. *The reason is that, in dimension* $d = \text{rank}(X) - 1 \geq 2$, *the convex hull of the columns of X is in general not contained in the convex hull of a subset of $d + 1$ columns. In the worst case, all the columns are vertices (take for example points on the unit circle in two dimensions).*

Exact NMF looks for a polytope that contains the data completely. The vertices of this polytope are the columns of W and can be any elements of the nonnegative orthant. Separable NMF is a restriction of Exact NMF that will be discussed thoroughly in Chapter 7. It also looks for this polytope but requires that its vertices are points from the data set. At times this means that more vertices are needed in a separable NMF solution than are needed for an Exact NMF solution, unless $\text{rank}(X) \leq 2$. *In other words, for* $\text{rank}(X) \geq 3$, *there is not always a separable NMF solution where the number of columns of W is equal to the nonnegative rank.*

Before proving the second result of Thomas [450], let us provide two useful lemmas and a corollary about Exact NMF when $\text{col}(X) = \text{col}(W)$.

Lemma 2.7. *Let $X \in \mathbb{R}_+^{m \times n}$, and let $X = WH$ be an Exact NMF of X of size $r = \text{rank}(X)$. Then $\text{rank}(W) = \text{rank}(X)$ and $\text{col}(W) = \text{col}(X)$.*

Proof. Since $X = WH$ is an Exact NMF of size r, $W \in \mathbb{R}^{m \times r}$ has r columns, and

$$r = \text{rank}(X) \leq \min\big(\text{rank}(W), \text{rank}(H)\big) \leq \text{rank}(W) \leq \min(m, r) \leq r.$$

Thus $\text{rank}(X) = \text{rank}(W) = r$. Moreover, $X = WH$ implies $\text{col}(X) \subseteq \text{col}(W)$. Since the dimensions of $\text{col}(X)$ and $\text{col}(W)$ are equal to one another, namely to $\text{rank}(X) = \text{rank}(X)$, we know that $\text{col}(X)$ is not strictly contained in $\text{col}(W)$, and thus $\text{col}(X) = \text{col}(W)$ as desired. □

Lemma 2.8. *Let $X \in \mathbb{R}_+^{m \times n}$ and $W \in \mathbb{R}_+^{m \times r}$ be matrices whose columns have unit ℓ_1 norm and be such that $\text{col}(X) = \text{col}(W)$. Then*

$$\text{aff}(W) = \text{aff}(X).$$

Proof. We have

$$\begin{aligned}
\text{aff}(W) &= \text{col}(W) \cap \{x \mid e^\top x = 1\} \\
&= \text{col}(X) \cap \{x \mid e^\top x = 1\} \\
&= \text{aff}(X),
\end{aligned}$$

where the first and third equalities follow from Lemma 2.4 and the second by the assumption that $\text{col}(X) = \text{col}(W)$. □

Corollary 2.9. *Let $X \in \mathbb{R}_+^{m \times n}$ and $W \in \mathbb{R}_+^{m \times r}$ be matrices whose columns have unit ℓ_1 norm and be such that* $\mathrm{col}(X) = \mathrm{col}(W)$. *Then,*

$$\mathrm{col}\left(\overline{X}\right) = \mathrm{col}\left(\overline{W}\right) \quad and \quad \mathrm{aff}\left(\overline{X}\right) = \mathrm{aff}\left(\overline{W}\right).$$

Proof. Since $\mathrm{col}(X) = \mathrm{col}(W)$ and by the definition of $\overline{X}$ which contains the first $m-1$ rows of X, $\mathrm{col}\left(\overline{X}\right) = \mathrm{col}\left(\overline{W}\right)$. More precisely, $\mathrm{col}(X) \subseteq \mathrm{col}(W)$ if and only if there exists $B \in \mathbb{R}^{r \times n}$ such that $X = WB$. Since $X = WB$ implies $\overline{X} = \overline{W}B$, we obtain $\mathrm{col}\left(\overline{X}\right) \subseteq \mathrm{col}\left(\overline{W}\right)$. Using the same argument, $\mathrm{col}(X) \subseteq \mathrm{col}(W)$ implies $\mathrm{col}\left(\overline{W}\right) \subseteq \mathrm{col}\left(\overline{X}\right)$.

By Lemma 2.8, $\mathrm{aff}(X) = \mathrm{aff}(W)$, while $\mathrm{aff}\left(\overline{X}\right) = \mathrm{aff}\left(\overline{W}\right)$ follows using the same argument as for the column spaces, by simply adding the constraint that $B^\top e = e$ in the argument above. □

With these tools we can now present the counterexample of Thomas demonstrating that $\mathrm{rank}_+(X)$ may not be equal to $\mathrm{rank}(X)$ when $\mathrm{rank}(X) \geq 3$.

Theorem 2.10. *[450] The nonnegative rank of*

$$X = \frac{1}{2} \begin{pmatrix} 1 & 1 & 0 & 0 \\ 1 & 0 & 1 & 0 \\ 0 & 1 & 0 & 1 \\ 0 & 0 & 1 & 1 \end{pmatrix} \tag{2.5}$$

is equal to 4, *while* $\mathrm{rank}(X) = 3$.

Proof. First, let us check that $\mathrm{rank}(X) = 3$:

- $X(2:4, 1:3)$ is an upper triangular matrix with its diagonal entries equal to $1/2$ implying that $\mathrm{rank}(X) \geq 3$, and

- $X(:,1) = X(:,2) + X(:,3) - X(:,4)$ implying that $\mathrm{rank}(X) \leq 3$.

It remains to show that $\mathrm{rank}_+(X) = 4$. Note that the columns of X have unit ℓ_1 norm, that is, $X = \theta(X)$. The factorizations $X = XI_4 = I_4X$ are Exact NMFs of size 4; hence $\mathrm{rank}_+(X) \leq 4$. Assume $\mathrm{rank}_+(X) = 3$ so that there exists a solution (W, H) of Exact NMF of size $r = 3$. By Lemma 2.1, we can assume w.l.o.g. that the columns of W and H have unit ℓ_1 norm, since the columns of X have unit ℓ_1 norm. By Theorem 2.2 and Lemma 2.3, we consider the equivalent geometric problem of finding $W \in \mathbb{R}_+^{4 \times 3}$ such that

$$\mathrm{conv}\left(\overline{X}\right) \subseteq \mathrm{conv}\left(\overline{W}\right) \subseteq \mathcal{S}^3.$$

By Lemma 2.7, $\mathrm{col}(W) = \mathrm{col}(X)$ since $r = \mathrm{rank}(X)$. By Corollary 2.9,

$$\mathrm{aff}\left(\overline{X}\right) = \mathrm{aff}\left(\overline{W}\right).$$

Therefore, if $\mathrm{rank}_+(X) = 3$, there exists $W \in \mathbb{R}_+^{4 \times 3}$ such that

$$\mathrm{conv}\left(\overline{X}\right) \subseteq \mathrm{conv}\left(\overline{W}\right) \subseteq \mathcal{S}^3 \cap \mathrm{aff}(\overline{X}),$$

since $\mathrm{conv}\left(\overline{W}\right) \subseteq \mathrm{aff}\left(\overline{W}\right) = \mathrm{aff}\left(\overline{X}\right)$. Note that since $\overline{W}$ has three columns and $\mathrm{rank}(W) = 3$, $\mathrm{conv}\left(\overline{W}\right)$ is a triangle.

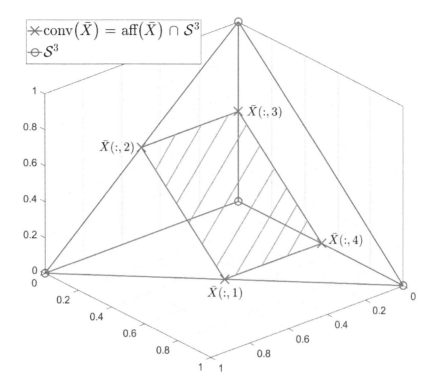

Figure 2.4. *Geometric illustration of Exact NMF for the 4-by-4 matrix X of Thomas from (2.5). The matrix $\overline{X}$ is obtained by discarding the last row of matrix X. The columns of $\overline{X}$ are the vectors (0.5,0.5,0), (0.5,0,0.5), (0,0.5,0), and (0,0,0.5).*

The vertices of $\mathrm{conv}(\overline{X})$ are given by (0.5,0.5,0), (0.5,0,0.5), (0,0.5,0), and (0,0,0.5); see Figure 2.4 for an illustration. Figure 2.4 shows that $\mathcal{S}^3 \cap \mathrm{aff}(\overline{X})$ is a square which coincides with $\mathrm{conv}(\overline{X})$, that is, $\mathrm{conv}(\overline{X}) = \mathcal{S}^3 \cap \mathrm{aff}(\overline{X})$. Since $\mathrm{conv}(\overline{W})$ must be nested between $\mathrm{conv}(\overline{X})$ and $\mathcal{S}^3 \cap \mathrm{aff}(\overline{X})$, this implies that

$$\mathrm{conv}(\overline{X}) = \mathrm{conv}(\overline{W}) = \mathcal{S}^3 \cap \mathrm{aff}(\overline{X}).$$

This is a contradiction since a square has four vertices. In other words, the triangle $\mathrm{conv}(\overline{W})$ cannot contain and be contained within the same square; hence $\mathrm{rank}_+(X) \geq 4$. □

Note that there are many other ways to prove that $\mathrm{rank}_+(X) = 4$; see Section 3.4.

2.1.4 ▪ On the necessity of $\mathrm{rank}(W) > \mathrm{rank}(X)$: illustration with nested hexagons

In this section, we present examples of Exact NMFs that correspond to a geometric problem with nested hexagons; these examples were presented in [353, 190]. It will help us gain more insight into the geometric interpretation of Exact NMF. In particular, they provide an instance of Exact NMF for which $\mathrm{rank}(W) > \mathrm{rank}(X)$ is necessary to obtain a decomposition of minimum size, that is, of size $\mathrm{rank}_+(X)$.

Given a scalar parameter $a > 1$, let us define the nonnegative matrix $X_a \in \mathbb{R}^{6 \times 6}$ whose columns have unit ℓ_1 norm by

$$X_a = \frac{1}{6a} \begin{pmatrix} 1 & a & 2a-1 & 2a-1 & a & 1 \\ 1 & 1 & a & 2a-1 & 2a-1 & a \\ a & 1 & 1 & a & 2a-1 & 2a-1 \\ 2a-1 & a & 1 & 1 & a & 2a-1 \\ 2a-1 & 2a-1 & a & 1 & 1 & a \\ a & 2a-1 & 2a-1 & a & 1 & 1 \end{pmatrix}. \tag{2.6}$$

The aim of this section is to determine the nonnegative rank of X_a depending on the parameter a via the geometric interpretation of the nonnegative rank.

Geometry of $\mathrm{conv}(X_a)$ and $\mathrm{col}(X_a) \cap \Delta^6$

One can check[10] that $\mathrm{rank}(X_a) = 3$ for $a > 1$, and that there exists a decomposition of X_a of the form

$$X_a = \frac{1}{6a} \begin{pmatrix} 1 & 2 & 0 \\ 0 & 1 & 0 \\ 0 & 0 & 1 \\ 1 & 0 & 2 \\ 2 & 1 & 2 \\ 2 & 2 & 1 \end{pmatrix} \begin{pmatrix} -1 & a-2 & -1 & 1-2a & 2-3a & 1-2a \\ 1 & 1 & a & 2a-1 & 2a-1 & a \\ a & 1 & 1 & a & 2a-1 & 2a-1 \end{pmatrix}. \tag{2.7}$$

This shows that the matrices X_a for $a > 1$ share the same column space because the basis in (2.7) does not depend on the parameter a. By Lemma 2.5, $\mathrm{conv}(X_a)$ is a two-dimensional polytope. Moreover, one can check[11] that no column of X_a is contained within the convex hull of the other columns, hence $\mathrm{conv}(X_a)$ has six vertices, that is, $\mathrm{conv}(X_a)$ is a hexagon. This hexagon $\mathrm{conv}(X_a)$ is contained in $\mathrm{aff}(X_a) \cap \Delta^6$: $\mathrm{conv}(X_a) \subseteq \mathrm{aff}(X_a)$ by definition while $\mathrm{conv}(X_a) \subseteq \Delta^6$ because the columns of X_a are nonnegative and have unit ℓ_1 norm. By Lemma 2.4, $\mathrm{aff}(X_a) = \mathrm{col}(X_a) \cap \{x \mid e^\top x = 1\}$, and taking the intersection with the nonnegative orthant, we obtain $\mathrm{aff}(X_a) \cap \mathbb{R}^6_+ = \mathrm{col}(X_a) \cap \Delta^6$. By Lemma 2.5, the polytope $\mathrm{col}(X_a) \cap \Delta^6$ has dimension 2; hence it is also a polygon. More precisely, it is a hexagon: the vertices of the hexagon $\mathrm{col}(X_a) \cap \Delta^6$ are given by the columns of the matrix

$$X = \lim_{a \to +\infty} X_a = \frac{1}{6} \begin{pmatrix} 0 & 1 & 2 & 2 & 1 & 0 \\ 0 & 0 & 1 & 2 & 2 & 1 \\ 1 & 0 & 0 & 1 & 2 & 2 \\ 2 & 1 & 0 & 0 & 1 & 2 \\ 2 & 2 & 1 & 0 & 0 & 1 \\ 1 & 2 & 2 & 1 & 0 & 0 \end{pmatrix}. \tag{2.8}$$

To see this, observe that $\mathrm{col}(X_a) = \mathrm{col}(X)$ for all $a > 1$ since three columns of X form a basis of $\mathrm{col}(X_a)$; see (2.7). Then, observe that the columns of X belong to

$$\mathrm{col}(X_a) \cap \Delta^6 = \mathrm{col}(X) \cap \Delta^6$$
$$= \{z \mid z = Xy, y \in \mathbb{R}^6, z \geq 0, e^\top z = 1\},$$

---

[10]The determinant of $X_a(1:3, 1:3)$ is equal to $\frac{1-a}{108a^2}$.

[11]This can be done by checking that the linear system of equalities and inequalities $X(:,k) = X(:,\bar{k})h$ and $h \in \Delta^5$ has no solution for $k = 1, 2, \ldots, 6$, where $\bar{k} = \{1, 2, \ldots, 6\} \backslash \{k\}$.

and, for each column of X, two of the six inequalities $z_i \geq 0$ ($1 \leq i \leq 6$) are active. In a polygon, vertices are the intersections of two adjacent segments. This implies that, for all $a > 1$,

$$\mathrm{col}(X_a) \cap \Delta^6 = \mathrm{conv}(X).$$

Another way to see this is to observe that, for all $a > 1$,

$$X_a = \frac{a-1}{a}X + \frac{1}{6a}ee^\top = X\left(\frac{a-1}{a}I_6 + \frac{1}{6a}ee^\top\right),$$

since $Xe = e$. This means that the columns of X_a are nonnegative linear combinations of the columns of X. As a increases, the columns of X_a get closer to the columns of X, and $\lim_{a\to+\infty} X_a = X$. In fact,

$$\mathrm{conv}(X_a) \subset \mathrm{conv}(X_b) \quad \text{for } 1 < a < b.$$

This follows from the equality

$$X_a = X_b\left(\frac{b(a-1)}{a(b-1)}I_6 + \frac{b-a}{6a(b-1)}ee^\top\right), \tag{2.9}$$

which implies that the columns of X_a are convex combinations of the columns of X_b when $b \geq a > 1$. The parameter a acts on X_a by scaling the size of the hexagon associated with $\mathrm{conv}(X_a)$. This hexagon grows as a increases with $\lim_{a\to\infty} \mathrm{conv}(X_a) = \mathrm{conv}(X)$. In Figures 2.5 and 2.6, we see the cases of $a = 2$ and $a = 3$, respectively.

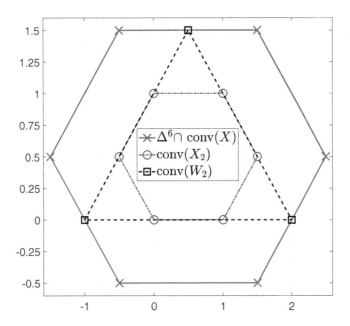

Figure 2.5. *Geometric illustration of Exact NMF for the 6-by-6 matrix X_2 from (2.6) representing a triangle nested between two hexagons.* [Matlab file: ExactNMF_nested_hexagons.m].

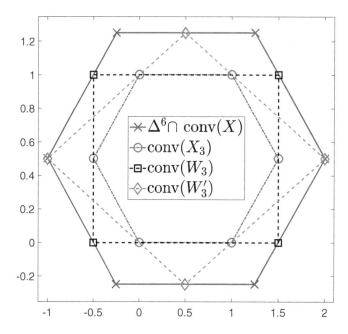

Figure 2.6. *Geometric illustration of Exact NMF for the 6-by-6 matrix X_3 from (2.6) representing two different quadrilaterals nested between the same two hexagons. Each quadrilateral corresponds to a different Exact NMF.* [Matlab file: ExactNMF_nested_hexagons.m].

Remark 2.3 (Figures 2.5 and 2.6). *Figures 2.5 and 2.6 are obtained by representing the columns of X_a in the subspace spanned by their columns (which has dimension 3), while the fact that their columns have unit ℓ_1 norm allows us to reduce the dimension by one, as in Section 2.1.2.1. For example, for $a = 2$, we use the basis $U = X_2(:, [4, 1, 5])$, for which we obtain*

$$X_2 = U \begin{pmatrix} 0 & 1 & 1.5 & 1 & 0 & -0.5 \\ 1 & 1 & 0.5 & 0 & 0 & 0.5 \\ 0 & -1 & -1 & 0 & 1 & 1 \end{pmatrix}$$

and

$$X = U \begin{pmatrix} -0.5 & 1.5 & 2.5 & 1.5 & -0.5 & -1.5 \\ 1.5 & 1.5 & 0.5 & -0.5 & -0.5 & 0.5 \\ 0 & -2 & -2 & 0 & 2 & 2 \end{pmatrix},$$

while the displayed solution W_2 is given by

$$W_2 = U \begin{pmatrix} -1 & 0.5 & 2 \\ 0 & 1.5 & 0 \\ 2 & -1 & -1 \end{pmatrix}.$$

Figure 2.5 represents the above matrices within the subspace spanned by U using only the first two indices (the entries of all the columns of the second factors shown above sum to one). For example, the three columns of W_2 are represented by $(-1, 0)$, $(0.5, 1.5)$, and $(2, 0)$. For Figure 2.6 with $a = 3$, we used the basis $U = X_3(:, [4, 1, 5])$.

Such geometric constructions can be obtained in a systematic way. They will be described in detail in the proof of Theorem 2.11 in Section 2.2 and can be obtained for any matrix of rank 3 via [Matlab file: NPPrank3matrix.m].

What is the nonnegative rank of X_a?   Let (W_a, H_a) be a solution of Exact NMF for X_a of size $\text{rank}_+(X_a)$, and we assume w.l.o.g. that the columns of W_a have unit ℓ_1 norm. Exact NMF is equivalent to finding W_a such that

$$\text{conv}(X_a) \quad \subseteq \quad \text{conv}(W_a) \quad \subseteq \quad \Delta^6;$$

see Theorem 2.2. Since $\text{conv}(X_a) \subseteq \text{conv}(X_b)$ for $a \leq b$ by (2.9), $\text{rank}_+(X_a) \leq \text{rank}_+(X_b)$ for $a \leq b$.

We have that $\text{rank}_+(X_a) \geq \text{rank}(X_a) = 3$. If $\text{rank}_+(X_a) = \text{rank}(X_a) = 3$, Lemma 2.7 implies $\text{col}(W_a) = \text{col}(X_a)$. When this equality is combined with Lemma 2.4 (applied on W_a), we have

$$\text{conv}(W_a) \quad \subseteq \quad \text{col}(X_a) \cap \Delta^6.$$

Since W_a has three columns and $\text{rank}(W_a) = 3$, $\text{conv}(W_a)$ is a triangle.

So far, these are the same observations as for the nested square problem discussed in the previous section. Geometrically, the problem of checking whether $\text{rank}_+(X_a) = 3$ reduces to checking whether there exists a triangle nested between two hexagons, namely $\text{conv}(X_a)$ and $\text{col}(X_a) \cap \Delta^6 = \text{conv}(X)$. For $a = 2$ shown in Figure 2.5, four triangles fit between the two hexagons: the one shown in the figure and its rotation by 60 degrees, and the two triangles made of three nonadjacent vertices of $\text{conv}(X)$. This implies that for any $1 < a \leq 2$, $\text{rank}_+(X_a) = 3$.

To compute an Exact NMF, the vertices of $\text{conv}(W_2)$ can be obtained by averaging two consecutive vertices of $\text{conv}(X)$, where X is given in (2.8); see Figure 2.5. We have

$$X_2 = \frac{1}{12} \begin{pmatrix} 1 & 2 & 3 & 3 & 2 & 1 \\ 1 & 1 & 2 & 3 & 3 & 2 \\ 2 & 1 & 1 & 2 & 3 & 3 \\ 3 & 2 & 1 & 1 & 2 & 3 \\ 3 & 3 & 2 & 1 & 1 & 2 \\ 2 & 3 & 3 & 2 & 1 & 1 \end{pmatrix}$$

$$= \frac{1}{12} \underbrace{\begin{pmatrix} 1 & 4 & 1 \\ 0 & 3 & 3 \\ 1 & 1 & 4 \\ 3 & 0 & 3 \\ 4 & 1 & 1 \\ 3 & 3 & 0 \end{pmatrix}}_{W_2} \begin{pmatrix} 2/3 & 2/3 & 1/3 & 0 & 0 & 1/3 \\ 0 & 1/3 & 2/3 & 2/3 & 1/3 & 0 \\ 1/3 & 0 & 0 & 1/3 & 2/3 & 2/3 \end{pmatrix},$$

where

$$W_2 = \frac{1}{12} \begin{pmatrix} 1 & 4 & 1 \\ 0 & 3 & 3 \\ 1 & 1 & 4 \\ 3 & 0 & 3 \\ 4 & 1 & 1 \\ 3 & 3 & 0 \end{pmatrix} = X \begin{pmatrix} 1/2 & 0 & 0 \\ 1/2 & 0 & 0 \\ 0 & 1/2 & 0 \\ 0 & 1/2 & 0 \\ 0 & 0 & 1/2 \\ 0 & 0 & 1/2 \end{pmatrix}.$$

As explained above, there are four distinct solutions in the case $a = 2$, each corresponding to a triangle nested between $\text{conv}(X_a)$ and $\text{conv}(X)$. Permutation and scaling of the columns of W and the rows of H does not change the geometry of the solutions. The W factors of the other

Exact NMFs of X_2 are given by

$$\frac{1}{12}\begin{pmatrix} 0 & 3 & 3 \\ 1 & 1 & 4 \\ 3 & 0 & 3 \\ 4 & 1 & 1 \\ 3 & 3 & 0 \\ 1 & 4 & 1 \end{pmatrix}, \quad \frac{1}{6}\begin{pmatrix} 0 & 2 & 1 \\ 0 & 1 & 2 \\ 1 & 0 & 2 \\ 2 & 0 & 1 \\ 2 & 1 & 0 \\ 1 & 2 & 0 \end{pmatrix}, \quad \frac{1}{6}\begin{pmatrix} 1 & 2 & 0 \\ 0 & 2 & 1 \\ 0 & 1 & 2 \\ 1 & 0 & 2 \\ 2 & 0 & 1 \\ 1 & 1 & 0 \end{pmatrix}.$$

The first one is obtained by symmetry of the problem and corresponds to the triangle containing the three other segments of the inner hexagon. The other two are made of three nonadjacent columns of X. Up to permutation and scaling, these are the only solutions of Exact NMF of X_2. Note that, for $1 < a < 2$, there are infinitely many solutions as infinitely many triangles fit between the two nested hexagons. The related nonuniqueness of Exact NMF solutions is the topic of Chapter 4.

The observation above also implies that $\text{rank}_+(X_a) \geq 4$ for any $a > 2$, as no triangle fits between the two hexagons[12] for $a > 2$. For $a = 3$, a quadrilateral fits between the two hexagons; see Figure 2.6. (By symmetry of the problem, there is a total of six solutions when $a = 3$; see Figure 2.6.) Therefore, $\text{rank}_+(X_a) = 4$ for $2 < a \leq 3$.

Lemma 2.7 does not apply when $\text{rank}_+(X_a) \geq 4$, so we cannot take as a given that $\text{col}(W_a) = \text{col}(X_a)$. Thus even though no quadrilateral fits between the two hexagons[12] when $a > 3$, it is not straightforward to conclude that $\text{rank}_+(X_a) \geq 5$ for $a > 3$. Fortunately, Corollary 2.16 in Section 2.2 will allow us to reach this conclusion. It states that for a matrix X_a such that $\text{rank}_+(X_a) \leq \text{rank}(X_a) + 1$ and such that X_a is symmetric up to permutation of its rows and columns ($X_a([2, 1, 6, 5, 4, 3], :)$ is symmetric), there always exists a solution W of Exact NMF such that $\text{col}(W) = \text{col}(X_a)$. Hence, if $\text{rank}_+(X_a) = 4$ for $a > 3$, then a quadrilateral should fit between the two hexagons but this is not the case. This implies that $\text{rank}_+(X_a) \geq 5$ for $a > 3$.

Is $\text{rank}_+(X_a) = 5$ or $\text{rank}_+(X_a) = 6$ for $a > 3$? For a sufficiently close to 3, there is a pentagon between the two hexagons implying that $\text{rank}_+(X_a) = 5$. Let us consider the case $a \to +\infty$ for which $X_a = X$; hence the two nested hexagons coincide since $\text{conv}(X) = \text{col}(X) \cap \Delta^6$. Therefore, the only polygon that fits between $\text{conv}(X)$ and $\text{col}(X) \cap \Delta^6$ is the hexagon $\text{conv}(X)$ itself. One would be tempted to conclude that $\text{rank}_+(X) = 6$. This would be true if we were to impose that[13] $\text{rank}(W_a) = \text{rank}(X) = 3$. All our previous observations were made under the condition $\text{col}(W_a) = \text{col}(X)$ (Lemma 2.7). However, it turns out that there exists a higher dimensional solution $W \in \mathbb{R}^{6 \times 5}$ with five vertices and with $\text{rank}(W) = 4 > \text{rank}(X)$ so that $\text{conv}(W)$ has dimension 3, namely

$$X = \frac{1}{6}\begin{pmatrix} 1 & 0 & 0 & 0 & 1 \\ 2 & 0 & 0 & 1 & 0 \\ 1 & 0 & 1 & 0 & 0 \\ 0 & 1 & 1 & 0 & 0 \\ 0 & 2 & 0 & 1 & 0 \\ 0 & 1 & 0 & 0 & 1 \end{pmatrix}\begin{pmatrix} 0 & 0 & 0 & 1 & 1 & 0 \\ 1 & 1 & 0 & 0 & 0 & 0 \\ 1 & 0 & 0 & 0 & 1 & 2 \\ 0 & 0 & 1 & 0 & 0 & 1 \\ 0 & 1 & 2 & 1 & 0 & 0 \end{pmatrix}. \tag{2.10}$$

In other words, there exists a three-dimensional polytope within the five-dimensional polytope Δ^6 that contains the two-dimensional polytope $\text{conv}(X)$; see Figure 2.7 for an illustration. This implies that $\text{rank}_+(X_a) = 5$ for any $a > 3$, and therefore $\text{rank}_+(X) = 5$ too.

[Matlab file: ExactNMF_nested_hexagons.m] can be used to compute Exact NMFs of X_a for any $a > 1$.

---

[12]This can be proved rigorously using arguments presented in Section 2.3.1.

[13]This is the topic of the next section. This quantity is referred to as the restricted nonnegative rank of X.

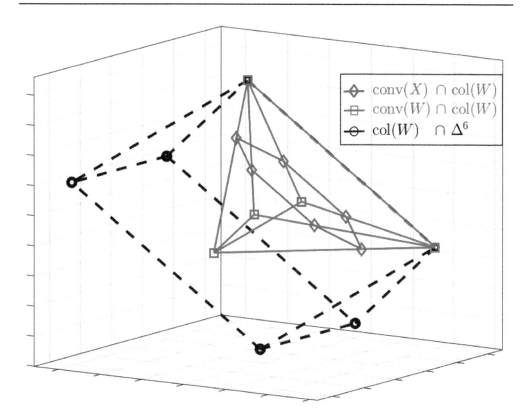

Figure 2.7. *Geometric illustration of the Exact NMF* (2.10) *of the 6-by-6 matrix X from* (2.8) *representing a hexagon contained within a three-dimensional polytope with five vertices within the unit simplex Δ^6. Figure adapted from [190].*

2.2 ▪ Restricted Exact NMF

The RE-NMF problem is obtained by imposing that $\mathrm{rank}(W) = \mathrm{rank}(X)$ in the Exact NMF problem:

Problem 2.2 (Restricted Exact NMF). *Given a nonnegative matrix $X \in \mathbb{R}_+^{m \times n}$ and a factorization rank r, compute, if possible, two nonnegative matrices $W \in \mathbb{R}_+^{m \times r}$ and $H \in \mathbb{R}_+^{r \times n}$ such that*

$$\mathrm{rank}(W) = \mathrm{rank}(X) \quad and \quad X = WH.$$

It is crucial to recall that $\mathrm{rank}(W) = \mathrm{rank}(X)$ and $X = WH$ imply that $\mathrm{col}(W) = \mathrm{col}(X)$. In fact, $X = WH$ implies $\mathrm{col}(X) \subseteq \mathrm{col}(W)$, while the dimension of these two linear subspaces is the same since $\mathrm{rank}(W) = \mathrm{rank}(X)$; hence $\mathrm{col}(X) = \mathrm{col}(W)$ (this is the same argument as in Lemma 2.7). As we will see, this restricts the search space significantly.

The smallest r such that an RE-NMF of X exists is the restricted nonnegative rank of X, denoted $\mathrm{rank}_+^*(X)$, which was introduced in [199]. Note that $\mathrm{rank}_+^*(X) \leq n$ since $X = XI$ is a feasible solution to RE-NMF with $r = n$. However, as opposed to the nonnegative rank, it is possible that $\mathrm{rank}_+^*(X) > m$. We refer the interested reader to [199, 88] for more properties on the restricted nonnegative rank.

The reason to consider RE-NMF is twofold:

1. In most data analysis applications, it is assumed that W has full column rank r and that the rank of the noiseless input matrix is also r. In other terms, in most real-world data analysis problems, $\tilde{X} = X + N = WH + N$ where $\text{rank}(W) = \text{rank}(X) = r$ and N is the noise. It is therefore interesting to better understand this particular scenario of Exact NMF.

2. As we will show, RE-NMF is equivalent to the NPP [114, 112, 135], a widely studied problem in the computational geometry literature. Hence any result for NPP applies to RE-NMF. Since RE-NMF and Exact NMF are closely related, this will also imply results for Exact NMF, for example when $r = \text{rank}(X)$ in which case RE-NMF and Exact NMF coincide.

This section is organized as follows. We define the NPP in Section 2.2.1 and prove its equivalence with RE-NMF in Section 2.2.2 (Theorem 2.11). In Section 2.2.3, we discuss the implications of this result for Exact NMF.

2.2.1 ▪ The nested polytope problem

Let us introduce the NPP.

Problem 2.3 (Nested polytope problem). *Let $\mathcal{A} \subseteq \mathcal{B} \subset \mathbb{R}^d$ be two full-dimensional nested polytopes, that is, the dimension of $\mathcal{A}$ and $\mathcal{B}$ is equal to d. The polytope $\mathcal{A}$, referred to as the inner polytope, is given via the convex hull of n points*

$$\mathcal{A} = \text{conv}\left(\{v_1, v_2, \dots, v_n\}\right), \ v_j \in \mathbb{R}^d \text{ for } j = 1, 2, \dots, n,$$

and the polytope $\mathcal{B}$, referred to as the outer polytope, via m inequalities

$$\mathcal{B} = \{x \in \mathbb{R}^d \mid Fx + g \geq 0\},$$

where $F \in \mathbb{R}^{m \times d}$ and $g \in \mathbb{R}^m$. Given k, find, if possible, a polytope $\mathcal{E}$ with k vertices, referred to as the nested polytope, such that

$$\mathcal{A} \subseteq \mathcal{E} \subseteq \mathcal{B}.$$

Note that since $\mathcal{A}$ and $\mathcal{B}$ are full dimensional, we must have $k \geq d + 1$; otherwise it is never possible to find $\mathcal{E}$ nested between $\mathcal{A}$ and $\mathcal{B}$. Similarly, $r \geq \text{rank}(X)$ is a necessary condition for RE-NMF to admit a solution. As we will see in Theorem 2.11, there is a one-to-one correspondence between these two problems.

2.2.2 ▪ Equivalence between RE-NMF and NPP

We now show that NPP is equivalent to RE-NMF, and they can be reduced to each other in polynomial time. This result was first proved by Vavasis in the case of RE-NMF with $r = \text{rank}(X)$ and NPP with $k = d + 1$ [465]. In [199], the reduction from RE-NMF to NPP was fully developed, while the reduction from NPP to RE-NMF was incomplete. A comprehensive proof was later provided by Chistikov et al. [88].

Theorem 2.11. *[465, 199, 88] There are polynomial-time reductions from RE-NMF to NPP and from NPP to RE-NMF.*

Proof. Let us construct explicitly polynomial-time reductions from RE-NMF to NPP and from NPP to RE-NMF. This means that given any instance of RE-NMF, we construct with polynomially many operations an instance of NPP such that solving the NPP instance solves the RE-NMF instance and vice versa.

From RE-NMF to NPP. The main idea behind the reduction from RE-NMF to NPP is the geometric interpretation described in the previous section: Given an RE-NMF instance, we construct an NPP instance with $d = \text{rank}(X) - 1$ and $k = r$ where the inner polytope corresponds to $\text{conv}(X)$ and the outer polytope to $\text{col}(X) \cap \Delta^m$; in fact, the dimension of $\text{col}(X) \cap \Delta^m$ is $\text{rank}(X) - 1$ as shown in Lemma 2.5. This part of the proof follows [199, Theorem 1].

First, we remove zero columns of X and replace X with its normalization $\theta(X)$. The corresponding RE-NMF problem has a solution if and only if it has a solution (W, H) where the columns of W and H have unit ℓ_1 norm; see Lemma 2.1.

Let us construct the NPP instance corresponding to RE-NMF. Let $d = \text{rank}(X) - 1$ and $(U, V) \in \mathbb{R}^{m \times (d+1)} \times \mathbb{R}^{(d+1) \times n}$ be an unconstrained factorization of X which can be computed, for example, by taking U as a subset of $d + 1$ linearly independent columns of X and computing V which solves a linear system of equations. Since the entries in each column of X sum to one, and U is formed as a subset of its columns, the entries of each column of U also sum to one. By Lemma 2.1, the entries in each column of V must sum to one. The NPP instance is defined as follows:[14]

$$\mathcal{A} = \text{conv}\left(\overline{V}\right) \quad \text{with} \quad \overline{V} = V(1:d, :),$$

$$\mathcal{B} = \left\{ x \in \mathbb{R}^d \mid U(:, 1:d)x + U(:, d+1)\left(1 - \sum_{i=1}^d x_i\right) \geq 0 \right\},$$

where $d = \text{rank}(X) - 1$, and $k = r$. The nonnegativity of $X = UV \geq 0$ implies that $\mathcal{A} \subseteq \mathcal{B}$: every column of $\overline{V}$ belongs to $\mathcal{B}$ since, for $j = 1, 2, \ldots, n$,

$$U(:, 1:d)V(1:d, j) + U(:, d+1)\left(1 - \sum_{i=1}^d V(i, j)\right)$$
$$= U(:, 1:d)V(1:d, j) + U(:, d+1)V(d+1, j)$$
$$= UV(:, j) = X(:, j) \geq 0,$$

since the sum of the entries of $V(:, j)$ is equal to one for all j. To verify that this is an NPP instance, we must show that $\mathcal{A}$ and $\mathcal{B}$ are of dimension d and that $\mathcal{B}$ is bounded (that is, it is a polytope). Clearly, the dimension of $\mathcal{A}$ is d since V has rank $d + 1$ because $X = UV$ where $\text{rank}(X) = d + 1$. The inclusion $\mathcal{A} \subseteq \mathcal{B}$ implies that $\mathcal{B}$ is full dimensional. To prove that $\mathcal{B}$ is bounded, define $F \in \mathbb{R}^{m \times d}$ with $F(:, \ell) = U(:, \ell) - U(:, d+1)$ for all $1 \leq \ell \leq d$, and denote $g = U(:, d+1) \in \mathbb{R}^m$. Thus the constraints on the set $\mathcal{B}$ can be written as $Fx + g \geq 0$. Note that F is full column rank since U is and that the entries in each column of F sum to zero since the entries in each column of U sum to one.

Now, assume $\mathcal{B}$ is unbounded, that is, assume there exists a point $y \in \mathcal{B}$ and a direction $q \neq 0$ such that $y + \lambda q \in \mathcal{B}$ for all $\lambda \geq 0$. This implies $F(y + \lambda q) + g \geq 0$ for all $\lambda \geq 0$, hence $Fq \geq 0$. Observe that (i) $Fq \neq 0$ since F is full column rank and $q \neq 0$, and (ii) the entries of Fq sum

---

[14]The NPP instance depends on the factorization of $X = UV$ where the entries in the columns of U and V sum to one. This decomposition is highly nonunique; hence the NPP associated to RE-NMF is not unique. In fact, any invertible matrix Q such that $e^\top Q = e^\top$ leads to another acceptable factorization $X = (UQ)(Q^{-1}V)$ since $e^\top UQ = e^\top Q = e^\top$, and hence to another NPP. Geometrically, the polytopes in the NPP can be translated, rotated, and dilated, and any feasible solution after these transformations remains feasible. However, any NPP constructed in this way is equivalent to the corresponding RE-NMF.

to zero (since the entries in each column of F do). This implies that at least one entry in Fq is negative, a contradiction.

It remains to be shown that the RE-NMF instance has a solution if and only if the above NPP instance has a solution.

($\Rightarrow$) Let (W, H) be a solution of RE-NMF where the columns of W and H have unit ℓ_1 norm, such that $X = WH$ and $\text{rank}(X) = \text{rank}(W)$. Since $\text{rank}(W) = \text{rank}(X) = d + 1$, we have $\text{col}(W) = \text{col}(X)$ (Lemma 2.7), and there exists C such that $W = UC$. Note that $W = UC \geq 0$ since W is a solution of RE-NMF. By Lemma 2.1, the entries in each column of C must sum to one. We have

$$X = UV = WH = UCH,$$

where the first equality is by construction, the second by assumption, and the third since $W = CH$. The left inverse of U exists since it is full column rank; hence $V = CH$. Moreover, the entries in the columns of V and C sum to one, which implies

$$V = CH \quad \Longleftrightarrow \quad \overline{V} = \overline{C}H;$$

see Lemma 2.3. This implies that $\mathcal{A} = \text{conv}(\overline{V}) \subseteq \text{conv}\left(\overline{C}\right)$. Recall that $W = UC \geq 0$ so that

$$UC = U(:, 1:d)\overline{C} + U(:, d+1)C(d+1, :) \geq 0,$$

where $C(d + 1, j) = 1 - \sum_{i=1}^{d} C(i, j)$ for all j. This implies $\overline{C}(:, j) \in \mathcal{B}$ for all j; hence $\text{conv}(\overline{C}) \in \mathcal{B}$. Therefore, the columns of $\overline{C}$ form the $k = r$ vertices of a polytope which is a solution of the NPP since

$$\mathcal{A} = \text{conv}\left(\overline{V}\right) \subseteq \text{conv}\left(\overline{C}\right) \subseteq \mathcal{B}. \tag{2.11}$$

($\Leftarrow$) Let $\overline{C}$ be the matrix whose columns are the vertices of a solution of the NPP so that $\overline{C}$ satisfies (2.11). This implies that there exists H whose columns have unit ℓ_1 norm such that

$$\overline{V} = \overline{C}H,$$

and $\overline{C}(:, j) \in \mathcal{B}$ for all j. This implies $V = CH$ (see Lemma 2.3), and multiplying on both sides by U gives $X = UV = UCH$ where $UC \geq 0$ by definition of $\mathcal{B}$.

Before providing the reduction from NPP to RE-NMF, let us illustrate the reduction from RE-NMF to NPP with an example.

Example 2.12. Let us consider the matrix

$$X = \frac{1}{3}\begin{pmatrix} 1 & 1 & 1/2 & 1/2 \\ 1 & 1/2 & 1 & 1/2 \\ 1/2 & 1 & 1/2 & 1 \\ 1/2 & 1/2 & 1 & 1 \end{pmatrix}. \tag{2.12}$$

The rank of X is equal to three: $X(:, 4) = -X(:, 1) + X(:, 2) + X(:, 3)$ implies $\text{rank}(X) \leq 3$, while $\det(X(1:3, 1:3)) = -1/72$ implies $\text{rank}(X) \geq 3$. Let us construct an NPP equivalent to the RE-NMF of X following the reduction of the proof of Theorem 2.11 above. We take U as the first three columns of X which are linearly independent, that is, $U = X(:, 1:3)$. The matrix V such that $X = UV$ is given by

$$V = \begin{pmatrix} 1 & 0 & 0 & -1 \\ 0 & 1 & 0 & 1 \\ 0 & 0 & 1 & 1 \end{pmatrix}.$$

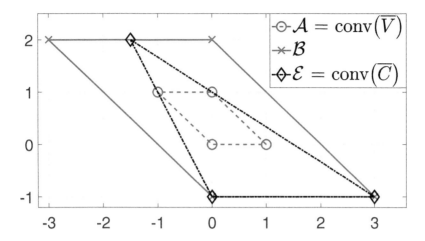

Figure 2.8. *Geometric illustration of the construction of the NPP corresponding to the RE-NMF of X from (2.12).*

Note that the entries in the columns of X, U, and V sum to one. The NPP corresponding to the RE-NMF of X is given by

$$\mathcal{A} = \text{conv}\left(V(1:2,:)\right) = \text{conv}\left(\overline{V}\right) = \text{conv}\left(\{(1,0),(0,1),(0,0),(-1,1)\}\right)$$

and

$$\begin{aligned}
\mathcal{B} &= \left\{x \in \mathbb{R}^2 \mid U(:,1)x_1 + U(:,2)x_2 + (1-x_1-x_2)U(:,3) \geq 0\right\} \\
&= \left\{x \in \mathbb{R}^2 \mid x_1 + x_2 + \frac{1}{2}(1-x_1-x_2) \geq 0, x_1 + \frac{1}{2}x_2 + (1-x_1-x_2) \geq 0,\right. \\
&\qquad \left. \frac{1}{2}x_1 + x_2 + \frac{1}{2}(1-x_1-x_2) \geq 0, \frac{1}{2}x_1 + \frac{1}{2}x_2 + (1-x_1-x_2) \geq 0\right\} \\
&= \left\{x \in \mathbb{R}^2 \mid x_1 + x_2 \geq -1, x_2 \leq 2, x_2 \geq -1, x_1 + x_2 \leq 2\right\} \\
&= \text{conv}\left(\{(0,2),(0,-1),(-3,2),(3,-1)\}\right);
\end{aligned}$$

see Figure 2.8 for an illustration. We observe that $\mathcal{A}$ and $\mathcal{B}$ are quadrilaterals.[15] We also observe that there is a triangle nested between these two quadrilaterals; hence the NPP has a solution with three vertices (note that this solution is not unique). The vertices of this solution are given by two vertices of $\mathcal{B}$, namely $(0,-1)$ and $(3,-1)$, and the middle point between the other two vertices of $\mathcal{B}$, namely $(-1.5, 2)$. In other words, $\mathcal{E} = \text{conv}\left(\overline{C}\right)$, where

$$\overline{C} = \begin{pmatrix} 0 & 3 & -1.5 \\ -1 & -1 & 2 \end{pmatrix}$$

is a solution of this NPP. Let us construct the corresponding RE-NMF solution as described in Theorem 2.11. Given $\overline{C}$, we construct

$$C = \begin{pmatrix} \overline{C} \\ e^\top - e^\top \overline{C} \end{pmatrix} = \begin{pmatrix} 0 & 3 & -1.5 \\ -1 & -1 & 2 \\ 2 & -1 & 0.5 \end{pmatrix}.$$

---

[15]Using the basis $U = X(:,[1,4,2])$ instead of $U = X(:,[1,2,3])$, we obtain another equivalent NPP instance (see footnote 14) made of nested squares; see Example 2.18.

Since conv $\left(\overline{C}\right)$ is a solution of the NPP, $\mathcal{A} = \text{conv}\left(\overline{V}\right) \subseteq \text{conv}\left(\overline{C}\right)$ (see Figure 2.8), and there exists a nonnegative matrix H whose columns have unit ℓ_1 norm such that $\overline{V} = \overline{C}H$, which is given by

$$H = \frac{1}{6} \begin{pmatrix} 1 & 0 & 3 & 2 \\ 3 & 2 & 1 & 0 \\ 2 & 4 & 2 & 4 \end{pmatrix}.$$

Taking

$$W = UC = 4 \begin{pmatrix} 0 & 2 & 1 \\ 2 & 2 & 0 \\ 0 & 0 & 2 \\ 2 & 0 & 1 \end{pmatrix}$$

gives $X = UV = UCH = WH$, while $\text{rank}(W) = \text{rank}(X) = 3$, and hence (W, H) is a solution of RE-NMF of size $r = 3$.

[Matlab file: `NPPrank3matrix.m`] allows us to display the NPP instance for any matrix of rank three, such as the one above. ∎

We now proceed to the reduction from NPP to RE-NMF.

From NPP to RE-NMF   This part of the proof follows [88, Appendix A]. Given the polytopes $\mathcal{A} \subseteq \mathcal{B}$ of an NPP as defined in Problem 2.3, let us construct the corresponding RE-NMF as follows: Define a matrix

$$X(:, j) = Fv_j + g \geq 0 \text{ for } j = 1, 2, \dots, n, \tag{2.13}$$

which is nonnegative because $\mathcal{A} \subseteq \mathcal{B}$ implies $Fv_j + g \geq 0$, and choose a factorization rank $r = k$ (the number of vertices in the nested polytope). Each column of X corresponds to a vertex of $\mathcal{A}$ and each row to a facet of $\mathcal{B}$. Note that since $\mathcal{A}$ is full dimensional, $\text{col}(X) = \text{col}([F, g])$. Let us show that the above RE-NMF instance has a solution if and only if the corresponding NPP has one.

($\Rightarrow$) Let $X = WH$ be a solution to RE-NMF with inner dimension r, and w.l.o.g., assume W does not contain a zero column. Since $\text{rank}(W) = \text{rank}(X)$, $\text{col}(W) = \text{col}(X) = \text{col}([F, g])$; see the proof of Lemma 2.7. Hence, there exists $C \in \mathbb{R}^{d+1 \times r}$ such that $W = [F, g]C$. Let us denote the last row of C by $c^\top$, so that $C = [\overline{C}; c^\top]$ and $W = F\overline{C} + gc^\top \geq 0$. Since $\mathcal{B}$ is bounded, $Fq \not\geq 0$ for all $q \neq 0$ (see the argument above). Let us use this observation to show that $c > 0$:

- If $c_i = 0$ for some i, then $F\overline{C}(:, i) \geq 0$ while $\overline{C}(:, i) \neq 0$ since $W(:, i) \neq 0$, a contradiction.

- If $c_i < 0$ for some i, let us divide the inequality $F\overline{C}(:, i) + c_i g \geq 0$ by $-c_i$ to obtain

$$-F\overline{C}(:, i)/c_i - g \geq 0.$$

For all $x \in \mathcal{B} \backslash \{\overline{C}(:, i)/c_i\}$, $Fx + g \geq 0$ and $x - \overline{C}(:, i)/c_i \neq 0$. Therefore,

$$F(x - \overline{C}(:, i)/c_i) = (Fx + g) + (-F\overline{C}(:, i)/c_i - g) \geq 0,$$

but this a contradiction, and thus we must have $c > 0$.

Let $D = \text{diag}(c)$ be the diagonal matrix with elements taken from c, and define $H' = DH$. We have

$$X = [F, g] \begin{pmatrix} v_1 & v_2 & \cdots & v_n \\ 1 & 1 & \cdots & 1 \end{pmatrix}$$
$$= WH = [F, g]CH = [F, g]CD^{-1}H'$$
$$= [F, g] \begin{pmatrix} \overline{C}(:, 1)/c_1 & \overline{C}(:, 2)/c_2 & \cdots & \overline{C}(:, r)/c_r \\ 1 & 1 & \cdots & 1 \end{pmatrix} H'.$$

Since $[F, g]$ is full column rank, applying its left inverse gives us

$$\begin{pmatrix} v_1 & v_2 & \cdots & v_n \\ 1 & 1 & \cdots & 1 \end{pmatrix} = \begin{pmatrix} \overline{C}(:, 1)/c_1 & \overline{C}(:, 2)/c_2 & \cdots & \overline{C}(:, r)/c_r \\ 1 & 1 & \cdots & 1 \end{pmatrix} H'.$$

The bottom row of the equality, $e^\top = e^\top H'$, implies that the columns of H' have unit ℓ_1 norm, and hence the v_i's belong to the convex hull of the $\overline{C}(:, i)/c_i$ for $1 \leq i \leq r$. Therefore the columns of CD^{-1} form the vertices of a solution to the NPP as they belong to $\mathcal{B}$ since $W(:, i) = F\overline{C}(:, i) + gc_i \geq 0$ and $c_i > 0$ for all i.

($\Leftarrow$) Let the columns of $C \in \mathbb{R}^{d \times r}$ be the vertices of a solution of the NPP. Let us define $W = FC + ge^\top$ where e is the vector of all ones. Since $C(:, j) \in \mathcal{B}$ for all j, $W \geq 0$. Since $\mathcal{A} \subseteq \text{conv}(C)$, there exists $H \geq 0$ whose columns have unit ℓ_1 norm such that $v_j = CH(:, j)$ for $1 \leq j \leq n$. For all $1 \leq j \leq n$,

$$X(:, j) = Fv_j + g = FCH(:, j) + g = (FC + ge^\top)H(:, j) = WH(:, j);$$

hence $X = WH$. Since $\text{col}(X) = \text{col}([F, g])$, $\text{col}(W) \subseteq \text{col}([F, g])$ (by construction) and $X = WH$, $\text{col}(X) = \text{col}(W)$ implying that (W, H) is an RE-NMF of X. $\qquad \square$

Note that the reduction from RE-NMF to NPP has been used in particular within the literature on SMCR to describe the set of feasible solutions; see Section 1.4.1. For example, when $r = \text{rank}(X) = 3$, it amounts to finding all possible triangles nested between the two-dimensional polytopes $\text{conv}(\theta(X))$ and $\text{col}(X) \cap \Delta^m$. For higher dimensions, the problem becomes quickly intractable; see for example [366] and the references therein.

Example 2.13 (Example 2.12, backward). Let us take Example 2.12 backward, that is, let us consider the NPP with

$$\mathcal{A} = \text{conv}\left(\{(1, 0), (0, 1), (0, 0), (-1, 1)\}\right)$$

and

$$\mathcal{B} = \{x \in \mathbb{R}^2 \mid x_1 + x_2 \geq -1, x_2 \leq 2, x_2 \geq -1, x_1 + x_2 \leq 2\},$$

and construct the matrix X whose RE-NMF is equivalent. The set $\mathcal{B}$ can be written in the form $\mathcal{B} = \{x \in \mathbb{R}^2 \mid Fx + g \geq 0\}$ where

$$F = \begin{pmatrix} 1 & 1 \\ 0 & -1 \\ 0 & 1 \\ -1 & -1 \end{pmatrix} \quad \text{and} \quad g = \begin{pmatrix} 1 \\ 2 \\ 1 \\ 2 \end{pmatrix}.$$

Let us denote

$$\overline{V} = \begin{pmatrix} 1 & 0 & 0 & -1 \\ 0 & 1 & 0 & 1 \end{pmatrix}$$

so that $\mathcal{A} = \text{conv}\left(\overline{V}\right)$. To construct the matrix X whose RE-NMF is equivalent to the above NPP, the reduction from NPP to RE-NMF simply requires us to take

$$X(:,j) = F\overline{V}(:,j) + g \;\text{ for }\; j = 1, 2, 3, 4,$$

which gives

$$X = \begin{pmatrix} 2 & 2 & 1 & 1 \\ 2 & 1 & 2 & 1 \\ 1 & 2 & 1 & 2 \\ 1 & 1 & 2 & 2 \end{pmatrix}$$

and is equal to the matrix in Example 2.12 multiplied by 6. (Note that we may have arbitrary scalings of the rows and columns of X, which does not modify its rank or its restricted nonnegative rank.) ∎

Other reductions from NPP to RE-NMF will be used later in this book: a square nested with itself in Section 3.6.3.1, an octagon nested with itself in Section 3.6.3.3, and a rectangle inside a square on page 129.

2.2.3 ▪ Implications for Exact NMF

For $r = \text{rank}(X)$, Exact NMF and RE-NMF are the same problems since $\text{rank}(W) = r = \text{rank}(X)$ in any factorization of $X = WH$; see Lemma 2.7.

Moreover, it turns out that in the case when $r = \text{rank}(X) + 1$, Exact NMF and RE-NMF also coincide in the sense that Exact NMF has a solution if and only if RE-NMF has a solution.

Theorem 2.14. *[199, Corollary 2] If RE-NMF admits a solution for $r \le \text{rank}(X) + 1$, that is, $\text{rank}_+^*(X) \le \text{rank}(X) + 1$, then $\text{rank}_+(X) = \text{rank}_+^*(X)$.*

Proof. If RE-NMF admits a solution for $r = \text{rank}(X)$, then $r = \text{rank}(X) = \text{rank}_+(X) = \text{rank}_+^*(X)$, and the proof is complete. Otherwise, $\text{rank}_+^*(X) = \text{rank}(X) + 1$. Let us assume $\text{rank}_+(X) \ne \text{rank}_+^*(X)$, hence $\text{rank}_+(X) = \text{rank}(X) = r_+$, because $\text{rank}(X) \le \text{rank}_+(X) \le \text{rank}_+^*(X) = \text{rank}(X) + 1$. Let $W \in \mathbb{R}_+^{m \times r_+}$ and $H \in \mathbb{R}_+^{r_+ \times n}$ be an Exact NMF of X. Since $X = WH$, $r_+ = \text{rank}(W) = \text{rank}(X)$ (Lemma 2.7), and hence (W, H) is an RE-NMF of X of size $r_+ < \text{rank}_+^*(X)$, a contradiction. □

Considering the transpose of X, we obtain the following corollary (which is, as far as we know, not present in the literature).

Corollary 2.15. *If the RE-NMF of X or of $X^\top$ admits a solution for $r \le \text{rank}(X) + 2$, that is,*

$$\min\left(\text{rank}_+^*(X), \text{rank}_+^*\left(X^\top\right)\right) \le \text{rank}(X) + 2, \tag{2.14}$$

then $\text{rank}_+(X) = \min(\text{rank}_+^*(X), \text{rank}_+^*(X^\top))$.

Proof. First note that $\text{rank}_+(X) = \text{rank}_+(X^\top)$, which follows from the symmetry of the problem, that is, $X = WH$ if and only if $X^\top = H^\top W^\top$; see Theorem 3.1(ii). Note, however, that we do not necessarily have $\text{rank}_+^*(X) = \text{rank}_+^*\left(X^\top\right)$ unless X is symmetric [199].

By Theorem 2.14, the theorem holds true for $\text{rank}^*_+(X) \le \text{rank}(X) + 1$ or $\text{rank}^*_+(X^\top) \le \text{rank}(X) + 1$. It remains to consider the case where the inequality in (2.14) is an equality. By symmetry of the problem, let us assume w.l.o.g. that $\text{rank}^*_+(X^\top) \ge \text{rank}^*_+(X) = \text{rank}(X) + 2$, and let us prove the result by contradiction, assuming

$$\text{rank}_+(X) \; < \; \text{rank}^*_+(X) \; = \; \text{rank}(X) + 2 \; \le \; \text{rank}^*_+(X^\top).$$

There are two cases:

Case 1. $\text{rank}_+(X) = \text{rank}(X)$: by Lemma 2.7, $\text{rank}_+(X) = \text{rank}^*_+(X)$, a contradiction.

Case 2. $\text{rank}_+(X) = \text{rank}(X) + 1$: let (W, H) be an Exact NMF of X of size $\text{rank}_+(X)$. We must have $\text{rank}(W) = \text{rank}(X) + 1$; otherwise, $\text{rank}^*_+(X) = \text{rank}_+(X)$ because (W, H) would be an RE-NMF of X of size $\text{rank}(X) + 1$, a contradiction. This implies that W admits a left inverse $W^\dagger$, and hence $H = W^\dagger X$ so that $\text{rank}(H) \le \text{rank}(X)$, implying $\text{rank}(H) = \text{rank}(X)$ since $\text{rank}(X) \le \text{rank}(H)$ as $X = WH$. Therefore $\text{rank}^*_+(X^\top) \le \text{rank}_+(X)$ as $H^\top W^\top$ is an RE-NMF of $X^\top$ of size $\text{rank}_+(X)$, which is a contradiction because $\text{rank}^*_+(X) = \text{rank}(X) + 2 \le \text{rank}^*_+(X^\top)$. $\qquad\square$

For a symmetric matrix, or a matrix that can be made symmetric after permutations and scalings of its rows and columns (which does not influence the restricted nonnegative rank), we have $\text{rank}^*_+(X) = \text{rank}^*_+(X^\top)$. In this case, if $\text{rank}^*_+(X) \le r + 2$, then $\text{rank}_+(X) = \text{rank}^*_+(X)$. For example, the 6-by-6 matrix X from (2.8) with $\text{rank}(X) = 3$ is symmetric after a proper permutation of its rows and columns. We have $\text{rank}^*_+(X) = 6$ because the corresponding NPP instance is such that $\mathcal{A} = \mathcal{B}$ is a hexagon. We showed in Section 2.1.4 that $\text{rank}(X) = 3 < \text{rank}_+(X) = 5 < \text{rank}^*_+(X) = 6$ and, by Corollary 2.15, this is the smallest possible example of a symmetric matrix with its nonnegative rank strictly smaller than its restricted nonnegative rank (recall that for $\text{rank}(X) = 2$, both ranks coincide).

We also have the following corollary, which we used in Section 2.1.4 to prove that the nonnegative rank of X_3 is equal to 4 and that of X_a for $a > 3$ is strictly larger than 4.

Corollary 2.16. *Let X be a symmetric matrix up to permutation and scaling of its rows and columns. If* $\text{rank}_+(X) \le \text{rank}(X) + 1$, *then* $\text{rank}_+(X) = \text{rank}^*_+(X)$.

Proof. The same proof as that of Corollary 2.15 can be used, considering the two cases $\text{rank}_+(X) = \text{rank}(X)$ and $\text{rank}_+(X) = \text{rank}(X) + 1$. $\qquad\square$

2.3 ▪ Computational complexity of RE-NMF and Exact NMF

As we have seen in Section 2.1, Thomas [450] showed that Exact NMF is easily solvable when $\text{rank}(X) \le 2$ since W can be constructed by picking two columns of X, namely the vertices of $\text{conv}(\theta(X))$; see Section 4.1 and Algorithm 4.1 for such a construction. In this particular case, Exact NMF and RE-NMF coincide since $r = \text{rank}_+(X) = \text{rank}(X) = 2$. In this section, we discuss the computational complexity of RE-NMF and Exact NMF when $\text{rank}(X) \ge 3$.

This section is organized as follows. In Section 2.3.1, we discuss the computational complexity of RE-NMF. As we will see, when $\text{rank}(X) = 3$, there exists a polynomial-time algorithm to tackle RE-NMF. This algorithm is derived via the equivalence between RE-NMF and NPP, and via a polynomial-time algorithm for the two-dimensional NPP (Section 2.3.1.1). When $\text{rank}(X) = 4$, it is NP-hard to find the minimal r such that an RE-NMF exists (Section 2.3.1.2). We also briefly discuss the case when r is assumed to be a fixed constant in RE-NMF, that is, when r is not part of the input (Section 2.3.1.3). In Section 2.3.2, we discuss the computational complexity of Exact NMF. It is NP-hard to check whether $\text{rank}(X) = \text{rank}_+(X)$. As we will see, even when $\text{rank}(X) \geq 3$, the computational complexity of Exact NMF is unknown. However, when r is not part of the input, Exact NMF can be solved in polynomial time; see Section 2.3.2.2. We briefly mention the practical implications of these complexity results in Section 2.3.2.3. Finally, we mention yet another important complexity result, namely that Exact NMF and RE-NMF are not in NP as there exist rational input matrices X that need factors (W, H) with irrational entries to be decomposed with an Exact NMF of size $\text{rank}_+(X)$; see Section 2.3.3.

2.3.1 ▪ RE-NMF

In the following three sections, we discuss the complexity of RE-NMF in three different cases: $\text{rank}(X) \leq 3$ (Section 2.3.1.1), $\text{rank}(X) \geq 4$ (Section 2.3.1.2), and r is not part of the input (Section 2.3.1.3).

2.3.1.1 ▪ Case $\text{rank}(X) \leq 3$

For $\text{rank}(X) = 3$, by Theorem 2.11, RE-NMF is equivalent to a two-dimensional NNP where one has to find a convex polygon $\mathcal{E}$ with minimum number of vertices nested in between two given convex polygons $\mathcal{A} \subseteq \mathcal{B}$. This problem was studied by Silio [420] (1979), who proposed an algorithm running in $\mathcal{O}(v_1 v_2 + v \log(v))$ operations, where v_1 (resp. v_2) is the number of vertices of $\mathcal{A}$ (resp. of $\mathcal{B}$) and $v = v_1 + v_2$. Note that if X is an m-by-n matrix then $v_1 \leq m$ and $v_2 \leq n$ since, for a polygon, the number of facets (that is, segments) is equal to the number of vertices. Aggarwal et al. [3] (1989) later proposed a very similar algorithm with improved complexity (they were not aware of Silio's work), running in $\mathcal{O}(v \log(k^*))$ operations,[16] where $k^* = \text{rank}_+^*(X) \leq n$ is the number of vertices of the minimal nested polygon $\mathcal{E}$; see below for a description of this algorithm.

Theorem 2.17. *For* $\text{rank}(X) \leq 3$, *RE-NMF can be solved in polynomial time.*

Proof. The case $\text{rank}(X) \leq 2$ follows from the result of Thomas (Theorem 2.6); see also Section 4.1. The case $\text{rank}(X) = 3$ follows from Theorem 2.11 and the polynomial-time algorithms of Silio [420] and Aggarwal et al. [3] for the two-dimensional NPP.   □

Algorithm for the two-dimensional NPP   Let us give the main ideas of the algorithms of Silio [420] and Aggarwal et al. [3]. They first make the following observations:

O1 Any vertex of a solution $\mathcal{E}$ can be assumed w.l.o.g. to belong to the boundary of the polygon $\mathcal{B}$. If this is not the case, it can be replaced by a point on the boundary of $\mathcal{B}$ in such a way that the new solution contains the previous one.

O2 Consider a segment $[p_1, p_2]$ whose vertices are on the boundary of $\mathcal{B}$ and which is tangent to $\mathcal{A}$. This segment $[p_1, p_2]$ defines a polygon D with the boundary of $\mathcal{B}$. This polygon D must contain a vertex of any feasible solution $\mathcal{E}$. If this were not the case, the point p_A at the intersection of the segment and the boundary of $\mathcal{A}$ could not be contained in the nested polygon $\mathcal{E}$. This is illustrated in Figure 2.9, where the triangle D_1 defined by the points p_1, p_2, and c_1 must contain at least one point in any nested polygon $\mathcal{E}$.

---

[16]Wang generalized the result for nonconvex polygons [474]. Bhadury and Chandrasekaran propose an algorithm for computing all the solutions [43].

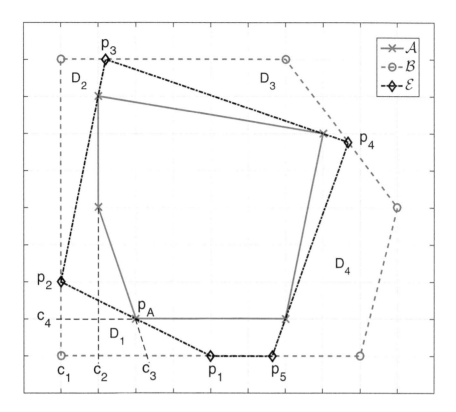

Figure 2.9. *Illustration of the algorithm of Silio [420] and Aggarwal et al. [3]: $\mathcal{A}$ is the inner polytope, $\mathcal{B}$ is the outer polytope, and $\mathcal{E}$ is a nested polytope between $\mathcal{A}$ and $\mathcal{B}$. The solution $\mathcal{E}$ is constructed starting from the point p_1 located on an edge of $\mathcal{B}$ which leads to a solution with five vertices. The polygons D_i ($1 \leq i \leq 4$) are delimited by the boundary of $\mathcal{B}$ and the segment $[p_i, p_{i+1}]$. Starting instead from the vertex of $\mathcal{B}$ between p_1 and p_2, namely the contact change point c_1, leads to an optimal solution $\mathcal{E}^*$ with only four vertices; see Figure 2.10.*

Then, given a point x on the boundary of $\mathcal{B}$, they define the point $f(x)$ as the intersection of the boundary of $\mathcal{B}$ and the line that goes through x, is tangent to $\mathcal{A}$, and goes clockwise. For example, in Figure 2.9, we have $p_2 = f(p_1)$, $p_3 = f(p_2)$, and so on. Letting p_1 be any point on the boundary of $\mathcal{B}$, the polynomial-time algorithm of Silio [420] and Aggarwal et al. [3] constructs iteratively a feasible solution $\mathcal{E}$ that contains p_1 as follows. First compute the two points $p_2 = f(p_1)$ and $p_3 = f(p_2)$. (Note that any nested polygon containing $\mathcal{A}$ must have at least three vertices.) Then, let $k = 3$, and, as long as the line going from p_k to p_1 goes through the interior of $\mathcal{A}$, set $p_{k+1} = f(p_k)$, and $k \leftarrow k + 1$. This construction terminates with the feasible solution $\mathcal{E} = \text{conv}\left(\{p_1, p_2, \ldots, p_k\}\right)$. Figure 2.9 illustrates such a construction of a feasible solution $\mathcal{E} = \text{conv}\left(\{p_1, p_2, \ldots, p_5\}\right)$.

Because of the two observations O1 and O2 above, we have the following:

- The solution $\mathcal{E}$ constructed as described above with k vertices has at most one more vertex than an optimal solution. In fact, this solution defines at least $k - 1$ disjoint polygons that have two endpoints on the boundary of $\mathcal{B}$ and that are tangent to $\mathcal{A}$. Since each of these polygons must contain a vertex of an optimal solution, $k^* = \text{rank}_+^*(X) \geq k - 1$, where k^* is the minimum number of vertices of a nested polygon between $\mathcal{A}$ and $\mathcal{B}$.

Figure 2.9 illustrates this observation where $\mathcal{E}$ with five vertices defines four disjoint polygons, namely D_i for $1 \leq i \leq 4$, that must contain a vertex of any feasible solution.[17]

- There must be a point on $\mathcal{B}$ between p_1 and p_2 that belongs to an optimal solution. In fact, any solution must contain a point in D_1 (O2), and this point can be assumed w.l.o.g. to be on the boundary of $\mathcal{B}$ (O1).

The last step of the algorithm is to identify whether the constructed solution $\mathcal{E}$ is optimal and, if it is not, to construct a solution with one less vertex. To do so, let us define the contact change points between p_1 and p_2: they are either (i) the vertices of $\mathcal{B}$ between p_1 and p_2, or (ii) the points that are located at the intersection of the boundary of $\mathcal{B}$ and a line containing an edge of $\mathcal{A}$. Figure 2.9 illustrates these contact change points between p_1 and p_2; there are four of them, namely c_1, c_2, c_3, and c_4.

Silio [420] and Aggarwal et al. [3] were able to prove that one only needs to consider solutions generated from these contact change points (instead of p_1) to possibly generate a solution that is better than the one constructed starting from p_1 (meaning a solution with one less vertex). In the example of Figure 2.9, starting the procedure to construct $\mathcal{E}$ from the (only) vertex of $\mathcal{B}$ between p_1 and p_2, namely c_1 (instead of p_1) generates an optimal solution of this NPP instance, since it reduces the original solution from five to four vertices; see Figure 2.10. (Another optimal solution is also obtained starting from c_2.)

Algorithm 2.1 summarizes the algorithm of Silio [420] and Aggarwal et al. [3].

Algorithm 2.1 The two-dimensional NPP [420, 3]

Input: Two nested polygons $\mathcal{A} \subseteq \mathcal{B}$.
Output: A polygon $\mathcal{E}$ nested between $\mathcal{A}$ and $\mathcal{B}$ with minimum number of vertices.

1: Pick a point p_1 on the boundary of $\mathcal{B}$.
2: Let $p_2 = f(p_1)$ and $p_3 = f(p_2)$, where the function $f(x)$ is defined as the intersection of the boundary of $\mathcal{B}$ and the line that goes through x, is tangent to $\mathcal{A}$, and goes clockwise.
3: Let $k = 3$.
4: **while** the line between p_1 and p_k goes through the interior of $\mathcal{A}$ **do**
5:        $p_{k+1} = f(p_k)$.
6:        $k = k + 1$.
7: **end while**
8: Let $\mathcal{E} = \text{conv}\left(\{p_1, p_2, \ldots, p_k\}\right)$ be a feasible solution of the NPP: $\mathcal{A} \subseteq \mathcal{E} \subseteq \mathcal{B}$.
9: Compute the contact change points between p_1 and p_2.
10: For each contact change point, construct the corresponding feasible solution for the NPP, exactly as done above when starting from p_1. If such a feasible solution has $k - 1$ vertices, replace $\mathcal{E}$ by this solution, and return.

When $\mathcal{A}$ and $\mathcal{B}$ intersect   In the case in which the inner polygon intersects the outer polygon, say at a point p, Silio [420] makes the following interesting observation: any solution $\mathcal{E}$ has to contain p. Hence one can assume w.l.o.g. that either one of the vertices of an optimal solution is p or two of the vertices of an optimal solution are two points on the segment containing p (in which case these two points can be assume w.l.o.g. to be the endpoint of this segment since this leads to a new solution containing the previous one). Therefore, using the above

---

[17]To be more precise, the polygons D_i are not disjoint since two consecutive polygons intersect in one point on the boundary of $\mathcal{B}$. However, if this intersection point is selected in a nested polygon, then a minimal solution containing this point is the one obtained by the procedure described above, and it contains k vertices.

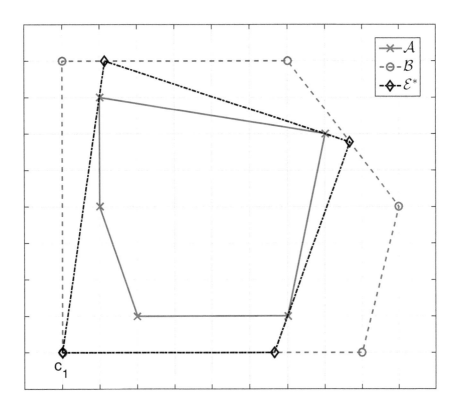

Figure 2.10. *Illustration of Algorithm 2.1: $\mathcal{A}$ is the inner polytope, $\mathcal{B}$ is the outer polytope and $\mathcal{E}^*$ is a nested polytope, between $\mathcal{A}$ and $\mathcal{B}$. The optimal solution $\mathcal{E}^*$ with four vertices is constructed starting from the point c_1.*

construction starting from p or from one of the endpoints of the segment containing p leads to an optimal solution.

Remark 2.4 (Generalization to higher dimensions). *The observation that the sets D_i (see Figure 2.9) must contain a vertex of any feasible solution can be generalized in higher dimensions: in any dimension, any hyperplane tangent to $\mathcal{A}$ defines with $\mathcal{B}$ a polytope that contains the vertex of any feasible solution. Unfortunately, it is not possible to use this observation to design an efficient algorithm in higher dimensions (the problem is NP-hard in dimension higher than two; see Theorem 2.19). However, it has been recently used to provide new lower bounds for the nonnegative rank [128].*

Example 2.18 (Nested squares). Before we go on with the complexity of Exact NMF, let us illustrate the notions covered in this section with a simple example.[18] Let

$$X = \begin{pmatrix} 1 & 1 & \epsilon & \epsilon \\ 1 & \epsilon & 1 & \epsilon \\ \epsilon & 1 & \epsilon & 1 \\ \epsilon & \epsilon & 1 & 1 \end{pmatrix} \quad \text{with} \quad \epsilon \in [0, 1]. \tag{2.15}$$

---

[18]This example has been used several times in the literature; see for example [51], [292, Example 2.1]. Example 2.12 (page 38) corresponds to the case $\epsilon = 1/2$. A more general example appears in [155, Proposition 4], where the authors consider a rectangle nested inside a square. Another similar example is presented in Section 4.2.5.

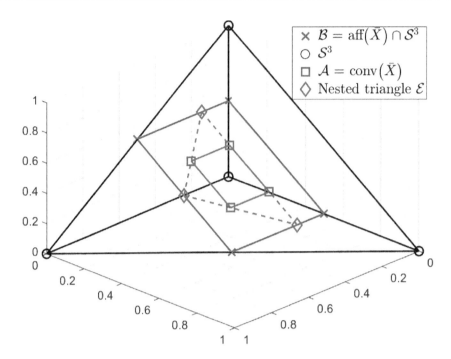

Figure 2.11. *Geometric illustration of Exact NMF for the 4-by-4 matrix X from* (2.15) *with $\epsilon = \sqrt{2} - 1$. The matrix $\overline{X}$ is the scaled matrix X where the last coordinate is discarded. The columns of $\mathrm{col}(\overline{X}) \cap \mathcal{S}^3$ are the points $(0.5, 0.5, 0)$, $(0.5, 0, 0.5)$, $(0, 0.5, 0)$, and $(0, 0, 0.5)$, while the columns of $\overline{X}$ are $(0.5 - \delta, 0.5 - \delta, \delta)$, $(0.5 - \delta, \delta, 0.5 - \delta)$, $(\delta, 0.5 - \delta, \delta)$, and $(\delta, \delta, 0.5 - \delta)$, where $\delta = \frac{\epsilon}{2(1+\epsilon)} = \frac{2 - \sqrt{2}}{4} \approx 0.1464$.*

In the following, we address the following question: What is the nonnegative rank of X depending on the parameter $\epsilon \in [0, 1]$?

For $\epsilon = 0$, this is the matrix of Thomas from (2.5), and hence $\mathrm{rank}_+(X) = 4$ (Theorem 2.10). For $\epsilon = 1$, the problem is trivial since $\mathrm{rank}(X) = \mathrm{rank}_+(X) = 1$. For $\epsilon \in (0, 1)$, $\mathrm{rank}(X) = 3$. In fact, the sum of the first and fourth columns of X equals the sum of the other two, hence $\mathrm{rank}(X) \leq 3$, while the determinant of $X(1 : 3, 1 : 3)$ equals $-(-\epsilon)^2(1 + \epsilon)$ and is different from zero, and hence $\mathrm{rank}(X) \geq 3$. Since $\mathrm{rank}(X) = 3$ and X is a 4-by-4 matrix, $\mathrm{rank}_+(X) \in \{3, 4\}$. In the following, we show that $\mathrm{rank}_+(X) = 4$ for $\epsilon \in [0, \sqrt{2} - 1)$ and $\mathrm{rank}_+(X) = 3$ for $\epsilon \in [\sqrt{2} - 1, 1)$. To do so, we follow the same strategy used for the matrix of Thomas in Theorem 2.10 to visualize the problem and its solutions in three dimensions[19] (see page 29).

We scale the columns of X so that their entries sum to one, and we discard the last coordinate to obtain $\overline{X}$. We then draw the columns of $\overline{X}$ in three dimensions. The intersection of the affine hull of $\overline{X}$ with $\mathcal{S}^3$ is the same as that for the matrix of Thomas regardless of the value of ϵ; thus the outer polytope $\mathcal{B}$ does not change as ϵ takes different values. If $\mathrm{rank}_+(X) = 3$, there must exist a triangle between $\mathcal{A} = \mathrm{conv}(\overline{X})$ and $\mathcal{B} = \mathrm{aff}(\overline{X}) \cap \mathcal{S}^3$; this is the case, for example, in Figure 2.11 with $\epsilon = \sqrt{2} - 1$. To simplify the representation, let us work in the two-dimensional space obtained by projecting $\mathrm{col}(\overline{X}) \cap \mathcal{S}^3$ onto the span of the first two canonical basis vectors;

---

[19]It would be simpler to use the reduction from RE-NMF to the NPP, which would give an equivalent problem. However, we present here the more intuitive approach using the description from Section 2.1. This allows the reader who has skipped the proof of Theorem 2.11 to follow the line of thought.

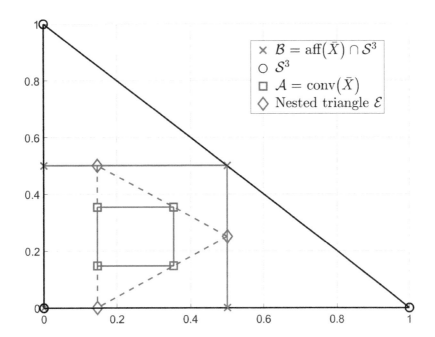

Figure 2.12. *Geometric illustration of Exact NMF for the 4-by-4 matrix X from (2.15) for $\epsilon = \sqrt{2} - 1$ in two dimensions via a projection on the (x, y) plane from Figure 2.11. The vertices of the outer square are the points $(0.5, 0.5)$, $(0.5, 0)$, $(0, 0.5)$, and $(0, 0)$, while the columns of the inner square are $(0.5 - \delta, 0.5 - \delta)$, $(0.5 - \delta, \delta)$, $(\delta, 0.5 - \delta)$, and (δ, δ) where $\delta = \frac{\epsilon}{2(1+\epsilon)} = \frac{2-\sqrt{2}}{4} \approx 0.1464$.*

see Figure 2.12 for an illustration. This projection is equivalent to viewing Figure 2.11 from above. In this two-dimensional projection, we have an NPP with two nested squares where the size of the inner square depends on the value of ϵ. Figure 2.13 depicts another example of this projection with $\epsilon = 0.25$; in this case, $\mathrm{rank}_+(X) = 4$ because no triangle can be nested between the two squares. How can we find the smallest value of ϵ such that $\mathrm{rank}_+(X) = 3$? This question can be answered using the arguments provided by Silio [420] and Aggarwal et al. [3] and using Algorithm 2.1.

Consider the example presented in Figure 2.13. We would like to find the smallest value of ϵ for which $\mathrm{rank}_+(X) = 3$. If we define the parameter $\delta = \frac{\epsilon}{2(1+\epsilon)}$, then by the observation O2 (page 44), the convex hull of the points $\{(0, 0), (0, 0.5), (\delta, 0.5), (\delta, 0)\}$ must contain a vertex of any polygon nested between the two squares $\mathcal{A}$ and $\mathcal{B}$.

If Algorithm 2.1 is initialized with $p_1 = (\delta, 0)$, the second vertex generated by the method is $f(p_1) = p_2 = (\delta, 0.5)$. In order to have $\mathrm{rank}_+(X) = 3$, we must be able to contain $\mathcal{A}$ within a triangle, and by the symmetry of the example the final vertex must be beyond $p_3 = (0.5, 0.25)$. If the secant line between p_1 and p_3 (similarly, between p_2 and p_3) intersects $\mathcal{A}$ at more than one point, the triangle will not contain $\mathcal{A}$. Thus if the secant line between $(\delta, 0.5)$ and $(0.5, 0.25)$ intersects the line $x = 0.5 - \delta$ at a value of $y \geq 0.5 - \delta$, the triangle will contain $\mathcal{A}$. One can check that this is possible for any $\frac{1}{4}(2 - \sqrt{2}) = 0.1464 \leq \delta < \frac{1}{4}$.

Now, let us show that the solution obtained by starting from $p_1 = (\delta, 0)$ cannot be improved upon, that is, it is an optimal solution of the NPP. If a better solution exists, it can be constructed using the contact change points between $(\delta, 0)$ and $(\delta, 0.5)$; this is the result from Silio [420] and Aggarwal et al. [3]; see Section 2.3.1.1. There are four contact change points, labeled as c_1, c_2,

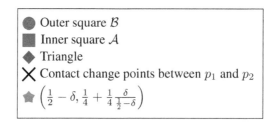

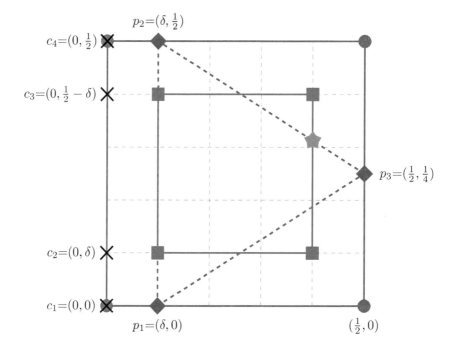

Figure 2.13. *Illustration of the nested squares problem for $\delta = 0.1$. In order for the triangle to contain the inner square, one needs the orange point with star shape to be located at $(\frac{1}{2} - \delta, \frac{1}{2} - \delta)$, that is, $\frac{1}{2} - \delta = \frac{1}{4}(1 + \frac{\delta}{0.5-\delta})$, which is achieved for $\delta^* = \frac{1}{4}(2 - \sqrt{2}) = 0.1464$.*

c_2, and c_4 in Figure 2.13. Two of them, namely, $(0, 0.5 - \delta)$ and $(0, \delta)$, lead to solutions which are rotations of the solution initialized at $(\delta, 0)$ by symmetry of the problem. Initializing with the point $(0,0)$ leads to a solution with more vertices than $(\delta, 0)$; for example, in Figure 2.12, starting from $(0,0)$ we obtain a solution which has four vertices instead of three when starting from $(\delta, 0)$. Thus the NPP has a solution with three vertices if and only if the algorithm initialized at $(\delta, 0)$ leads to a solution with three vertices. As previously noted, this is possible for $\delta \in [\frac{1}{4}(2 - \sqrt{2}), \frac{1}{4})$, and finally we see that $\mathrm{rank}_+(X) = 4$ for $\epsilon \in [0, \sqrt{2} - 1)$ and $\mathrm{rank}_+(X) = 3$ for $\epsilon \in [\sqrt{2} - 1, 1)$ because $\delta = \frac{\epsilon}{2(1+\epsilon)}$. ∎

2.3.1.2 ▪ Case $\mathrm{rank}(X) \geq 4$.

For $\mathrm{rank}(X) = 4$, RE-NMF reduces to a three-dimensional NPP. This problem has been studied in the computational geometry literature by Das [112] and Das and Joseph [114] among others and has been shown to be NP-hard when minimizing the number of facets of $\mathcal{E}$ (the reduction is from planar 3-SAT).[20] From this result, one can deduce (using a duality argument) that minimizing the number of vertices of $\mathcal{E}$ is NP-hard as well [113, 99].

Theorem 2.19. *Let $X \in \mathbb{R}_+^{m \times n}$. For $\mathrm{rank}(X) \geq 4$, finding the minimum r such that RE-NMF has a solution is NP-hard.*

Proof. This is a consequence of Theorem 2.11 and the NP-hardness results of Das and others [114, 112, 113]. $\square$

Note, however, that several approximation algorithms have been proposed in the literature. For example, Mitchell and Suri [345] propose an algorithm running in $\mathcal{O}(p^3)$ to compute $\mathrm{rank}_+^*(X)$ in case $\mathrm{rank}(X) = 4$ within an $\mathcal{O}(\log(p))$ factor, where $p = m + n$. Clarkson [99] proposes a randomized algorithm for finding $\mathcal{E}$ with at most $r_+^* \mathcal{O}(5d \ln(r_+^*))$ vertices which runs in $\mathcal{O}(r_+^{*2} p^{1+\delta})$ expected time, where $r_+^* = \mathrm{rank}_+^*(X)$, $d = \mathrm{rank}(X) - 1$, and δ is any positive fixed value.

In his seminal paper on the complexity of NMF, Vavasis [465] considers RE-NMF with $r = \mathrm{rank}(X)$. Note that, in this case, RE-NMF and Exact NMF are equivalent (Lemma 2.7). He refers to the corresponding NPP as the intermediate simplex problem where $k = d + 1$. Vavasis proves that this problem is NP-hard using a reduction of 3-SAT (see footnote 20) to the intermediate simplex problem [465, Theorem 4].

Theorem 2.20. *[465] It is NP-hard to check whether $\mathrm{rank}(X) = \mathrm{rank}_+(X)$.*

Theorem 2.20 implies that RE-NMF and Exact NMF are NP-hard.

2.3.1.3 ▪ Case r is not part of the input.

Theorem 2.20 considers that r is part of the input in RE-NMF, while Theorem 2.19 does not fix r a priori. However, if we assume instead that r is a fixed constant, then solving RE-NMF can be done in polynomial time in m and n, namely in time $\mathcal{O}\big((mn)^{cr^2}\big)$ for some constant c; see Theorem 2.21 in Section 2.3.2 and the discussion that follows.

2.3.2 ▪ Exact NMF

Let us now discuss the computational complexity of Exact NMF. We consider two cases: r part of the input and r not part of the input.

2.3.2.1 ▪ Case when r is part of the input

When $\mathrm{rank}(X) \leq 2$, Exact NMF coincides with RE-NMF and $\mathrm{rank}_+(X) = \mathrm{rank}_+^*(X) = \mathrm{rank}(X)$. Thus Exact NMF can also be solved in polynomial time in this situation.

When $\mathrm{rank}(X) = 3$, Theorem 2.17 shows that the minimal r for which an RE-NMF exists can be found in polynomial time; however, the computational complexity of finding the minimal

---

[20]SAT, or satisfiability, is an instrument problem in computational complexity used to prove NP-completeness results. It refers to the problem of deciding whether a set of clauses composed of Boolean variables, or their negation, can be satisfied. SAT is the first problem ever proved to be NP-complete. 3-SAT refers to the subset of satisfiability problems where each clause contains exactly three Boolean variables. It can be shown that SAT reduces to 3-SAT and vice versa; see [180] for more details. The incidence graph of a 3-SAT instance is defined as follows. The vertices of the incidence graph are the variables and the clauses, while the edges connect variables to clauses containing them. Planar 3-SAT requires the incidence graph to be planar.

r for which an Exact NMF exists is still unknown to the best of our knowledge. In the next section, we explain that Exact NMF can be solved in polynomial time when r is a fixed constant. However, this does not allow us to conclude that Exact NMF can be solved in polynomial time when $\mathrm{rank}(X) = 3$ because the nonnegative rank of rank-three matrices cannot be bounded above as a function of the rank. In Chapter 3 we will see some examples of this. In particular page 92 describes a class of n-by-n matrices of rank three whose nonnegative rank is $\Omega(\sqrt{n})$. Note that if $\min\left(\mathrm{rank}_+^*(X), \mathrm{rank}_+(X^\top)\right) \leq \mathrm{rank}(X) + 2 = 5$, then Algorithm 2.1 can be used to compute $\mathrm{rank}_+(X)$ since the RE-NMF of X or of $X^\top$ provides a solution for Exact NMF; see Corollary 2.15.

Deciding whether there exists a solution to Exact NMF for $r = \mathrm{rank}(X)$ is NP-hard, which was proved by Vavasis [465]; see Theorem 2.20. Note that $r = \mathrm{rank}(X)$ is part of the input meaning that, unless P=NP, there is no algorithm polynomial in r and in the size of X that solves Exact NMF. We refer the reader to [414] for a different proof using algebraic arguments (instead of geometric ones) and the references to earlier works that implied NP-hardness of NMF. Moreover, Arora et al. [15, 16] showed that there is no algorithm for solving Exact NMF that runs in time $(mn)^{o(r)}$ unless 3-SAT can be solved in time $2^{o(n)}$ (which is believed to be impossible unless P=NP).

2.3.2.2 ▪ Case when r is not part of the input

In practice, r is usually small, so it is meaningful to wonder whether Exact NMF can be solved in polynomial time in m and n for a small, fixed r (in other words, the complexity is polynomial in m and n but not in r). It turns out that the answer is yes, that is, Exact NMF can be solved in polynomial time when r is not considered part of the input. This can be proved using the seminal result of Basu, Pollack, and Roy [26] based on quantifier elimination. Their result shows that finding a point in a semialgebraic set, that is, a set defined via polynomial equations and strict polynomial inequalities, with p constraints on polynomials of degree at most d with q variables can be done in time $\mathcal{O}\left((pd)^{cq}\right)$ for some constant c.

In its original form, Exact NMF is written with $mr + nr$ variables (the entries of W and H) as mn equalities of degree two, namely $X = WH$. Hence Exact NMF can be decided in time $\mathcal{O}\left((mn)^{c(mr+nr)}\right)$ for some constant c. Note that the nonnegativity constraints can be achieved by using auxiliary variables whose squares are equal to the entries of W and H, leading to equations of degree four but no inequality constraints, which does not change the overall complexity.

In the particular case of $r = \mathrm{rank}(X)$, the number of variables to formulate Exact NMF can be drastically reduced.

Theorem 2.21. *[16, Lemma 2.2], [16, Corollary 3.20] Deciding whether there exists an Exact NMF of an m-by-n matrix X with $r = \mathrm{rank}(X)$ can be done in time $\mathcal{O}\left((mn)^{cr^2}\right)$ for some constant c.*

Proof. Let $C = X(:, \mathcal{K})$ with $|\mathcal{K}| = r$ be a subset of linearly independent columns of X, and let $R = X(\mathcal{L}, :)$ with $|\mathcal{L}| = r$ be a subset of linearly independent rows of X. In any Exact NMF of $X = WH$ with $r = \mathrm{rank}(X)$, $\mathrm{col}(X) = \mathrm{col}(W)$ and $\mathrm{col}(X^\top) = \mathrm{col}(H^\top)$; see Lemma 2.7. Hence there exists an Exact NMF (W, H) of X of size r with $X = WH$ if and only if there exist $S \in \mathbb{R}^{r \times r}$ and $T \in \mathbb{R}^{r \times r}$ such that

$$W = CS \geq 0, \quad H = TR \geq 0, \quad \text{and} \quad X = CSTR.$$

The reformulation in terms of the variables S and R has $2r^2$ variables and $\mathcal{O}(mn)$ constraints of degree at most two. $\square$

Theorem 2.21 also applies to RE-NMF since $r = \text{rank}(X)$, hence $\text{rank}(W) = r$ (Lemma 2.7).

Arora et al. were able to obtain a reformulation of Exact NMF in the general case, that is, when $\text{rank}(W)$ is possibly larger than $\text{rank}(X)$, using only $\mathcal{O}(r^2 2^r)$ variables and hence making it solvable in polynomial time for r fixed [15, 16]. Later, Moitra [351] was able to reduce it to $\mathcal{O}(r^2)$ variables, essentially as small as that for the case $r = \text{rank}(X)$. These constructions are rather complicated; we refer the reader to [15, 351, 16] for the details.

2.3.2.3 ▪ Practical implications

Although r is usually small in practice, the approach based on quantifier elimination described above cannot currently be used to solve practical problems because of its high computational cost: although the term $\mathcal{O}((mn)^{cr^2})$ is a polynomial in m and n for r fixed, it grows extremely fast. Even to factorize a 4-by-4 matrix with $r = 3$, current such algorithms fail to produce a solution; we have $(mn)^{r^2} = 16^9 > 10^{10}$ (which does not consider c or the hidden constants in the Big O notation). In contrast, most heuristic NMF algorithms run in $\mathcal{O}(mnr)$ operations; see Section 8. Hence designing a practical algorithm for Exact NMF when r is small is a direction for future research. Note that, on top of the complexity of the problem, there is another obstacle to designing such an algorithm: even for a rational input matrix, Exact NMF might require factors with irrational entries; see Section 2.3.3.

Algorithms for Exact NMF Not many algorithms exist that are designed specifically for Exact NMF. Most numerical algorithms are developed for approximation problems, such as

$$\min_{W \geq 0, H \geq 0} \|X - WH\|_F^2.$$

If such an algorithm obtains a reconstruction error equal to zero, that is, $X = WH$, then it has solved the corresponding Exact NMF problem. We refer the reader to Chapter 8 for the description of such algorithms.

Let us mention a few heuristic approaches (that is, they do not come with theoretical guarantees) that focus on Exact NMF:

- In [460], Vandaele et al. specifically designed globalization heuristics to tackle Exact NMF. These include a simulated annealing approach and a greedy randomized adaptive search procedure. These heuristics were able to compute solutions for a wide class of nonnegative matrices of small size (with dimensions up to around 100).

- Dong, Lin, and Chu [137] consider an algorithm that sequentially computes rank-one factors in the case $\text{rank}(X) = \text{rank}_+(X)$. At each step, a nonnegative rank-one factor is subtracted from the current residual in order to reduce its rank by one. Their approach is based on the Wedderburn rank reduction formula [479].

- For $r = \text{rank}(X)$, the number of variables in the problem can be drastically reduced from $mr + nr$ to $2r^2$; see the proof of Theorem 2.21. This approach has been used in particular in the SMCR literature; see [366] and the references therein. Other works have used the same idea, for example in [251] for symNMF where $W = H^\top$.

2.3.3 ▪ Are RE-NMF and Exact NMF in NP?

In 1993, Cohen and Rothblum [100] asked whether there always exists a rational Exact NMF of size $\text{rank}_+(X)$ for an input matrix X with rational entries (rational Exact NMF means that the entries of W and H are rational). This question is related to the problem of whether Exact NMF is in NP which requires that its solution be able to be written as a polynomial in the size of the input. It turns out that the answer to Cohen and Rothblum's question is no. Explicit counterexamples were obtained independently by Shitov [412, 414] and Chistikov et al. [89]. Shitov's counterexample is a 21-by-21 matrix with entries in $\{0, 1, 2\}$ whose nonnegative rank is less than 19, while there is no rational NMF of size smaller than 20. Chistikov et al. created a counterexample with a 6-by-11 rational matrix whose nonnegative rank is 5 while over the rationals it is 6. The same result holds for RE-NMF, as shown by Chistikov et al. [88]. In the case $r = \text{rank}(X) = \text{rank}_+(X)$, the problem is still open as the above counterexamples satisfy $\text{rank}(X) < \text{rank}_+(X)$. However, Shitov [416] showed that there exists a nonnegative matrix whose entries belong to a subfield of $\mathbb{R}$ with $r = \text{rank}(X) = \text{rank}_+(X)$, while its nonnegative rank over that subfield (meaning that the entries of W and H are required to belong to that subfield) is strictly larger than r.

Another closely related result is the fact that Exact NMF is $\exists\mathbb{R}$-complete, that is, Exact NMF is equivalent to deciding if a given system of polynomial equations with integral coefficients has a real solution [413]. The same holds for RE-NMF [135].

2.4 ▪ Take-home messages

The three main take-home messages from this chapter are as follows:

1. Exact NMF can be reduced to finding a polytope, $\text{conv}\left(\theta(W)\right)$, nested between two given polytopes, $\text{conv}\left(\theta(X)\right) \subseteq \Delta^m$. The dimension of the convex hull of $\theta(X)$ is $\text{rank}(X) - 1$ (Lemma 2.5) and the dimension of Δ^m is $m - 1$, while the dimension of $\text{conv}\left(\theta(W)\right)$ is unknown in advance but belongs to the interval $\left[\text{rank}(X) - 1, r - 1\right]$, since $\text{rank}(W) \in [\text{rank}(X), r]$.

2. RE-NMF imposes that the dimension of $\text{conv}\left(\theta(W)\right)$ is the same as that of $\text{conv}\left(\theta(X)\right)$ by imposing $\text{rank}(W) = \text{rank}(X)$. The smallest size of an RE-NMF is referred to as the restricted nonnegative rank of X and denoted $\text{rank}_+^*(X)$. RE-NMF is equivalent to the NPP (where the inner and outer polytopes have the same dimension); see Theorem 2.11. RE-NMF can be solved in polynomial time when $\text{rank}(X) = 3$ using an algorithm for the two-dimensional NPP (Algorithm 2.1). For $\text{rank}(X) \geq 4$, it is NP-hard to compute $\text{rank}_+^*(X)$ (Theorem 2.19).

3. Exact NMF is NP-hard. If r is fixed and not part of the input (that is, r is considered as a fixed constant), it can be solved in polynomial time, namely in time $\mathcal{O}\left(mn^{\mathcal{O}(r^2)}\right)$. However, this is not very useful in practice because of the high computational cost and, as far as we know, all available existing algorithms are heuristics and come with no global optimality guarantee.

Chapter 3

Nonnegative rank

This chapter discusses the nonnegative rank and is oriented toward theoretical results. Recall that the nonnegative rank of a nonnegative matrix X is the smallest r such that an Exact NMF of X exists and is denoted $\mathrm{rank}_+(X)$. In other words, the nonnegative rank of X is the smallest r such that X can be decomposed as the sum of r nonnegative rank-one terms $X = \sum_{p=1}^{r} W(:,p)H(p,:)$, where $W \geq 0$ and $H \geq 0$. By convention, the nonnegative rank of a matrix containing only zero entries is equal to zero.

If your interests are rather geared toward data analysis applications, this chapter can be skipped.

Organization of the chapter   In Section 3.1, several properties of the nonnegative rank are presented. In Section 3.2, we show how the nonnegative rank is impacted by small and by rank-one perturbations of the input matrix. In Section 3.3, we discuss what the nonnegative rank of randomly generated matrices is. In Sections 3.4 and 3.5, several techniques for computing lower and upper bounds for the nonnegative rank are presented, respectively. Finally, we discuss the link between the nonnegative rank and several closely related problems, namely the compact representations of polytopes which led to recent breakthrough results in combinatorial optimization (Section 3.6), and communication complexity (Section 3.7) as well as other applications (Section 3.8). The chapter is concluded with take-home messages (Section 3.9).

3.1 ▪ Some properties of the nonnegative rank

The paper by Cohen and Rothblum [100] (1993) is the first to thoroughly investigate the nonnegative rank, its properties, and its applications. Let us start with some properties that were proved in their paper.

Theorem 3.1. *Let* $X \in \mathbb{R}_+^{m \times n}$. *The following properties hold:*

(i) $\mathrm{rank}(X) \leq \mathrm{rank}_+(X) \leq \min(m, n)$, *where the inequalities can be strict.*

(ii) $\mathrm{rank}_+(X) = \mathrm{rank}_+(X^\top)$.

(iii) *For any* $Y \in \mathbb{R}_+^{m \times n}$, $\mathrm{rank}_+(X + Y) \leq \mathrm{rank}_+(X) + \mathrm{rank}_+(Y)$.

(iv) *For any* $Y \in \mathbb{R}_+^{n \times p}$, $\mathrm{rank}_+(XY) \leq \min\big(\mathrm{rank}_+(X), \mathrm{rank}_+(Y)\big)$.

(v) *For any* $Y \in \mathbb{R}_+^{p \times q}$, $\text{rank}_+ \begin{pmatrix} X & 0 \\ 0 & Y \end{pmatrix} = \text{rank}_+(X) + \text{rank}_+(Y)$.

(vi) *If* $\text{rank}(X) \leq 2$ *or* $\min(m,n) \leq 3$ *or* $\min(m,n) = \text{rank}(X)$, *then*

$$\text{rank}_+(X) = \text{rank}(X).$$

Proof. (i) The rank of an m-by-n matrix X can be defined as the smallest r such that there exists $W \in \mathbb{R}^{m \times r}$ and $H \in \mathbb{R}^{r \times n}$ with $X = WH$. Since the nonnegative rank is defined in the same way with the additional constraint that U and V are componentwise nonnegative, $\text{rank}(X) \leq \text{rank}_+(X)$. The second inequality is implied by the trivial factorizations $X = XI_n = I_m X$. We have proved in Section 2.1 that the 6-by-6 matrix X from Equation (2.8) corresponding to the nested hexagon problem satisfies

$$\text{rank}(X) = 3 < \text{rank}_+(X) = 5 < \min(m,n) = 6,$$

which provides an example where the inequalities are strict.
(ii) This follows directly from $X = WH \iff X^\top = H^\top W^\top$.
(iii) Let (W, H) be an Exact NMF of X with $r_X = \text{rank}_+(X)$ rank-one factors, and let (U, V) be an Exact NMF of Y with $r_Y = \text{rank}_+(Y)$ rank-one factors. We have

$$X + Y = WH + UV = (W \ U) \begin{pmatrix} H \\ V \end{pmatrix},$$

which provides an Exact NMF of $X + Y$ with $r_X + r_Y$ factors, which gives the result.
(iv) Let (W, H) be an Exact NMF of X with $r_X = \text{rank}_+(X)$ rank-one factors, and let (U, V) be an Exact NMF of Y with $r_Y = \text{rank}_+(Y)$ rank-one factors. We have

$$XY = WHUV = W \underbrace{(HUV)}_{A \geq 0} = \underbrace{(WHU)}_{B \geq 0} V.$$

Hence WA is an Exact NMF of XY with r_X rank-one factors, and BV is an Exact NMF of XY with r_Y rank-ones factors, which gives the result.
(v) Let us consider the following Exact NMF:

$$Z = \begin{pmatrix} X & 0 \\ 0 & Y \end{pmatrix} = WH = \begin{pmatrix} W_1 \\ W_2 \end{pmatrix} \begin{pmatrix} H_1 & H_2 \end{pmatrix} = \begin{pmatrix} W_1 H_1 & W_1 H_2 \\ W_2 H_1 & W_2 H_2 \end{pmatrix}.$$

We must have $W_1 H_2 = 0$ and $W_2 H_1 = 0$. Since $W_i \geq 0$ and $H_i \geq 0$ for $i = 1, 2$, $W_1(:, k) \neq 0$ implies $H_2(k, :) = 0$ and vice versa. Similarly, $W_2(:, k) \neq 0$ implies $H_1(k, :) = 0$ and vice versa. This means that each rank-one factor $W(:, k)H(k, :)$ can have nonzero entries in positions corresponding to X or Y but not for both simultaneously. Therefore we need at least $\text{rank}_+(X) + \text{rank}_+(Y)$ rank-one factors in any Exact NMF of Z. If $W_X H_X$ is an Exact NMF of X of size $\text{rank}_+(X)$, and $W_Y H_Y$ is an Exact NMF of Y of size $\text{rank}_+(Y)$, then we can easily construct an Exact NMF of Z of size $\text{rank}_+(X) + \text{rank}_+(Y)$ as

$$Z = \begin{pmatrix} X & 0 \\ 0 & Y \end{pmatrix} = \begin{pmatrix} W_X & 0 \\ 0 & W_Y \end{pmatrix} \begin{pmatrix} H_X & 0 \\ 0 & H_Y \end{pmatrix} = \begin{pmatrix} W_X H_X & 0 \\ 0 & W_Y H_Y \end{pmatrix},$$

which concludes the proof.
(vi) The case $\text{rank}(X) \leq 2$ is the result of Thomas [450]; see Theorem 2.6. The case $\min(m,n) = \text{rank}(X)$ follows from the trivial decompositions $X = XI_n = I_m X$. The case $\min(m,n) \leq 3$ follows from the two previous cases: since $\text{rank}(X) \leq \min(m,n)$, either $\text{rank}(X) \leq 2$ or $\text{rank}(X) = 3 = \min(m,n)$, and in both cases $\text{rank}_+(X) = \text{rank}(X)$. $\qquad\square$

3.2 ▪ The nonnegative rank under perturbations

In this section, we discuss how the nonnegative rank of a nonnegative matrix X is modified under small and rank-one perturbations of X in Sections 3.2.1 and 3.2.2, respectively.

3.2.1 ▪ Lower semicontinuity

An important property of the nonnegative rank is that it is lower semicontinuous, that is, it can only increase in a neighborhood of a matrix.

Theorem 3.2. *[51, Theorem 3.1] Let $X \in \mathbb{R}_+^{m \times n}$ without zero columns and $\mathrm{rank}_+(X) = k$. Then there exists a ball $B(X, \epsilon) = \{Y \mid \|X - Y\|_F \leq \epsilon\}$ for some $\epsilon > 0$ such that $\mathrm{rank}_+(Y) \geq k$ for all $Y \in B(X, \epsilon) \cap \mathbb{R}_+^{m \times n}$.*

Proof. We refer the interested reader to [51, Theorem 3.1] for a formal proof which is based on the geometric interpretation of NMF: in the problem of finding a nested polytope between an inner and an outer polytope, perturbing slightly the inner and outer polytopes cannot lead to a nested polytope with fewer vertices. □

We have already encountered matrices such that $\mathrm{rank}_+(X) > \mathrm{rank}(X)$; see for example (2.5) and (2.6). Together with Theorem 3.2, this implies that the set of matrices of rank r with $3 \leq r < \min(m, n)$ and nonnegative rank strictly larger than r, that is, the set

$$\left\{ X \in \mathbb{R}_+^{m \times n} \mid \mathrm{rank}(X) = r < \mathrm{rank}_+(X) \right\},$$

has a positive measure within the set $\{X \in \mathbb{R}_+^{m \times n} \mid \mathrm{rank}(X) = r\}$ of matrices of rank r; see also Section 3.3 for a discussion.

3.2.2 ▪ Rank-one perturbations

Given a matrix $X \in \mathbb{R}^{m \times n}$ and any rank-one perturbation $yz^\top$ with $y \in \mathbb{R}^m$ and $z \in \mathbb{R}^n$, it is well-known that

$$\mathrm{rank}(X) - 1 \leq \mathrm{rank}\left(X + yz^\top\right) \leq \mathrm{rank}(X) + 1.$$

One may wonder whether this property holds for the nonnegative rank. For $X \in \mathbb{R}_+^{m \times n}$, $y \in \mathbb{R}_+^m$, and $z \in \mathbb{R}_+^n$,

$$\mathrm{rank}_+\left(X + yz^\top\right) \leq \mathrm{rank}_+(X) + 1,$$

which follows from Theorem 3.1(iii).

What about lower bounds for $\mathrm{rank}_+\left(X + yz^\top\right)$? Can the nonnegative rank of $X + yz^\top$ be smaller than the nonnegative rank of X minus one? The answer is yes. This result is a consequence of the following theorem;[21] see Corollary 3.5 below.

Theorem 3.3. *For any nonnegative matrix $X \in \mathbb{R}_+^{m \times n}$, there exists $y \in \mathbb{R}_+^m$ and $z \in \mathbb{R}_+^n$ such that*

$$\mathrm{rank}_+\left(X + yz^\top\right) = \mathrm{rank}(X).$$

---

[21]Although it is relatively easy to prove, Theorem 3.3 is a result not present in the literature to the best of our knowledge.

Proof. First, w.l.o.g., let us

- discard the zero columns of X: it does not impact its rank or its nonnegative rank, while taking the entries of z corresponding to zero columns of X equal to zero, that is, $z(i) = 0$ for all i such that $X(:, i) = 0$, does not impact the rank or the nonnegative rank of $X + yz^\top$ (the corresponding columns of $X + yz^\top$ are zero);

- normalize the columns of X to unit ℓ_1 norm. This amounts to multiplying X on the right by a diagonal matrix D with positive diagonal elements. Scaling the vector z in the same way, that is, $z \leftarrow Dz$, does not influence the rank or the nonnegative rank of X and $X + yz^\top$.

Then, let us denote $r = \operatorname{rank}(X)$, and let us construct an unconstrained factorization of $X = UV$ with $U \in \mathbb{R}^{m \times r}$ and $V \in \mathbb{R}^{r \times n}$ as follows. Pick r linearly independent columns of X to form U; hence $U \geq 0$ and the columns of U have unit ℓ_1 norm. Since the columns of U are linearly independent, there must exist V (not necessarily nonnegative) such that $X = UV$ with $\operatorname{rank}(V) = r$, and the entries in each column of V sum to one, that is, $e^\top V = e^\top$; see Lemma 2.1. Note that this is the same construction as in the proof of Theorem 2.11.

Now, let $y = Ue \geq 0$ and $z = \alpha e$ where $\alpha = |\min_{i,j} V(i,j)|$ so that $V' = V + \alpha ee^\top \geq 0$. We have

$$X + yz^\top = UV + \alpha Uee^\top = U\left(V + \alpha ee^\top\right) = UV',$$

which provides an Exact NMF of $X + yz^\top = AV'$ of size $r = \operatorname{rank}(X)$, which proves $\operatorname{rank}_+\left(X + yz^\top\right) \leq r$.

To prove $\operatorname{rank}_+\left(X + yz^\top\right) \geq r$, it suffices to prove that $\operatorname{rank}(X + yz^\top) = r$ since the rank is a lower bound for the nonnegative rank; see Theorem 3.1(i). To prove $\operatorname{rank}(X + yz^\top) = r$, it is sufficient to prove that V' has rank r, since $X + yz^\top = UV'$ where $U \in \mathbb{R}^{m \times r}$ and $\operatorname{rank}(U) = r$. Since $V \in \mathbb{R}^{r \times n}$ has rank r, we can construct a matrix $B \in \mathbb{R}^{r \times r}$ with r linearly independent columns of V. Let us show that $C = B + \alpha ee^\top$ has rank r, implying that V' has rank r since C is made of r columns of V'. Assume $\operatorname{rank}(C) < r$ so that there exists $x \neq 0$ such that

$$Cx = (B + \alpha ee^\top)x = 0.$$

Multiplying by $e^\top$ on both sides, we obtain

$$e^\top(B + ee^\top)x = (e^\top B + \alpha(e^\top e)e^\top)x = (1 + \alpha r)e^\top x = 0,$$

and hence $e^\top x = 0$ since $\alpha \geq 0$. This implies that $Bx = 0$, which is a contradiction since B has rank r.   □

Let us illustrate the construction of the proof of Theorem 3.3.

Example 3.4 (Illustration of Theorem 3.3). In Section 2.1.4, it was proved that the matrix

$$X = \begin{pmatrix} 0 & 1 & 2 & 2 & 1 & 0 \\ 0 & 0 & 1 & 2 & 2 & 1 \\ 1 & 0 & 0 & 1 & 2 & 2 \\ 2 & 1 & 0 & 0 & 1 & 2 \\ 2 & 2 & 1 & 0 & 0 & 1 \\ 1 & 2 & 2 & 1 & 0 & 0 \end{pmatrix}$$

satisfies $\mathrm{rank}(X) = 3$ and $\mathrm{rank}_+(X) = 5$. Let us follow the construction of Theorem 3.3 to find a rank-one perturbation to X that makes its nonnegative rank equal to $\mathrm{rank}(X) = 3$ and hence reduces its nonnegative rank by two. The first three columns of X are linearly independent, so let us take $U = X(:, 1:3)$ for which V, such that $X = UV$, is given by

$$V = \begin{pmatrix} 1 & 0 & 0 & 1 & 2 & 2 \\ 0 & 1 & 0 & -2 & -3 & -2 \\ 0 & 0 & 1 & 2 & 2 & 1 \end{pmatrix}.$$

Taking $\alpha = |\min_{i,j} V(i,j)| = 3$, we obtain

$$X + 3(Ue)e^\top = X + 3 \begin{pmatrix} 3 & 1 & 1 & 3 & 5 & 5 \end{pmatrix}^\top \begin{pmatrix} 1 & 1 & 1 & 1 & 1 & 1 \end{pmatrix}$$

$$= \begin{pmatrix} 9 & 10 & 11 & 11 & 10 & 9 \\ 3 & 3 & 4 & 5 & 5 & 4 \\ 4 & 3 & 3 & 4 & 5 & 5 \\ 11 & 10 & 9 & 9 & 10 & 11 \\ 17 & 17 & 16 & 15 & 15 & 16 \\ 16 & 17 & 17 & 16 & 15 & 15 \end{pmatrix}$$

$$= UV + 3Uee^\top = U(V + 3ee^\top)$$

$$= \begin{pmatrix} 0 & 1 & 2 \\ 0 & 0 & 1 \\ 1 & 0 & 0 \\ 2 & 1 & 0 \\ 2 & 2 & 1 \\ 1 & 2 & 2 \end{pmatrix} \begin{pmatrix} 4 & 3 & 3 & 4 & 5 & 5 \\ 3 & 4 & 3 & 1 & 0 & 1 \\ 3 & 3 & 4 & 5 & 5 & 4 \end{pmatrix},$$

which shows that the nonnegative rank of $X + 3(Ue)e^\top$ is smaller than three. By Theorem 3.3, we know it is equal to three since $\mathrm{rank}(U) = \mathrm{rank}(V + 3ee^\top) = 3$. ∎

Geometric interpretation of Theorem 3.3 Given a nonnegative matrix X, let y and z be as described in the proof of Theorem 3.3. Let us denote $C = X + yz^\top$, and let us provide the geometric intuition why $\mathrm{rank}_+(C) = \mathrm{rank}(X)$. Since $\mathrm{rank}(X) = \mathrm{rank}(C)$ (see the proof of Theorem 3.3), the nonnegative rank and the restricted nonnegative rank of C coincide, that is, $\mathrm{rank}_+^*(C) = \mathrm{rank}_+(C)$; see Theorem 2.14. By Theorem 2.11, computing $\mathrm{rank}_+^*(C)$ is equivalent to finding a nested polytope between two given full-dimensional polytopes $\mathcal{A} \subseteq \mathcal{B}$ in dimension $d = \mathrm{rank}(C) - 1$. Adding the constant vector αUe to all columns of X in the RE-NMF instance turns out to be equivalent to moving all points of $\mathcal{A}$ toward a point in the interior of $\mathcal{A}$. In particular, for $\alpha \to \infty$ and after normalization of the columns of X (which is required in the reduction from RE-NMF to NPP), all columns of X collide to a single point, namely $\frac{Ue}{\|Ue\|_1}$, so that $\mathcal{A}$ becomes a single point. In other words, the transformation $X + \alpha Uee^\top$ shrinks the inner polytope $\mathcal{A}$ in the NPP instance, while the outer polytope $\mathcal{B}$ is unchanged as it only depend on the column space of X. Therefore, for α sufficiently large, there always exists a nested polytope with $d + 1$ vertices nested between $\mathcal{A}$ and $\mathcal{B}$, and hence $\mathrm{rank}_+^*(C) = \mathrm{rank}(X)$.

Interestingly, a similar behavior was observed in Section 2.1.4: we analyzed the matrix X_a whose corresponding NPP is made of two nested hexagons, and the inner hexagon is shrunk as a decreases. If the vertices of the inner hexagon are moved sufficiently close to the center, that is, if a is sufficiently small (namely for $a \leq 2$), then a triangle fits between the two hexagons and $\mathrm{rank}_+(X_a) = 3$; see Figure 2.5. Note that in this example the perturbations are not of rank one.

Rank-one perturbations can modify the nonnegative rank by more than one   As we will see later in this chapter, there exist matrices whose rank is significantly smaller than their nonnegative rank. For example, there exist

- n-by-n matrices of rank 3 and nonnegative rank larger than $\log n$ (namely linear Euclidean distance matrices (EDMs); see Section 3.4.2);

- n-by-n matrices of rank 3 and nonnegative rank in $\Omega(\sqrt{n})$ (namely the slack matrices of generic n-gons; see Section 3.6.3.5); and

- 2^n-by-2^n matrices of rank $\frac{n(n+1)}{2} + 1$ and nonnegative rank larger than $\left(\frac{3}{2}\right)^n$ (namely some unique disjointness (UDISJ) matrices; see Section 3.7.1).

We refer the interested reader to the recent paper [295] that discusses in depth the separation between the rank and the nonnegative rank. In particular, the authors provide a family of n-by-n matrices with rank $n^{o(1)}$ and nonnegative rank $n^{1-o(1)}$. This means that there are n-by-n matrices whose ratio between the nonnegative rank and the rank is of the order of the dimension of the matrix, namely $n^{1-o(1)}$.

By Theorem 3.3, there exists nonnegative rank-one perturbations that reduce the nonnegative rank of the three classes of matrices listed above by a quantity proportional to $\log(n)$, $\sqrt{n}$, and $\left(\frac{3}{2}\right)^n$, respectively.

Also, using, for example, the 6-by-6 matrix X of Example 3.4 which satisfies $\operatorname{rank}(X) = 3$ and $\operatorname{rank}_+(X) = 5$, we can construct a $6p$-by-$6p$ matrix Y with rank $3p$ and nonnegative rank $5p$ for any $p \geq 1$. This can be done by taking Y as a block diagonal matrix with p diagonal blocks equal to X; see Theorem 3.1(v).

We therefore have the following corollary.

Corollary 3.5. *For any positive integer r, there exist nonnegative matrices X and nonnegative rank-one perturbations $yz^\top$ such that*

$$\operatorname{rank}_+(X + yz^\top) \leq \operatorname{rank}_+(X) - r.$$

Proof. This follows from Theorem 3.3 and the matrices mentioned above for which $\operatorname{rank}_+(X)$ is arbitrarily larger than $\operatorname{rank}(X)$. $\qquad\square$

Theorem 3.3 also implies that if y and z are not required to be nonnegative but only that $X + yz^\top \geq 0$, then rank-one perturbations of X can increase its nonnegative rank by more than one.

Corollary 3.6. *For any positive integer r, there exist nonnegative matrices X such that*

$$\operatorname{rank}_+(X + yz^\top) \geq \operatorname{rank}_+(X) + r,$$

where $X + yz^\top \geq 0$, $y \in \mathbb{R}^m$, and $z \in \mathbb{R}^n$ (y and z are not required to be nonnegative).

Proof. Let Y be a nonnegative matrix such that $\operatorname{rank}_+(Y) \geq \operatorname{rank}(Y) + r$; see the discussion before Corollary 3.5 for examples. By Theorem 3.3, there exists a, b such that $\operatorname{rank}_+(X) = \operatorname{rank}(Y)$ where $X = Y + ab^\top$. Now take $y = -a$ and $z = b$ so that $Y = X + yz^\top$ for which

$$\operatorname{rank}_+\left(X + yz^\top\right) = \operatorname{rank}_+(Y) \geq \operatorname{rank}(Y) + r = \operatorname{rank}_+(X) + r. \qquad\square$$

3.3 ▪ Generic values of the nonnegative rank

In this section, we address the following question: given a nonnegative matrix X randomly generated, what is the probability that $\text{rank}(X) = \text{rank}_+(X)$? To the best of our knowledge, this question has not been discussed much in the literature, and this chapter contains some new results. Of course, the answer to this question depends on how the matrix is generated.

3.3.1 ▪ Independently distributed entries

If all entries of X are independently distributed using a continuous distribution (such as the uniform distribution), then $\text{rank}(X) = \text{rank}_+(X) = \min(m, n)$ with probability one (this does not hold for discrete distributions such as the Bernoulli distribution). In fact, the probability for an m-by-n matrix whose entries are independently distributed using a continuous distribution to have rank $\min(m, n)$ is one.

If the entries of $W \in \mathbb{R}_+^{m \times r}$ and $H \in \mathbb{R}_+^{r \times n}$ are independently distributed using a continuous distribution (for example using the uniform distribution in [0,1]) and we let $X = WH$, then $\text{rank}_+(X) = \text{rank}(X) = \text{rank}(W) = \text{rank}(H) = r$ with probability one. This is closely related to the assumptions made in most data analysis applications, such as hyperspectral imaging, text mining, or audio source separation. It is usually assumed that the observed data matrix $\tilde{X} \in \mathbb{R}_+^{m \times n}$ is generated as follows:

$$\tilde{X} = X + N \quad \text{with} \quad X = WH,$$

where N is the noise, and $W \in \mathbb{R}_+^{m \times r}$ and $H \in \mathbb{R}_+^{r \times n}$ are nonnegative matrices whose rank is equal to $r = \text{rank}(X)$. In other words, it is implicitly assumed that $\text{rank}(X) = \text{rank}_+(X) = r$.

3.3.2 ▪ Geometric distribution

In some applications, in particular in the study of extended formulations (see Section 3.6), $\text{rank}(X) = \text{rank}_+(X)$ usually does not hold.

In this section, we investigate another way to randomly generate X based on the geometric interpretation of NMF. Let us start with the simplest nontrivial case, that is, when $\min(m, n) = 4$ and $\text{rank}(X) = 3$. In fact, for $\min(m, n) < 4$ and/or $\text{rank}(X) < 3$, $\text{rank}(X) = \text{rank}_+(X)$; see Theorem 3.1(vi). Without loss of generality, let us consider the case $m = 4$ and $\text{rank}(X) = 3$. In this case, $\text{rank}_+(X)$ is equal to three or four, hence $\text{rank}_+(X) = \text{rank}_+^*(X)$ (Theorem 2.14), so that deciding whether $\text{rank}_+(X) = 3$ or $\text{rank}_+(X) = 4$ is equivalent to deciding whether the corresponding NPP instance has a solution with three or four vertices.

A possible way to generate X is to first generate randomly its column space, spanned by some matrix $U \in \mathbb{R}^{m \times r}$, making sure it has a nontrivial intersection with the nonnegative orthant. This can be achieved, for example, using independently and uniformly distributed entries in the interval $[0, 1]$. Then, the columns of X are picked uniformly at random in the polytope $\text{col}(U) \cap \Delta^m$. However, such a construction is difficult to analyze. To simplify the analysis, let us consider a similar construction using the geometric interpretation of NMF and instead generate an NPP instance from which we will construct the matrix X using the one-to-one correspondence between these two problems; see Theorem 2.11. Recall that the dimension of the NPP is given by $\text{rank}(X) - 1 = 2$. We proceed as follows:

- Generate at random four points in the two-dimensional unit disk, that is, within the unit circle. The outer polygon $\mathcal{B}$ in the NPP is defined as the convex hull of these four points. Note that $\mathcal{B}$ will be either a triangle or a quadrilateral, with probability one. This is similar to the step of generating the basis U for the span of X since there is a one-to-one correspondence between $\mathcal{B}$ and the column space of X; see the proof of Theorem 2.11.

- Generate n points uniformly at random in $\mathcal{B}$ to obtain the points whose convex hull defines the inner polytope $\mathcal{A}$ in the NPP. This step is equivalent to the step of picking points uniformly at random in $\mathrm{col}(U) \cap \Delta^m$, since the points whose convex hull is $\mathcal{A}$ correspond to the columns of X; see the proof of Theorem 2.11.

- Given the NPP instance defined by $\mathcal{A} \subseteq \mathcal{B}$, generate the matrix X from the corresponding RE-NMF instance; see Theorem 2.11.

This data generation can be easily analyzed via a well-known result in geometry which gives the probability that four points generated at random in a disk are the vertices of a convex quadrilateral or a convex triangle. This is the so-called Sylvester's four-point problem. The probability of obtaining a quadrilateral is equal[22] to $1 - \frac{35}{12\pi^2} \approx 70.45\%$ [281, 223, 381]. In terms of the original NMF problem, this means that the probability for the randomly generated hyperplane corresponding to $\mathrm{col}(U) \cap \Delta^m$ to intersect the unit simplex in four segments is 70.45% and in three segments is 29.65%. In fact, any hyperplane which intersects the three-dimensional simplex in its interior has to intersect exactly three or four of its edges; see Figure 2.4 for an example with four intersections.

Lemma 3.7. *Let the n columns of the matrix X be generated as described in the previous paragraph. Then* $\mathrm{rank}(X) = 3$ *with probability one for $n \geq 3$, and*

$$\lim_{n \to \infty} \mathbb{P}\left(\mathrm{rank}_+(X) = 4\right) = 1 - \frac{35}{12\pi^2}.$$

Proof. With probability one, four points generated at random within the unit disk are not aligned, hence $\mathcal{B}$ is a two-dimensional polytope, that is, a polygon. Since the columns of $\mathcal{A}$ are generated uniformly at random within $\mathcal{B}$, they are also not aligned with probability one, as long as $n \geq 3$. This implies $\mathrm{rank}(X) = 3$; see Theorem 2.11.

As explained above, the probability that $\mathcal{B}$ has four vertices is $1 - \frac{35}{12\pi^2}$. In the NPP instance, as n goes to infinity, $\mathcal{A} \to \mathcal{B}$ as the points in $\mathcal{A}$ are generated uniformly at random within $\mathcal{B}$. Since the solution $\mathcal{E}$ must satisfy $\mathcal{A} \subset \mathcal{E} \subset \mathcal{B}$, this implies $\mathcal{E} \to \mathcal{B}$, and hence, for n sufficiently large, $\mathcal{E}$ must have four vertices. $\square$

The above result can be easily generalized in the case $\mathrm{rank}(X) = m - 1$ because the probability for the convex hull of $m + 2$ points picked uniformly at random in the m-dimensional unit ball to have $m + 2$ vertices is also known [281, 223, 381] and given by

$$1 - \frac{(n+2)\left(\frac{1}{2}(n+1)\right)^{n+1}}{2^n \left(\frac{(n+1)^2}{\frac{1}{2}(n+1)^2}\right)^{n+1}}.$$

3.4 ▪ Lower bounds on the nonnegative rank

In some applications, it is key to compute bounds on the nonnegative rank. In particular, lower bounds are particularly useful for the study of extended formulations as they correspond to a lower bound on the size of linear programs over a given polytope; see Section 3.6. These bounds are also useful in other contexts, such as communication complexity; see Sections 3.7 and 3.8.

Many lower bounds are based on the sparsity pattern of the input matrix X. The zero pattern of X implies a zero pattern on any of its Exact NMF $X = WH$. This can be leveraged to bound the nonnegative rank. In particular, if $X(i,j) = 0$ for some (i,j), then for all p, we must have

---

[22]See also http://mathworld.wolfram.com/SylvestersFour-PointProblem.html (consulted June 24, 2019).

$W(i,p) = 0$ or $H(p,j) = 0$, otherwise, $(WH)_{i,j} > 0$. In the following, we review several lower bounding techniques for the nonnegative rank.

3.4.1 ▪ Fooling sets

One of the first lower bounding techniques for the nonnegative rank is based on the identification of a fooling set, which was first introduced in [100] (1993) and referred to as a set of pairwise independent entries; see also [162] for more details.

Definition 3.8 (Pairwise independent entries). *Given a matrix $X \in \mathbb{R}_+^{m \times n}$, two entries (i,j) and (k,ℓ) are pairwise independent if*

(i) *$X(i,j) > 0$ and $X(k,\ell) > 0$, and*

(ii) *$X(i,\ell) = 0$ or $X(k,j) = 0$.*

The two conditions in Definition 3.8 imply that $i \neq k$ and $j \neq \ell$; otherwise, we have a contradiction (an entry would have to be positive and equal to zero simultaneously). Note that (i) can be equivalently formulated as $X(i,j)X(k,\ell) > 0$, and (ii) as $X(i,\ell)X(k,j) = 0$.

Let us illustrate this notion with the matrix of Thomas [450]:

$$X = \begin{pmatrix} 1 & 1 & 0 & 0 \\ 1 & 0 & 1 & 0 \\ 0 & 1 & 0 & 1 \\ 0 & 0 & 1 & 1 \end{pmatrix}.$$

The entries of X at positions (1,1) and (2,3) are pairwise independent since $X(1,1) = X(2,3) = 1$ and $X(1,3) = 0$. The entries at positions (1,1) and (3,4) are also pairwise independent. As a matter of fact, any two entries (i,j) and (k,ℓ) of X such that $X(i,j)X(k,\ell) > 0$, $i \neq k$, and $j \neq \ell$, are pairwise independent.

The following lemma provides a simple observation about pairwise independent entries.

Lemma 3.9. *Let $X = WH$ for some nonnegative matrices W and H, and let the two entries of X at positions (i,j) and (k,ℓ) be pairwise independent. Then no rank-one factor $W(:,p)H(p,:)$ for any p can be positive simultaneously on these two entries: for all p,*

$$W(i,p)H(p,j) = 0 \quad or \quad W(k,p)H(p,\ell) = 0.$$

Proof. Assume there exists p such that

$$W(i,p)H(p,\ell) > 0 \text{ and } W(k,p)H(p,j) > 0.$$

This contradicts $X = WH$ since $X(i,\ell) = 0$ or $X(k,j) = 0$ as (i,j) and (k,ℓ) are pairwise independent, while W and H are nonnegative (no cancellation is possible). □

We can now define a fooling set. A fooling set of a nonnegative matrix X is a set of entries that are all pairwise independent. It leads to a lower bound for the nonnegative rank.

Lemma 3.10. *If $X \in \mathbb{R}_+^{m \times n}$ is a nonnegative matrix with a fooling set of size f, then*

$$\mathrm{rank}_+(X) \geq f.$$

Proof. Let (W, H) be an Exact NMF of X of size $\text{rank}_+(X)$. Since the rank-one factors $W(:, p)H(p, :)$ cannot be positive simultaneously for two entries in the fooling set (Lemma 3.9), and since $X = WH$ implies that WH must be positive at least once in each entry in the fooling set, this implies that $\text{rank}_+(X) \geq f$.   □

The size of the largest fooling set of X, denoted $\omega(X)$, is referred to as the *fooling set bound*.

Lemma 3.10 leads to a simple and direct proof that the matrix of Thomas [450] has nonnegative rank four: $\{(1, 1), (2, 3), (3, 2), (4, 4)\}$ is a fooling set of size 4, represented as follows using framed entries:

$$\begin{pmatrix} \boxed{1} & 1 & 0 & 0 \\ 1 & 0 & \boxed{1} & 0 \\ 0 & \boxed{1} & 0 & 1 \\ 0 & 0 & 1 & \boxed{1} \end{pmatrix}.$$

Note that it is not possible to have a larger fooling set, that is, $\omega(X) = 4$, since X is a 4-by-4 matrix.

Rectangle graph and link with the clique number

Let us define the rectangle graph of a matrix X.

Definition 3.11. *The rectangle graph of X, denoted $G(X) = (V, E)$, is constructed as follows. The set of vertices corresponds to the positive entries of X, that is, $V = \{(i, j) \mid X(i, j) > 0\}$, and an edge connects two vertices (i, j) and (k, ℓ) in V if and only if they are pairwise independent, that is, $X(i, \ell) = 0$ or $X(k, j) = 0$.*

By construction, the clique number of $G(X)$, that is, the size of the largest subset of vertices that are all connected, is equal to the fooling set bound of X. Note that the clique number is hard to compute in general [180].

Limitations of the fooling set bound

The fooling set bound does not always provide strong lower bounds, especially for matrices with only a few zero entries. In particular, Fiorini et al. showed the following.

Lemma 3.12. *[162, Lemma 5.5] If every row or if every column of X contains at most s zero entries, then $\omega(X) \leq 2s + 1$.*

Proof. Consider a fooling set of X of size $f = \omega(X)$. The f-by-f submatrix of X induced by the rows and columns of the entries belonging to this fooling set contains at least $\binom{f}{2} = f(f - 1)/2$ zeros, because every pair of entries in the fooling set must correspond to at least one zero in X. This implies that one row of the submatrix has at least $(f - 1)/2$ zeros. In the case when every row of X contains at most s zero entries (the same result holds for the columns by symmetry), we obtain $(f - 1)/2 \leq s$, that is, $f \leq 2s + 1$.   □

Example 3.13 (Linear EDMs). Throughout this chapter, we will illustrate lower bounds for the nonnegative rank on the class of linear EDMs. Linear EDMs of size n-by-n are defined as follows:

$$X(i, j) = (x_i - x_j)^2 \quad \text{for } i, j = 1, 2, \ldots, n,$$

where $x \in \mathbb{R}^n$ and[23] $x_i \neq x_j$ for all $i \neq j$. For example, for $x_i = i$ for all i and $n = 6$, we obtain

$$X_6 = \begin{pmatrix} 0 & 1 & 4 & 9 & 16 & 25 \\ 1 & 0 & 1 & 4 & 9 & 16 \\ 4 & 1 & 0 & 1 & 4 & 9 \\ 9 & 4 & 1 & 0 & 1 & 4 \\ 16 & 9 & 4 & 1 & 0 & 1 \\ 25 & 16 & 9 & 4 & 1 & 0 \end{pmatrix}. \tag{3.1}$$

Linear EDMs X of size n-by-n have zeros only on their diagonal, with a single zero entry on each row, hence the size of the largest fooling set is at most 3, that is, $\omega(X) \leq 3$ (Lemma 3.12). We have that $\omega(X) = 3$ for $n \geq 3$; take for examples the entries at positions (1,2), (2,3), and (3,1) to form a fooling set of size 3.

What is the nonnegative rank of linear EDMs? First, their rank is at most three [27]: since $(x_i - x_j)^2 = x_i^2 - 2x_i x_j + x_j^2$,

$$X = \begin{bmatrix} e & x & x \circ x \end{bmatrix} \begin{bmatrix} x \circ x & -2x & e \end{bmatrix}^\top,$$

where $x \circ x$ is the componentwise (Hadamard) product of x with itself. One can easily check that it is exactly three for $n \geq 3$, showing that any 3-by-3 linear EDM has three linearly independent columns.

Let us try to understand the geometry of the NPP instance corresponding to linear EDMs. This will allow us to determine their restricted nonnegative rank. First, since the rank of these matrices is equal to three, the corresponding NPP has dimension 2, and $\mathcal{A}$ and $\mathcal{B}$ are polygons. Second, recall that the entries of matrix X of the Exact NMF instance are given by $X(:,j) = Fv_j + g$ for all j, where $\mathcal{A} = \text{conv}(\{v_1, v_2, \ldots, v_n\})$ and $\mathcal{B} = \{x \in \mathbb{R}^r | Fx + g \geq 0\}$ with $v_j \in \mathbb{R}^2$ for all j, $F \in \mathbb{R}^{n \times 2}$, and $g \in \mathbb{R}^n$. The equality $X(i,j) = F(i,:)v_j + g(i) = 0$ implies that the jth point of $\mathcal{A}$, that is, v_j, is located on the ith segment of $\mathcal{B}$. For a rank-three matrix X with only zeros on its diagonal, the jth point of $\mathcal{A}$ is located on the jth segment of $\mathcal{B}$. In other words, there is a one-to-one correspondence between the n points of $\mathcal{A}$ and the n edges of $\mathcal{B}$. Therefore, $\mathcal{B}\backslash\mathcal{A}$ is made of n disjoint regions which implies that $\text{rank}_+^*(X) = n$ [199]: any solution of this NPP instance must contain a point in each region; see Section 2.3.1, and see Figure 3.1 for an illustration of this geometric observation on the matrix X_6 from (3.1).

What about the nonnegative rank of such matrices? It is not possible to obtain a fooling set of size larger than 3; hence we can only obtain the lower bound $\text{rank}_+(X) \geq \text{rank}(X) = 3$ using fooling sets. Using the facts that $\text{rank}_+^*(X) = \text{rank}_+(X)$ for $\text{rank}_+^*(X) \leq \text{rank}(X) + 1$ (Theorem 2.14) and that linear EDMs are submatrices of larger linear EDMs, we conclude that, for $n \geq 4$, the nonnegative rank of linear EDMs is larger than four. ∎

Remark 3.1. *For any nonnegative m-by-n rank-three matrix whose columns have supports not contained in one another (the support is the index set of nonzero entries), $\text{rank}_+^*(X) = n$ [199, Theorem 8]. The proof follows the same reasoning as that for linear EDMs.*

3.4.2 ▪ Counting argument

In Goemans preprint that appeared on his website in 2008 (although only published in 2015) and that was presented at ISMP 2009 in Chicago under the title "Smallest Compact Formulation for the Permutahedron," he [215] makes the following observation.

---

[23]Usually linear EDMs are not required to satisfy $x_i \neq x_j$ for all $i \neq j$. However, in terms of rank computations, one can assume w.l.o.g. that it holds. If $x_i = x_j$ for some $i \neq j$, then $X(:,i) = X(:,j)$ and $X(i,:) = X(j,:)$; hence removing one of these two columns and the corresponding row does not modify its rank or its nonnegative rank.

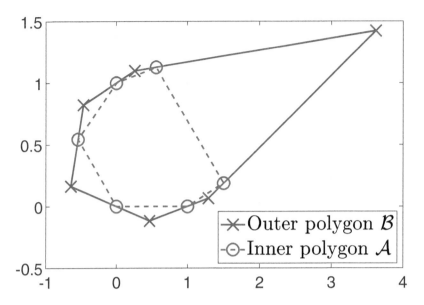

Figure 3.1. *NPP instance corresponding to the linear EDM X_6 from (3.1).* [Matlab file: NPPrank3matrix.m].

Lemma 3.14. *[215] Let X be a nonnegative matrix with p columns or p rows with distinct supports. Then,*

$$\text{rank}_+(X) \geq \log_2(p).$$

Proof. Let p be the number of columns of X with distinct supports. Let $X = WH$ be an Exact NMF of size $r = \text{rank}_+(X)$. Since

$$X(:,j) = \sum_{k=1}^{r} W(:,k)H(k,j) \quad \text{for all } j, \tag{3.2}$$

and H is nonnegative, each of the p different supports of the columns of X must be generated using the union of a subset of the supports of the r columns of W. In other words, denoting by $\text{supp}(x)$ the support of the vector x, we have for all j that

$$\text{supp}(X(:,j)) = \bigcup_{k \in \mathcal{K}_j} \text{supp}(W(:,k)), \quad \text{where } \mathcal{K}_j = \{k | H(k,j) > 0\}.$$

The r columns of W can generate at most 2^r different supports, which implies that

$$2^r \geq p \quad \Longleftrightarrow \quad r \geq \log_2(p).$$

By symmetry the same observation applies to the rows, which concludes the proof. $\qquad\square$

We were recently told by an anonymous reviewer that the following stronger result was hidden in the paper of Yannakakis [494], or, rather, it is an implication of the discussions in his paper; see Remark 3.2 below.

Lemma 3.15. *[494] Let X be a nonnegative matrix, and let p be the maximum number of distinct supports that can be obtained as the union of supports of the columns of X. Then,*

$$\text{rank}_+(X) \geq \log_2(p).$$

By symmetry the same result applies to the rows of X.

Proof. Let $X = WH$ be an Exact NMF of size $r = \text{rank}_+(X)$. For any vector $v \in \mathbb{R}^n_+$,

$$Xv = WHv = Wv', \quad \text{where} \quad v' = Hv \geq 0.$$

This implies that

$$\text{supp}(Xv) = \bigcup_{j \text{ s.t. } v(j)>0} \text{supp}(X(:,j))$$
$$= \text{supp}(Wv')$$
$$= \bigcup_{k \text{ s.t. } v'(j)>0} \text{supp}(W(:,k)).$$

Since v can be chosen as any nonnegative vector, this means that the union of the supports of any subset of the columns of X must be equal to the union of the supports of a subset of the columns of W. (This generalizes the observation of Lemma 3.14, taking $v = e_i$ for all i.) Since at most 2^r supports can be generated using the columns of W, the result follows; this is the same argument as in Lemma 3.14. □

Lemma 3.15 requires us to compute the maximum number of distinct supports that can be obtained as the union of supports of the columns of X, which is not necessarily trivial, but any lower bound for this number leads to a lower bound for the nonnegative rank; Lemma 3.14 provides such a lower bound.

Example 3.16 (Linear EDMs, Example 3.13 continued). For an n-by-n linear EDM X, all the columns of X have a different support. Hence Lemma 3.14 implies that

$$\text{rank}_+(X) \geq \lceil \log_2(n) \rceil.$$

Although the rank is fixed and equal to three, the nonnegative rank of n-by-n linear EDMs grows at least as fast as $\log_2(n)$. Note, however, that for $n = 6$, this bound gives $\text{rank}_+(X) \geq \log_2(6) = 2.585$ and hence does not improve upon the fooling set bound.

Observe that only $n + 2$ distinct supports can be generated using linear combinations of the n columns of X since the sum of any two columns has a support containing all elements $\{1, 2, \ldots, m\}$, while the empty set is also a valid support (corresponding to the choice $v = 0$ in Lemma 3.15). Hence Lemma 3.15 implies that for an n-by-n linear EDM X,

$$\text{rank}_+(X) \geq \lceil \log_2(n + 2) \rceil.$$

To improve the lower bound for the nonnegative rank of linear EDMs we have derived so far, which is 4 for $n \geq 4$ using Theorem 2.14, n needs to be larger than 15 (so that $n + 2 > 16 = 2^4$), in which case Lemma 3.15 leads to a lower bound for their nonnegative rank which is larger than 5. ∎

Remark 3.2 (Geometric interpretation of Lemma 3.15). *Given a nonnegative matrix X, let us consider the NPP instance $\mathcal{A} \subseteq \mathcal{B}$ corresponding to the RE-NMF instance of X. Each column of X corresponds to an element of $\mathcal{A}$ and each row to a facet of $\mathcal{B}$, implying that $X(i, j) = 0$ if and only if the jth element of $\mathcal{A}$ is located on the ith facet of $\mathcal{B}$; see Theorem 2.11. Taking linear combinations of columns of X is equivalent to taking convex combinations of elements of $\mathcal{A}$. Let us refer to the support of a point within $\mathcal{B}$ as the set of facets of $\mathcal{B}$ that it does not belong to. With this definition, and the observation above, the support of each column of X coincides with the* *supports of the corresponding element of $\mathcal{A}$. Since points in the relative interior of the same face of $\mathcal{B}$ have the same support since a face is defined as the intersection of facets, one can show that the maximum number p of distinct supports that can be obtained as the union of supports of columns of X is equal to the number of faces of $\mathcal{B}$ that $\mathcal{A}$ intersects. In the particular case when $\mathcal{A} = \mathcal{B}$ (this will correspond to slack matrices; see Section 3.6), p is the number of faces of $\mathcal{B}$ since $\mathcal{A}$ touches all faces of $\mathcal{B}$ as $\mathcal{A} = \mathcal{B}$. This is the observation made by Yannakakis [493]. For example, taking $\mathcal{A} = \mathcal{B}$ as a polygon with n vertices gives $p = 2n + 2$ because any polygon has $2n + 2$ faces, and there is a one-to-one correspondence with the linear combinations of the columns of X that generate distinct supports, namely*

- *n vertices (zero-dimensional faces): the original columns of X with two zero entries (each vertex is located on two segments);*

- *n edges (one-dimensional faces): the linear combinations of two columns of X corresponding to adjacent vertices of the polygon with one zero entry (any other combinations of two vertices lead to a point in the interior of the polygon);*

- *the polygon itself (a two-dimensional face): the linear combinations of three columns of X (or two columns that do not correspond to adjacent vertices) that do not contain any zero entry; and*

- *the empty set: the trivial combination of the columns of X with n zero entries.*

Another example is n-by-n linear EDMs X: the NPP instance is such that $\mathcal{B}$ is a polygon, and the elements of $\mathcal{A}$ are located on each segment of $\mathcal{B}$; see Figure 3.1 for an illustration in the case $n = 6$. For such matrices, $n + 2$ distinct supports can be generated using linear combinations of the columns of X; see Example 3.16. Geometrically, $\mathcal{A}$ intersects $n + 2$ faces of $\mathcal{B}$: the n segments and the interior, to which we need to add the empty set; see Figure 3.1 for the case $n = 6$.

3.4.3 ▪ Antichain

Instead of assuming that there are p columns of X with different supports as in the previous subsection, let us make the stronger assumption that there are p columns whose supports are not contained in one another. This stronger assumption is particularly meaningful as, in many cases, the supports of the columns of X are not contained in one another, for example for linear EDMs and slack matrices discussed in Section 3.6. This assumption means that there exist p columns of X whose supports form a Sperner family of size p, also known as an antichain of size p, which is a family of p sets that are not contained in one another [429]. For example, for linear EDMs of size n, $p = n$ since they have zeros only on their diagonal.

This subsection is organized as follows. First, we show that if the supports of p columns of X form an antichain of size p, then the supports of the corresponding p columns of H in any Exact NMF (W, H) of X must also form an antichain of size p (Lemma 3.17). Then, we recall the well-known Sperner theorem on the size of the largest antichain over r elements (Theorem 3.18). Together with Lemma 3.17, this implies a lower bound for the nonnegative rank (Theorem 3.20).

Lemma 3.17. *Let X be a nonnegative matrix with p columns whose supports are not contained in one another, that is, they form an antichain of size p. Then, in any Exact NMF $X = WH$ of X of size r, the supports of the corresponding p columns of H form an antichain of size p.*

Proof. Let (W, H) be an Exact NMF of X of size r. Let us denote the supports of the rows of W and columns of H as follows:

$$u_i = \text{supp}\left(W(i,:)\right) = \{k \mid W_{ik} \neq 0\} \text{ for } 1 \leq i \leq m$$

and

$$v_j = \text{supp}\left(H(:,j)\right) = \{k \mid H_{kj} \neq 0\} \text{ for } 1 \leq j \leq n.$$

Since $X_{ij} = W(i,:)H(:,j)$,

$$X_{ij} = 0 \quad \Longleftrightarrow \quad u_i \cap v_j = \emptyset. \tag{3.3}$$

If $v_{j_1} \subseteq v_{j_2}$ for some j_1, j_2, (3.3) implies that the support of the j_2th column of X contains the support of the j_1th column of X, that is, $\text{supp}(X(:,j_1)) \subseteq \text{supp}(X(:,j_2))$. In fact, $X(:,j_1)$ is a linear combination of a subset of the columns of W used to reconstruct $X(:,j_2)$. In other words, for $v_{j_1} \subseteq v_{j_2}$ and any set u_ℓ, $u_\ell \cap v_{j_1} \neq \emptyset \Rightarrow u_\ell \cap v_{j_2} \neq \emptyset$. Therefore, if X contains p columns whose supports are not contained in one another, then the subset containing the corresponding p elements of $\{v_j\}_{j=1}^n$ are not contained in one another and hence form an antichain of size p. $\square$

Let us now recall the well-known result by Sperner that bounds the size of an antichain over r elements. We provide the proof from [330] for completeness and because it is very elegant. Combined with Lemma 3.17, this will lead to a new lower bound for the nonnegative rank (Theorem 3.20).

Theorem 3.18 (Sperner). *Let $S = \{s_1, s_2, \dots, s_n\}$ be a set of n subsets of $\{1, 2, \dots, r\}$. Also let S be an antichain, that is, no subset in S is contained in another subset in S. Then,*

$$n \leq \binom{r}{\lfloor r/2 \rfloor}, \tag{3.4}$$

and the bound is tight (take all subsets of size $\lfloor r/2 \rfloor$).

Proof. [330] This proof is based on a counting argument observing that there are $r!$ permutations of $\{1, 2, \dots, r\}$. Given $s_i \in S$ with k elements, there are $k!(r-k)!$ permutations of $\{1, 2, \dots, r\}$ whose first k elements are in s_i. Because the subsets in S are not contained in one another, two permutations generated using two different subsets s_i and s_j cannot coincide (otherwise, this would imply that $s_i \subset s_j$ or $s_j \subset s_i$). Let us also denote by c_k the number of sets with k elements contained in S, that is, $c_k = |\{s \in S \mid |s| = k\}|$; hence $n = \sum_{k=0}^r c_k$. We have

$$\sum_{k=0}^r c_k k!(r-k)! \quad \leq \quad r!.$$

The left-hand side counts the different permutations generated using the subsets in S as described above; the right-hand side is the total number of different permutations of the set $\{1, 2, \dots, r\}$. Therefore, dividing both sides by $r!$, we obtain

$$\frac{n}{\binom{r}{\lfloor r/2 \rfloor}} = \sum_{k=0}^r \frac{c_k}{\binom{r}{\lfloor r/2 \rfloor}} \quad \leq \quad \sum_{k=0}^r \frac{c_k}{\binom{r}{k}} = \sum_{k=0}^r c_k \frac{k!(r-k)!}{r!} \quad \leq \quad 1,$$

since $\binom{r}{\lfloor r/2 \rfloor} \geq \binom{r}{k}$ for all k. This completes the proof. $\square$

Lemma 3.17 and Theorem 3.18 were used in [70] to lower bound the Boolean rank of binary matrices. Given a binary matrix $X \in \{0,1\}^{m \times n}$, its Boolean rank is the smallest r such that there exists $W \in \{0,1\}^{m \times r}$ and $H \in \{0,1\}^{r \times n}$ with $(WH)_{i,j} = 0$ if and only if $X(i,j) = 0$. The corresponding (W, H) is referred to as a Boolean factorization of X. The Boolean rank only considers the location of the positive entries in X, W, and H and requires that the supports of X and WH coincide. Let us denote by $\mathrm{bin}(X)$ the binarization of X, that is,

$$\mathrm{bin}(X)_{i,j} = \begin{cases} 1 & \text{if } X(i,j) > 0, \\ 0 & \text{if } X(i,j) = 0. \end{cases}$$

Let us also denote by $\mathrm{rank}_{01}(X)$ the Boolean rank of the binarization of X. We have the following result.

Lemma 3.19. *For $X \in \mathbb{R}_+^{m \times n}$,*

$$\mathrm{rank}_+(X) \geq \mathrm{rank}_{01}(X).$$

Proof. Any Exact NMF (W, H) of X leads to a Boolean factorization of X by binarizing W and H. □

Finally, Lemma 3.17, which only focuses on the supports of X, W, and H, also applies to Boolean factorizations of X.

Theorem 3.20. *[70] Let X be a matrix having p rows or p columns whose supports form an antichain. Then,*

$$\mathrm{rank}_+(X) \ \geq \ \mathrm{rank}_{01}(X) \ \geq \ \min\left\{ r \ \middle| \ \binom{r}{\lfloor r/2 \rfloor} \geq p \right\}.$$

Proof. The first inequality is from Lemma 3.19. The second inequality follows from Lemma 3.17 and Theorem 3.18. In fact, let (W, H) be a Boolean factorization of $\mathrm{bin}(X)$. Using exactly the same argument as in the proof of Lemma 3.17, this implies that there are p subsets of $\{1, 2, \ldots, r\}$ corresponding to the supports of p columns of H that are not contained in one another, that is, they form an antichain of size p. Theorem 3.18 allows us to conclude the proof for the case when p columns of X form an antichain. By symmetry, the same argument holds if X has p rows with supports not contained in one another. □

Example 3.21 (Linear EDMs, Example 3.13 continued). The above result implies that the nonnegative rank of linear EDMs is larger than the minimum r such that $n \leq \binom{r}{\lfloor r/2 \rfloor}$ [27]. For linear EDMs, this means that to obtain a lower bound of 5 for the nonnegative rank, one only needs $n \geq 7$ since $\binom{4}{2} = 6$, which significantly improves the bound using the counting argument (Lemma 3.15) which requires $n \geq 15$; see Example 3.16. ∎

Making additional assumptions on the supports of the rows/columns of X, one can use similar arguments to improve the above antichain-based lower bound for the nonnegative rank [459, Theorem 2].

3.4.4 ▪ Rectangle covering bound

The rectangle covering bound (RCB) for a matrix X is the minimum number of rectangles necessary to cover all positive entries of X. A rectangle is a subset of rows and columns of X such that the corresponding submatrix of X contains only positive entries (see below for an example). More precisely, let

$$\mathrm{supp}(X) = \{(i,j) \mid X(i,j) > 0\}$$

be the support of X. The RCB of X, denoted $\mathrm{rc}(X)$, is the minimum number of nonnegative rank-one matrices R_k $(1 \leq k \leq \mathrm{rc}(X))$ such that

$$\mathrm{supp}(X) \;=\; \cup_{k=1}^{\mathrm{rc}(X)} \mathrm{supp}(R_k).$$

The support of a rank-one matrix corresponds to a rectangle, that is, the support of each nonzero row/column is the same (hence the name).

It is easy to see that the RCB is equal to the Boolean rank of the binarization of X, that is,

$$\mathrm{rc}(X) \;=\; \mathrm{rank}_{01}(X) \;\leq\; \mathrm{rank}_+(X),$$

since any rectangle covering provides a Boolean factorization and vice versa.

Example 3.22 (Linear EDMs, Example 3.13 continued). For the 6-by-6 linear EDM X_6 from (3.1), $\mathrm{rc}(X_6) = 4$ with the following four rectangles:

$$
\begin{pmatrix}
0 & 0 & 0 & 0 & 0 & 0 \\
0 & 0 & 0 & 0 & 0 & 0 \\
0 & 0 & 0 & 0 & 0 & 0 \\
1 & 1 & 1 & 0 & 0 & 0 \\
1 & 1 & 1 & 0 & 0 & 0 \\
1 & 1 & 1 & 0 & 0 & 0
\end{pmatrix}
+
\begin{pmatrix}
0 & 0 & 0 & 0 & 0 & 0 \\
1 & 0 & 0 & 1 & 0 & 1 \\
1 & 0 & 0 & 1 & 0 & 1 \\
0 & 0 & 0 & 0 & 0 & 0 \\
1 & 0 & 0 & 1 & 0 & 1 \\
0 & 0 & 0 & 0 & 0 & 0
\end{pmatrix}
+
\begin{pmatrix}
0 & 1 & 0 & 0 & 1 & 1 \\
0 & 0 & 0 & 0 & 0 & 0 \\
0 & 1 & 0 & 0 & 1 & 1 \\
0 & 1 & 0 & 0 & 1 & 1 \\
0 & 0 & 0 & 0 & 0 & 0 \\
0 & 0 & 0 & 0 & 0 & 0
\end{pmatrix}
$$

$$
+
\begin{pmatrix}
0 & 0 & 1 & 1 & 1 & 0 \\
0 & 0 & 1 & 1 & 1 & 0 \\
0 & 0 & 0 & 0 & 0 & 0 \\
0 & 0 & 0 & 0 & 0 & 0 \\
0 & 0 & 0 & 0 & 0 & 0 \\
0 & 0 & 1 & 1 & 1 & 0
\end{pmatrix}
=
\begin{pmatrix}
0 & 1 & 1 & 1 & 2 & 1 \\
1 & 0 & 1 & 2 & 1 & 1 \\
1 & 1 & 0 & 1 & 1 & 2 \\
1 & 2 & 1 & 0 & 1 & 1 \\
2 & 1 & 1 & 1 & 0 & 1 \\
1 & 1 & 2 & 1 & 1 & 0
\end{pmatrix};
\tag{3.5}
$$

see [Matlab file: `rec_cov_linEDM.m`]. ∎

Link with the biclique covering number　Let us interpret the matrix $X \in \mathbb{R}^{m \times n}$ as the biadjacency matrix of a bipartite graph $G_b = (V_1 \times V_2, E)$ where $V_1 = \{s_1, s_2, \ldots, s_m\}$, $V_2 = \{t_1, t_2, \ldots, t_n\}$, and $(s_i, t_j) \in E \iff X(i,j) > 0$. Then a rectangle whose support is contained in the support of X corresponds to a so-called biclique of G_b, that is, a subset of vertices from V_1 and V_2 that are all connected. The biclique covering number of G_b is the minimum number of such bicliques needed to cover all edges of G_b. The biclique covering number of G_b equals the RCB of X; see [419] and the references therein for more details.

Link with the chromatic number of $G(X)$　Using the rectangle graph $G(X)$ of X (Definition 3.11, page 64), one can check that the RCB is equal to the chromatic number of $G(X)$, that is, the smallest number of colors needed to color the vertices of $G(X)$ so that no two adjacent vertices share the same color. In fact, in the rectangle graph, two vertices are not connected if the corresponding 2-by-2 submatrix of X only has positive entries. Hence a rectangle in X corresponds to a set of non-connected vertices in G, which can be assigned the same color; see [162] for more details.

Linear combinatorial formulation　Unfortunately, the RCB is hard to compute in general. However, it can be formulated as a linear combinatorial optimization problem, that is, a linear optimization problem with binary variables [368]. A formulation can be obtained as follows. First, generate all rectangles R_p for $1 \leq p \leq q$ whose support is contained in the support of X and whose support is not contained in the support of any other such rectangle (these are referred to as maximal rectangles). There are at most $\min(2^m, 2^n) - 1$ such rectangles. Once the support of the rows of a rectangle is fixed, it is easy to select the support of the columns to obtain the corresponding unique maximal rectangle: select all columns that have positive entries on these rows. By symmetry, the same observation holds when the support of the columns is fixed. This allows us to generate all maximal bicliques of the bipartite graph corresponding to X; see for example [6] and the references therein for more sophisticated approaches.

Then, given the q maximal rectangles $R_p \in \{0, 1\}^{m \times n}$ for $1 \leq p \leq q$, let us define the binary variables

$$y_p = \begin{cases} 1 & \text{if the } p\text{th rectangle is selected,} \\ 0 & \text{otherwise.} \end{cases}$$

A feasible covering of X corresponds to a vector $y \in \{0, 1\}^q$ such that $\sum_{p=1}^q y_p R_p \geq \text{bin}(X)$. The RCB can therefore be computed using a linear integer formulation over the variable y. This is similar to a minimum set covering problem:

$$\text{rc}(X) \quad = \quad \min_{y \in \{0,1\}^q} \sum_{p=1}^q y_p \text{ such that } \sum_{p=1}^q y_p R_p \geq \text{bin}(X). \tag{3.6}$$

The RCB can therefore be computed for matrices of small size and/or for which the number of maximal rectangles q is small. For example, for the slack matrices of four-dimensional 0/1 polytopes and for the slack matrices of n-gons (see Section 3.6 for the details on these matrices), the RCB can be computed on a standard laptop for $\min(m, n) \leq 13$ [368, 459]. Note, however, that the RCB was derived for the slack matrices of n-gons for n up to 33 via computer-assisted proofs [23].

Example 3.23 (Linear EDMs, Example 3.13 continued). With our naive implementation of the above strategy to compute the RCB, we were able to compute the RCB for the linear EDMs for $n \leq 10$ within a few seconds (it took about 20 seconds for $n = 10$, while for $n = 11$ we stopped the program after 5 minutes); see [Matlab file: rec_cov_bound.m]. For $4 \leq n \leq 6$, the RCB of n-by-n linear EDMs is 4; see (3.5) for a solution for $n = 6$. For $n \in [7, 10]$, the RCB is 5. In this case, the RCB coincides with the lower bound based on antichains (Section 3.4.3) that required $n \geq 7$ to have a lower bound of 5 for the nonnegative rank of n-by-n linear EDMs. ∎

Bounding the rectangle covering number
Since in many cases the RCB cannot be computed, one usually resorts to using lower bounds for the RCB itself [162].

The first example is the fooling set bound which is a lower bound for the RCB:

$$\omega(X) \leq \text{rc}(X).$$

Interestingly, this bound corresponds to a well-known bound in graph theory: the clique number of the rectangle graph $G(X)$ is smaller than its chromatic number, since every vertex in a clique must be assigned a different color.

Similarly, the lower bounds using the counting argument and antichains (Sections 3.4.2 and 3.4.3) also apply to the RCB since they only take into account the support of X.

Another lower bound for the RCB is the ratio between the number of positive entries in X and the size of the largest rectangle whose support is contained in the support of X, that is,

$$\text{rc}(X) \geq \frac{|\text{supp}(X)|}{\text{size of largest rectangle}}. \tag{3.7}$$

However, finding the largest rectangle is also a difficult problem in general as it corresponds to finding the largest biclique in the corresponding bipartite graph, referred to as the maximum-edge biclique problem, which is NP-complete [380]. However, it can be computed or upper bounded accurately in some cases. We will discuss an important use of this bound in Section 3.7, on the so-called UDISJ matrices.

Example 3.24 (Linear EDMs, Example 3.13 continued). Let us illustrate the bound (3.7) on n-by-n linear EDMs which have zero entries only on their diagonal. The number of positive entries is equal to $n^2 - n$, while the size of the largest rectangle is $\lceil \frac{n}{2} \rceil \lfloor \frac{n}{2} \rfloor$, taking for example the bottom left rectangle; see (3.5) for an example when $n = 6$. For simplicity, assume n is even, in which case

$$\mathrm{rc}(X) \geq \frac{n^2 - n}{n^2/4} = 4 - 4/n,$$

where X is any n-by-n linear EDM. This is not a very strong bound since it only provides the lower bound of 4 for the nonnegative rank of X when $n \geq 5$. ∎

Looking back at the combinatorial formulation (3.6) of the RCB, it is possible to relax the combinatorial problem, replacing the constraints $y \in \{0,1\}^r$ with $0 \leq y \leq 1$ (this is the first relaxation in a standard branch and bound procedure). This gives a lower bound for $\mathrm{rc}(X)$, hence a lower bound for $\mathrm{rank}_+(X)$, which is referred to as the fractional RCB. For the same reason as for the RCB, it can only be computed if the number of rectangles is not too large.

Note that the RCB is the strongest lower bound among the bounds that solely rely on the support of X. We refer the reader to [162] for other examples, discussions, and limitations of the RCB.

3.4.5 ▪ Refined rectangle covering bound

The RCB can be refined by taking into account the values of the positive entries of X instead of just their location. Oelze, Vandaele, and Weltge [368] introduced the so-called refined RCB (RRCB) of a matrix X, denoted $\mathrm{rrc}(X)$, which requires that the diagonal entries of any 2-by-2 rank-two submatrix of X must be covered by at least two rectangles. Using the notation of the formulation (3.6), for any pair (i, j) and (k, ℓ) such that $X(i,j)X(k,l) - X(i,\ell)X(k,j) \neq 0$, the RRCB imposes the constraints

$$A(i,j) \geq 2 \text{ and } A(k,\ell) \geq 2, \quad \text{where } A = \sum_{p=1}^{r} y_p R_p.$$

Note that these constraints are satisfied by the solution of (3.6) when $X(i,\ell) = 0$ or $X(k,j) = 0$ since in that case (i,j) and (k,ℓ) are pairwise independent and can only be covered using two rectangles; see Lemma 3.9. Compared to the RCB, the RRCB adds the constraints that require the diagonal entries of any 2-by-2 positive rank-two submatrix of X to be covered by at least two rectangles. This implies that the RRCB is always larger than the RCB, that is,

$$\mathrm{rc}(X) \leq \mathrm{rrc}(X) \leq \mathrm{rank}_+(X),$$

for any nonnegative matrix X [368, Theorem 3.4]. For example, for the matrix

$$X = \begin{pmatrix} 2 & 1 \\ 1 & 1 \end{pmatrix},$$

we have $\mathrm{rc}(X) = 1 < \mathrm{rrc}(X) = 2$. In particular, the RRCB improves the RCB and is tight for the slack matrices of all 4-dimensional 0/1 polytopes [368] and of the regular 9-gon and 13-gon (see Section 3.6.3).

3.4.6 ▪ Geometric bound

In this section, we present the geometric lower bound proposed in [199] which is based on the restricted nonnegative rank.

Theorem 2.2 states that Exact NMF of matrix X is equivalent to finding a polytope with r vertices, namely $\mathrm{conv}\,(\theta(W))$, nested between $\mathrm{conv}\,(\theta(X))$ and the unit simplex Δ^m. RE-NMF is the same problem as Exact NMF except that it requires $\mathrm{col}(W) = \mathrm{col}(X)$. Moreover, the number of vertices of a nested polytope that satisfies this condition has at least $\mathrm{rank}_+^*(X)$ vertices, by definition of the restricted nonnegative rank.

Let $X = WH$ be an Exact NMF of size $\mathrm{rank}_+(X)$. Observe that the polytope

$$P = \mathrm{conv}\,(\theta(W)) \cap \mathrm{col}(X)$$

is a feasible solution for RE-NMF of $\theta(X)$. Hence we know that P has at least $\mathrm{rank}_+^*(X)$ vertices. This observation can be used to lower bound the nonnegative rank. Let us define the quantity $\mathrm{faces}(n, d, k)$ to be the maximal number of k-faces (that is, faces of dimension k) of a polytope with n vertices in dimension d; see (3.9) below for an explicit formula. For example, $\mathrm{faces}(n, d, 0) = n$ for any $d \geq 2$ since 0-faces are vertices. We have the following result.

Theorem 3.25. *[199, Theorem 5] The restricted nonnegative rank of a nonnegative matrix X with $r = \mathrm{rank}(X)$ and $r_+ = \mathrm{rank}_+(X)$ can be bounded above by*

$$\mathrm{rank}_+^*(X) \;\leq\; \max_{r \leq r_w \leq r_+} \mathrm{faces}(r_+, r_w - 1, r_w - r). \tag{3.8}$$

Proof. This follows from the discussion above. Let (W, H) be a solution to Exact NMF of X of size $r_+ = \mathrm{rank}_+(X)$. The polytope $T = \mathrm{conv}\,(\theta(W)) \subseteq \Delta^m$ with r_+ vertices has dimension $r_w - 1 \in \{r - 1, \ldots, r_+ - 1\}$, where $r_w = \mathrm{rank}(W)$ (see Lemma 2.5). If we intersect T with $Q = \mathrm{col}(X)$ of dimension r, we obtain a feasible solution for RE-NMF which must have more than $\mathrm{rank}_+^*(X)$ vertices, that is,

$$\mathrm{rank}_+^*(X) \;\leq\; \#\,\mathrm{vertices}(T \cap Q).$$

It remains to show that the number of vertices of $P = T \cap Q$ is at most $\mathrm{faces}(r_+, r_w - 1, r_w - r)$. Since the linear subspace Q intersected with Δ^m has dimension $r - 1$ (Lemma 2.5), the number of vertices of $T \cap Q \subseteq \Delta^m$ is upper bounded by the number of $(r_w - r)$-faces of T, which is upper bounded by $\mathrm{faces}(r_+, r_w - 1, r_w - r)$. $\qquad\square$

Note that when $r_w = r$, $\mathrm{faces}(r_+, r - 1, 0) = r_+$. This corresponds to the case $\mathrm{rank}(W) = \mathrm{rank}(X)$ for which $\mathrm{rank}_+(X) = \mathrm{rank}_+^*(X)$.

A tight bound for the function $\mathrm{faces}(n, d, k)$ exists and is attained by cyclic polytopes [517, Corollary 8.28, p. 257]:

$$\mathrm{faces}(n, d, k - 1) = \sum_{i=0}^{\frac{d}{2}}{}^{*} \left(\binom{d - i}{k - i} + \binom{i}{k - d + i} \right) \binom{n - d - 1 + i}{i}, \tag{3.9}$$

where $\sum^*$ denotes a sum where only half of the last term is taken for $i = \frac{d}{2}$ if d is even, and the whole last term is taken for $i = \lfloor \frac{d}{2} \rfloor = \frac{d-1}{2}$ if d is odd. Let us introduce for easier reference a function ϕ corresponding to the upper bound in Theorem 3.25:

$$\phi(r, r_+) = \max_{r \leq r_w \leq r_+} \mathrm{faces}(r_+, r_w - 1, r_w - r).$$

When $r = \mathrm{rank}(X)$ is fixed, ϕ is an increasing function of its second argument r_+, since $\mathrm{faces}(n, d, k)$ increases with n. Therefore the inequality $\mathrm{rank}_+^*(X) \leq \phi(r, r_+)$ from Theorem 3.25 implicitly provides a lower bound on the nonnegative rank r_+ that depends on both rank r and restricted nonnegative rank $\mathrm{rank}_+^*(X)$.

Corollary 3.26. *[199] If X is a nonnegative matrix, then*

$$\text{rank}_+(X) \geq \min_k \left\{ k \mid \phi(\text{rank}(X), k) \geq \text{rank}_+^*(X) \right\}.$$

The function ϕ can be upper bounded in the following way.

Lemma 3.27. *[199, Theorem 6] The upper bound $\phi(r, r_+)$ on the restricted nonnegative rank of a nonnegative matrix X with $r = \text{rank}(X)$ and $r_+ = \text{rank}_+(X)$ satisfies*

$$\phi(r, r_+) = \max_{r \leq r_w \leq r_+} \text{faces}(r_+, r_w - 1, r_w - r)$$

$$\leq \max_{r \leq r_w \leq r_+} \binom{r_+}{r_w - r + 1} \leq \binom{r_+}{\lfloor r_+/2 \rfloor} \leq 2^{r_+} \sqrt{\frac{2}{\pi r_+}} \leq 2^{r_+}.$$

Proof. The first inequality follows from $\text{faces}(n, d, k - 1) \leq \binom{n}{k}$, since any set of k distinct vertices defines at most one $(k - 1)$-face. The second follows from the maximality of central binomial coefficients. The third is a standard upper bound on central binomial coefficients, and the fourth is an even cruder upper bound.   □

Of course, the above lower bound requires knowledge of the restricted nonnegative rank, which is hard to compute in general for $\text{rank}(X) \geq 4$ (Theorem 2.19). However, in some cases, it can be computed easily.

Example 3.28 (Linear EDMs, Example 3.13 continued). For an n-by-n linear EDMs X, $\text{rank}_+^*(X) = n$; see the discussion on page 65. For example, for $n = 6$, the geometric bound gives $\text{rank}_+(X_6) \geq 4$, while for $n = 7$, it gives $\text{rank}_+(X_7) \geq 5$, which coincides with the RCB.

The geometric bound can be improved in several ways, for example when the input matrix is symmetric, or when the input matrix has rank three; see [199] and [185, Chapter 3.6] for more details, and see [Matlab file: geometric_bound]. It can also be improved by using the f-vector of polytopes whose entries are the number of k-faces for all k, instead of just the number of vertices (which are the 0-faces) [127]. For the 6-by-6 linear EDM $X(i, j) = (i - j)^2$ for all $1 \leq i, j \leq 6$, we obtain $\text{rank}_+(X) \geq 5$ with the improved geometric bound of [199]. It turns out this bound is tight, since the nonnegative rank of X is at most 5, as proved by the following decomposition of X:

$$\begin{pmatrix} 0 & 1 & 4 & 9 & 16 & 25 \\ 1 & 0 & 1 & 4 & 9 & 16 \\ 4 & 1 & 0 & 1 & 4 & 9 \\ 9 & 4 & 1 & 0 & 1 & 4 \\ 16 & 9 & 4 & 1 & 0 & 1 \\ 25 & 16 & 9 & 4 & 1 & 0 \end{pmatrix} = \begin{pmatrix} 5 & 0 & 4 & 1 & 0 \\ 3 & 0 & 1 & 0 & 1 \\ 1 & 0 & 0 & 1 & 4 \\ 0 & 1 & 0 & 1 & 4 \\ 0 & 3 & 1 & 0 & 1 \\ 0 & 5 & 4 & 1 & 0 \end{pmatrix} \begin{pmatrix} 0 & 0 & 0 & 1 & 3 & 5 \\ 5 & 3 & 1 & 0 & 0 & 0 \\ 0 & 0 & 1 & 1 & 0 & 0 \\ 0 & 1 & 0 & 0 & 1 & 0 \\ 1 & 0 & 0 & 0 & 0 & 1 \end{pmatrix}. \quad \blacksquare$$

Comparison with the antichain-based bound   As for linear EDMs, the restricted nonnegative rank of the slack matrices that will be discussed in Section 3.6 is also maximal and equal to the number of columns (see page 85 for an explanation of this result). Interestingly, in these two cases, Lemma 3.27 shows that the geometric bound (Corollary 3.26) provides a tighter bound than the bound based on antichains (Sections 3.4.3), which itself is stronger than the bound based on counting arguments (Sections 3.4.2); see Figure 3.3 in Section 3.6.3.4 (page 91) for an illustration on the slack matrices of regular n-gons.

3.4.7 • Hyperplane separation

In personal communication with Thomas Rothvoss, Samuel Fiorini provided the following lemma.
It was implicitly present in the paper [57] later published as the journal version [58].

Lemma 3.29. *Let* $X \in \mathbb{R}_+^{m \times n}$. *Let also* $Z \in \mathbb{R}^{m \times n}$, *and define*

$$\alpha(Z) \;=\; \max_{R \in \{0,1\}^{m \times n}} \langle Z, R \rangle \text{ such that } \operatorname{rank}(R) = 1. \tag{3.10}$$

Then,

$$\operatorname{rank}_+(X) \;\geq\; \frac{\langle Z, X \rangle}{\alpha(Z) \|X\|_\infty},$$

where $\|X\|_\infty = \max_{i,j} |X(i,j)|$.

Proof. [400, Lemma 1] First, let us show that $\langle Z, R \rangle \leq \alpha(Z)$ holds for any rank-one matrix
$R \in [0,1]^{m \times n}$ and not just for the binary rank-one matrices. For this, let us write $R = xy^\top$
with $x \geq 0$ and $y \geq 0$. After scaling, one can assume w.l.o.g. that $x \in [0,1]^m$ and $y \in [0,1]^n$.
Consider the optimization problem

$$\max_{x \in [0,1]^m, y \in [0,1]^n} \langle Z, xy^\top \rangle,$$

and suppose the optimal solution (x^*, y^*) is not binary. Assume x^* is not binary. For $y = y^*$
fixed, the above optimization problem is a linear program over the cube $[0,1]^n$ so that there exists
a binary optimal solution,[24] and hence x^* can be assumed to be binary w.l.o.g. By symmetry, the
same reasoning holds for y^*. Geometrically speaking, this means that

$$\operatorname{conv}\{R \in [0,1]^{m \times n} \mid \operatorname{rank}(R) \leq 1\} = \operatorname{conv}\{R \in \{0,1\}^{m \times n} \mid \operatorname{rank}(R) \leq 1\},$$

even though the set of matrices of rank at most one is not a convex set itself.

  Now, let $X = \sum_{k=1}^r R_k$ with $R_k \geq 0$ and $\operatorname{rank}_+(R_k) = 1$ (this corresponds to an Exact
NMF of X of size r). We have

$$\langle Z, X \rangle = \sum_{k=1}^r \|R_k\|_\infty \left\langle Z, \frac{R_k}{\|R_k\|_\infty} \right\rangle$$

$$\leq \alpha(Z) \sum_{k=1}^r \|R_k\|_\infty$$

$$\leq \alpha(Z) r \|X\|_\infty.$$

The first inequality follows from the result above since $\frac{R_k}{\|R_k\|_\infty} \in [0,1]^{m \times n}$ for all k; the second
inequality follows from the nonnegativity of the R_k's.   □

  This lower bound for the nonnegative rank is interesting in two aspects:

  1. It does not depend directly on the support of X and hence may lead to nontrivial bounds
     for positive matrices.

---

[24]The vertices of $[0,1]^n$ are the binary vectors $\{0,1\}^n$, while there exists at least one optimal vertex for any linear
program over a polyhedron containing at least one vertex.

2. It actually leads to an infinite number of lower bounds depending on the choice of Z. Appropriate choices of Z may lead to strong lower bounds. In particular, it is via the hyperplane separation bound that Rothvoss was able to prove his seminal result, namely that the perfect matching polytope has exponential extension complexity [400, 401]; see Section 3.6.

Note that it is always best to take $Z(i, j)$ as small as possible for all (i, j) such that $X(i, j) = 0$ because it leads to the best possible lower bound for the nonnegative rank of X. In fact, this choice does not affect the numerator $\langle Z, X \rangle$, while it can only decrease $\alpha(Z)$. Also, it limits the number of rectangles to consider when computing $\alpha(Z)$ via (3.10): the objective takes the value $-\infty$ for any rectangle containing an entry equal to $-\infty$, and hence such rectangles can be discarded.

Still, the main drawback of the hyperplane separating bound is the computation of $\alpha(Z)$ which requires solving a difficult combinatorial problem. Using the factorization $R = xy^\top$ for x and y binary, and assuming x is fixed, the optimal y can be easily computed:

$$\left[y^*(x) \right]_j = \begin{cases} 0 & \text{if } x^\top Z(:, j) < 0, \\ 1 & \text{otherwise.} \end{cases}$$

The same result holds when y is fixed, by symmetry. This means that the quantity $\alpha(Z)$ can be computed in $\mathcal{O}\left(mn2^{\min(m,n)} \right)$ operations, as done in [Matlab file: hyperplane_separation_bound.m]. This is impractical when $\min(m, n)$ is large. However, $\alpha(Z)$ can be replaced with any valid upper bound as done, for example, in [400].

Example 3.30. Consider the matrix of Thomas [450],

$$X = \begin{pmatrix} 1 & 1 & 0 & 0 \\ 1 & 0 & 1 & 0 \\ 0 & 1 & 0 & 1 \\ 0 & 0 & 1 & 1 \end{pmatrix},$$

and let $Z = 2X - 1$, meaning that Z is obtained from X by replacing its zero entries with -1 (they could also be chosen as $-\infty$; see above). One can check by inspection that $\alpha(Z) = 2$ (an optimal solution of (3.10) is obtained, for example, with the rectangle containing the first row and first two columns of X), and hence

$$\text{rank}_+(X) \geq \frac{\langle Z, X \rangle}{\alpha(Z) \|X\|_\infty} = \frac{8}{2} = 4.$$

This is yet another proof that the nonnegative rank of the matrix of Thomas is four. This example can be run with [Matlab file: bound_nnr_Thomas.m] that also computes the other bounds presented in this section. All bounds, from the fooling set bound to the bounds presented hereafter, provide a lower bound of 4.

Note that generating Z randomly, for example using the normal distribution randn(m,n), usually leads to poor lower bounds. Out of 100,000 such generated Z's, the best lower bound we have obtained for the matrix of Thomas is 2.81. ∎

3.4.8 ▪ Nonnegative nuclear norm

Let us present the bound proposed by Fawzi and Parrilo [154] and explain its connections with the nuclear norm, a widely used convex relaxation for the rank [394]. This bound has some similarities with the hyperplane separation bound presented in the previous section but relies on the Frobenius norm, instead of the infinity norm. It also relies on the set of copositive matrices $\mathfrak{C}^n$ defined as

$$\mathfrak{C}^n = \{A \in \mathbb{R}^{n \times n} \mid x^\top A x \geq 0 \text{ for all } x \geq 0\}.$$

Theorem 3.31. *[154, Theorem 1] Let* $X \in \mathbb{R}_+^{m \times n}$, *and let*

$$\nu_+(X) = \max_{Z \in \mathbb{R}^{m \times n}} \langle X, Z \rangle \text{ such that } \begin{pmatrix} I & -Z \\ -Z^\top & I \end{pmatrix} \in \mathfrak{C}^{m+n}.$$

Then,

$$\text{rank}_+(X) \geq \left(\frac{\nu_+(X)}{\|X\|_F} \right)^2. \tag{3.11}$$

Proof. Let $X = WH = \sum_{k=1}^r W(:,k)H(k,:)$ be an Exact NMF of X of size $r = \text{rank}_+(X)$. Let us assume w.l.o.g. that $\|W(:,k)\|_2 = \|H(k,:)\|_2$ for all k; this can be achieved by a scaling of the rank-one factors in the decomposition. Using the inequality $\|x\|_1 \leq \sqrt{r}\|x\|_2$ for the vector $x \in \mathbb{R}^r$ with $x_k = \|W(:,k)\|_2\|H(k,:)\|_2$ for all k, we obtain

$$\frac{\sum_{k=1}^r \|W(:,k)\|_2 \|H(k,:)\|_2}{\sqrt{\sum_{k=1}^r \|W(:,k)\|_2^2 \|H(k,:)\|_2^2}} \leq \sqrt{r} = \sqrt{\text{rank}_+(X)}.$$

Let us show that the numerator of the left-hand side is smaller than $\nu_+(X)$ and that the denominator is larger than $\|X\|_F$, which will prove the result.

- For any matrix Z such that $\begin{pmatrix} I & -Z \\ -Z^\top & I \end{pmatrix}$ is copositive, we have for all k that

$$\begin{pmatrix} W(:,k) \\ H(k,:)^\top \end{pmatrix}^\top \begin{pmatrix} I & -Z \\ -Z^\top & I \end{pmatrix} \begin{pmatrix} W(:,k) \\ H(k,:)^\top \end{pmatrix} \geq 0$$

since W and H are nonnegative. Expanding the product, we obtain

$$W(:,k)^\top Z H(k,:)^\top \leq \frac{1}{2}(\|W(:,k)\|_2^2 + \|H(k,:)\|_2^2) = \|W(:,k)\|_2\|H(k,:)\|_2$$

since $\|W(:,k)\|_2 = \|H(k,:)\|_2$ for all k. Therefore,

$$\langle X, Z \rangle = \left\langle \sum_{k=1}^r W(:,k)H(k,:), Z \right\rangle = \sum_{k=1}^r W(:,k)^\top Z H(k,:)^\top$$

$$\leq \sum_{k=1}^r \|W(:,k)\|_2\|H(k,:)\|_2.$$

- We have

$$\|X\|_F^2 = \left\| \sum_{k=1}^r W(:,k)H(k,:) \right\|_F^2$$

$$= \sum_{k=1}^r \left\| W(:,k)H(k,:) \right\|_F^2 + 2\sum_{k<\ell} \langle W(:,k)H(k,:), W(:,\ell)H(\ell,:) \rangle$$

$$\geq \sum_{k=1}^r \left\| W(:,k) \right\|_2^2 \left\| H(k,:) \right\|_2^2,$$

since $\|W(:,k)H(k,:)\|_F^2 = \|W(:,k)\|_2^2\|H(k,:)\|_2^2$, and W and H are nonnegative. $\square$

Link with the nuclear norm   The nuclear norm of a matrix A, denoted $\|A\|_*$, is the sum of its singular values. It can be shown that [257]

$$\|A\|_* \;=\; \min_{U,V} \sum_{k=1}^{\min(m,n)} \|U(:,k)\|_2 \|V(k,:)\|_2 \text{ such that } A = UV.$$

Moreover, using the inequality $\|x\|_1 \leq \sqrt{r}\|x\|_2$ for any vector x,

$$r = \operatorname{rank}(A) \geq \left(\frac{\sigma_1(A) + \sigma_2(A) + \cdots + \sigma_r(A)}{\sqrt{\sigma_1^2(A) + \sigma_2^2(A) + \cdots + \sigma_r(A)}} \right)^2 = \left(\frac{\|A\|_*}{\|A\|_F} \right)^2. \tag{3.12}$$

For $A \geq 0$, the quantity $\nu_+(A)$ can be interpreted as a nonnegative nuclear norm: one can show that [154, Theorem 4]

$$\nu_+(A) \;=\; \min_{W \geq 0, H \geq 0} \sum_k \|W(:,k)\|_2 \|H(k,:)\|_2 \text{ such that } X = WH.$$

Therefore, the lower bound (3.11) for the nonnegative rank can be interpreted as the analogue of the lower bound (3.12) for the rank. We refer the interested reader to [154] for more details.

Computation   It is hard to optimize over the copositive cone, that is, the set of copositive matrices [140]. However, it can be approximated using semidefinite programming. Let us introduce the cone of completely positive matrices, denoted $\mathfrak{C}_+^n$, which is the dual cone[25] of $\mathfrak{C}^n$:

$$\mathfrak{C}_+^n \;=\; \left\{ A \in \mathbb{R}^{n \times n} \mid A = BB^\top \text{ for some } B \geq 0 \right\}.$$

In other words, completely positive matrices admit an Exact NMF with $W = H^\top$; this will be referred to later as the symNMF problem in Section 5.3. The minimal r such that such a decomposition exists is referred to as the completely positive rank (cp-rank). The cp-rank behaves rather differently from the nonnegative rank. In particular, the cp-rank can be higher than the dimension of the matrix (up to $n(n+1)/2 - 1$ [231, 24]) or might not exist. For example, the matrix

$$X = \begin{pmatrix} 1 & 1 \\ 1 & 0 \end{pmatrix}$$

is not completely positive, that is, its cp-rank is infinite. Let us explain why, but first let us define the set $\mathbb{S}_+^n$ of positive semidefinite (PSD) matrices: a symmetric matrix A is PSD if $x^\top A x \geq 0$ for all x. Equivalently, a symmetric matrix A is PSD if all its eigenvalues are nonnegative. Since $x^\top (BB^\top) x = \|Bx\|_2^2 \geq 0$ for any B and x, $\mathfrak{C}_+^n \subseteq \mathbb{S}_+^n$. The matrix X above has one negative eigenvalue (its eigenvalues are $(1 \pm \sqrt{5})/2$), so that it does not belong to $\mathbb{S}_+^n$ and hence does not belong to $\mathfrak{C}_+^n$. We refer the reader to [39] for more information on the cp-rank.

Using duality, one can show that [154, Theorem 4]

$$\nu_+(X) \;=\; \min_{P \in \mathbb{S}^m, Q \in \mathbb{S}^n} \left\{ \frac{1}{2}(\operatorname{tr}(P) + \operatorname{tr}(Q)) \;\middle|\; \begin{pmatrix} P & X \\ X^\top & Q \end{pmatrix} \in \mathfrak{C}_+^{m+n} \right\}.$$

Assuming we have a lower bound for $\nu_+(X)$, we can derive a new lower bound for the nonnegative rank; this follows from (3.11). To do so, instead of optimizing over $\mathfrak{C}_+^{m+n}$, one can optimize over sets that contain $\mathfrak{C}_+^{m+n}$ so that the minimum is decreased. In particular, there exists a hierarchy of such sets. The first level of the hierarchy is $\mathfrak{C}_{[0]}^n = \mathbb{S}_+^n \cap \mathbb{R}_+^{n \times n} \supseteq \mathfrak{C}_+^n$: as explained above, any completely positive matrix must be componentwise nonnegative and

---
[25] See page 106 for a definition.

PSD. Then, without going into detail, they showed that there exists a hierarchy of sets $\mathfrak{C}^n_{[k]}$ that are representable using PSD matrices (with an increasing number of variables) such that $\mathfrak{C}^n_{[k+1]} \subset \mathfrak{C}^n_{[k]}$ and that tend to $\mathfrak{C}^n_+$ as k increases; see [154] for more details. We denote

$$\nu^{[0]}_+(X) = \min_{\substack{P \in \mathbb{S}^m \\ Q \in \mathbb{S}^n}} \left\{ \frac{1}{2}(\operatorname{tr}(P) + \operatorname{tr}(Q)) \,\middle|\, \begin{pmatrix} P & X \\ X^\top & Q \end{pmatrix} \in \mathbb{S}^{m+n}_+ \cap \mathbb{R}^{(m+n)\times(m+n)}_+ \right\},$$

for which

$$\operatorname{rank}_+(X) \geq \left(\frac{\nu^{[0]}_+(X)}{\|X\|_F} \right)^2.$$

Even the computation of $\nu^{[0]}_+(X)$ is relatively expensive since it requires solving a semidefinite program with matrices of size $(m+n)$ by $(m+n)$. On a standard laptop and using interior-point methods, $\nu^{[0]}_+(X)$ can be computed for m and n up to about 100; see [Matlab file: nonneg_nuclear_norm_bound.m].

For linear EDMs, the bound (3.11) is rather weak; for example it is equal to 2.27 for $n = 6$. Similar behavior will be observed for the slack matrices of n-gons; see Section 3.6.3.

An advantage of the bound (3.11) is that, as for the hyperplane separation bound, it does not rely on the support of the input matrix. However, for some matrices, it can be rather poor. For linear EDMs, the bound based on $\nu^{[0]}_+(X)$ is below 3 for all $n \leq 16$ (we have not tried higher values). As we will see in Section 3.6.3, this bound is also poor for the slack matrices of regular n-gons. Note, however, that this bound can be improved by adding some degrees of freedom in the formulation (including the scaling of the rows and columns of X, which does not affect its nonnegative rank). We refer the reader to [154] for details and more examples.

3.4.9 ▪ Self-scaled bound and sum of squares

In the paper [155], Fawzi and Parrilo develop a general approach to lower bound cone ranks, that is, ranks corresponding to factorizations where the factors belong to specific cones (such as the nonnegative orthant); see the subsection on cone factorizations (page 93). We present in this section their result only in terms of the nonnegative rank, and we refer the reader to the paper for more details. For example, their result also applies to the cp-rank.

Given a nonnegative matrix $X \in \mathbb{R}^{m\times n}_+$, let us define the set of nonnegative rank-one matrices which are componentwise smaller than X:

$$\mathcal{X}_+(X) = \{R \in \mathbb{R}^{m\times n} \mid \operatorname{rank}(R) \leq 1 \text{ and } 0 \leq R \leq X\}.$$

The supports of the matrices in $\mathcal{X}_+(X)$ correspond to the rectangles in the RCB.

Lemma 3.32. *[155] Let $L(.)$ be a linear functional such that*

$$L(R) \leq 1 \text{ for all } R \in \mathcal{X}_+(X). \tag{3.13}$$

Then, $L(X) \leq \operatorname{rank}_+(X)$.

Proof. Let $X = \sum_{k=1}^{r_+} R_k$ correspond to an Exact NMF for X with $r_+ = \operatorname{rank}_+(X)$ where $R_k \geq 0$ and $\operatorname{rank}(R_k) = 1$ for all k. This implies that $R_k \leq X$ and hence $R_k \in \mathcal{X}_+(X)$ so that $L(R_k) \leq 1$ for all k, and therefore

$$L(X) = \sum_{k=1}^{r_+} L(R_k) \leq r_+ = \operatorname{rank}_+(X). \qquad \square$$

Given X, one can try to find the linear functional satisfying the condition (3.13) that leads to the largest lower bound for the nonnegative rank of X, namely

$$\tau_+(X) = \max_{L \text{ linear}} L(X) \quad \text{such that} \quad L(R) \leq 1 \text{ for all } R \in \mathcal{X}_+(X).$$

We have the following result.

Theorem 3.33. *[155, Theorem 2] For any $X \in \mathbb{R}_+^{m \times n}$,*

$$\tau_+(X) \leq \operatorname{rank}_+(X).$$

Proof. This follows directly from Lemma 3.32. □

Generalization of the hyperplane separation bound Lemma 3.32 can be used to derive the hyperplane separation bound: it corresponds to the choice

$$L(X) = \left\langle \frac{Z}{\alpha(Z)}, X \right\rangle,$$

where $Z \in \mathbb{R}^{m \times n}$ and $\alpha(Z)$ is given in (3.10), and where X is normalized, that is, $X \leftarrow \frac{X}{\|X\|_\infty}$. This implies that $\tau_+(X)$ is always larger than the hyperplane separation bound.

Note that the hyperplane separation bound implicitly requires $L(R) \leq 1$ for *all* rectangles (not only those satisfying $R \leq A$); see the definition of $\alpha(Z)$ in (3.10). However, choosing Z such that $Z(i,j) = -\infty$ for all (i,j) such that $X(i,j) = 0$ (which is the best possible choice; see the discussion in Section 3.4.7) means that one can only consider the rectangles R such that $\operatorname{supp}(R) \subseteq \operatorname{supp}(X)$.

Computation of $\tau_+(X)$ The quantity $\tau_+(X)$ is difficult to compute. Note that $\tau_+(X)$ is larger than the fractional RCB [155, Theorem 4], which itself is hard to compute.

As for the quantity $\nu_+(X)$, it is possible to derive a semidefinite relaxation to lower bound $\tau_+(X)$. This is based on sum of squares, and we refer the reader to [155] for more details. Here is the formulation:

$$\tau_+(X) \geq \tau_+^{\text{sos}}(X) = \min_{t \in \mathbb{R}, Y \in \mathbb{R}^{mn \times mn}} t$$

$$\text{such that} \begin{pmatrix} t & \operatorname{vec}(X)^\top \\ \operatorname{vec}(X) & Y \end{pmatrix} \succeq 0,$$

$$Y_{ij,ij} \leq X(i,j)^2 \text{ for all } i,j,$$

$$Y_{ij,k\ell} \leq Y_{i\ell,kj} \text{ for all } i < k \text{ and } j < \ell.$$

The variable Y has mn columns and rows, hence this optimization problem cannot be solved when mn is large using a standard interior-point solver. On a standard laptop, one can handle m and n up to around 15 [155]; see also [Matlab file: self_scaled_bound.m]. However, some simplifications are possible, in particular to reduce the number of variables to $|\operatorname{supp}(X)|$ [155].

Example 3.34 (Linear EDMs, Example 3.13 continued). For the 6-by-6 linear EDM defined as $X(i,j) = (i-j)^2$ for $1 \leq i, j \leq 6$, $\tau^{sos}(X) = 3.82$, which provides a lower bound of 4 for the nonnegative rank of X. For $n = 7$, $\tau^{sos}(X_7) = 4.19$, which provides a lower bound of 5. These coincide with the antichain bound, the RCB, and the geometric bound.

Starting from the RCB presented in Section 3.4.4, all bounds can be compared using [Matlab file: bound_nnr_linEDM.m]. ∎

3.4.10 ▪ Further readings

In the paper [161], the authors propose highly efficient implementations of several lower bounds for the nonnegative rank in[26] C++, namely the fooling set bound, the RCB and the fractional RCB, the RRCB and the fractional RRCB (which is obtained in the same way as for the RCB), the hyperplane separation bound, and a nonnegative variant.

Let us briefly mention two other powerful approaches for lower bounding the nonnegative rank. Introducing the concepts needed to understand these approaches and showing explicitly how they can be used in the context of the nonnegative rank is beyond the scope of this book.

The first approach is based on information-theory concepts. Intuitively, the idea is to use the connection between the nonnegative rank and the decomposition of a distribution P into independent "components"; see Section 1.4.6. These bounds lead to strong lower bounds for the nonnegative rank, especially in the context of (approximate) extended formulations. We refer the reader to [61] and the references therein for the details.

The second approach [219] uses noncommutative polynomial optimization to formulate hierarchies of semidefinite programming lower bounds on matrix factorization ranks. This approach is rather general and applies to the nonnegative rank, the PSD rank (see the discussion in Section 3.6.4 on cone factorizations), and their symmetric analogues: the CP-rank (see page 79) and the completely PSD rank. This approach shares some similarities with the work of Fawzi and Parrilo [155] presented in Section 3.4.9. We refer the reader to the paper [219] for complete details and some numerical examples.

3.5 ▪ Upper bounds for the nonnegative rank

Upper bounds are usually obtained by providing explicit factorizations. This has been done, for example, for

- linear EDMs of the form $X(i,j) = (i-j)^2$ for $i,j = 1, 2, \ldots, n$ leading to the upper bound of $2\log_2(n)$ when n is a power of two [199, 244]; and

- the slack matrices of regular n-gons [165, 459]; see Section 3.6.3.

However, there exist a few general upper bounds for the nonnegative rank. Of course, the restricted nonnegative rank is such an upper bound since for any nonnegative matrix X,

$$\mathrm{rank}_+(X) \leq \mathrm{rank}_+^*(X).$$

Shitov [411] was able to prove the following result.

Theorem 3.35. *[411, Theorem 3.2] The nonnegative rank of a rank-three matrix $X \in \mathbb{R}_+^{m \times n}$ does not exceed $\frac{6\min(m,n)}{7}$.*

This results implies for example that any 7-gon can be represented as the projection of a three-dimensional polytope with at most six facets; see the next section for more details. Shitov also proved a sublinear upper bound for the nonnegative rank of rank-three matrices [410], later improved to the following.

Theorem 3.36. *[417] The nonnegative rank of a rank-three matrix $X \in \mathbb{R}_+^{m \times n}$ does not exceed $147\min(m,n)^{2/3}$.*

---

[26]The code is available from https://bitbucket.org/matthias-walter/nonnegrank/.

3.6 ▪ Lower bounds on extended formulations via the nonnegative rank

Lower bounding techniques for the nonnegative rank have had a tremendous impact in the study of extended formulations (see also Section 1.4.7 in the introduction). Let us quote Braun and Pokutta [61]:

> Nonnegative matrix factorizations and lower bounds for those are the main (arguably even the only) strong tools to establish lower bounds on the extension complexity.

In a nutshell (see the details in Section 3.6.1 below), the minimum number of inequalities needed to represent the feasible set of a linear program, namely, a polyhedron, is equal to the nonnegative rank of a particular matrix, referred to as the slack matrix of that polyhedron. The relatively recent breakthrough results in the field of linear integer programming bounding the size of such compact representations were obtained by bounding the nonnegative rank of the corresponding slack matrices. Examples include the proof that there does not exist a polynomial-size linear formulation for the TSP polytope [163] (2012) and the matching polytope [400] (2014).

In this section, we review this connection and present several examples. It is organized as follows. In Section 3.6.1, we introduce the setup and define an extended formulation and the extension complexity of a polytope. In Section 3.6.2, we prove the seminal paper of Yannakakis (Theorem 3.39), namely that the nonnegative rank of the slack matrix of a polytope is equal to the extension complexity of that polytope. In Section 3.6.3, we illustrate this result on the n-gons. This allows us to review the different lower bounding techniques for the nonnegative rank presented in Section 3.4 on several examples. We conclude the section by mentioning two generalizations of extended formulations for polytopes: approximate and conic extended formulations (Section 3.6.4), and by discussing the extension complexity of the Cartesian product of two polytopes (Section 3.6.5).

3.6.1 ▪ Introduction

A standard approach for tackling combinatorial optimization problems is to use linear optimization. By representing the convex hull of the set of feasible solutions via linear equalities and inequalities, the corresponding optimization problem can be solved via linear optimization. An important problem is therefore the compact description of polytopes. In particular, if one is able to represent the convex hull of the set of feasible solutions using a polynomial number of inequalities, then the problem can be solved in polynomial time via linear optimization.

In this section we focus on bounded polyhedra, that is, polytopes, but the results can be extended to unbounded polyhedra; see the discussion in [162]. A key issue is representing these polytopes compactly, that is, with as few inequalities (that is, facets) as possible, possibly using additional variables. Such reformulations are referred to as extended formulations.

Definition 3.37 (Extended formulation). *Let*

$$\mathcal{P} = \{x \in \mathbb{R}^d \mid Ax \leq b\} \ \text{where} \ A \in \mathbb{R}^{m \times d} \ \text{and} \ b \in \mathbb{R}^m$$

be a polytope. An extended formulation of $\mathcal{P}$ is a polyhedron

$$\mathcal{Q} = \{(x, y) \in \mathbb{R}^{d+p} \mid Cx + Dy \leq g \ \text{and} \ Ex + Fy = h\}$$

in variables $(x, y) \in \mathbb{R}^{d+p}$ such that $x \in \mathcal{P}$ if and only if there exists y such that $(x, y) \in \mathcal{Q}$.

In the definition above, the projection of $\mathcal{Q}$ onto the variable x, which we denote $\mathrm{proj}_x(\mathcal{Q})$, is equal to $\mathcal{P}$. Optimizing a linear function $f(x)$ over $\mathcal{P}$ is therefore equivalent to optimizing the same function $f(x)$ over $\mathcal{Q}$. The size of the extended formulation $\mathcal{Q}$ of $\mathcal{P}$ is the number of inequalities in $\mathcal{Q}$, that is, the number of rows of the matrices C and D.

Most polytopes arising in combinatorial optimization have an exponential number of facets in their "natural" formulations, that is, in terms of their original variables. However, some of them can be reformulated using only a polynomial number of inequalities via appropriate extended formulations; these are referred to as compact extended formulations.

Example 3.38 (Permutahedron). Let us illustrate the notion of compact extended formulations on the permutahedron; see [376] and the references therein for more details. The permutahedron Π_n in dimension n is the convex hull of the set of points corresponding to all permutations of $\{1, 2, \ldots, n\}$. For example, for $n = 3$,

$$\Pi_3 = \mathrm{conv}\left((1, 2, 3), (1, 3, 2), (2, 1, 3), (3, 1, 2), (2, 3, 1), (3, 2, 1)\right).$$

The permutahedron has $n!$ vertices and can be written as follows using $2^n - 2$ inequalities (it has $2^n - 2$ facets):

$$\left\{ x \in \mathbb{R}^n \mid \sum_{i=1}^{n} x_i = \frac{n(n+1)}{2}, \sum_{i \in S} x_i \geq \frac{|S|(|S|+1)}{2} \text{ for all } \emptyset \neq S \subset \{1, 2, \ldots, n\} \right\}.$$

The inequalities require that any nonempty subset of $|S|$ entries of x must sum to at least $\frac{|S|(|S|+1)}{2}$. Introducing the n^2 variables $Z \in \mathbb{R}^{n \times n}$, the permutahedron can be written as follows:

$$\Pi_n = \left\{ x \in \mathbb{R}^n \mid \text{there exists } Z \in \mathbb{R}_+^{n \times n} \text{ such that } \sum_{i=1}^{n} iZ(i, j) = x_j \text{ for all } j, \right.$$

$$\sum_{i=1}^{n} Z(i, j) = 1 \text{ for all } i,$$

$$\left. \sum_{j=1}^{n} Z(i, j) = 1 \text{ for all } j \right\}.$$

This reformulation has $n^2 + n$ variables and n^2 inequalities (namely $Z \geq 0$). Hence this is a compact extended formulation of size n^2 of the permutahedron.

There exists a more compact formulation of Π_n of size $n \log(n)$ using sorting networks [215]. Since the slack matrix of the permutahedron has $n!$ columns with different supports, its nonnegative rank is lower bounded by $\log(n!) = \Theta(n \log(n))$ (because $(n/2)^{n/2} \leq n! \leq n^n$); see Lemma 3.14 in Section 3.4.2. This proves that there cannot exist more compact formulations (see Theorem 3.39). ∎

We refer the reader to the papers [105, 462, 267, 483, 375, 153] for more details on extended formulations, including more examples of compact extended formulations.

The smallest size of an extended formulation of $\mathcal{P}$ is called the extension complexity of $\mathcal{P}$ and denoted $\mathrm{xc}(\mathcal{P})$. As explained above, describing the polytope corresponding to a combinatorial problem allows us to solve it using linear optimization. Hence knowing whether such a polytope has an extension complexity polynomial in the number of variables is crucial as it indicates whether the corresponding combinatorial problem can be solved in polynomial time via linear optimization.

3.6.2 ▪ Yannakakis's result

In 1988, Yannakakis unraveled a key result: the extension complexity of a polytope is equal to the nonnegative rank of its slack matrix; see Theorem 3.39 below. The slack matrix $S \in \mathbb{R}_+^{m \times n}$ of the polytope

$$\mathcal{P} = \left\{ x \in \mathbb{R}^d \mid a_i^\top x \leq b_i, i = 1, 2, \ldots, m \right\}$$

is defined as

$$S(i, j) = b_i - a_i^\top v_j \geq 0 \quad \text{for all } i, j,$$

where $v_1, v_2, \ldots, v_n$ are the vertices of $\mathcal{P}$. Before stating and proving the seminal result of Yannakakis, let us make a few remarks:

- The slack matrix of a polytope is not unique, since its representation in terms of inequalities is not (for example by multiplying an inequality by a positive number, or by adding redundant inequalities). However, the nonnegative rank is invariant with respect to the choice of the representation of $\mathcal{P}$. For example, adding rows (resp. columns) corresponding to valid inequalities for $\mathcal{P}$ (resp. points in $\mathcal{P}$) to the slack matrix does not change its nonnegative rank since any valid inequality is a conic combination of the inequalities defining $\mathcal{P}$ (resp. any point inside $\mathcal{P}$ is a convex combination of its vertices).

- The slack matrix S is equal to the matrix X of the RE-NMF instance obtained by using the reduction from the NPP with

$$\mathcal{A} = \mathcal{P} \quad \subseteq \quad \mathcal{B} = \mathcal{P};$$

  see the paragraph "From NPP to RE-NMF" (page 40) and in particular (2.13). In the NPP, since the inner polytope $\mathcal{A}$ and the outer polytope $\mathcal{B}$ coincide, the only solution is $\mathcal{E} = \mathcal{A} = \mathcal{B}$, implying that the restricted nonnegative rank of a slack matrix X is equal to the number of vertices of $\mathcal{P}$, that is, $\operatorname{rank}_+^*(X) = n$ [199, Theorem 7].

- In his original paper [492] (later published as a journal paper [493]), Yannakakis defined the extension complexity as the number of variables plus the number of constraints in the extended formulation and showed that extension complexity and nonnegative rank are within a factor of two of each other; see also the discussion in [162].

Theorem 3.39. *[492, 493] Let $\mathcal{P}$ be a polytope of dimension larger than one, and let S be its slack matrix. Then*

$$\operatorname{rank}_+(S) = \operatorname{xc}(\mathcal{P}).$$

Proof. [163, Theorem 3], [400, Theorem 2] Let us first show that $\operatorname{xc}(\mathcal{P}) \leq \operatorname{rank}_+(S)$. For this, given $\mathcal{P} = \{x \in \mathbb{R}^d \mid b - Ax \geq 0\}$ with $A \in \mathbb{R}^{m \times d}$ and $b \in \mathbb{R}^m$, let us show that any Exact NMF of $S = WH$ of size r with $W \geq 0$ and $H \geq 0$ provides the following explicit extended formulation of $\mathcal{P}$ of size r (with some redundant equalities):

$$Q = \{(x, y) \mid b - Ax = Wy \text{ and } y \geq 0\};$$

that is, let us show that $\operatorname{proj}_x(Q) = \{x \mid \exists y \text{ s.t. } (x, y) \in Q\} = \mathcal{P}$. We have

- $\operatorname{proj}_x(Q) \subseteq \mathcal{P}$ since $W \geq 0$ and $y \geq 0$, and hence $b - Ax = Wy \geq 0$ for all $(x, y) \in Q$;

- $\mathcal{P} \subseteq \operatorname{proj}_x(Q)$ because all vertices v_j ($1 \leq j \leq n$) of $\mathcal{P}$ belong to $\operatorname{proj}_x(Q)$: by construction, $(v_j, H(:, j)) \in Q$ since $S(:, j) = b - Av_j = WH(:, j)$ and $H(:, j) \geq 0$.

It remains to show that $\mathrm{rank}_+(S) \leq \mathrm{xc}(\mathcal{P})$. Let

$$\mathcal{Q} \;=\; \{(x,y) \in \mathbb{R}^{d+p} \mid Cx + Dy \leq g \text{ and } Ex + Fy = q\}$$

be an extended formulation of $\mathcal{P}$ with r inequalities, that is, C and D have r rows. From this extended formulation, let us construct the r-dimensional nonnegative vectors w_i and h_j such that $S(i,j) = w_i^\top h_j$ for $1 \leq i \leq m$ and $1 \leq j \leq n$, which will imply $\mathrm{rank}_+(S) \leq r$. Since $\mathcal{Q}$ is an extended formulation of $\mathcal{P}$, there exists z_j such that $(v_j, z_j) \in \mathcal{Q}$ for all j. Let $h_j = g - Cv_j - Dz_j$ be the slack of (v_j, z_j) with respect to the inequalities of $\mathcal{Q}$. By linear programming (LP) duality, we know that each valid constraint $a_i^\top x + 0^\top y \leq b_i$ $(1 \leq i \leq m)$ of $\mathcal{Q}$ can be obtained by a conic combination of its defining constraint $Cx + Dy \leq g$ (this requires the assumption that the dimension of $\mathcal{P}$ is larger than one), that is, for all i, there exists $w_i \in \mathbb{R}_+^r$ such that

$$w_i^\top (C \; D \; g) = (a_i^\top \; 0 \; b_i).$$

We obtain

$$w_i^\top h_j = w_i^\top (g - Cv_j - Dz_j) = b_i - a_i^\top v_j = S(i,j). \qquad \square$$

Many results were obtained later (starting around 2010) to provide bounds on the extended formulations of difficult combinatorial problems using Theorem 3.39. In particular, Fiorini et al. [163, 164] proved that the extension complexity of the TSP polytope has exponential extension complexity, and Rothvoss proved that the perfect matching polytope has exponential extension complexity [401].

3.6.3 ▪ Application to the slack matrices of n-gons

In this section, we apply the different notions covered in this chapter to the slack matrices of n-gons. We first review the cases of the square and the regular hexagon that we have actually already investigated in Section 2.1. We then present the case of the regular octagon, in particular to illustrate the use of the different lower bounds for the nonnegative rank on this 8-by-8 matrix. We present the general case of regular n-gons and describe an application: approximating the second-order cone with a polytope. Finally, we briefly discuss the case of generic n-gons.

3.6.3.1 ▪ Square

Let us start with the square with four facets

$$\mathcal{P} \;=\; \{x \in \mathbb{R}^2 \mid x_1 \geq 0, x_2 \geq 0, x_1 \leq 1, x_2 \leq 1\}$$

and four vertices

$$(0,0), (1,0), (1,1) \text{ and } (0,1).$$

This leads to the following facet-by-vertex slack matrix:

	$(1,1)$	$(1,0)$	$(0,1)$	$(0,0)$
$x_1 \geq 0$	1	1	0	0
$x_2 \geq 0$	1	0	1	0
$x_2 \leq 1$	0	1	0	1
$x_1 \leq 1$	0	0	1	1

.

This matrix equals the matrix of Thomas (2.5) (we have ordered the facets and vertices in order to achieve this). Looking back at Figure 2.4 (page 29), this is not surprising: the inner and outer polytopes in the corresponding NPP instance are the same square. We have seen that the nonnegative rank of the matrix of Thomas is 4; it is therefore not possible to represent the square with fewer than four inequalities.

3.6.3.2 ▪ Hexagon

We have considered in Section 2.1 the matrix

$$X = \begin{pmatrix} 0 & 1 & 2 & 2 & 1 & 0 \\ 0 & 0 & 1 & 2 & 2 & 1 \\ 1 & 0 & 0 & 1 & 2 & 2 \\ 2 & 1 & 0 & 0 & 1 & 2 \\ 2 & 2 & 1 & 0 & 0 & 1 \\ 1 & 2 & 2 & 1 & 0 & 0 \end{pmatrix}.$$

Doing the same exercise as for the square, one can check that this matrix is a slack matrix of the regular hexagon. In fact, we have proved in Section 2.1 that the corresponding NPP instance has the inner and outer polytopes being the same regular hexagon. Moreover, we have seen that the nonnegative rank of this matrix is five, implying that the extension complexity of the regular hexagon is five. We have also seen that the columns of X belong to the convex hull of a three-dimensional polytope with only five vertices, because $X = WH$ where W has five columns and $\mathrm{rank}(W) = 4$; see Figure 2.7 (page 35). Equivalently (Theorem 3.39), the hexagon can be represented as the projection of a polytope with five facets (Theorem 3.39); see Figure 3.2.

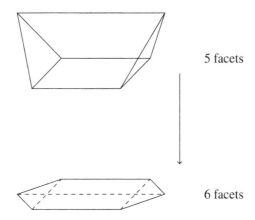

5 facets

6 facets

Figure 3.2. *Minimum-size extended formulation of the regular hexagon with five facets. Figure taken from [459].*

3.6.3.3 ▪ Octagon

Let $\mathcal{P} = \{x \in \mathbb{R}^2 \mid Ax \le b\}$ be the regular octagon centered at the origin with

$$A = \begin{pmatrix} 1 & a & 0 & -a & -1 & -a & 0 & a \\ 0 & a & 1 & a & 0 & -a & -1 & -a \end{pmatrix}^\top, \text{ and } b(i) = 1 + a \text{ for all } i,$$

where $a = \frac{\sqrt{2}}{2}$. The vertices of $\mathcal{P}$ are the columns of

$$V = \begin{pmatrix} 1+a & a & -a & -(1+a) & -(1+a) & -a & a & (1+a) \\ a & 1+a & 1+a & a & -a & -(1+a) & -(1+a) & -a \end{pmatrix}.$$

We obtain the slack matrix

$$S = AV - [b \; b \; \ldots \; b]$$

$$= \begin{pmatrix}
0 & 1 & 1+2a & 2+2a & 2+2a & 1+2a & 1 & 0 \\
0 & 0 & 1 & 1+2a & 2+2a & 2+2a & 1+2a & 1 \\
1 & 0 & 0 & 1 & 1+2a & 2+2a & 2+2a & 1+2a \\
1+2a & 1 & 0 & 0 & 1 & 1+2a & 2+2a & 2+2a \\
2+2a & 1+2a & 1 & 0 & 0 & 1 & 1+2a & 2+2a \\
2+2a & 2+2a & 1+2a & 1 & 0 & 0 & 1 & 1+2a \\
1+2a & 2+2a & 2+2a & 1+2a & 1 & 0 & 0 & 1 \\
1 & 1+2a & 2+2a & 2+2a & 1+2a & 1 & 0 & 0
\end{pmatrix}.$$

Let us show that $\operatorname{rank}_+(S) = 6$. We have $\operatorname{rank}_+(S) \leq 6$ since $S = WH$ where

$$W = \begin{pmatrix}
1 & 0 & 0 & 1 & 0 & 1+2a \\
1+2a & 0 & 0 & 0 & 1 & 2+2a \\
1 & 1 & 0 & 0 & 0 & 1+2a \\
0 & 2a & 1 & 0 & 0 & 1 \\
0 & 1 & 1+2a & 0 & 1 & 0 \\
1 & 0 & 2+2a & 0 & 1+2a & 0 \\
0 & 0 & 1+2a & 1 & 1 & 0 \\
0 & 0 & 1 & 2a & 0 & 1
\end{pmatrix}$$

and

$$H = \begin{pmatrix}
0 & 0 & 0 & 1 & 0 & 0 & 1 & 0 \\
1 & 0 & 0 & 0 & 0 & 1 & 1+2a & 1+2a \\
1 & 1 & 0 & 0 & 0 & 0 & 0 & 0 \\
0 & 1 & 1+2a & 1+2a & 1 & 0 & 0 & 0 \\
0 & 0 & 1 & 0 & 0 & 0 & 0 & 1 \\
0 & 0 & 0 & 0 & 1 & 1 & 0 & 0
\end{pmatrix},$$

with $\operatorname{rank}(W) = 4$ and $\operatorname{rank}(H) = 5$. Such factorizations can be computed using the code from [460], [Matlab file: exactNMFheur.m], or using explicit factorizations from [165, 459]. Note that one can write down explicitly the extended formulation of $\mathcal{P}$ of size 6 using the Exact NMF given above. It suffices to follow the first part of the proof of Theorem 3.39:

$$\mathcal{Q} = \{(x, y) \in \mathbb{R}^{2+6} \mid b = Ax + Wy, y \geq 0\}$$

is an extended formulation for $\mathcal{P}$ with six inequalities. Since $\operatorname{rank}([A, W]) = 4$ while $b \in \operatorname{col}([A, W])$, four variables can be discarded. Hence we can write the regular 8-gon as the projection of a polytope in four dimensions with six inequalities.

Let us review the different lower bounds presented in this chapter.

• **Rank**. We have $\operatorname{rank}(S) = 3 \leq \operatorname{rank}_+(S)$.

• **Fooling set**. The largest fooling of S set has size 4. The largest fooling set for the slack matrix of any n-gon has size 4 for $n \geq 6$. This is because such matrices have essentially the same pattern of zeros: only two zeros per column and row (each vertex intersects exactly two facets, and vice versa) and the pattern of zeros is circulant (it is shifted by one on the right between each row). To prove this result, first observe that such circulant matrices do not contain any 5-by-5 submatrix with (at least) two zeros per row and column, and then apply Lemma 3.12. (Note that applying Lemma 3.12 directly on S only provides an upper bound of size 5 for any fooling set since S has two zeros per row and columns.)

Remark 3.3. *Surprisingly, the slack matrices of 5-gons have a fooling set of size 5:*

$$X = \begin{pmatrix} 0 & \boxed{*} & * & * & 0 \\ 0 & 0 & \boxed{*} & * & * \\ * & 0 & 0 & \boxed{*} & * \\ * & * & 0 & 0 & \boxed{*} \\ \boxed{*} & * & * & 0 & 0 \end{pmatrix},$$

*where * indicates a positive entry.*

- **Counting argument**. The matrix S has eight columns with different supports, hence the counting argument of Goemans leads to $\mathrm{rank}_+(S) \geq \log_2(8) = 3$ (Lemma 3.14). Using the refined argument of Yannakakis (Lemma 3.15) and observing that one can generate $2n + 2$ columns with distinct supports using linear combinations of the n columns of S (see Remark 3.2), we obtain $\mathrm{rank}_+(S) \geq \log_2(18) = 4.17$, which leads to a lower bound of 5.

- **Antichain**. The matrix S has eight columns with supports not contained in one another, as for linear EDMs. The antichain bound gives the lower bound $\mathrm{rank}_+(S) \geq 5$ since

$$5 \;=\; \min_r r \text{ such that } \binom{r}{\lfloor r/2 \rfloor} \geq 8.$$

- **RCB and RRCB**. The RCB gives $\mathrm{rc}(S) = 6$ and hence is tight. As explained in Section 3.4.4, it can be computed via the resolution of a combinatorial problem. Since the RCB is tight, the RRCB leads to the same bound.

- **Geometric**. The restricted nonnegative rank of a slack matrix of a polytope $\mathcal{P}$ is always equal to the number of vertices of $\mathcal{P}$ [199, Theorem 7]. This is because in the corresponding NPP instance the inner and outer polytopes coincide with $\mathcal{P}$, and hence the only nested polytope is $\mathcal{P}$ itself. Now, we need to find the smallest k such that

$$\phi(\mathrm{rank}(S), k) = \phi(3, k) \geq \mathrm{rank}_+^*(S) = 8;$$

see Corollary 3.26. Since $\phi(3, 4) = 6$ while $\phi(3, 5) = 10$, the geometric lower bound gives $\mathrm{rank}_+(S) \geq 5$.

Note that by using the improved geometric bound for rank-three matrices proposed in [199, Corollary 4], we obtain $\mathrm{rank}_+(S) \geq 6$ which is tight.

- **Hyperplane separation**. The best lower bound we were able to obtain with the hyperplane separation bound is 2.712, using

$$Z(i, j) = \begin{cases} -\infty & \text{if } S(i, j) = 0, \\ 0 & \text{if } S(i, j) = 1, \\ S(i, j) & \text{otherwise.} \end{cases}$$

For this matrix, $\alpha(Z) = 20.14$ where the best rectangle is the 3-by-4 bottom left rectangle; hence

$$\mathrm{rank}_+(X) \geq \frac{\langle Z, S \rangle}{\alpha(Z)\|S\|_\infty} = 2.712.$$

Generating 10000 matrices Z whose entries are generated uniformly at random in $[0, 1]$ (except the ones corresponding to the zero entries of S which are set to $-\infty$) did not improve this bound.

- **Nonnegative nuclear norm.** We have $\nu_+^{[0]}(S) = 29.564$, and hence

$$\text{rank}_+(S) \geq \left(\frac{\nu_+^{[0]}(S)}{\|S\|_F} \right)^2 = 2.95.$$

- **Self-scaled bound.** We have $\tau_+^{\text{sos}}(S) = 5.15 \leq \text{rank}_+(S)$ which implies $\text{rank}_+(S) \geq 6$, which is tight.

In summary, the extension complexity of the regular octagon is equal to six. This can be proved by computing an explicit Exact NMF of rank six and providing tight lower bounds for the nonnegative rank of this matrix. For the regular octagon, four lower bounds are tight: the RCB, the RRCB, the improved geometric bound, and $\tau_+^{\text{sos}}(S)$. Except for the RRCB, these bounds can be computed using [Matlab file: bound_nnr_octagon.m], along with the hyperplane separation bound and the bound based on the nonnegative nuclear norm.

3.6.3.4 ▪ Regular n-gons

Ben-Tal and Nemirovski [32] provided extended formulations of the regular n-gons when n is a power of two ($n = 2^k$ for some k) with $2\log_2(n) + 4$ inequalities. They used this construction to approximate the unit disk with regular n-gons. This allowed them to approximate, up to any given accuracy ϵ, second-order cone optimization problems (where inequalities have the form $\|Ax - b\|_2 \leq c^\top x - d$ for some matrix A, vectors b and c, and scalar d) with linear optimization problems of polynomial size in $\ln(1/\epsilon)$ and in the dimension and the number of constraints of the second-order cone. This is interesting because it essentially shows that, from a theoretical point of view, the expressiveness of second-order cone optimization is not much more powerful than linear optimization; see also [214, 25] for some numerical experiments and applications. Note that the lower bound based on a counting argument of Lemma 3.15 provides the bound $\log_2(2n + 2)$ since $2n + 2$ distinct supports can be generated using the n columns of the slack matrix; see also Remark 3.2. For $n = 2^k$ for some integer k, the construction of Ben-Tal and Nemirovski was slightly reduced to size $2\log_2(n)$ by Glineur [214]. Kaibel and Pashkovich [268, 269] proposed a general construction for arbitrary n of size $2\lceil\log_2(n)\rceil + 2$. Fiorini, Rothvoss, and Tiwary [165] improved the construction, leading to an extended formulation of size $2\lceil\log_2(n)\rceil$. Finally, the authors in [459] improved the bound to $2\lceil\log_2(n)\rceil - 1$ when $2^{k-1} < n \leq 2^{k-1} + 2^{k-2}$ for some integer k. They used explicit Exact NMFs of these slack matrices. They conjectured that the bound is tight, which holds for $n \leq 13$, $21 \leq n \leq 24$, and $31 \leq n \leq 32$ as it matches the best known lower bound; see Figure 3.3.

Figure 3.3 compares the different lower and upper bounds for the nonnegative rank of the slack matrix or regular n-gons for $3 \leq n \leq 33$. It displays three upper bounds:

- Shitov: the bound $\lceil \frac{6\min(m,n)}{7} \rceil$ of Shitov [411].

- Reflection: the bound of [165] that uses a geometric construction based on reflections to obtain explicit extended formulations of the regular n-gons.

- Factorization: the bound of [459] that proposed explicit Exact NMFs of the slack matrices of regular n-gons.

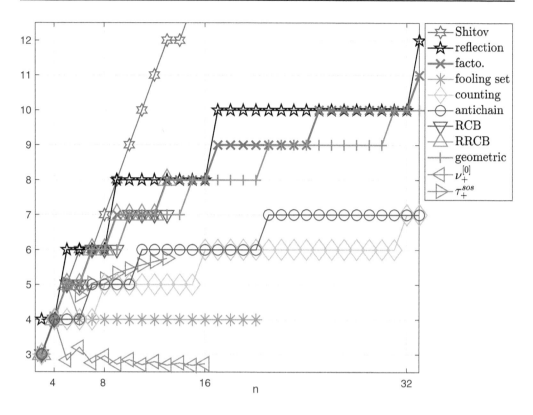

Figure 3.3. *Lower and upper bounds for the nonnegative rank of the slack matrices of regular n-gons. Figure adapted from [459, Figure 2].*

It displays eight lower bounds presented in this chapter: the ones based on the counting argument (Section 3.4.2) and on antichains (Section 3.4.3), the RCB (Section 3.4.4), the RRCB (Section 3.4.5), the geometric bound (Section 3.4.6), $\nu_+^{[0]}$ (Section 3.4.8), and τ_+^{sos} (Section 3.4.9).

The following observations can be made:

- Some bounds are computationally expensive to compute (namely, RCB, RRCB, $\nu_+^{[0]}$, and τ_+^{sos}) and cannot be computed for all values of n. For these matrices, we know the fooling set bound is always 4 (see the discussion in subsection 3.6.3.3 about the regular octagon). However, in general, it requires solving a difficult combinatorial problem (corresponding to a maximum clique problem that we solved explicitly for $n \leq 20$).

- The RRCB improves upon the RCB, by one, for $n = 9$ and $n = 13$. The RRCB is the only tight lower bound in these two cases and, together with the explicit factorizations, proves that the extension complexity of the regular 9-gon is 7 and of the regular 13-gon is 8.

- The bound $\nu_+^{[0]}$, and to a lesser extent $\tau_+^{sos}(X)$, is relatively poor for these matrices (it is smaller than 3 for $n \geq 7$). Also, neither can be computed for large n.

- The geometric bound is the improved version for rank-three matrices [199, Corollary 7]. It performs rather well, being tight for quite a few values of n; it is the only tight (as it matches the value of explicit factorizations) and computable bound for $n = 15, 16, 21, 22, 23, 24, 31, 32$.

3.6.3.5 ▪ Generic n-gons

As opposed to regular n-gons, generic n-gons have large extension complexity in $\Omega(\sqrt{n})$ [165], and hence cannot be represented compactly in general. Shitov provided a sublinear upper bound of $147n^{2/3}$ for the slack matrices of n-gons [417]; see Theorem 3.36.

3.6.4 ▪ Approximate and conic extended formulations

There are two important generalizations of the bounds for extended formulations based on the nonnegative rank. Let us briefly mention them.

3.6.4.1 ▪ Approximate extended formulations

As we have mentioned previously, many combinatorial problems do not admit compact extended formulations, such as the TSP and the matching polytopes. However, one may wonder whether this holds when the extended formulations only need to approximate the given polytope up to some accuracy.

An approach to tackle this question is to consider the slack matrix of a pair of polytopes. Let us look back at the NPP: given the polytope $\mathcal{A}$ defined via its vertices $\{v_1, v_2, \ldots, v_n\}$ and the polytope $\mathcal{B}$ defined via its facets $Fx + g \geq 0$ with $\mathcal{A} \subseteq \mathcal{B}$, find, if possible, a polytope $\mathcal{E}$ with p vertices nested between $\mathcal{A}$ and $\mathcal{B}$, that is, $\mathcal{A} \subseteq \mathcal{E} \subseteq \mathcal{B}$. We have seen that this problem is equivalent to RE-NMF of the matrix X defined as $X(:, j) = Fv_j + g$ for all j; see Theorem 2.11. Let us define the matrix X using this construction as the slack matrix of the pair $(\mathcal{A}, \mathcal{B})$. One can show the following.

Theorem 3.40. *[57, Theorem 1] The minimum extension complexity over all polytopes $\mathcal{E}$ such that $\mathcal{A} \subseteq \mathcal{E} \subseteq \mathcal{B}$ is equal to the nonnegative rank of the slack matrix of the pair of polytopes $(\mathcal{A}, \mathcal{B})$. This quantity is referred to as the extension complexity of the pair $(\mathcal{A}, \mathcal{B})$ and denoted* $\mathrm{xc}(\mathcal{A}, \mathcal{B})$. *The corresponding extended formulation of $\mathcal{E}$ is referred to as an extended formulation of $(\mathcal{A}, \mathcal{B})$.*

Note that $\mathrm{xc}(\mathcal{P}) = \mathrm{xc}(\mathcal{P}, \mathcal{P})$ for any polytope $\mathcal{P}$, since the slack matrix of the pair $(\mathcal{P}, \mathcal{P})$ is the slack matrix of $\mathcal{P}$.

Theorem 3.40 allowed Braunn et al. [57, 58] to prove that $\mathcal{O}(n^{1/2-\epsilon})$-approximations of the clique polytope require a linear program of exponential size, where n is the number of nodes in the graph and ϵ is a small fixed constant. This was later improved by showing that the same bound applies to $\mathcal{O}(n^{1-\epsilon})$-approximations [62]; see also [59, 61]. Braun and Pokutta [60] established that for all fixed $0 < \epsilon < 1$, every linear program approximating the matching polytope by a factor $(1 + \epsilon/n)$ must have exponential size, where n is the number of nodes in the graph.

Let us briefly explain the idea behind the proofs of these results. For simplicity, let us consider a maximization problem over a polytope of the form

$$\mathcal{P} = \{x \in \mathbb{R}^r \mid Ax \leq b\} \text{ with } b \geq 0.$$

The generalization to any polyhedron and to whether maximization or minimization is considered can be found in [57]. For any $\rho \geq 1$, let us define

$$\rho\mathcal{P} = \{x \in \mathbb{R}^r \mid Ax \leq \rho b\},$$

where $\rho\mathcal{P}$ is a dilatation of $\mathcal{P}$ by a factor ρ so that $\mathcal{P} \subseteq \rho\mathcal{P}$ since $0 \leq b \leq \rho b$. Note that $x \in \mathcal{P}$ if and only if $\rho x \in \rho\mathcal{P}$. Hence, for any $c \in \mathbb{R}^d$,

$$\max_{x \in \rho\mathcal{P}} c^\top x = \rho \max_{x \in \mathcal{P}} c^\top x.$$

Now, in order to bound the extension complexity of an approximation of a polytope $\mathcal{P}$, we can consider the extension complexity of the pair of polytopes $\mathcal{P}$ and $\rho\mathcal{P}$. Any polytope in between $\mathcal{P}$ and $\rho\mathcal{P}$ provides a ρ-approximation for $\mathcal{P}$. In other words, finding an extended formulation for the pair $(\mathcal{P}, \rho\mathcal{P})$ allows us to approximate any linear program over $\mathcal{P}$ by a factor of at most ρ. Finally, the proofs rely on bounding the nonnegative rank of the slack matrix of the pair $(\mathcal{P}, \rho\mathcal{P})$.

3.6.4.2 ▪ Cone factorizations

In an Exact NMF of size r of a matrix $X \in \mathbb{R}^{m \times n}_+$, we are looking for m vectors $w_i \in \mathbb{R}^r_+$ ($1 \le i \le m$, the rows of W) and n vectors $h_j \in \mathbb{R}^r_+$ ($1 \le j \le n$, the columns of H) such that

$$X(i, j) = \langle w_i, h_j \rangle = w_i^\top h_j \text{ for all } i, j.$$

The nonnegative rank of X is the smallest r such that such a decomposition exists. Replacing the constraints

$$w_i \in \mathbb{R}^r_+ \text{ and } h_j \in \mathbb{R}^r_+ \text{ for all } i, j$$

with

$$w_i \in \mathcal{K} \text{ and } h_j \in \mathcal{K}^* \text{ for all } i, j$$

for some cone $\mathcal{K}$ and its dual[27] $\mathcal{K}^*$ leads to a generalization of the notion of Exact NMF and hence of the notion of the nonnegative rank.

Let us focus on the cone of r-by-r PSD matrices (which is self-dual like the nonnegative orthant). The PSD rank of a nonnegative m-by-n matrix X is the smallest r such that there exists $w_i, h_j \in \mathbb{S}^r_+$ with

$$X(i, j) = \langle w_i, h_j \rangle \text{ for all } i, j.$$

Note that w_i, h_j are r-by-r PSD matrices, and this factorization is referred to as a PSD factorization. The power of this generalization is that the theorem of Yannakakis [217, Theorem 4] also generalizes leading to conic extended formulation. In particular, the PSD rank of the slack matrix of a polytope is equal to the size of its smallest PSD extended formulation, which is the size of the PSD matrices involved in the formulation. This can be generalized to sets other than polytopes but then requires dealing with infinite dimensional matrices (the slack matrix is infinite since the number of extreme points and facets is not finite). We refer to the surveys [152, 451, 153] for more details.

Note that the PSD rank can be significantly smaller than the nonnegative rank: there exists a family of 2^n-by-2^n matrices with PSD rank in $\mathcal{O}(1)$ and nonnegative rank in $2^{\Omega(n)}$ [415]. However, similarly as for (linear) extended formulations, these results allowed them to show that there are no polynomial-size PSD extended formulations for the cut, TSP, and stable set polytopes [306].

These ideas also recently allowed Hamza Fawzi to prove that the PSD cone cannot be represented using the second-order cone [151]; the proof relies on the second-order cone rank of the cone of 3-by-3 PSD matrices being infinite.

3.6.5 ▪ Extension complexity of the Cartesian product of polytopes

To conclude this section on extended formulations, we mention an important open question about the extension complexity of the Cartesian product of polytopes and its link with the nonnegative rank. Given two polytopes $\mathcal{P}$ and $\mathcal{Q}$, the open question is to determine the extension complexity

---

[27]See page 106 for a definition.

of their Cartesian product

$$\mathcal{P} \times \mathcal{Q} = \{(x, y) \mid x \in \mathcal{P} \text{ and } y \in \mathcal{Q}\}.$$

It is easy to show (see below) that $\mathrm{xc}(\mathcal{P} \times \mathcal{Q}) \leq \mathrm{xc}(\mathcal{P}) + \mathrm{xc}(\mathcal{Q})$, and it is conjectured that equality holds; indeed, it has been proved to hold when $\mathcal{P}$ or $\mathcal{Q}$ is a pyramid [454]. Let $S \in \mathbb{R}^{m \times n}$ and $T \in \mathbb{R}^{p \times q}$ be the slack matrices of $\mathcal{P}$ and $\mathcal{Q}$, respectively. The above conjecture is equivalent to showing that the nonnegative rank of the slack matrix of $\mathcal{P} \times \mathcal{Q}$ of dimension $(m + p) \times nq$, given by

$$S \times T = \left(\begin{array}{ccc|ccc|c|ccc} S(:,1) & \ldots & S(:,1) & S(:,2) & \ldots & S(:,2) & \ldots & S(:,n) & \ldots & S(:,n) \\ \hline & T & & & T & & \ldots & & T & \end{array} \right),$$

is equal to the sum of the nonnegative ranks of S and T, that is,

$$\mathrm{rank}_+(S \times T) = \mathrm{rank}_+(S) + \mathrm{rank}_+(T).$$

It is easy to prove that $\mathrm{rank}_+(S \times T) \leq \mathrm{rank}_+(S) + \mathrm{rank}_+(T)$, decomposing S and T separately. Note that the inequality can be strict for matrices which are not slack matrices (take for example $T = S = e$).

3.7 ▪ Link with communication complexity

In its simplest variant, communication complexity addresses the following problem: Alice and Bob have to compute the function

$$f \colon \{0, 1\}^m \times \{0, 1\}^n \mapsto \{0, 1\}.$$

Alice knows $a \in \{0, 1\}^m$ and Bob $b \in \{0, 1\}^n$, and the aim is to minimize the number of bits exchanged between Alice and Bob to compute $f(a, b)$. Nondeterministic communication is a variant where Bob and Alice first receive a message z before starting their communication. The nondeterministic communication complexity of f, denoted $\mathrm{NCC}(f)$, is the minimum of the length of the message z plus the communication in the deterministic protocol in order to be able to compute f. The communication matrix $X \in \{0, 1\}^{2^m \times 2^n}$ is equal to the function f for all possible combinations of inputs. Let us use the notation $X(a, b)$ for the entry of the communication matrix corresponding to $a \in \{0, 1\}^n$ and $b \in \{0, 1\}^n$ so that $f(a, b) = X(a, b)$ for all a and b.

Theorem 3.41. *The NCC of f is equal to the logarithm of the RCB of X rounded from above:*

$$\lceil \log_2(\mathrm{rc}(X)) \rceil = \mathrm{NCC}(f). \tag{3.14}$$

Proof. [307, Theorem 1] Let us first show that $\mathrm{NCC}(f) \leq \lceil \log_2(\mathrm{rc}(X)) \rceil$. Alice (resp. Bob) knows the row a (resp. column b) of X where the answer is located. Assume a rectangle covering of X into $r = \mathrm{rc}(X)$ rectangles exists. If $X(a, b) = 1$, let the message z be the string identifying the rectangle in X containing (a, b). The length of z is $\lceil \log_2(\mathrm{rc}(X)) \rceil$ bits. Both Alice and Bob can verify that the rectangle contains the row a and the column b, respectively. If $f(a, b) = 0$, no rectangle contains (a, b) so that Alice or Bob output $f(a, b) = 0$.

Let us now show that $\mathrm{NCC}(f) \geq \lceil \log_2(\mathrm{rc}(X)) \rceil$. For this, let us introduce another definition of the NCC. The NCC of the function f is the minimum k such that there exist two functions

$A : \{0,1\}^m \times \{0,1\}^k \mapsto \{0,1\}$ and $B : \{0,1\}^n \times \{0,1\}^k \mapsto \{0,1\}$ satisfying

$f(a,b) = 1 \quad \Rightarrow \quad$ there exists $z \in \{0,1\}^k$ such that $A(a,z) = 1$ and $B(b,z) = 1$, and

$f(a,b) = 0 \quad \Rightarrow \quad$ for all $z \in \{0,1\}^k$ such that $A(a,z) = 0$ or $B(b,z) = 0$.

The function A (resp. B) is the function Alice (resp. Bob) uses to compute the value of $f(a,b)$, given $z \in \{0,1\}^k$. By definition of the NCC, Alice (resp. Bob) needs such a function to compute $f(a,b)$ given a (resp. b) and the input z (which depends on a and b). Suppose $\text{NCC}(f) = k$, and let A and B be the functions as described above. For a given $z \in \{0,1\}^k$, define a rectangle $C \times D$ using $C = \{a \mid A(a,z) = 1\}$ and $D = \{b \mid B(b,z) = 1\}$. The union of all such rectangles has to cover the communication matrix X by definition of A and B, and there are 2^k such rectangles, which concludes the proof. $\qquad\square$

Theorem 3.41 implies that the NCC of f is upper bounded by the logarithm of the nonnegative rank of the communication matrix, since $\text{rank}_+(X) \geq \text{rc}(X)$, as noted by Yannakakis [493]. We refer the reader to [308, 402] and the references therein for more details.

3.7.1 ▪ Unique disjointness problem

Let us discuss a key example, namely the UDISJ problem. In this problem, both Alice and Bob receive a subset of $\{1, 2, \ldots, n\}$, and they have to determine whether the two subsets have zero or one element in common. In the other cases, every output is considered correct. Equivalently, Alice is given $a \in \{0,1\}^n$ and Bob is given $b \in \{0,1\}^n$, and they have to output 1 if $a^\top b = 0$, 0 if $a^\top b = 1$, and any nonnegative real otherwise. The communication matrix corresponding to this problem is referred to as the UDISJ matrix (because the function f can take any value if the two subsets intersect in strictly more than one element, this is actually a class of matrices). Let us use the notation $X(a,b)$ for the entry of a UDISJ matrix corresponding to some $a \in \{0,1\}^n$ and $b \in \{0,1\}^n$. Any UDISJ matrix has the form

$$X(a,b) = \begin{cases} 1 & \text{if } a^\top b = 0, \\ 0 & \text{if } a^\top b = 1, \\ \geq 0 & \text{otherwise.} \end{cases}$$

Let us stress that entries corresponding to subsets with strictly more than one element in common, that is, $X(a,b)$ with $a^\top b > 1$, are not of interest, and any nonnegative output is considered correct. The UDISJ matrix appears as a submatrix in many combinatorial problems and has been instrumental in proving the lower bounds of (approximate) extended formulations; see for example [61] and the references therein.

The NCC of this problem is $\Omega(n)$ because the RCB of X is lower bounded by $\left(\frac{3}{2}\right)^n$ [163, 164]. This implies that the nonnegative rank of any such matrix is $2^{\Omega(n)}$ because

$$\left(\frac{3}{2}\right)^n \leq \text{rc}(X) \leq \text{rank}_+(X).$$

The proof, which relies on lower bounding $\text{rc}(X)$, uses the same idea as the simple lower bound for the RCB based on the ratio of the largest rectangle contained in X to its total number of positive entries; see Equation (3.7), page 72. There are 3^n entries of X such that $a^\top b = 0$ for which $X(a,b) = 1$. In fact, when $a^\top b = 0$ there are three possibilities for each $1 \leq i \leq n$: (1) $a(i) = b(i) = 0$, (2) $a(i) = 0$ and $b(i) = 1$, or (3) $a(i) = 1$ and $b(i) = 0$. Moreover, one can prove that every rectangle can cover at most 2^n of these entries of X; see [402, Lemma 5.10] and the references therein. Therefore, $\text{rc}(X) \geq \left(\frac{3}{2}\right)^n$. Note that this bound holds regardless of the values of $X(a,b)$ for $a^\top b > 1$.

Open question on a UDISJ matrix An example of a UDISJ matrix is given by
$U_n(a, b) = (1 - a^\top b)^2$ where $a, b \in \{0, 1\}^n$. For example, for $n = 3$,

$$U_3 = \begin{pmatrix} 1 & 1 & 1 & 1 & 1 & 1 & 1 & 1 \\ 1 & 0 & 1 & 0 & 1 & 0 & 1 & 0 \\ 1 & 1 & 0 & 0 & 1 & 1 & 0 & 0 \\ 1 & 0 & 0 & 1 & 1 & 0 & 0 & 1 \\ 1 & 1 & 1 & 1 & 0 & 0 & 0 & 0 \\ 1 & 0 & 1 & 0 & 0 & 1 & 0 & 1 \\ 1 & 1 & 0 & 0 & 0 & 0 & 1 & 1 \\ 1 & 0 & 0 & 1 & 0 & 1 & 1 & 4 \end{pmatrix},$$

and $\mathrm{rank}(U_3) = 7$. One can check that all the lower bounds presented in this chapter do not improve upon the bound $\mathrm{rank}(U_3) = 7 \leq \mathrm{rank}_+(U_3)$. It is an open problem to decide whether $\mathrm{rank}_+(U_3) = 7$ or $\mathrm{rank}_+(U_3) = 8$. It is strongly believed that $\mathrm{rank}_+(U_3) = 8$. For example, the code from [460] cannot compute any factorization with small error (for $r = 7$, the best solution found has relative error $\|U_3 - WH\|_F/\|U_3\|_F = 8.36\%$); see [Matlab file: UDISJ.m].

More generally, the rank of the above UDISJ matrix is $\mathrm{rank}(U_n) = \frac{n(n+1)}{2} + 1$ (see Remark 3.4 below), while the nonnegative rank is larger than $(3/2)^n$; see the discussion in Section 3.7. This is another example of a large gap between the rank and the nonnegative rank. Can one prove stronger lower bounds? In particular, it has been conjectured that $\mathrm{rank}_+(U_n) = 2^n$.

Remark 3.4 (Rank of U_n). *Let us show that* $\mathrm{rank}(U_n) = 1 + \frac{n(n+1)}{2}$ *by induction. (We prove this result because it is a nice exercise, and we were not able to find it in the literature.) It suffices to show that* $\mathrm{rank}(U_1) = 2$ *and* $\mathrm{rank}(U_{n+1}) = \mathrm{rank}(U_n) + n$. *The matrix*

$$U_1 = \begin{pmatrix} 1 & 1 \\ 0 & 1 \end{pmatrix} \text{ so that } \mathrm{rank}(U_1) = 2.$$

It remains to prove that $\mathrm{rank}(U_{n+1}) = \mathrm{rank}(U_n) + n$. *For that, let us define recursively the matrix* $F_n \in \{0, 1\}^{2^n \times n}$ *as follows:*

$$F_1 = \begin{pmatrix} 0 \\ 1 \end{pmatrix} \text{ and } F_n = \begin{pmatrix} F_{n-1} & 0 \\ F_{n-1} & e \end{pmatrix}.$$

Since the rows of F_n *contain all possible 0/1 patterns, including the rows of the identity matrix,* $\mathrm{rank}(F_n) = n$. *Let us also denote* $V_n = F_n F_n^\top$, *with* $\mathrm{rank}(V_n) = n$ *since* $\mathrm{rank}(F_n) = n$. *Note that* $\mathrm{col}(V_n) = \mathrm{col}(F_n)$. *We have*

$$U_n = (ee^\top - V_n)^{\circ 2} = ee^\top - 2V_n + V_n^{\circ 2},$$

and hence $V_n^{\circ 2} = U_n - ee^\top + 2V_n$. *We obtain*

$$U_{n+1} = \left(ee^\top - \begin{pmatrix} F_n & 0 \\ F_n & e \end{pmatrix} \begin{pmatrix} F_n & 0 \\ F_n & e \end{pmatrix}^\top\right)^{\circ 2}$$

$$= \left(ee^\top - \begin{pmatrix} V_n & V_n \\ V_n & V_n + ee^\top \end{pmatrix}\right)^{\circ 2}$$

$$= \begin{pmatrix} U_n & U_n \\ U_n & U_n \end{pmatrix} - \begin{pmatrix} 0 & 0 \\ 0 & ee^\top \end{pmatrix} + 2 \begin{pmatrix} 0 & 0 \\ 0 & V_n \end{pmatrix}.$$

We have that $\binom{0}{e} \in \operatorname{col}\left(\begin{smallmatrix} U_n \\ U_n \end{smallmatrix}\right)$ which follows from $\binom{e}{e}$ and $\binom{e}{0}$ belonging to $\operatorname{col}\left(\begin{smallmatrix} U_n \\ U_n \end{smallmatrix}\right)$, which itself follows from the form of U_n (see above) and because 0 is the first row of F_n and $\binom{e}{0}$ is the last column of F_n. This implies that

$$\operatorname{rank}(U_{n+1}) = \operatorname{rank}\left(\begin{array}{cc} U_n & 0 \\ U_n & V_n \end{array} \right) = \operatorname{rank}\left(\begin{array}{cc} U_n & 0 \\ U_n & F_n \end{array} \right) = \operatorname{rank}(U_n) + \operatorname{rank}(F_n),$$

where the second equality follows from $\operatorname{col}(V_n) = \operatorname{col}(F_n)$ and the third because the matrix is block triangular. Since $\operatorname{rank}(F_n) = n$, this concludes the proof.

3.8 ▪ Other applications of the nonnegative rank

The nonnegative rank is related to other problems in the literature; however, for these problems, it did not have as much impact as for extended formulations. Let us recall some of them:

- The nonnegative rank is an upper bound for the minimum biclique cover number of a bipartite graph; see Section 3.4.4.

- Let P be a probability matrix P (where $\sum_{i,j} P(i,j) = 1$) corresponding to the distribution of two random variables (X, Y) so that

$$\mathbb{P}(X = i, Y = j) = P(i,j).$$

  The nonnegative rank of P is equal to the minimum number of pairs of independent variables that can explain this distribution; see Section 1.4.6.

- The nonnegative rank is closely related to minimal state representations of stochastic sequential machines and the minimal cover of labeled Markov chains; see Section 1.4.3.

3.9 ▪ Take-home messages

The nonnegative rank of a nonnegative matrix is an intriguing quantity. Although it shares some properties with the usual rank (see in particular Theorem 3.1), it may behave rather differently. For example, we have seen that rank-one perturbations may increase or decrease the nonnegative rank by more than one unit (Corollaries 3.5 and 3.6). Moreover, the nonnegative rank is NP-hard to compute; see Chapter 2. Although the nonnegative rank has been used in various contexts, such as probability (Section 1.4.6) and communication complexity (Section 3.7), it has had the most impact in the study of extended formulations of polytopes in combinatorial optimization. Yannakakis proved that the nonnegative rank of the slack matrix of a polytope is equal to its extension complexity (the minimum number of inequalities needed to represent it); see Theorem 3.39. This result has been used to provide lower bounds on the extension complexity of polytopes, which in turn allowed providing limits of LP for solving combinatorial optimization problems. A notable example is the proof that the perfect matching polytope cannot be represented via a polynomial number of inequalities [400, 401].

Chapter 4

Identifiability

There are several issues when using NMF in practice, including for example the choice of the factorization rank r and of additional constraints that W and H should satisfy depending on the application at hand; see Chapter 5. However, the two main issues are arguably the *NP-hardness* of computing solutions and the *nonuniqueness* of the solutions. The NP-hardness of Exact NMF was discussed in Chapter 2, while NP-hardness of NMF will be discussed in Chapter 6. In this chapter, we discuss the nonuniqueness issue of the solutions of Exact NMF, also known as the identifiability issue. In other words, we discuss conditions under which the factors W and H in an Exact NMF decomposition $X = WH$ are unique (up to scaling and permutation ambiguities) and hence correspond to the true factors that generated the data. This is crucial in many applications. For example, in blind HU (Section 1.3.2), it ensures that the recovered matrix W corresponds to the spectral signatures of the true endmembers and that the matrix H corresponds to the abundances of the endmembers within the pixels of the image. In audio source separation (Section 1.3.2), it ensures that the matrix W corresponds to the frequency response of the sources and that the matrix H corresponds to the activation of the sources over time. In Chapter 2, when studying the geometric interpretation of Exact NMF, we have encountered several matrices whose Exact NMF are not unique. As we will see, the geometric interpretation plays a crucial role in characterizing the solutions of Exact NMF.

Most theoretical results on identifiability focus on Exact NMF. The reason is twofold. First, analyzing identifiability in noisy settings is much more difficult. Second, identifiability of NMF can be decomposed into two subproblems:

1. The first subproblem is the uniqueness of the low-rank approximation $\tilde{X} = WH$, that is, the uniqueness of the solution to the problem

$$\min_{\tilde{X} \in \mathcal{N}_+^r} D\big(X, \tilde{X}\big), \tag{4.1}$$

where $\mathcal{N}_+^r$ is the set of nonnegative matrices of nonnegative rank at most r, that is,

$$\mathcal{N}_+^r = \{\tilde{X} \mid \text{there exist } W \in \mathbb{R}_+^{m \times r} \text{ and } H \in \mathbb{R}_+^{r \times n} \text{ such that } \tilde{X} = WH\}.$$

Although the solution to the problem (4.1) is not necessarily unique, this type of nonuniqueness is usually not a problem in practice, as long as $\tilde{X}$ is a good approximation of X. In fact, assume there are two solutions $\tilde{X}_1$ and $\tilde{X}_2$ such that $D(X, \tilde{X}_1) = D(X, \tilde{X}_2) = \epsilon$ for some small ϵ. If D is a distance (for example, the Frobenius, ℓ_1, or ℓ_∞ norm), using the triangle inequality, we obtain

$$D\big(\tilde{X}_1, \tilde{X}_2\big) \leq D\big(\tilde{X}_1, \tilde{X}\big) + D\big(\tilde{X}, \tilde{X}_2\big) \leq 2\epsilon,$$

hence $\tilde{X}_1$ and $\tilde{X}_2$ are close to one another, given that ϵ is small. In other words, even if there are several solutions, they are close to one another, given that they are sufficiently close to X. We do not discuss this type of nonuniqueness in this book. We refer the interested reader to, for example, the discussion in [224] under the relaxed assumptions that $\tilde{X} \geq 0$ and $\text{rank}(\tilde{X}) = r$.

2. What is more important to practitioners is that *the recovered factors are close to the true underlying factors that generated the data.* It was shown in [298, Theorem 6] that if $\tilde{X} - X$ is sufficiently small[28] and the Exact NMF (W, H) of X is unique, then the optimal solution $(\tilde{W}, \tilde{H})$ of the NMF problem (1.1) for $\tilde{X}$ is close to the Exact NMF (W, H) of X, up to permutation and scaling of the rank-one factors. In other words, if the Exact NMF (W, H) of X is unique and the noise N is small, the estimation error on the factors W and H is small when solving the NMF for $\tilde{X} = X + N$. Hence, assuming the noisy input data $\tilde{X}$ is close to X, the problem of identifiability of NMF reduces to the identifiability of Exact NMF of X. This is the problem we discuss in this chapter.

Let us formally define the uniqueness of an Exact NMF decomposition.

Definition 4.1 (Uniqueness of Exact NMF). *The Exact NMF (W, H) of $X = WH$ of size r is unique if and only if for any other Exact NMF (W', H') of $X = W'H'$ of size r, there exists a permutation matrix[29] $\Pi \in \{0, 1\}^{r \times r}$ and a diagonal scaling matrix D with positive diagonal elements such that*

$$W' = W\Pi D \quad \text{and} \quad H' = D^{-1}\Pi^\top H.$$

In other words, an Exact NMF (W, H) of X of size r is unique if and only if the only other Exact NMFs (W', H') of X of size r have the form

$$W'H' = \sum_{k=1}^{r} W'(:, k)H'(k, :) = \sum_{k=1}^{r} \underbrace{\alpha_k W(:, \pi_k)}_{W'(:,k)} \underbrace{\frac{1}{\alpha_k}H(\pi_k, :)}_{H'(k,:)}, \tag{4.2}$$

for some permutation π of $\{1, 2, \ldots, r\}$, and some positive scalars α_k $(1 \leq k \leq r)$. This definition of uniqueness is also often referred to as *essential uniqueness* in the literature, and (W, H) is said to be essentially unique. Another terminology says that (W, H) is unique up to permutation and scaling. These terminologies are more precise since we can always permute and scale[30] the rank-one factors to obtain equivalent factorizations; see (4.2). However, for simplicity, we prefer to use the term "unique" as defined above.

In this chapter, the main question we address is the following: Given an Exact NMF (W, H) of X of size r, when is it unique? We will mostly focus on the case $r = \text{rank}(X)$, as done in the literature.

Before delving into the details, let us point out that in some applications, identifiability is not a concern. For example,

- in the study of extended formulations discussed in Section 3.6, any factorization provides an extended formulation, and there is no reason to look for a particular one;

---

[28]No explicit bounds are provided, and such bounds would probably be rather weak. It would be interesting to investigate further the stability of NMF solutions under small perturbations.

[29]A permutation matrix is a square binary matrix that has exactly one entry equal to 1 in each row and each column and has zeros everywhere else. In other words, permutation matrices are obtained by permuting the columns of the identity matrix.

[30]There is a slight abuse of language here since we scale not the rank-one factors $W(:, k)H(k, :)$ $(1 \leq k \leq r)$ but rather the vectors $W(:, k)$ and $H(k, :)$ that compose them. We will use this abuse of language throughout and sometimes refer to it as the scaling degree of freedom.

- in some data analysis applications, the main goal is linear dimensionality reduction, and there is no ground truth to recover. For example, in facial feature extraction (Section 1.3.1), it is difficult to argue that there exists a unique set of features that one is desperately looking for. In this context, computing any meaningful set of features with low reconstruction error is satisfactory. These features are then used, for example, within a facial recognition system [227].

Organization of the chapter We first briefly discuss the case rank$(X) \leq 2$ (Section 4.1). Then we present the two main parts of this chapter. In the first part (Section 4.2), we discuss the identifiability of Exact NMF, providing geometric and algebraic characterizations of the set of solutions. We also present sufficient and necessary conditions for Exact NMF to have a unique solution. As we will see, in most cases, we should not expect Exact NMF to have a unique solution. In the second part (Section 4.3), we discuss regularized Exact NMF models, that is, we consider Exact NMF models where the factors (W, H) satisfy additional properties, namely, separability, orthogonality, minimum volume, or sparsity. Under these additional properties, we will see that Exact NMFs are unique under much milder conditions. This is crucial in practice: taking into account prior knowledge allows us to recover the ground truth factors (W, H); we will discuss this further in Section 5.3.

4.1 ▪ Case rank$(X) \leq 2$

Let us start with the simple case rank$(X) \leq 2$ for which rank$(X) = \text{rank}_+(X)$ (Theorem 2.6). For rank$(X) = 1$, the Exact NMF of X is always unique, as the single column of $W \in \mathbb{R}^{m \times 1}$ has to be taken as a multiple of any nonzero column of X which are multiples of one another.

For rank$(X) = 2$, the uniqueness of Exact NMF can be characterized as follows.

Theorem 4.2. *Let* $X \in \mathbb{R}_+^{m \times n}$ *with* rank$(X) = 2$*. The Exact NMF of* X *of size* 2 *is unique if and only if there exists* i, j, k, ℓ *such that*

$$X(\{i, j\}, \{k, \ell\}) = \begin{pmatrix} \alpha & 0 \\ 0 & \beta \end{pmatrix}$$

for some $\alpha > 0$ *and* $\beta > 0$*, that is, if and only if* X *contains a 2-by-2 diagonal matrix with positive diagonal elements as a submatrix.*

Proof. To prove this result, let us use the geometric interpretation of Exact NMF using nested convex hulls described in Section 2.1.2, which we briefly recall here. W.l.o.g., let us remove the zero columns of X and normalize its columns to unit ℓ_1 norm, and let (W, H) be an Exact NMF of X of size two where the columns of W and H have unit ℓ_1 norm (Lemma 2.1). Since rank$(X) = 2$, rank$_+(X) = 2$ (Theorem 2.6) and hence for any Exact NMF of X of size 2, col$(W) = \text{col}(X)$; see Lemma 2.7. This implies that

$$\text{conv}(X) \subseteq \text{conv}(W) \subseteq \Delta^r \cap \text{col}(X),$$

where conv(X) and $\Delta^r \cap \text{col}(X)$ are segments, since their dimension is equal to rank$(X) - 1$; see Lemma 2.5. Therefore, the nested segment conv(W) is unique if and only if the two segments conv(X) and $\Delta^r \cap \text{col}(X)$ coincide. Note that, given W, H is uniquely determined since rank$(W) = 2$.

Let us now show that X contains a 2-by-2 diagonal matrix with positive diagonal elements as a submatrix (which for simplicity we refer to as a diagonal submatrix for the remainder of the proof) if and only if conv$(X) = \Delta^r \cap \text{col}(X)$, which will conclude the proof. For this, let us construct the feasible solution (W, H) of the Exact NMF of X of size 2 such that $\text{conv}(W) = \text{conv}(X)$, that is, the two columns of $W \in \mathbb{R}^{m \times 2}$ are the two vertices of $\text{conv}(X)$. This means that $W = X(:, \mathcal{K})$ for some index set $\mathcal{K}$ with $|\mathcal{K}| = 2$. This implies that for all $1 \le j \le n$, there exists $H(:, j) \in \Delta^2$ such that

$$X(:, j) = WH(:, j) = W(:, 1)H(1, j) + W(:, 2)H(2, j).$$

Therefore the support of each column of X is either $\text{supp}\left(W(:, 1)\right)$, $\text{supp}\left(W(:, 2)\right)$, or $\text{supp}\left(W(:, 1)\right) \cup \text{supp}\left(W(:, 2)\right)$. Hence X contains a diagonal submatrix if and only if W contains a diagonal submatrix. Now observe that $\text{conv}(W) = \Delta^r \cap \text{col}(X)$ if and only if for $(i, j) \in \{(1, 2), (2, 1)\}$ we have

$$W(:, i) + \alpha(W(:, i) - W(:, j)) \notin \Delta^r \text{ for any } \alpha > 0,$$

that is, when starting from $W(:, i)$ and going in the direction opposite to $W(:, j)$, we go out of the nonnegative orthant. This happens if and only if $W(:, i) - W(:, j)$ has one negative entry where $W(:, i)$ is zero, that is, there must exist k such that $W(k, i) = 0$ and $W(k, j) > 0$. Since this must hold for $(i, j) = (1, 2)$ and $(i, j) = (2, 1)$, this is equivalent to requiring that W contains a diagonal submatrix. $\qquad\square$

Figure 4.1 provides an example of a nonunique Exact NMF of a matrix X, since $\text{conv}\left(\theta(X)\right) \subset \Delta^r \cap \text{col}(X)$. (Recall that $\theta(X)$ is the matrix X whose columns have been normalized to unit ℓ_1 norm.)

The two vertices of $\text{conv}\left(\theta(X)\right)$ can be identified in different ways. For example, the first vertex can be identified as the column of $\theta(X)$ with the largest ℓ_2 norm,[31] and the second can be identified as the column of $\theta(X)$ furthest away from the first identified column. Moreover, one can easily characterize the set of feasible solutions: $W(:, 1)$ and $W(:, 2)$ can be chosen as any point on the segment $\text{col}(X) \cap \Delta^m$ (which is the intersection between the

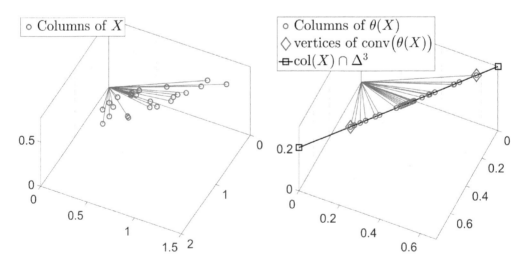

Figure 4.1. *Geometric illustration of rank-two Exact NMF for $m = 3$ and $n = 25$ with a nonunique solution. Both figures represent the same data set, up to scaling. The figure on the right is the normalization to unit ℓ_1 norm of the columns of X from the figure on the left.*

---

[31] A strongly convex function such as $f(x) = \|x\|_2^2$ always attains its maximum over a polytope at one of its vertices; see Section 7.4.1 for more details.

line containing the segment $\mathrm{conv}\left(\theta(X)\right)$ and the nonnegative orthant) but not in the relative interior of $\mathrm{conv}\left(\theta(X)\right)$, $W(:,1)$ and $W(:,2)$ being on a different side of $\mathrm{conv}\left(\theta(X)\right)$; see Algorithm 4.1. Such constructions have been known for a long time, in particular in the SMCR literature [301]; see Section 1.4.1 and also [355].

Algorithm 4.1 Exact NMF of size $r = 2$ [Matlab file: `Rank2NMF.m`]

Input: A rank-two nonnegative matrix $X \in \mathbb{R}_+^{m \times n}$.
Output: Exact NMF (W, H) of X of size $r = 2$.

1: Remove the zero columns of X.
2: Scale X, that is, $X \leftarrow \theta(X)$.
3: *% Identify the two vertices of* $\mathrm{conv}(X)$
4: Let $j_1 = \mathrm{argmax}_j \|X(:,j)\|_2$.
5: Let $j_2 = \mathrm{argmax}_j \|X(:,j) - X(:,j_1)\|_2$.
6: *% Construct a feasible solution* $\mathrm{conv}(X) \subseteq \mathrm{conv}(W) \subseteq \mathrm{col}(X) \cap \Delta^m$
7: $W(:,1) = X(:,j_1) + \alpha_1(X(:,j_1) - X(:,j_2))$ where

$$\alpha_1 \in \left[0, \min_{\{k \mid X(k,j_2) > X(k,j_1)\}} \frac{X(k,j_1)}{X(k,j_2) - X(k,j_1)}\right],$$

so that $W(:,1) \geq 0$.
% The choice for α_1 *is unique and equal to zero if there exists k such that $X(k,j_1) = 0$ and $X(k,j_2) > 0$*
8: $W(:,2) = X(:,j_2) + \alpha_2(X(:,j_2) - X(:,j_1))$ where

$$\alpha_2 \in \left[0, \min_{\{k \mid X(k,j_1) > X(k,j_2)\}} \frac{X(k,j_2)}{X(k,j_1) - X(k,j_2)}\right],$$

so that $W(:,2) \geq 0$.
% The choice for α_2 *is unique and equal to zero if there exists k such that $X(k,j_2) = 0$ and $X(k,j_1) > 0$*
9: Let $H \in \mathbb{R}_+^{2 \times n}$ be such that $X = WH$.

When $\mathrm{rank}(X) \geq 3$, the problem is much more complicated since we are facing NPPs in dimensions higher than 2 for which uniqueness is difficult to characterize. This is the topic of the remainder of this chapter.

4.2 ▪ Exact NMF with $r = \mathrm{rank}(X)$

Most results on identifiability of Exact NMF focus on the case $r = \mathrm{rank}_+(X) = \mathrm{rank}(X)$. This is because most applications for which identifiability is a key aspect come from data analysis: in this context, $\mathrm{rank}_+(X) = \mathrm{rank}(X)$ is a standard assumption. The factors W and H and the data matrix $X = WH$ have rank r; this is the case, for example, for the four applications presented in Section 1.3 (namely facial feature extraction, hyperspectral unmixing, text mining, and audio source separation).

In this section, we first characterize the set of solutions of the Exact NMF problem with $r = \mathrm{rank}(X)$ in three different ways, namely using

- its equivalence with the NPP (Section 4.2.1.1),

- an algebraic characterization (Section 4.2.1.2), and

- its interpretation in terms of nested convex cones (Section 4.2.1.3).

We then provide two sufficient conditions for W and H to obtain unique Exact NMF solutions for the matrix $X = WH$, namely

- the separability condition (Section 4.2.2) and

- the sufficiently scattered condition (SSC; Section 4.2.3).

In Section 4.2.4, we present a necessary condition based on the support of W and H to have a unique Exact NMF. Finally, we discuss the sparsity of the input matrix X; in particular, we explain why sparsity of the input matrix is not a necessary condition to have a unique Exact NMF, and we illustrate this observation with several examples of positive matrices that have a unique Exact NMF (Section 4.2.5).

4.2.1 ▪ Characterizations of the set of solutions of Exact NMF

In this section, we provide three different ways to represent the set of solutions of Exact NMF.

4.2.1.1 ▪ Geometric characterization I: NPP

The case $r = \operatorname{rank}(X)$ can be characterized more easily, as Exact NMF coincides with RE-NMF which is equivalent to the NPP where the inner and outer polytopes have the same dimension $r - 1$. We have the following result.

Theorem 4.3. *The Exact NMF (W, H) of X of size $r = \operatorname{rank}(X)$ is unique if and only if the solution to the corresponding NPP problem constructed using Theorem 2.11 is unique.*

Proof. For $r = \operatorname{rank}(X)$, Exact NMF coincides with RE-NMF, so that the result follows from the one-to-one relationship between RE-NMF and NPP instances; see Theorem 2.11.

Note that in the NPP, there is no permutation nor scaling ambiguity, and hence a unique solution is truly unique.   □

Theorem 4.3 has been known for a long time; see Section 1.4 and also, for example, [138, 298, 465] and the references therein. Theorem 4.3 is particularly useful when $\operatorname{rank}(X) = 3$ as the corresponding NPP is two-dimensional and can be solved efficiently; see Section 2.3.1.1. It will be used to construct interesting examples in Section 4.2.5.

4.2.1.2 ▪ Algebraic characterization

In an Exact NMF (W, H) of X of size $r = \operatorname{rank}(X)$, $\operatorname{col}(W) = \operatorname{col}(X)$ (Lemma 2.7). This is not true in general when $r > \operatorname{rank}(X)$; see Sections 2.1.4 and 3.6.3 for some examples. Therefore, when $r = \operatorname{rank}(X)$, as we have already seen, the space of solutions of Exact NMF can be parametrized with an invertible r-by-r matrix Q (Theorem 2.21). Let us recall this result stated in a different way.

Lemma 4.4. *Let (W, H) be an Exact NMF of X of size $r = \operatorname{rank}(X)$. Then, any Exact NMF (W', H') of X of size r has the form*

$$W' = WQ \geq 0 \quad \text{and} \quad H' = Q^{-1}H \geq 0$$

for some invertible r-by-r matrix Q.

Proof. Since $X = W'H' = WH$ are Exact NMFs of size $r = \mathrm{rank}(X)$, we have, by Lemma 2.7, that $\mathrm{col}(X) = \mathrm{col}(W) = \mathrm{col}(W')$ of dimension r. Therefore there exists an invertible r-by-r matrix Q such that

$$W' = WQ.$$

Moreover, since W has rank r, it admits a left inverse $W^\dagger$ such that $W^\dagger W = I_r$. Hence

$$W'H' = WQH' = X = WH \quad \Rightarrow \quad QH' = H,$$

where the left-hand side was multiplied by $W^\dagger$ on both sides to obtain the right-hand side. $\qquad\square$

The only nonnegative matrices Q whose inverse is nonnegative are diagonal matrices with positive diagonal elements, up to permutations of rows and columns [38]. Such matrices lead to equivalent factorizations $(WQ, Q^{-1}H)$ corresponding to permutations and scalings of the rank-one factors and hence do not destroy uniqueness of Exact NMF (Definition 4.1).

Theorem 4.5. *The Exact NMF (W, H) of X of size $r = \mathrm{rank}(X)$ is unique if and only if the only invertible matrices Q such that*

$$WQ \geq 0 \quad and \quad Q^{-1}H \geq 0 \tag{4.3}$$

are diagonal matrices with positive diagonal elements, up to permutations of rows and columns.

Proof. This follows directly from Lemma 4.4 and the fact that Q and Q^{-1} are both nonnegative if Q is a diagonal matrix, up to permutations of its rows and columns. $\qquad\square$

4.2.1.3 ▪ Geometric characterization II: nested cones

The geometric characterization in Theorem 4.3 is based on the construction of an equivalent geometric problem, the NPP, that must have a unique solution. However, this reformulation does not tell us much about the conditions that W and H must satisfy for uniqueness to hold. In this section, we derive another equivalent reformulation which can be traced back from the work of Thomas [450]; see also the discussion in [298]. This characterization will be useful to derive necessary and sufficient conditions for the uniqueness of Exact NMF.

To derive this second geometric characterization, we use the notions of cones and their dual, which we recall here. A cone $\mathcal{W}$ is a set such that $x \in \mathcal{W}$ implies that $\alpha x \in \mathcal{W}$ for all $\alpha \geq 0$. Let us recall that the cone generated by the columns of a matrix W is defined as

$$\mathrm{cone}(W) = \{x \mid x = Wh \text{ for } h \geq 0\}.$$

The extreme rays of a cone $\mathcal{W}$ are the points in $\mathcal{W}$ that cannot be represented as nonnegative linear combinations of other points in $\mathcal{W}$; more precisely, $x \in \mathcal{W}$ is an extreme ray of $\mathcal{W}$ if $x \notin \mathcal{W} \setminus \mathrm{cone}(x)$. An extreme direction $\mathcal{D}$ of a convex cone $\mathcal{W}$ is defined as

$$\mathcal{D} = \{\alpha x \mid \alpha \geq 0, x \text{ is an extreme ray of } \mathcal{W}\}.$$

A cone $\mathcal{W}$ is polyhedral (or finitely generated) if there exists a matrix W such that $\mathrm{cone}(W) = \mathcal{W}$. The order of a polyhedral cone is the minimum number of columns in such a W; it is equal to the number of extreme directions of $\mathrm{cone}(W)$. A simplicial cone is a polyhedral cone of the form $\mathrm{cone}(W)$ where W has full column rank; such a cone is of order r if W has r columns. Given two matrices $A \in \mathbb{R}^{r \times m}$ and $B \in \mathbb{R}^{r \times n}$, it is easy to verify that

$$\mathrm{cone}(A) \subseteq \mathrm{cone}(B) \quad \Longleftrightarrow \quad A(:, j) \in \mathrm{cone}(B) \text{ for all } j.$$

Given a cone $\mathcal{W}$, its dual is denoted by $\mathcal{W}^*$ and defined as

$$\mathcal{W}^* = \left\{ y \mid x^\top y \geq 0 \text{ for all } x \in \mathcal{W} \right\}.$$

Lemma 4.6. *Given a matrix W, the dual of* $\mathrm{cone}(W)$ *is given by*

$$\mathrm{cone}^*(W) = \left\{ y \mid W^\top y \geq 0 \right\}.$$

Proof. By definition,

$$
\begin{aligned}
\mathrm{cone}^*(W) &= \left\{ y \mid x^\top y \geq 0 \text{ for all } x \in \mathrm{cone}(W) \right\} \\
&= \left\{ y \mid x^\top y \geq 0 \text{ for all } x = Wh, h \geq 0 \right\} \\
&= \left\{ y \mid h^\top W^\top y \geq 0 \text{ for all } h \geq 0 \right\} \\
&= \left\{ y \mid W^\top y \geq 0 \right\}.
\end{aligned}
$$

The last equality follows from the fact that a vector z has nonnegative inner product with all nonnegative vectors if and only if it is nonnegative, that is,

$$z \geq 0 \quad \Longleftrightarrow \quad z^\top h \geq 0 \text{ for all } h \geq 0.$$

In other words, the nonnegative orthant is a self-dual cone: $\left(\mathbb{R}^n_+ \right)^* = \mathbb{R}^n_+$. $\qquad\square$

We have the following corollary.

Corollary 4.7. *Let $W \in \mathbb{R}^{m \times r}$ and $H \in \mathbb{R}^{r \times n}$. Then $WH \geq 0$ if and only if*

$$\mathrm{cone}\left(W^\top \right) \subseteq \mathrm{cone}^*(H).$$

Proof. We have

$$
WH \geq 0 \iff H^\top W^\top \geq 0 \iff W(i,:)^\top \in \mathrm{cone}^*(H) \text{ for all } i
$$
$$
\iff \mathrm{cone}(W^\top) \subseteq \mathrm{cone}^*(H),
$$

where the second equivalence follows from Lemma 4.6. $\qquad\square$

Before we provide the second geometric characterization of the uniqueness of Exact NMF, let us provide an elementary result from convex geometry. We refer the interested reader to [399] for more on convex cones.

Lemma 4.8. *Let $Q \in \mathbb{R}^{r \times r}$ be invertible. Then $\mathrm{cone}^*(Q^\top) = \mathrm{cone}(Q^{-1})$.*

Proof. We have

$$
\begin{aligned}
\mathrm{cone}^*\left(Q^\top \right) &= \left\{ y \mid Qy \geq 0 \right\} \\
&= \left\{ Q^{-1}x \mid Q(Q^{-1}x) \geq 0 \right\} \\
&= \left\{ Q^{-1}x \mid x \geq 0 \right\} = \mathrm{cone}(Q^{-1}),
\end{aligned}
$$

where the first equality follows from Lemma 4.6 and the second from the change of variable $y = Q^{-1}x$ which is made w.l.o.g. since Q is invertible. $\qquad\square$

Theorem 4.9. *[298, Theorem 1] The Exact NMF (W, H) of X of size $r = \operatorname{rank}(X)$ is unique if and only if the only simplicial cone $\mathcal{T}$ of order r such that*

$$\operatorname{cone}\left(W^{\top}\right) \subseteq \mathcal{T} \subseteq \operatorname{cone}^{*}(H) \tag{4.4}$$

is the nonnegative orthant $\mathbb{R}_{+}^{r}$.

Proof. First observe that for an r-by-r matrix Q, $\operatorname{cone}(Q)$ is the nonnegative orthant, that is, $\operatorname{cone}(Q) = \mathbb{R}_{+}^{r}$, if and only if Q is a permutation of a diagonal matrix with positive diagonal elements. In fact, the unit vectors e_k ($1 \leq k \leq r$) are extreme rays of $\mathbb{R}_{+}^{r}$ and hence have to be among the columns of any matrix Q (up to scaling) given that $\operatorname{cone}(Q) = \mathbb{R}_{+}^{r}$ (see also Lemma 4.11 for other equivalent characterizations).

Using this observation, let us rephrase Theorem 4.5: (W, H) is a unique Exact NMF of X of size r if and only if the only invertible r-by-r matrices Q satisfying $WQ^{-1} \geq 0$ and $QH \geq 0$ are such that $\operatorname{cone}(Q) = \mathbb{R}_{+}^{r}$.

Moreover, since $r = \operatorname{rank}(W)$ in any Exact NMF (W, H) of X of size r, the dimension of $\mathcal{T}$ in (4.4) is at least r, and hence there must exist an invertible r-by-r matrix P such that $\mathcal{T} = \operatorname{cone}(P)$. Let us denote $Q = P^{-\top}$ so that $\mathcal{T} = \operatorname{cone}(Q^{-\top})$ (we use this transformation to have this Q coincide with the Q of Theorem 4.5).

Below, we show that for an invertible r-by-r matrix Q,

(i) $\operatorname{cone}(W^{\top}) \subseteq \mathcal{T} = \operatorname{cone}(Q^{-\top}) \iff WQ \geq 0$, and

(ii) $\mathcal{T} = \operatorname{cone}(Q^{-\top}) \subseteq \operatorname{cone}^{*}(H) \iff Q^{-1}H \geq 0$.

Together with the observations above, this concludes the proof, because the condition (4.4) in Theorem 4.9 is equivalent to the condition (4.3) in Theorem 4.5.

It remains to prove (i) and (ii):

(i) By Lemma 4.8, $\operatorname{cone}(Q^{-\top}) = \operatorname{cone}^{*}(Q)$. By Lemma 4.6, we obtain

$$\operatorname{cone}(W^{\top}) \subseteq \operatorname{cone}(Q^{-\top}) = \operatorname{cone}^{*}(Q) \iff Q^{\top}W^{\top} \geq 0 \iff WQ \geq 0.$$

(ii) By Lemma 4.6,

$$\operatorname{cone}(Q^{-\top}) \subseteq \operatorname{cone}^{*}(H) \iff H^{\top}Q^{-\top} \geq 0 \iff Q^{-1}H \geq 0. \qquad \square$$

We will see in the next two sections that Theorem 4.9 is particularly useful for deriving sufficient conditions on W and H for the Exact NMF (W, H) of X to be unique.

4.2.2 ▪ Sufficient condition I: separability

Let us define a *separable matrix* as follows.

Definition 4.10 (Separable matrix). *The matrix $H \in \mathbb{R}_{+}^{r \times n}$ is separable if $\operatorname{cone}(H) = \mathbb{R}_{+}^{r}$.*

Lemma 4.11. *Let $H \in \mathbb{R}_{+}^{r \times n}$. The following conditions are equivalent:*

(i) *H is separable, that is, $\operatorname{cone}(H) = \mathbb{R}_{+}^{r}$.*

(ii) *$\operatorname{cone}^{*}(H) = \mathbb{R}_{+}^{r}$.*

(iii) *H contains an r-by-r submatrix which is a permutation of a diagonal matrix with positive diagonal entries.*

(iv) *There exists an index set $\mathcal{K}$ of size r such that $\operatorname{cone}\left(H(:, \mathcal{K})\right) = \mathbb{R}_{+}^{r}$.*

Proof. (i) $\iff$ (ii) follows from the nonnegative orthant being self-dual.

(iii) $\iff$ (iv) follows directly since the cone spanned by an r-by-r diagonal matrix with positive diagonal entries is $\mathbb{R}^n_+$.

(iv) $\Rightarrow$ (i) follows directly, while (i) $\Rightarrow$ (iv) follows from the unit vectors being extreme rays of the nonnegative orthant (they are not conic combinations of other vectors within the nonnegative orthant). □

Combining the definition of a separable matrix and Theorem 4.9, we obtain the following sufficient conditions on W and H for the Exact NMF (W, H) of X to be unique.

Theorem 4.12. *If X admits an Exact NMF (W, H) of size $r = \mathrm{rank}(X)$ where $W^\top$ and H are separable, then the Exact NMF (W, H) of X is unique.*

Proof. This follows directly from Theorem 4.9: by separability of $W^\top$ and H,

$$\mathrm{cone}(W^\top) = \mathbb{R}^r_+ = \mathrm{cone}^*(H),$$

so that the unique simplicial cone of order r nested between $\mathrm{cone}(W^\top)$ and $\mathrm{cone}^*(H)$ is $\mathbb{R}^r_+$. □

Let us state two simple corollaries.

Corollary 4.13. *The r-by-r identity matrix I_r admits a unique NMF of size r, namely $W = I_r$ and $H = I_r$.*

Proof. This result follows from Theorem 4.12.

A simpler, more straightforward, proof follows from the fact that the unit vectors are the extreme rays of the nonnegative orthant. □

Corollary 4.14. *Let X be a nonnegative matrix with $\mathrm{rank}(X) = \mathrm{rank}_+(X)$, and let the matrix X contain a diagonal matrix with positive diagonal entries as an r-by-r submatrix. Then the Exact NMF (W, H) of X of size r is unique.*

Proof. This follows from Theorem 4.12 and Corollary 4.13. The diagonal submatrix in X must be factorized with diagonal matrices (Corollary 4.13). This implies that W and H contain diagonal matrices as submatrices, and hence the Exact NMF of X is unique (Theorem 4.12). □

Looking back at the case $\mathrm{rank}(X) = 2$, it is interesting to observe that the separability condition of $W^\top$ and H is also necessary; see Theorem 4.2.

Relaxed sufficient conditions based on separability   The separability condition was first introduced by Donoho and Stodden [138] (2004). In their paper, Donoho and Stodden derive sufficient conditions for Exact NMF to be unique. In short, they require H to be separable while they require a milder nontrivial condition on the support of W. In Laurberg et al. [298], other sufficient conditions were derived, which are closely related to separability (they require a condition on H which is stronger than separability, while the condition on W is also based on its support but is milder).

We do not provide these conditions in detail here; we refer the interested reader directly to these papers, and also to the discussions in [251, 170]. In the next section, we present a much milder sufficient condition than separability for uniqueness.

Is separability of $W^\top$ and H a reasonable condition in practice? The requirement that $W^\top$ and H are both separable in Theorem 4.12 is unlikely to be satisfied in real-world settings. As we will see in detail in Chapter 7, it makes sense to assume separability of H *or* W in several applications. For example, in airborne hyperspectral images, which are images of the earth taken from an airplane or a drone (such as the one shown in Figure 1.6), H is separable if for each material present in the image, there exists a pixel containing only that material. This is called the *pure-pixel* assumption in the hyperspectral imaging literature which is reasonable in several scenarios. This notably requires a relatively high-resolution image; see Chapter 7 for more details. However, the matrix W is typically not separable as most of its entries are positive.

We will see in Section 4.3.1 how to require only one of the two factors to be separable and obtain a uniqueness result under additional constraints, namely imposing that $W = X(:, \mathcal{K})$ for some index set $\mathcal{K}$ of size r. In Chapter 7, we will see how to tackle this problem numerically, presenting polynomial-time algorithms that provably recover (W, H), even in the presence of noise.

4.2.3 ▪ Sufficient condition II: sufficiently scattered

In this section, we consider the SSC. As we will see, this is an instrumental condition when studying the identifiability of Exact NMF and of regularized Exact NMF, which will be discussed in Section 4.3.

Definition 4.15 (Sufficiently scattered condition). *The matrix $H \in \mathbb{R}_+^{r \times n}$ is sufficiently scattered if the following two conditions are satisfied:*

SSC1. $\mathcal{C} = \{x \in \mathbb{R}_+^r \mid e^\top x \geq \sqrt{r-1}\|x\|_2\} \subseteq \text{cone}(H)$.

SSC2. *There does not exist any orthogonal matrix Q such that $\text{cone}(H) \subseteq \text{cone}(Q)$, except for permutation matrices. (An orthogonal matrix Q is a square matrix such that $Q^\top Q = I$.)*

The SSC is a milder condition than separability: if H is separable, that is, $\text{cone}(H) = \mathbb{R}_+^r$, then

- SSC1 holds: $\mathcal{C} \subseteq \text{cone}(H)$ since $\mathcal{C} \subseteq \mathbb{R}_+^r$;

- SSC2 holds: the only orthogonal r-by-r matrices Q such that $\text{cone}(H) = \mathbb{R}_+^r \subseteq \text{cone}(Q)$ are permutation matrices; see the proof of Theorem 4.9.

Figure 4.2 provides an illustration that compares separability and the SSC in the case $r = 3$. The SSC is not an easy concept to grasp, and the aim of this section is to discuss the SSC and its implications on Exact NMF. It is organized as follows.

- Section 4.2.3.1 discusses the geometric interpretation of the SSC.

- Section 4.2.3.2 provides several lemmas that are used in Section 4.2.3.3 to prove the uniqueness of the Exact NMF (W, H) of X when H and $W^\top$ satisfy the SSC.

- Section 4.2.3.4 sheds some light on the question of whether the SSC is likely to be satisfied for a given matrix H. Unfortunately, checking whether a given matrix H satisfies the SSC is NP-hard in general. However, we provide a new necessary condition for SSC1 to be satisfied based on the sparsity of H. This new condition can be checked efficiently, which we illustrate with some numerical experiments.

- Finally, Section 4.2.3.7 discusses whether requiring both $W^\top$ and H to satisfy the SSC is reasonable in practice.

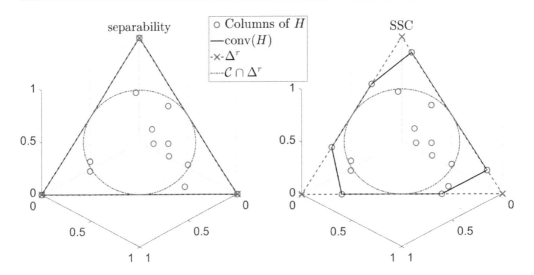

Figure 4.2. *Comparison of separability (left) and the SSC (right) for $r = 3$. The figure represents the columns of H, normalized to unit ℓ_1 norm. On the left, separability and the SSC are satisfied as $\mathrm{cone}(H)$ coincides with $\mathbb{R}^r_+$. On the right, only the SSC is satisfied since $\mathrm{cone}(H)$ contains $\mathcal{C}$ (SSC1) while it can be checked that SSC2 holds as well.*

4.2.3.1 ▪ Geometric interpretation of the SSC

Let us discuss the two conditions in the SSC.

Condition SSC1   SSC1 requires $\mathcal{C} \subseteq \mathrm{cone}(H)$. The set $\mathcal{C}$ is a second-order cone, also known as a Lorentz cone or an ice cream cone; see the green shape in Figure 4.3 for its representation for $r = 3$. The set $\mathcal{C} \cap \Delta^r$ is an $(r-1)$-dimensional sphere, namely

$$\left\{ x \in \Delta^r \mid \|x\|_2 \leq \frac{1}{\sqrt{r-1}} \right\} = \left\{ x \in \mathbb{R}^r_+ \mid \|x\|_2 \leq \frac{1}{\sqrt{r-1}} \text{ and } e^\top x = 1 \right\}$$

centered at $\frac{1}{r}e$ of radius $\frac{1}{\sqrt{r(r-1)}}$, within the affine subspace $\{x \mid e^\top x = 1\}$. This sphere goes through the points $\frac{1}{r-1}(e - e_k)$ for $k = 1, 2, \ldots, r$. For example, for $r = 3$, $\mathcal{C} \cap \Delta^3$ is the disk centred at $(1/3, 1/3, 1/3)$ of radius $\frac{1}{\sqrt{6}}$ that goes through the points $(1/2,1/2,0)$, $(1/2,0,1/2)$, and $(0,1/2,1/2)$; see Figure 4.3 for an illustration. This means that $\mathcal{C}$ is tangent to every facet of the nonnegative orthant (see Lemma 4.18 for a proof). Hence the condition SSC1, that is, $\mathcal{C} \subseteq \mathrm{cone}(H)$, implies that the columns of H are sufficiently spread in the nonnegative orthant since their conical hull contains $\mathcal{C}$; see Figures 4.2 and 4.4 for illustrations. In particular, $\mathrm{cone}(H)$ must contain the vectors $(e - e_k) \in \mathcal{C}$ for $k = 1, 2, \ldots, r$ whose kth entry is equal to zero, hence H must have some degree of sparsity. In fact, to reconstruct exactly a vector that has zero entries using nonnegative linear combinations of nonnegative basis vectors, the basis vectors must have zero entries in these positions; see Theorem 4.28 for a rigorous characterization.

Note also that the only simplicial cone of order r contained in the nonnegative orthant and containing $\mathcal{C}$ is the nonnegative orthant [137, Lemma 1]. This implies that the SSC boils down to the separability condition when $r = n$. However, in general, r is much smaller than n and SSC is a much milder condition than separability. However, for $r = \mathrm{rank}(X) = 2$, the SSC and the separability condition are equivalent since $\mathcal{C} = \mathbb{R}^2_+$ in that case.

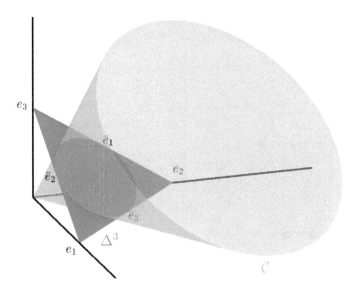

Figure 4.3. *Illustration of the sets Δ^3 and $\mathcal{C}$ and their intersection $\mathcal{C} \cap \Delta^3$. (This figure is similar to [251, Figure 2], and we are grateful to the authors for providing us with the code to generate it.)*

Condition SSC2   SSC2 requires that the only orthogonal r-by-r matrices Q such that $\mathrm{cone}(H) = \mathbb{R}^r_+ \subseteq \mathrm{cone}(Q)$ are permutation matrices. In a nutshell, this is a regularity condition that prevents $\mathrm{cone}(H)$ from being too small. A first intuition is that if there exists a scalar $q < \sqrt{r-1}$ such that

$$\{x \in \mathbb{R}^r_+ \mid e^\top x \geq q\|x\|_2\} \subseteq \mathrm{cone}(H), \tag{4.5}$$

then SSC2 holds; see [320]. Clearly, the condition (4.5) implies SSC1 because $\mathcal{C}$ is contained in $\{x \in \mathbb{R}^r_+ \mid e^\top x \geq q\|x\|_2\}$ for $q < \sqrt{r-1}$. This means that if SSC1 is satisfied, then SSC2 will in general be satisfied as well. Interestingly, for $q = 1$, we recover the separability condition since $e^\top x \geq \|x\|_2$ holds for any nonnegative vector, including the unit vectors.

The condition (4.5) was introduced in [320], where it replaces SSC1 and SSC2 which is more elegant but slightly less general.

Example 4.16. Let us provide an example that allows us to better understand SSC1 and SSC2; this example was presented in [298, Example 3] and [251, Example 2]. Let

$$H = \begin{pmatrix} \omega & 1 & 1 & \omega & 0 & 0 \\ 1 & \omega & 0 & 0 & \omega & 1 \\ 0 & 0 & \omega & 1 & 1 & \omega \end{pmatrix} \tag{4.6}$$

for $\omega \in [0, 1]$. Note that, for $\omega = 0$, H is separable since $H(:, \{2, 1, 4\}) = I_3$. This example is also closely related to the example with nested hexagons presented in Section 2.1.4, because the NPP instance corresponding to the RE-NMF of $X = H^\top H$ is nested hexagons.

For which values of ω does H satisfy the SSC? Figure 4.4 illustrates this problem geometrically. We have the following:

- For $\omega > 0.5$, H does not satisfy the SSC. More precisely, in Figure 4.4(a), we see that H does not satisfy SSC1, because $\mathrm{cone}(H)$ represented by the black squares does not contain $\mathcal{C}$, which is shown as the circle with the solid boundary. The matrix H does not satisfy

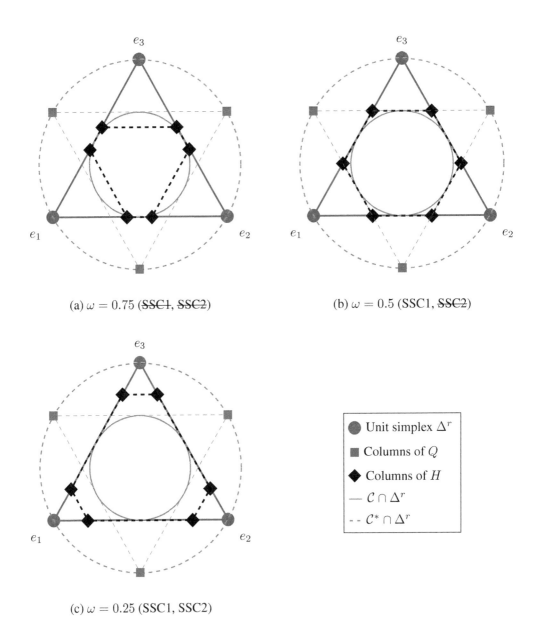

(a) $\omega = 0.75$ (~~SSC1~~, ~~SSC2~~)   (b) $\omega = 0.5$ (SSC1, ~~SSC2~~)

(c) $\omega = 0.25$ (SSC1, SSC2)

> ● Unit simplex Δ^r
> ■ Columns of Q
> ◆ Columns of H
> — $\mathcal{C} \cap \Delta^r$
> -- $\mathcal{C}^* \cap \Delta^r$

Figure 4.4. *Illustration of the SSC on the matrix H from (4.6), after projection onto* $\{x \mid e^\top x = 1\}$. *(It may be insightful to look at Figures 4.2 and 4.3 displaying $\mathcal{C} \cap \Delta^r$.) For* $\omega = 0.75$ *(top left), H does not satisfy SSC1 because $\mathcal{C} \nsubseteq \operatorname{cone}(H)$, nor SSC2 because there exists Q orthogonal such that $\operatorname{cone}(Q) \subseteq \operatorname{cone}(H) \subseteq \mathcal{C}^*$. For $\omega = 0.5$ (top right), H does not satisfy SSC2; note, however, that H satisfies SSC1. For $\omega = 0.25$ (bottom left), H satisfies the SCC: $\mathcal{C} \subseteq \operatorname{cone}(H)$ (SSC1), and there does not exist an orthogonal matrix Q within $\mathcal{C}^*$ containing $\operatorname{cone}(H)$, except the unit simplex (SSC2). Note that for $\omega = 0$, H is separable as its columns coincide with e_1, e_2 and e_3. (This figure is similar to [251, Figure 2].)*

SSC2 either, as we can see that there exists an orthogonal matrix Q such that the columns of H are contained within $\operatorname{cone}(Q)$, denoted by the blue dashed triangle, namely

$$Q = \frac{1}{3} \begin{pmatrix} -1 & 2 & 2 \\ 2 & -1 & 2 \\ 2 & 2 & -1 \end{pmatrix}.$$

In fact, we have $H = QB$, where

$$B = \frac{1}{3} \begin{pmatrix} 2-\omega & 2\omega-1 & 2\omega-1 & 2-\omega & 2\omega+2 & 2\omega+2 \\ 2\omega-1 & 2-\omega & 2\omega+2 & 2\omega+2 & 2-\omega & 2\omega-1 \\ 2\omega+2 & 2\omega+2 & 2-\omega & 2\omega-1 & 2\omega-1 & 2-\omega \end{pmatrix},$$

which is nonnegative for any $\omega \geq 0.5$.

- For $\omega = 0.5$, H satisfies SSC1 but not SSC2. In fact, we see in Figure 4.4(b) that the cone generated by the columns of H contains $\mathcal{C}$, while there exists an orthogonal matrix Q (the same as for the case above), different from a permutation matrix, whose convex cone contains the columns of H.

- For $\omega < 0.5$, H satisfies SSC1 and SSC2. In fact, we see in Figure 4.4(c) that the columns of H contain $\operatorname{cone}(C)$, while it can be shown that no orthogonal matrix Q exists that contains the columns of H, except for permutation matrices.

This example shows that the SSC is much milder than separability, which requires the columns of H to contain the unit vectors and hence holds only for $\omega = 0$. ∎

4.2.3.2 ▪ Useful lemmas for the identifiability of Exact NMF under the SSC

In this section, we provide four lemmas that will be key to further understanding the SSC and to proving identifiability of Exact NMF under the SSC. If you are less interested in theoretical aspects of NMF, you can skip this section and go directly to the identifiability result based on the SSC presented in Section 4.2.3.3.

The first lemma is well known in convex geometry.

Lemma 4.17. *Let $\mathcal{A}$ and $\mathcal{B}$ be two convex cones. If $\mathcal{A} \subseteq \mathcal{B}$, then $\mathcal{B}^* \subseteq \mathcal{A}^*$.*

Proof. Let $x \in \mathcal{B}^*$, hence $x^\top y \geq 0$ for all $y \in \mathcal{B}$. Since $\mathcal{A} \subseteq \mathcal{B}$, $x^\top y \geq 0$ for all $y \in \mathcal{A}$, hence $x \in \mathcal{A}^*$. □

The second lemma provides the dual cone of $\mathcal{C}$. Although this result is well known, we provide a proof here for completeness, and because such a proof is not so easily found in the literature. Understanding this proof is not necessary to follow the rest of this chapter and hence can be skipped. It is however a nice exercise.[32]

Lemma 4.18. *The dual cone of $\mathcal{C}$ is given by*

$$\mathcal{C}^* = \left\{ y \in \mathbb{R}^r \mid e^\top y \geq \|y\|_2 \right\}. \tag{4.7}$$

Moreover $\mathcal{C} \subseteq \mathbb{R}_+^r \subseteq \mathcal{C}^*$, and $\mathcal{C}$ is tangent to every facet of $\mathbb{R}_+^r$.

---

[32]There are simpler proofs; for example using the ice-cream cone $\{(t, x) \in \mathbb{R} \times \mathbb{R}^{r-1} \mid t \geq q\|x\|_2\}$ (a rotation of $\mathcal{C}$) and its dual $\{(s, y) \in \mathbb{R} \times \mathbb{R}^{r-1} \mid s \geq 1/q\|y\|_2\}$ (this relationship is much easier to prove).

Proof. For $r = 1$, $\mathcal{C} = \mathcal{C}^* = \mathbb{R}_+$, and the result is trivial, so let us consider $r \geq 2$.

Let us first show that $\mathcal{C} \subseteq \mathbb{R}_+^r$; see Figure 4.3 for an illustration in three dimensions. This will imply that $\mathbb{R}_+^r \subseteq \mathcal{C}^*$ (Lemma 4.17) since $\mathbb{R}_+^r$ is self-dual (see Figure 4.4 for an illustration). Assume $x \in \mathcal{C}$ has a negative entry; w.l.o.g. assume $x(1) < 0$. This implies that $e^\top x(2:r) > e^\top x$, hence

$$\|x(2:r)\|_1 \geq e^\top x(2:r) > e^\top x \geq \sqrt{r-1}\|x\|_2 > \sqrt{r-1}\|x(2:r)\|_2,$$

where the third inequality follows from the definition of $\mathcal{C}$. This is a contradiction, since $\|y\|_1 \leq \sqrt{r-1}\|y\|_2$ for any $(r-1)$-dimensional vector y. Using the same arguments but where all inequalities are replaced with equalities, $\mathcal{C}$ is tangent to every facet of $\mathbb{R}_+^r$. In other words, if $x(k) = 0$ for some k then $\|x\|_1 = \sqrt{r-1}\|x\|_2$, hence $x = e - e_k \in \mathcal{C}$, up to scaling. These are the only extreme rays of $\mathcal{C}$ tangent to the facets of $\mathbb{R}_+^r$. The inclusion $\mathbb{R}_+^r \subseteq \mathcal{C}^*$ follows by Lemma 4.17 and self-duality of $\mathbb{R}_+^r$.

Let us now show that $\mathcal{C}^*$ as defined in (4.7) is in fact the dual of $\mathcal{C}$. For simplicity, let us denote $\mathcal{A} = \{x \in \mathbb{R}^r \mid e^\top x \geq \|x\|_2\}$ which we need to prove is the dual of $\mathcal{C}$. Let us first show that $\mathcal{A} \subseteq \mathcal{C}^*$, that is, $y^\top x \geq 0$ for all $x \in \mathcal{C}$ and all $y \in \mathcal{A}$. Since $\mathcal{C}$ is a convex cone, it is sufficient to check that $y^\top x \geq 0$ for x on the boundary of $\mathcal{C}$, because any other point in $\mathcal{C}$ is a convex combination of these boundary points. The boundary of $\mathcal{C}$ consists of the points $\{x \mid e^\top x = \sqrt{r-1}\|x\|_2\}$ which we will denote as $\mathbf{bd}\,\mathcal{C}$. Since $e^\top y \geq \|y\|_2 > 0$ for $y \neq 0$, we can scale y w.l.o.g. such that $e^\top y = 1$ as it does not alter the inequality $y^\top x \geq 0$. Let us now consider the convex optimization problem

$$\min_{y} x^\top y \text{ such that } e^\top y = 1 \text{ and } \|y\|_2 \leq 1. \tag{4.8}$$

The feasible set of the optimization problem (4.8) is $\mathcal{A} \cap \{y \mid e^\top y = 1\}$. Showing that the optimal value of (4.8) is nonnegative for any $x \in \mathbf{bd}\,\mathcal{C} \cap \{x \mid e^\top x = 1\}$ is therefore equivalent to showing that $\mathcal{A} \subseteq \mathcal{C}^*$. The first-order optimality conditions of (4.8) are

$$x = \lambda e - \mu y, \qquad e^\top y = 1, \qquad \mu \geq 0, \qquad \|y\|_2 \leq 1, \qquad \mu(1 - \|y\|_2) = 0,$$

where λ and $\mu \geq 0$ are the Lagrangian variables associated with the constraint $e^\top y = 1$ and $\|y\|_2 \leq 1$, respectively. For a convex optimization problem admitting a Slater point (a point strictly inside the relative interior of the feasible set), which is the case here (for example, $y = e/r$), these conditions are necessary and sufficient for optimality. Let us find the solutions that satisfy these conditions. Either $\mu = 0$ or $\|y\|_2 = 1$. If $\mu = 0$, we must have $x = \lambda e \geq 0$, in which case $x^\top y = \lambda e^\top y \geq \lambda \|y\|_2 \geq 0$. Otherwise $\|y\|_2 = 1$, and the optimality conditions give

$$y = y(\lambda) = \frac{\lambda e - x}{\|\lambda e - x\|_2}.$$

We have $x^\top y \geq 0$ if and only if $\lambda x^\top e - \|x\|_2^2 \geq 0$ if and only if $\lambda \geq \frac{\|x\|_2^2}{x^\top e} = \frac{1}{r-1}$. It remains to show that, at optimality, $\lambda \geq \frac{1}{r-1}$. The value of λ that provides the optimal solution $y(\lambda)$ must satisfy the constraint $g(\lambda) = e^\top y(\lambda) = 1$. We need to find λ such that $g(\lambda) = 1$. We have

$$g(\lambda) = e^\top y(\lambda) = e^\top \frac{\lambda e - x}{\|\lambda e - x\|_2} = \frac{\lambda r - 1}{\|\lambda e - x\|_2} = \frac{\lambda r - 1}{(\lambda^2 r - 2\lambda + \frac{1}{r-1})^{1/2}},$$

where we used $\|\lambda e - x\|_2^2 = \lambda^2 r - 2\lambda + \frac{1}{r-1}$. Note that $g(1/r) = 0$, while $\lim_{\lambda \to \infty} g(\lambda) = \sqrt{r} > 1$. We have

$$g'(\lambda) = \frac{r - 1}{(\lambda^2 r - 2\lambda + \frac{1}{r-1})^{3/2}} > 0,$$

so that $g(\lambda)$ is increasing. This implies that $g(\lambda) = 1$ has a unique solution in $[\frac{1}{r}, +\infty]$: one can check that it is given by $\lambda = \frac{1}{r-1}$, for which $y(\lambda)^\top x = 0$. This concludes the first part of the proof.

It remains to show that $\mathcal{C}^* \subseteq \mathcal{A}$. First, let us show that for any $y \in \mathcal{C}^* \setminus \{0\}$, $e^\top y > 0$. Since $y^\top x \geq 0$ for all $x \in \mathcal{C} \subseteq \mathbb{R}_+$, $y \in \mathcal{C}^* \setminus \{0\}$ has at least one positive entry. Assume $e^\top y < 0$ and let $k = \text{argmax}_k y_k$ so that $y_k > 0$. Since $x = e - e_k \in \mathcal{C}$, $x^\top y < e^\top y < 0$, a contradiction. Now, assume there exists $z \in \mathcal{C}^* \setminus \mathcal{A}$, that is, $e^\top z < \|z\|_2$ while $x^\top z \geq 0$ for all $x \in \mathcal{C}$. Since $e^\top z > 0$, we can assume w.l.o.g. that $e^\top z = 1 < \|z\|_2$ (simply normalize z, and $e^\top z > \|z\|_2$ is not affected). Consider the optimization problem

$$\max_x z^\top x \text{ such that } e^\top x = 1 \text{ and } \|x\|_2^2 \leq \frac{1}{r-1},$$

where the feasible set is $\mathcal{C} \cap \{x \mid e^\top x = 1\}$. Using the same derivations as above, the optimal x has the form

$$x(\lambda) = \frac{1}{\sqrt{r-1}} \frac{\lambda e - z}{\|\lambda e - z\|_2}.$$

Note that $z^\top x(\lambda) \geq 0$ if and only if $\lambda \geq \|z\|_2 > 1$. Let us show that the value of λ at optimality, for which $e^\top x(\lambda) = 1$, satisfies $\lambda < \|z\|_2$, which leads to a contradiction. Let us define

$$h(\lambda) = e^\top x(\lambda) = \frac{1}{\sqrt{r-1}} \frac{\lambda r - 1}{\|\lambda e - z\|_2} = \frac{1}{\sqrt{r-1}} \frac{\lambda r - 1}{\sqrt{\lambda^2 r - 2\lambda + \|z\|_2^2}}.$$

As for $g(\lambda)$, one can show that $h(\lambda)$ is an increasing function of λ. Denoting $\alpha = \|z\|_2 > 1$, we have

$$h(\alpha) = \frac{1}{\sqrt{r-1}} \frac{\alpha r - 1}{\sqrt{\alpha^2 r - 2\alpha + \alpha^2}} > 1$$

since

$$(\alpha r - 1)^2 - (r-1)(\alpha^2 r - 2\alpha + \alpha^2) = (\alpha - 1)^2 > 0.$$

Since h is increasing, $h(\lambda) = 1$ if and only if $\lambda < \alpha = \|z\|_2^2$, leading to a contradiction. $\qquad\square$

The third lemma shows that a cone generated by an orthogonal matrix is self-dual.

Lemma 4.19. *Let Q be an orthogonal matrix. Then* $\text{cone}(Q) = \text{cone}^*(Q)$*, that is,* $\text{cone}(Q)$ *is self-dual.*

Proof. By Lemma 4.8, $\text{cone}^*(Q) = \text{cone}^*(Q^{-\top}) = \text{cone}(Q)$, since $Q^{-\top} = Q$ as $Q^\top Q = I$. $\square$

Lemmas 4.17 and 4.19 already shed some light on SSC2: under SSC1 ($\mathcal{C} \subseteq \text{cone}(H)$), $\text{cone}(H) \subseteq \text{cone}(Q)$ and Q orthogonal imply that

$$\mathcal{C} \subseteq \text{cone}(Q) \subseteq \mathcal{C}^*.$$

Figure 4.4 provides an example for $r = 3$, projected onto the subset $\{x \mid e^\top x = 1\}$, where the columns of Q belong to the border of $\mathcal{C}^*$. We observe that as long as the columns of H are sufficiently spread within the nonnegative orthant so that their conical hull contains $\mathcal{C}$ (SSC1), it is unlikely for SSC2 to be violated.

The last lemma is key to proving uniqueness of Exact NMF under the SSC. It shows that the only simplicial cones of order r nested between $\mathcal{C}$ and its dual $\mathcal{C}^*$ are the ones generated by orthogonal matrices (up to scaling of the columns). Intuitively, looking at Figure 4.4, the only triangles containing $\mathcal{C}$ and contained in $\mathcal{C}^*$ are equilateral triangles whose vertices belong to the border of $\mathcal{C}^*$.

Lemma 4.20. *[251, Lemma 1] Let $Q \in \mathbb{R}^{r \times r}$ and $\|Q(:,j)\|_2 = 1$ for all j. If*

$$\mathcal{C} \subseteq \mathrm{cone}(Q) \subseteq \mathcal{C}^*,$$

then

- *Q is orthogonal, that is, $Q^\top Q = I_r$, and*

- *$Q^\top e = e$, that is, $Q(:,j) \in \mathbf{bd}\,\mathcal{C}^* = \{x \in \mathbb{R}^r \mid e^\top x = \|x\|_2\}$ for all j.*

Proof. Since $\mathrm{cone}(Q) \subseteq \mathcal{C}^*$,

$$e^\top Q(:,j) \geq \|Q(:,j)\|_2 = 1 \text{ for } j = 1, 2, \ldots, r, \tag{4.9}$$

that is, $e^\top Q \geq e^\top$. Since $\mathcal{C} \subseteq \mathrm{cone}(Q)$ and $\mathcal{C}$ is full dimensional, $\mathrm{rank}(Q) = r$. Let $P = Q^{-\top}$ so that $P^\top Q = I_r$. By Lemma 4.8, $\mathrm{cone}(Q) = \mathrm{cone}^*(P)$ while $\mathcal{C} \subseteq \mathrm{cone}^*(P)$, and hence $\mathrm{cone}(P) \subseteq \mathcal{C}^*$ so that

$$e^\top P(:,j) \geq \|P(:,j)\|_2 \text{ for } j = 1, 2, \ldots, r. \tag{4.10}$$

Let us multiply and sum the inequalities (4.9) and (4.10), one by one, to obtain

$$e^\top Q P^\top e \geq \sum_{j=1}^{r} \|Q(:,j)\|_2 \|P(:,j)\|_2. \tag{4.11}$$

The left-hand side of (4.11) equals r since $Q P^\top = I_r$. Let us lower bound the right-hand side of (4.11) using the Cauchy–Schwarz inequality

$$\sum_{j=1}^{r} \|Q(:,j)\|_2 \|P(:,j)\|_2 \geq \sum_{j=1}^{r} Q(:,j)^\top P(:,j) = r \tag{4.12}$$

since $Q^\top P = I_r$. This implies that all inequalities (4.9)–(4.12) are equalities. In particular $Q^\top e = e$, and

$$\sum_{j=1}^{r} \|Q(:,j)\|_2 \|P(:,j)\|_2 = \sum_{j=1}^{r} Q(:,j)^\top P(:,j),$$

which is possible only if $\|Q(:,j)\|_2 \|P(:,j)\|_2 = Q(:,j)^\top P(:,j)$ for $j = 1, 2, \ldots, r$, implying that $P(:,j)$ is a positive scaling of $Q(:,j)$ for $j = 1, 2, \ldots, r$, that is, $P = DQ$ for a diagonal matrix D. Multiplying by $Q^\top$ on both sides of $P = DQ$, we obtain $PQ^\top = I_r = DQQ^\top$, and hence $D = I_r$ as $\|Q(:,j)\|_2 = 1$ for all j. Finally, $P = Q$, and hence Q is orthogonal. $\square$

4.2.3.3 ▪ Unique Exact NMF under the SSC

Let us now state the main result of this Section 4.2.3.

Theorem 4.21. *[251, Theorem 4] If $W^\top$ and H are sufficiently scattered, then the Exact NMF (W, H) of X of size $r = \mathrm{rank}(X)$ is unique.*

Proof. By Theorem 4.9, it suffices to prove that the unique simplicial cone $\mathcal{T}$ of order r such that

$$\text{cone}(W^\top) \subseteq \mathcal{T} \subseteq \text{cone}^*(H)$$

is the nonnegative orthant, that is, $\mathcal{T} = \mathbb{R}^r_+$. Since $\mathcal{C} \subseteq \text{cone}(W^\top)$ because $W^\top$ satisfies SSC1, and $\text{cone}^*(H) \subseteq \mathcal{C}^*$ because H satisfies SSC1 (combined with duality—Lemma 4.6),

$$\mathcal{C} \subseteq \mathcal{T} \subseteq \mathcal{C}^*.$$

By Lemma 4.20, $\mathcal{T} = \text{cone}(Q)$, where Q is an orthogonal matrix. This allows us to conclude since H satisfies SSC2, that is, the only orthogonal matrices Q such that $\text{cone}(H) \subseteq \text{cone}(Q)$ are permutations of the identity matrix, and hence $\mathcal{T} = \mathbb{R}^r_+$. $\square$

The last part of the proof of Theorem 4.21 does not rely on $W^\top$ satisfying SSC2. In fact, Theorem 4.21 remains valid if this condition is not satisfied for one of the two factor matrices, $W^\top$ or H. Let us state this observation formally.

Corollary 4.22. *Let $W^\top \in \mathbb{R}^{r \times m}_+$ and $H \in \mathbb{R}^{r \times n}_+$ satisfy SSC1 and $W^\top$ or H satisfy SSC2. Then the Exact NMF (W, H) of X of size $r = \text{rank}(X)$ is unique.*

Proof. The same proof as Theorem 4.21 applies. By symmetry, whether it is $W^\top$ or H that satisfies SSC2 allows us to conclude that $\mathcal{T} = \mathbb{R}^r_+$. $\square$

Let us now present an example to illustrate Theorem 4.21. After that, we will further discuss the SSC.

Example 4.23 (Example 4.16 continued). Let us take

$$W^\top = H = \begin{pmatrix} \omega & 1 & 1 & \omega & 0 & 0 \\ 1 & \omega & 0 & 0 & \omega & 1 \\ 0 & 0 & \omega & 1 & 1 & \omega \end{pmatrix} \tag{4.13}$$

for $\omega \in [0, 1]$. For $\omega < 0.5$, $H = W^\top$ satisfies the SSC; see Example 4.16. It turns out that, for this example, the SSC is tight in the sense that, for $\omega = 0.5$, the Exact NMF (W, H) of X of size 3 is not unique. The following pair (W', H') is also an Exact NMF of X:

$$W'^\top = H' = \begin{pmatrix} 0 & 0.5 & 1 & 1 & 0.5 & 0 \\ 0.5 & 0 & 0 & 0.5 & 1 & 1 \\ 1 & 1 & 0.5 & 0 & 0 & 0.5 \end{pmatrix};$$

(W', H') cannot be obtained by permutation and scaling of the columns of W and rows of H for $\omega = 0.5$. Interestingly, constructing the NPP instance corresponding to this RE-NMF leads to a nested hexagon problem, similar to that presented in Section 2.1.4.

This example also allows us to construct an example where Corollary 4.22 applies while Theorem 4.21 does not. Take H as in (4.6) for any $\omega < 0.5$ so that H satisfies the SSC, and take $W^\top$ as in (4.6) with $\omega = 0.5$ so that $W^\top$ satisfies SSC1 but not SSC2. Then the NMF of $X = WH$ is unique by Corollary 4.22. ∎

As we will see in Section 4.2.5 with some examples, the requirement that both $W^\top$ and H satisfy SSC1 is not a necessary condition to have a unique NMF.

4.2.3.4 ▪ Can we check the SSC efficiently?

Given a nonnegative matrix H, it is NP-hard to check in general whether H satisfies the SSC [251]. To show this, let us consider the following optimization problem:

$$\max_x \|x\|_2^2 \quad \text{such that} \quad H^\top x \geq 0 \text{ and } e^\top x = 1. \tag{4.14}$$

Solving (4.14) is NP-hard in general [168]. Note that the optimal value of (4.14) is always at least one since the unit vectors are feasible solutions. We have the following lemma.

Lemma 4.24. *The optimal value of* (4.14) *is strictly larger than one if and only if SSC1 fails. Hence checking SSC1 is NP-hard.*

Proof. The following statements are equivalent:

- The optimal value of (4.14) is strictly larger than one.

- There exists a point x such that $H^\top x \geq 0$ and $\|x\|_2 > e^\top x = 1$.

- There exists a point $x \in \text{cone}^*(H)$ and $x \notin \mathcal{C}^*$.

- $\text{cone}^*(H) \not\subseteq \mathcal{C}^*$.

- $\mathcal{C} \not\subseteq \text{cone}(H)$.

- H does not satisfy SSC1.

To show these equivalences, we simply need to recall the definitions of $\mathcal{C}$ and $\mathcal{C}^*$ and $\text{cone}^*(H) = \{x \mid H^\top x \geq 0\}$ (Lemma 4.6), and to use duality (Lemma 4.17).

Since it is NP-hard to solve (4.14) [168], it is NP-hard to check whether SSC1 holds. □

Lemma 4.24 is rather disappointing. In fact, if an Exact NMF algorithm outputs a solution (W, H), we cannot check in general whether $W^\top$ and H satisfy the SSC, hence we cannot check whether this solution is unique using the SSC. However, it does not prevent us from trying to understand better when the SSC is satisfied. In particular, it is useful to derive sufficient and necessary conditions for the SSC to hold that can be checked in polynomial time. In Section 4.2.3.5, we present a sufficient condition for the SSC that can be easily checked. In Section 4.2.3.6, we provide a new necessary condition for SSC1 to hold. Moreover, this new condition can be checked efficiently (see Algorithm 4.2). It is based on the geometric interpretation of the SSC and the sparsity pattern of the columns of H.

Note that separability of H can be easily checked (the columns of H must contain all the unit vectors, up to scaling) but is much stronger (it implies the SSC) and unlikely to be satisfied for both $W^\top$ and H; see Section 4.2.2.

4.2.3.5 ▪ Sufficient condition for the SSC

In [320], Lin et al. provide the following sufficient condition for the SSC to hold and that focuses on the 2-sparse columns of H, that is, the columns of H containing at most two nonzero entries.

Theorem 4.25. *[320, Theorem 4] Let $H \in \mathbb{R}_+^{r \times n}$ with $r \geq 3$. If for all $i \neq j \in \{1, 2, \ldots, r\}$, there exists $k, \ell \in \{1, 2, \ldots, n\}$ such that*

- $H(p, k) = 0$ *and* $H(p, \ell) = 0$ *for all* $p \neq i, j$*, that is,* $H(:, k)$ *and* $H(:, \ell)$ *are 2-sparse with nonzero entries at location i and j, and*

- $H(i, k) > \alpha(r)H(j, k)$ *and* $H(j, \ell) > \alpha(r)H(i, \ell)$ *where* $\alpha(3) = 2$ *and* $\alpha(r) = 1$ *for* $r \geq 4$,

then H satisfies the SSC.

Note that if there exists a 1-sparse column of H with a positive entry at position i, then the condition $H(i, k) > \alpha(r)H(j, k)$ needed for Theorem 4.25 is satisfied for any $j \neq i$ by taking k as the index of the 1-sparse column. In particular, if H is separable, then Theorem 4.25 guarantees that H satisfies the SSC (which is a nice sanity check).

Example 4.26. Let us take the matrix H from Example 4.16,

$$H = \begin{pmatrix} \omega & 1 & 1 & \omega & 0 & 0 \\ 1 & \omega & 0 & 0 & \omega & 1 \\ 0 & 0 & \omega & 1 & 1 & \omega \end{pmatrix}, \tag{4.15}$$

where $\omega \in [0, 1)$. As explained in Example 4.16, H satisfies the SSC if and only if $\omega < 0.5$; see Figure 4.4 for an illustration. For this matrix, the conditions of Theorem 4.25 are met for any $\omega < 0.5$, and hence these sufficient conditions are also necessary. For example, for the pair $(i, j) = (1, 2)$, take $(k, \ell) = (2, 1)$ so that

$$H(1, 2) = 1 > 2H(2, 2) = 2\omega, \text{ and } H(2, 1) = 1 > 2H(1, 1) = 2\omega. \qquad \blacksquare$$

If H does not contain 1-sparse columns, the condition of Theorem 4.25 requires H to have at least $2\binom{r}{2} = r(r-1)$ columns that are 2-sparse. This is a rather strong condition typically not met in practice because it requires two 2-sparse observations for each possible pair of columns of W. For example, in hyperspectral imaging, this would require each pair of materials to be observed together in pixels several times. However, this condition can be checked easily.

Providing milder sufficient conditions for the SSC to hold and that can be checked efficiently is a topic for future research.

4.2.3.6 ▪ Necessary condition for the SSC

Let us try to link the sparsity of H with the SSC, and let us first discuss the simplest case with $r = 3$. Let $H \in \mathbb{R}_+^{3 \times n}$ satisfy SSC1, and let us look at Figure 4.4 (page 112), which provides a nice interpretation of the SSC, after projection on the set $\{x \mid e^\top x = 1\}$. Let us denote $\bar{e}_j = \frac{1}{r-1}(e - e_j)$ for $j = 1, 2, \ldots, r$. If H satisfies SSC1, the three points $\bar{e}_1 = (0, 1/2, 1/2)$, $\bar{e}_2 = (1/2, 0, 1/2)$, and $\bar{e}_3 = (1/2, 1/2, 0)$ belong to $\mathcal{C}$, and hence to $\text{cone}(H)$.

Moreover, $\mathcal{C}$ is also tangent to $\text{cone}(H)$ at each of these points. Therefore, if any of these three points is an extreme ray of $\text{cone}(H)$, $\mathcal{C}$ will not be contained within $\text{cone}(H)$. Thus $\text{cone}(H)$ must contain a small interval around $\bar{e}_1$, $\bar{e}_2$, and $\bar{e}_3$ in order to contain $\mathcal{C}$. From Figure 4.4, we observe that for $\text{cone}(H)$ to contain $\mathcal{C}$, H must have two columns on each segment $[e_i, e_k]$ for $1 \leq i < k \leq 3$: one closer to e_i and the other closer to e_k. Therefore, H must contain two columns in the sets $\{x \mid x(j) = 0\}$ for $j = 1, 2, 3$; hence H has at least six zero entries. Note that if no column of H is a unitary vector (after scaling), then H must have at least six columns, that is, $n \geq 6$ (as in Figure 4.4). This observation is closely related to Theorem 4.25.

Let us generalize this result to higher dimensions, allowing us to link the SSC of H with its sparsity. We begin with a lemma.

Lemma 4.27. *Let $H \in \mathbb{R}_+^{r \times n}$ satisfy SSC1, and, for $k = 1, 2, \ldots, r$, let the index set*

$$\mathcal{I}_k = \{j \mid H(k, j) = 0\}$$

correspond to the columns of H on the kth facet of the nonnegative orthant which is defined as $\mathcal{F}_k = \{x \in \mathbb{R}_+^r \mid x(k) = 0\}$. Then

$$e \notin \mathbf{bd}\,\mathrm{cone}\,(H(\mathcal{K}, \mathcal{I}_k)) \quad \text{for} \ \ k = 1, 2, \ldots, r, \tag{4.16}$$

where $\mathcal{K} = \{1, 2, \ldots, r\} \backslash \{k\}$.

Proof. By SSC1, $e \in \mathrm{cone}\,(H(\mathcal{K}, \mathcal{I}_k))$ since $e - e_k \in \mathcal{C} \subseteq \mathrm{cone}(H)$ (note that the two e's in this expression do not have the same dimension). Assume there exists k such that (4.16) is not satisfied, that is, $e \in \mathbf{bd}\,\mathrm{cone}\,(H(\mathcal{K}, \mathcal{I}_k))$. This means that there exists a vector $v \in \mathbb{R}^{r-1}$ defining a valid, nontrivial inequality for $\mathrm{cone}\,(H(\mathcal{K}, \mathcal{I}_k))$ which goes through e. That is, there exists a $v \in \mathbb{R}^{r-1}$ such that $v^\top H(\mathcal{K}, \mathcal{I}_k) \geq 0$, $v \neq 0$, and $v^\top e = 0$.

W.l.o.g. assume $k = 1$, and let us choose α such that $\hat{v} = \binom{\alpha}{v} \in \mathbb{R}^r$ is a valid inequality for $\mathrm{cone}(H)$. By construction,

$$\delta = \min_{j \notin \mathcal{I}_k} H(k, j) > 0.$$

Let

$$\beta = \min_{j \notin \mathcal{I}_k} v^\top H(\mathcal{K}, j) \quad \text{and} \quad \alpha = \max\left(0, \frac{-\beta}{\delta}\right).$$

The inequality $\hat{v}^\top x \geq 0$ is valid for $\mathrm{cone}(H)$, that is,

$$\mathrm{cone}(H) \subseteq \mathcal{V} = \{x \mid \hat{v}^\top x \geq 0\},$$

because

$$\hat{v}^\top H(:, \mathcal{I}_k) = v^\top H(\mathcal{K}, \mathcal{I}_k) \geq 0, \quad \text{and} \quad \hat{v}^\top H(:, j) \geq \alpha \delta + \beta \geq 0 \text{ for } j \notin \mathcal{I}_k.$$

Moreover, $\bar{e}_k \in \mathbf{bd}\,\mathcal{V}$ since $\hat{v}^\top \bar{e}_k = v^\top e = 0$.

Note that $\bar{e}_k \in \mathbf{bd}\,\mathcal{C}$ since $e^\top \bar{e}_k = 1 = \sqrt{r-1} \|\bar{e}_k\|_2$. Now, the only nontrivial valid inequality for $\mathcal{C}$ that goes through $\bar{e}_k$ is its tangent ($\mathcal{C}$ is a second-order cone), namely $\mathcal{F}_k = \{x \mid e_k^\top x = x(k) = 0\}$. In fact, by Lemma 4.18, $\mathcal{C}$ is tangent to every facet of the nonnegative orthant; see Figure 4.3 for an illustration. This is a contradiction since $\mathcal{V}$ is a different valid inequality for $\mathcal{C}$ that goes through $\bar{e}_k$; in fact, $\mathcal{V} \neq \mathcal{F}_k$ and $\mathcal{C} \subseteq \mathrm{cone}(H) \subseteq \mathcal{V}$. □

We can now generalize our observations for $r = 3$ to higher dimensions.

Theorem 4.28. *Let $H \in \mathbb{R}_+^{r \times n}$ satisfy SSC1, and let $\mathcal{I}_k = \{j \mid H(k, j) = 0\}$ denote the index set of columns of H containing a zero in the kth entry for $k = 1, 2, \ldots, r$. Then, for all $k = 1, 2, \ldots, r$,*

- *$|\mathcal{I}_k| \geq r - 1$, that is, H has at least $r - 1$ zeros per row;*

- *$\mathrm{cone}\,(H(\mathcal{K}, \mathcal{I}_k))$ contains e in its relative interior, with $\mathcal{K} = \{1, 2, \ldots, r\} \backslash \{k\}$.*

Proof. Let us use the same notation as in Lemma 4.27. For e to belong to the interior of cone $(H(\mathcal{K}, \mathcal{I}_k))$, it is required that $\mathcal{I}_k \geq r - 1$. In fact, a cone in dimension $r - 1$ has a nonempty interior if and only if it has at least $r - 1$ extreme directions. The second claim follows directly from Lemma 4.27. $\square$

Theorem 4.28 implies that if $H \in \mathbb{R}_+^{r \times n}$ satisfies SSC1, then H has at least $r(r-1)$ zero entries. This implies that if each column of H is k-sparse, that is, each column has at most k nonzero entries, we must have $n \geq \frac{r(r-1)}{r-k}$ for H to satisfy SSC1. Moreover, columns of H containing more than two zero entries, say, the jth and other ones, are located on the border of the jth facet of the nonnegative orthant. Therefore, it is more likely for such columns to contain $\bar{e}_k$ in the interior of their conical hull.

Example 4.29. Let us consider the 4-by-6 matrix H containing all 2-sparse columns with nonzero entries equal to one:

$$H = \begin{pmatrix} 1 & 1 & 1 & 0 & 0 & 0 \\ 1 & 0 & 0 & 1 & 1 & 0 \\ 0 & 1 & 0 & 1 & 0 & 1 \\ 0 & 0 & 1 & 0 & 1 & 1 \end{pmatrix}. \tag{4.17}$$

This matrix satisfies the conditions of Theorem 4.28 (for example, sum the last three columns to obtain $2\bar{e}_1$) and has exactly $r(r-1) = 12$ zeros; see Figure 4.5 for an illustration. Note that it does not satisfy the sufficient condition from Theorem 4.25 based on the 2-sparse columns of H.

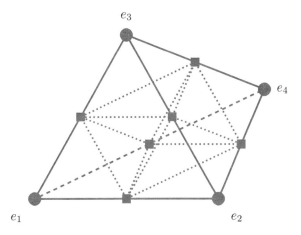

Figure 4.5. *Geometric illustration of the matrix H from (4.17): the squares represent the columns of H after projection onto $\{x \mid e^\top x = 1\}$. This matrix satisfies the assumptions of Theorem 4.28.*

This matrix satisfies SSC1: the polytope whose vertices are the squares in Figure 4.5 contains $\mathcal{C}$ intersected with $\{x \mid e^\top x = 1\}$. This can be checked by solving (4.14) via the enumeration of all the vertices of the feasible set (a convex function is always maximized at a vertex of a polytope); see [Matlab file: isSSC.m], and see [Matlab file: checkSSC_H46_2sparse.m] for its application on the matrix H of (4.17). This matrix, however, does not satisfy SSC2; it turns out that all vertices of the feasible set of (4.14) have their objective function equal to one.

Another way to see this is via the decomposition

$$
H = QH' = \begin{pmatrix} 0.5 & 0.5 & 0.5 & -0.5 \\ 0.5 & 0.5 & -0.5 & 0.5 \\ 0.5 & -0.5 & 0.5 & 0.5 \\ -0.5 & 0.5 & 0.5 & 0.5 \end{pmatrix} \begin{pmatrix} 1 & 1 & 0 & 1 & 0 & 0 \\ 1 & 0 & 1 & 0 & 1 & 0 \\ 0 & 1 & 1 & 0 & 0 & 1 \\ 0 & 0 & 0 & 1 & 1 & 1 \end{pmatrix},
$$

which shows that H does not satisfy SSC2 since $Q = Q^{-\top}$ is orthogonal and $\mathrm{cone}(H) \subseteq \mathrm{cone}(Q)$, since $Q^{-1}H = H' \geq 0$ implies $\mathrm{cone}(H) \in \mathrm{cone}^*(Q^{-\top}) = \mathrm{cone}^*(Q) = \mathrm{cone}(Q)$ (Lemmas 4.6 and 4.19). Moreover, the matrix

$$
X = H^\top H = \begin{pmatrix} 2 & 1 & 1 & 1 & 1 & 0 \\ 1 & 2 & 1 & 1 & 0 & 1 \\ 1 & 1 & 2 & 0 & 1 & 1 \\ 1 & 1 & 0 & 2 & 1 & 1 \\ 1 & 0 & 1 & 1 & 2 & 1 \\ 0 & 1 & 1 & 1 & 1 & 2 \end{pmatrix}
$$

does not admit a unique NMF as $X = H'^\top H'$ is another Exact NMF (which cannot be obtained via permutations of the rows of H). ∎

While SSC is NP-hard to check, the conditions of Theorem 4.28 can be checked in polynomial time; see Algorithm 4.2 which implements such a procedure. It could be used for example a posteriori, after an NMF solution has been computed, to check whether $W^\top$ and H satisfy the necessary condition of Theorem 4.28 for SSC1 to be satisfied.

Is it likely that a randomly generated sparse matrix H will satisfy the SSC?

Let us use Theorem 4.28 to lower bound the number of columns of H that are needed on average to have a chance to satisfy the SSC, when the nonzero entries of H are picked at random. Assume that the probability of an entry of H being equal to zero follows a Bernoulli distribution of parameter $\theta \in (0,1)$, that is, $\mathbb{P}(H(k,j) = 0) = \theta$ for all k, j. When a column of H is sampled, there is a probability of θ that it belongs to $\mathcal{F}_k = \{x \mid x(k) = 0\}$, and we need to have at least $r - 1$ columns on $\mathcal{F}_k$ for $k = 1, 2, \ldots, r$ (Theorem 4.28). How many samples are needed on average to satisfy this condition? This is a well-known question in dictionary learning, closely related to the coupon collector problem;[33] see for example Spielman, Wang, and Wright [431, Theorem 1]. The coupon collector problem tells us that, on average, about $\frac{1}{\theta} \ln(r)$ samples are needed to collect one point in each subspace, that is, to have a single point on each $\mathcal{F}_k$. Since we need $r - 1$ on each $\mathcal{F}_k$, about $\frac{1}{\theta} r \ln(r)$ samples are enough on average.[34] For example, if the columns are on average $(r - 1)$-sparse (that is, there is on average a single zero entry per column), that is, $\theta = \frac{1}{r}$, one requires on average $n \geq \mathcal{O}(r^2 \log(r))$ samples. Note that this is not significantly larger than the deterministic lower bound of $n = r(r - 1)$ when there is a single zero entry in each column of H (Theorem 4.28). We will illustrate this observation through a numerical example in the next subsection.

The numerical experiments in [251, 170] validate these observations (Huang and coworkers check the SSC by solving (4.14) with a heuristic), and the authors claim that

> If H is generated randomly and if every row has *at least* $r - 1$ zero elements, then the sufficiently scattered condition is satisfied with very high probability, which is a fairly mild condition.

---

[33] In the coupon collector problem, the goal is to obtain r distinct coupons via several purchases. Each purchase gives a coupon with some probability, and the contents of the purchases are independent of one another.

[34] This is a crude estimation; fewer samples are actually needed. However, a better estimate is nontrivial to obtain. The corresponding problem is referred to as the *double Dixie cup problem* [365].

Algorithm 4.2 Checking the necessary condition of Theorem 4.28 for SSC1 [Matlab file: SSC1_nec_cond.m]

Input: $H \in \mathbb{R}_+^{r \times n}$
Output: SSC1-nec = *yes* if H satisfies the conditions of Theorem 4.28,
         SSC1-nec = *no* otherwise.

1: **if** $\text{rank}(H) < r$ or $\min_{k,j} H(k,j) < 0$ **then**
2:    SSC1-nec = *no*. Return.
3: **end if**
4: Remove zero columns of H.
5: Normalize the columns of H: $H(:,j) \leftarrow \frac{H(:,j)}{e^\top H(:,j)}$ for $j = 1, 2, \ldots, n$.
6: Remove duplicated columns in H.
7: **for** $k = 1, 2, \ldots, r$ **do**
8:    Let $\mathcal{I}_k = \{i \mid H(k,i) = 0\}$, and denote $\mathcal{K} = \{1, 2, \ldots, r\} \backslash \{k\}$.
9:    % *We need* $\text{rank}\big(H(\mathcal{K}, \mathcal{I}_k)\big) \geq r-1$ *for* $\text{cone}\big(H(\mathcal{K}, \mathcal{I}_k)\big)$ *to have*
        % *a nonempty interior.*
10:    **if** $\text{rank}\big(H(:, \mathcal{I}_k)\big) < r - 1$ **then**
11:        SSC1-nec = *no*. Return.
12:    **end if**
        % *Extract the extreme rays of* $\text{cone}\big(H(\mathcal{K}, \mathcal{I}_k)\big)$.
13:    Let $\mathcal{I}'_k \subseteq \mathcal{I}_k$ be the indices corresponding to the extreme rays of $\text{cone}\big(H(\mathcal{K}, \mathcal{I}_k)\big)$, that is,
$$i \in \mathcal{I}'_k \iff i \in \mathcal{I}_k \text{ and } \min_{x \geq 0} \|H(\mathcal{K}, i) - H(\mathcal{K}, \mathcal{I}_k \backslash \{i\})x\|_2 > 0.$$
        % *If e is in the interior of* $\text{cone}\big(H(\mathcal{K}, \mathcal{I}'_k)\big)$, *then* $e = H(\mathcal{K}, \mathcal{I}_k)x$ *for some*
        % $x \geq 0$ *with the support of x being at least*[†] $r - 1$.
14:    Solve
$$x^* = \text{argmin}_{x \geq 0} \big\|e - H\big(\mathcal{K}, \mathcal{I}'_k\big)x\big\|_2.$$
15:    **if** $\big\|e - H\big(\mathcal{K}, \mathcal{I}'_k\big)x^*\big\|_2 > 0$ or $\text{supp}(x^*) < r - 1$ **then**
16:        SSC1-nec = *no*. Return.
17:    **end if**
18: **end for**
19: SSC1-nec = *yes*.

[†] It it possible that $\text{supp}(x^*) < r - 1$ while e belongs to the interior of $\text{cone}\big(H(\mathcal{K}, \mathcal{I}_k)\big)$. For example, let us consider a cone within the nonnegative orthant in three dimensions with four extreme rays. After the projection onto Δ^r (step 5), this cone is a quadrilateral and all the points between two opposite vertices are in the interior of the quadrilateral while being nonnegative linear combinations of only two vertices. However, the set of such points has a zero measure, hence this is unlikely to happen. Even if this were to happen, it is also unlikely for an algorithm to end up at the sparsest solution of the least squares problem (this subset of optimal solutions is of measure zero within the set of optimal solutions). For example, by construction, interior-point methods would avoid such solutions.

Although this statement might be a bit bold since having $r - 1$ zero rows is a necessary condition (Theorem 4.28), the experiments in their papers show that H satisfy the SSC with high probability given that it is sufficiently sparse.

Numerical experiments Let us perform an experiment to observe when the necessary condition of Theorem 4.28 is satisfied for randomly generated matrices H. Let us generate H using the `sprand(r,n,d)` function of MATLAB that produces an r-by-n matrix with density d (that is, the probability for an entry to be equal to zero is $1 - d$). We make sure that H does not have zero columns by resampling these columns (this is necessary when r and d are small).

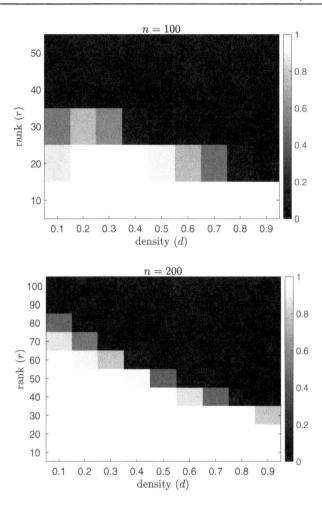

Figure 4.6. *Percentage of matrices H randomly generated using* `sprand(r,n,d)` *for $n = 100$ (top) and $n = 200$ (bottom) that pass the test of Algorithm 4.2 which checks whether the necessary condition of Theorem 4.28 for SSC1 is satisfied. (White squares indicate that all matrices pass the test, black squares that none do.) You can generate these figures using [*`Matlab file: SSC1_nec_cond_illus.m`*].*

We take $r \in \{10, 20, \ldots, 100\}$, which is a typical range in practice, $n \in \{100, 200\}$, which is smaller than the typical values observed in practice, and $d \in \{0.1, 0.2, \ldots, 0.9\}$. For each combination of values, we generate 100 such matrices, and Figure 4.6 reports how many times the necessary condition of Theorem 4.28 holds for $n = 100$ (on the top) and $n = 200$ (on the bottom).

As expected, the probability of passing the test decreases as the density or the dimension r increases. It is important to keep in mind that the value $n = 100$ is small as $n \leq r^2$ for all tested r. In practice, n is typically much larger, and hence it would be more likely for random matrices to pass the test; for example, it is the number of pixels in a hyperspectral image, the number of documents in a corpus, or the number of time windows in an audio signal. For $n = 100$, we observe that matrices generated for $r \geq 30$ and density $d \geq 0.5$ do not satisfy the SSC in most cases. Even when the density is as low as $d = 0.1$, some matrices with rank as low as $r = 20$ do not pass the test. This behavior is consistent with the results in [170, Figure S2]. These failures occur when there are not enough positive entries in each row of H, because $\bar{e}_k$ must be contained in the relative interior of the conical hull of the columns of H.

From $n = 100$ to $n = 200$, there is a drastic improvement. For example, for $n = 200$ and

- $r \leq 20$, all generated matrices pass the test;

- $r \leq 30$ and $d \leq 0.8$, $r \leq 40$ and $d \leq 0.5$, or $r \leq 50$ and $d \leq 0.4$, almost all generated matrices pass the test (only 4, 5, and 16 fail out of 2400, 2000, and 2000, respectively).

Let us shed some light on these numerical results using the observation above predicting the average value of n to have at least $r - 1$ zeros per row of H, namely $n \gtrsim \frac{1}{1-d} r \ln(r)$ or, equivalently, $d \lesssim \psi(r, n) = 1 - \frac{r \ln(r)}{n}$. Table 4.1 provides the values of $\psi(r, n)$ for the values of n and r from Figure 4.6.

Table 4.1. *Value of $\psi(r, n) = 1 - \frac{r \ln(r)}{n}$, which provides an upper bound on the density, d, for which the average $H = \texttt{sprand(r,n,d)}$ has $r - 1$ zeros per row.*

	$r = 10$	$r = 20$	$r = 30$	$r = 40$	$r = 50$	$r = 60$
$n = 100$	0.770	0.401	< 0	< 0	< 0	< 0
$n = 200$	0.885	0.700	0.490	0.262	0.022	< 0

The bound for the values of d in Table 4.1 is smaller than for the corresponding larger value of d in Figure 4.6 that allow us to pass the test of Algorithm 4.2. The reason is that $\psi(r, n)$ is larger than the actual average value of d needed to have $r - 1$ zero per rows in H (see footnote 34). For example, $\psi(30, 200) = 0.49$, which is smaller than the largest value of d where all matrices pass the test; see the bottom of Figure 4.6, where this happens for $d \leq 0.6$. This illustrates that the necessary condition for SSC is easily satisfied for n sufficiently large and d sufficiently small. For example, in practice, when $r \approx 40$, the number of samples n is typically much larger than 200 while the density is typically lower than 50%: most samples do not use more than half the basis elements when r is large. Similar observations were made in [251, 170]. In particular, [170, Figure S2] shows the same behavior as Figure 4.6 with a sharp phase transition.

A direction for further research would be to identify more precisely under which conditions the SSC is satisfied when H is randomly generated. This is a difficult question since checking the SSC is NP-hard (Lemma 4.24).

4.2.3.7 ▪ Is the SSC on $W^\top$ and H reasonable in practice?

The requirement that $W^\top$ and H both satisfy the SSC in Theorem 4.21 makes it difficult to be satisfied in real-world settings. For example, in airborne hyperspectral imaging, the matrix W typically has mostly positive entries (spectral signature are typically positive) and hence does not satisfy the SSC. We will see in Section 4.3.3 how to relax the condition that $W^\top$ and H both satisfy the SSC and require only one of the two factors to satisfy the SSC (this can be achieved by looking for minimum-volume solutions).

There are, however, some applications where $W^\top$ and H satisfy the SSC. This is the case for example in audio source separation of relatively simple signals (for example, a piano recording). The SSC for H requires that all sources are not active at all times, which is typically the case (during each time window, only a small subset of all piano notes are played simultaneously). The SSC for $W^\top$ requires that the frequency response of the sources does not overlap too much. This is also typically the case (this holds for piano notes). Such an example was presented in Section 1.3.4 for the piano recording of "Mary Had a Little Lamb," where the solution shown in Figure 1.8 was obtained with a standard NMF algorithm without any additional constraints. This is why plain NMF usually works well for simple source separation problems, as observed in Smaragdis et al. [425] but without any theoretical explanations.

4.2.4 ▪ Necessary condition: Sperner family

In this section, we discuss a simple to derive, yet important, necessary condition for an Exact NMF to be unique. This condition is presented in Theorem 4.34 and requires that the supports of the columns of W are not contained in one another, that is, they form a Sperner family (also known as an antichain). By symmetry, the same condition holds for the rows of H. Before doing so, we derive several intermediate results which will imply Theorem 4.34.

The following necessary condition for Exact NMF is relatively straightforward.

Theorem 4.30. *[298] If the Exact NMF (W, H) of X of size r is unique, then (W, I_r) is a unique Exact NMF of W of size r and (I_r, H) is a unique Exact NMF of H of size r.*

Proof. Let W admit another Exact NMF: $W = (WQ)(Q^{-1}I_r)$, where Q is not the permutation of a diagonal matrix (Theorem 4.5). This implies that $WQ \geq 0$ and $Q^{-1} \geq 0$. Therefore, $X = (WQ)(Q^{-1}H)$ is another Exact NMF of X since $Q^{-1}H \geq 0$ as $Q^{-1} \geq 0$ and $H \geq 0$, a contradiction. The same holds for H by symmetry. $\qquad\square$

It is interesting to observe that if H satisfies SSC1 then the Exact NMF (H, I_r) of H is unique.

Corollary 4.31. *Let $H \in \mathbb{R}_+^{r \times n}$ satisfy SSC1; then the Exact NMF (H, I_r) of H of size r is unique.*

Proof. The identity matrix I_r is separable and hence satisfies the SSC, while H satisfies SSC1 by assumption. This implies that HI_r is a unique Exact NMF of H of size r; see Corollary 4.22. $\qquad\square$

Let us show the following lemma.

Lemma 4.32. *Let $W \in \mathbb{R}_+^{m \times r}$ have a column whose support is contained in the support of another column; then the Exact NMF (W, I_r) of W of size r is not unique.*

Proof. Let us assume that the support of the kth column of W is contained in the support of the ℓth column of W. Let us construct another Exact NMF (W', H') of W which is not a permutation and scaling of (W, I_r). For simplicity, let us denote $H = I_r$. We proceed as follows.

- For $\epsilon > 0$ sufficiently small, we take

$$W'(:, \ell) = W(:, \ell) - \epsilon W(:, k) \geq 0$$

  since $\operatorname{supp}(W(:, k)) \subseteq \operatorname{supp}(W(:, \ell))$. The maximum possible value of ϵ such that $W'(:, \ell) \geq 0$ can be computed explicitly as follows:

$$\epsilon_{\max} = \min_{p \in \operatorname{supp}(W(:,k))} \frac{W(p, \ell)}{W(p, k)} > 0.$$

- $W'(:, j) = W(:, j)$ for $j \neq \ell$.

- $H'(i,:) = H(i,:)$ for $i \neq k$.

- $H(k,:) = H(k,:) + \epsilon H(\ell,:) \geq 0$ since $H \geq 0$ and $\epsilon > 0$.

Let us show that $W'H' = WH$ and hence that (W', H') is an Exact NMF of W of size r. Since only two rank-one factors have been modified, it suffices to show that

$$W'(:,k)H'(k,:) + W'(:,\ell)H'(\ell,:) = W(:,k)H(k,:) + W(:,\ell)H(\ell,:).$$

We have

$$
\begin{aligned}
W'(:,k)&H'(k,:) + W'(:,\ell)H'(\ell,:) \\
&= W(:,k)\big(H(k,:) + \epsilon H(\ell,:)\big) + \big(W(:,\ell) - \epsilon W(:,k)\big)H(\ell,:) \\
&= W(:,k)H(k,:) + \epsilon W(:,k)H(\ell,:) + W(:,\ell)H(\ell,:) - \epsilon W(:,k)H(\ell,:) \\
&= W(:,k)H(k,:) + W(:,\ell)H(\ell,:).
\end{aligned}
$$

Finally, one can check that $W' = WQ$ and $H' = Q^{-1}H = Q^{-1}$, where Q is the identity matrix where the zero entry $Q(k,\ell)$ is replaced by $-\epsilon$. The inverse of Q is the identity matrix where the zero entry at position (k,ℓ) is replaced by ϵ, hence $H' = Q^{-1} \geq 0$. Since Q is not the permutation of a diagonal matrix, (W, I_r) is not a unique Exact NMF of W (Theorem 4.5). $\quad\square$

Example 4.33. The matrix

$$W = \begin{pmatrix} 1 & 0 \\ 2 & 4 \\ 3 & 5 \end{pmatrix}$$

can never be part of a unique Exact NMF because

$$W' = WQ = \begin{pmatrix} 1 & 0 \\ 2 & 4 \\ 3 & 5 \end{pmatrix} \begin{pmatrix} 1 & 0 \\ -0.5 & 1 \end{pmatrix} = \begin{pmatrix} 1 & 0 \\ 0 & 1 \\ 0.5 & 1 \end{pmatrix} \geq 0,$$

where

$$Q^{-1} = \begin{pmatrix} 1 & 0 \\ 0.5 & 1 \end{pmatrix} \geq 0,$$

hence $Q^{-1}H \geq 0$ for any $H \geq 0$, while Q is not the permutation of a diagonal matrix. $\quad\blacksquare$

Now, we can derive an interesting necessary condition for the uniqueness of an Exact NMF, $X = WH$. Recall that a Sperner family, also known as an antichain, is a family of sets that are not contained in one another; see Section 3.4.3.

Theorem 4.34. *[355] If the Exact NMF (W, H) of X of size r is unique, then the supports of the columns of W form a Sperner family, and the supports of the rows of H form a Sperner family.*

Proof. This follows directly from Theorem 4.30, Lemma 4.32, and the symmetry of NMF ($X = WH \iff X^\top = H^\top W^\top$).

This result can also be found in [298, Theorem 3], [251, Theorem 3], [186, Remark 7], and [288, Theorem 3.4]. $\quad\square$

Given an NMF solution computed by an NMF algorithm, the necessary condition on the supports of W and H of Theorem 4.34 can be checked easily, allowing us to know whether there is a chance for this NMF to be unique.

Theorem 4.34 implies that each column of W and each row of H in a unique Exact NMF of size $r \geq 2$ must have a least one entry equal to zero. In fact, the support of a column of W (resp. a row of H) with only positive entries is $\{1, 2, \ldots, m\}$ (resp. $\{1, 2, \ldots, n\}$) and hence contains any other support.

Geometric interpretation   Theorem 4.34 can be interpreted geometrically as follow. For (W, I_r) to be a unique Exact NMF of W, the only simplicial cone of order r nested between $\text{cone}(W^\top)$ and $\text{cone}^*(I_r) = \mathbb{R}^r_+$ must be $\mathbb{R}^r_+$; see Theorem 4.9. For this to be possible, there must be at least one row of W on each facet of $\mathbb{R}^r_+$, that is, each column of W must have at least one entry equal to zero in a different position. Otherwise, we can easily move that facet of $\mathbb{R}^r_+$ toward the inside of $\mathbb{R}^r_+$ to generate a smaller simplicial cone of order r nested between $\text{cone}(W^\top)$ and $\mathbb{R}^r_+$ (this is actually what is implicitly done in the proof of Lemma 4.32). Moreover, if a row is located on several facets (that is, has several other zero entries), then another row must also be on that facet. Otherwise, there would be another simplicial cone of order r obtained by rotating that facet of $\mathbb{R}^r_+$ toward its interior, and nested between $\text{cone}(W^\top)$ and $\mathbb{R}^r_+$.

Looking back at the case rank$(X) = 2$ It is interesting to observe that, for $r = \text{rank}(X) = 2$, the two columns of W form a Sperner family if and only if W contains the identity as a submatrix (up to scaling), and this condition becomes sufficient; see Theorem 4.2.

Improved necessary condition   Krone and Kaie have provide an improved necessary condition using rigidity theory in [288, Proposition 4.8]. Let us try to summarize their contribution in a few words. A necessary condition for (W, H) to be a unique Exact NMF of $X = WH$ is that (W, H) is locally rigid. Intuitively, this means that (W, H) cannot be modified locally to generate another solution. For example, the triangle nested between two hexagons in Figure 2.5 (page 31) is locally rigid. Of course, this local condition does not imply uniqueness. For the example of Figure 2.5, there exist four locally rigid solutions by symmetry of the problem (see the discussion on page 34). A factorization that is locally rigid is either infinitesimally rigid or the Kruskal rank[35] of a certain matrix, denoted $Z(W, H)$, is not maximal. The matrix $Z(W, H)$ is constructed from (W, H) and the number of columns of this matrix is equal to the number of zeros in (W, H). If the matrix $Z(W, H)$ does not have maximal Kruskal rank, then the infinitesimally rigid condition requires W and H to have at least $r^2 - r + 1$ zeros in total. (Theorem 4.34 only requires at least $2r$ zeros.) We refer the interested reader to their paper for more details as these results are out of the scope of this book.

4.2.5 ▪ Sparsity of the input matrix

As we have seen in the previous section, it is necessary for W and H to have some degree of sparsity to be a unique Exact NMF. Interestingly, this observation was already made by Paatero and Tapper [371] (1994):

> If all columns of the correct unknown W and all rows of the correct unknown H contain a significant number of zeros, then the result by NMF is unique.

However, as we will see below with several examples, this is not a necessary condition for X. If the data points are strictly within the interior of the nonnegative orthant, one may expect to have more than one cone containing them, but this intuition is not correct. The following was even mentioned in [137]:

---

[35]The Kruskal rank of a matrix X is the maximum k such that any subset of k columns of X is linearly independent.

In short, we must look for situations where the data do not obey strict positivity in order to have uniqueness.

Of course, the presence of zero entries in the input matrix increases the chances to have a unique Exact NMF since it implies zeros in the factors W and H:

$$X(i,j) = \sum_k W(i,k)H(k,j) = 0 \;\Rightarrow\; W(i,k) = 0 \text{ or } H(k,j) = 0 \text{ for all } k;$$

see for example the discussion in [186]. However, this is not a necessary condition for X to have a unique Exact NMF. We now provide several examples of positive matrices that have a unique NMF. The second example also shows that the SSC is not a necessary condition to have a unique Exact NMF.

Matrix from Example 4.16 Looking back at Example 4.23 (page 117) with $\omega = 0.25$, we obtain

$$X = WH = \frac{1}{16} \begin{pmatrix} 17 & 8 & 4 & 1 & 4 & 16 \\ 8 & 17 & 16 & 4 & 1 & 4 \\ 4 & 16 & 17 & 8 & 4 & 1 \\ 1 & 4 & 8 & 17 & 16 & 4 \\ 4 & 1 & 4 & 16 & 17 & 8 \\ 16 & 4 & 1 & 4 & 8 & 17 \end{pmatrix},$$

which has a unique Exact NMF, since $W^\top$ and H are sufficiently scattered for $\omega = 0.25$.

Unique Exact NMF for which SSC is not necessary Let us construct another example where X has only positive entries and admits a unique NMF. This example will also show that W and H do not need to be sufficiently scattered to have a unique Exact NMF, showing that this condition is not necessary. This example is based on a triangle nested between a square and a rectangle. We derive this example step by step, allowing an interested reader to construct their own examples. The key is to use the equivalence between RE-NMF and NPP (Theorem 2.11) and the fact that the dimension of the NPP is equal to $\text{rank}(X) - 1$. For $\text{rank}(X) = 3$, the NPP has dimension 2 and can be solved efficiently (Theorem 2.17).

Let us take the square defined with the inequalities

$$f_1(x,y) = x \geq 0,$$
$$f_2(x,y) = y \geq 0,$$
$$f_3(x,y) = 1 - x \geq 0,$$
$$f_4(x,y) = 1 - y \geq 0$$

as the outer polytope $\mathcal{B}$ in the NPP. Let $\delta = 0.25$ and let us pick the four points

$$(x_1, y_1) = (\delta, 0.2),$$
$$(x_2, y_2) = (\delta, 0.2 + 1.6\delta),$$
$$(x_3, y_3) = (1 - \delta, 0.2 + 1.6\delta),$$
$$(x_4, y_4) = (1 - \delta, 0.2)$$

to be the vertices of the inner polytope $\mathcal{A}$ in the NPP, which is a rectangle; see Figure 4.7. It turns out that the triangle whose vertices are

$$(0, 0.2), (0.5, 1), \text{ and } (1, 0.2)$$

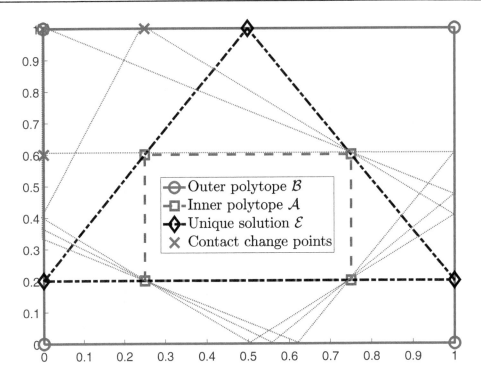

Figure 4.7. *Illustration of the NPP instance corresponding to the matrix X given in (4.18). The crosses are the contact change points between the vertices (0,0.2) and (0.5,1) of the unique nested triangle. The dotted lines are solutions of the two-dimensional NPP starting from these contact change points using Algorithm 2.1. All these solutions have four vertices, implying uniqueness of the displayed solution which cannot be locally modified.*

is the only triangle nested between the square and the rectangle. To prove this, one can use Algorithm 2.1 for the two-dimensional NPP problem; see Section 2.3.1. If there were other solutions, they could be constructed using Algorithm 2.1 by starting from

- one of the contact change points between (0,0.2) and (0.5,1); as shown in Figure 4.7, such solutions have four vertices;

- points in the neighborhood of (0,0.2). However, the solution constructed starting from (0,0.2) is locally unique because it ends up exactly on (0,0.2) while being tangent to the inner polygon. (More precisely, this solution is locally rigid; see [288] for more details.)

This implies the uniqueness of the triangle shown in Figure 4.7.

Using Theorem 2.11, the matrix X in RE-NMF corresponding to this NPP instance is given by

$$X(i,j) = f_i(x_j, y_j) \quad \text{for} \quad i = 1, \ldots, 4 \text{ and } j = 1, \ldots, 4,$$

so that

$$X = \begin{pmatrix} 0.25 & 0.25 & 0.75 & 0.75 \\ 0.2 & 0.6 & 0.6 & 0.2 \\ 0.75 & 0.75 & 0.25 & 0.25 \\ 0.8 & 0.4 & 0.4 & 0.8 \end{pmatrix}. \tag{4.18}$$

Up to permutation and scaling, the unique Exact NMF of X of size 3 is given by

$$X = WH = \begin{pmatrix} 0 & 0.5 & 1 \\ 0.2 & 1 & 0.2 \\ 1 & 0.5 & 0 \\ 0.8 & 0 & 0.8 \end{pmatrix} \begin{pmatrix} 0.75 & 0.5 & 0 & 0.25 \\ 0 & 0.5 & 0.5 & 0 \\ 0.25 & 0 & 0.5 & 0.75 \end{pmatrix}$$

and corresponds to the unique solution of the corresponding NPP displayed in Figure 4.7, and $W(1:2,:)$ are the vertices of the triangle.

Let us make a few observations:

- The factors $W^\top$ and H do not satisfy SSC1 because it requires that there are at least $r - 1 = 2$ zeros per row; see Theorem 4.28.

- This example is similar to the nested squares problem presented in Example 2.18. However, for the nested squares (where the squares have the same center), a triangle nested between the squares, if it exists, is never unique because of the symmetry (the same behavior appears for the nested hexagon problem; see Figure 2.5). Here the inner polygon is not a square (but a rectangle) and has a different center than the outer polygon, breaking this symmetry and allowing the solution to be unique.

- If δ is taken sufficiently large in the construction of the vertices of the inner rectangle, the solution of the corresponding Exact NMF problem is not unique: more than one triangle is nested between the rectangle and the square; see Figure 4.8 for an illustration with $\delta = 0.375$ for which

$$X = \begin{pmatrix} 0.375 & 0.375 & 0.625 & 0.625 \\ 0.2 & 0.8 & 0.8 & 0.2 \\ 0.625 & 0.625 & 0.375 & 0.375 \\ 0.8 & 0.2 & 0.2 & 0.8 \end{pmatrix}. \tag{4.19}$$

The value of $\delta = 0.375$ was chosen such that the solution of the NPP constructed using the first contact change point has three vertices, hence making the solution of the NPP nonunique. Using the symmetry of this NPP problem, one can check that there are four solutions. Figure 4.8 displays these four triangles, numbered from 1 to 4, which correspond, respectively, to W equal to

$$\begin{pmatrix} 0 & 0.5 & 1 \\ 0.2 & 1 & 0.2 \\ 1 & 0.5 & 0 \\ 0.8 & 0 & 0.8 \end{pmatrix}, \begin{pmatrix} 0 & 0.5 & 1 \\ 0.8 & 0 & 0.8 \\ 1 & 0.5 & 0 \\ 0.2 & 1 & 0.2 \end{pmatrix}, \frac{1}{8}\begin{pmatrix} 3 & 3 & 8 \\ 0 & 8 & 4 \\ 5 & 5 & 0 \\ 8 & 0 & 4 \end{pmatrix}, \frac{1}{8}\begin{pmatrix} 5 & 5 & 0 \\ 0 & 8 & 4 \\ 3 & 3 & 8 \\ 8 & 0 & 4 \end{pmatrix}.$$

- Given the matrix X, it is possible to compute an Exact NMF using NMF algorithms; see Chapter 8. For $\delta < 0.375$, the Exact NMF is unique, and so is the optimal NMF solution. However, as δ gets closer to 0.375, there are four regions in the search space with very small objective function values corresponding to four local minima close to the W's shown above. Hence we have observed numerically that as δ gets closer to 0.375, NMF algorithms need in general more (randomly generated) initial points to find the global optimum.

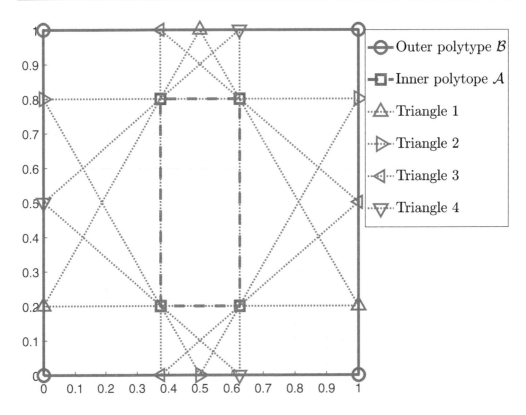

Figure 4.8. *Illustration of the NPP instance corresponding to the matrix X given in* (4.19) *with four distinct solutions nested between the square and the rectangle.*

Other examples from [288] Several examples of 5-by-5 positive matrices with a unique Exact NMF[36] of size $r = \mathrm{rank}(X) = 4$ can be found in [288]. In these examples, the matrices $W^\top$ and H also do not satisfy the SSC as $W^\top$ and H do not have $(r - 1) = 3$ zeros per row (Theorem 4.28); here is one of their examples:

$$
\begin{pmatrix}
104184 & 229176 & 94392 & 336996 & 77040 \\
94663 & 117528 & 485070 & 3404 & 7979 \\
535318 & 168896 & 1169348 & 255210 & 182576 \\
156494 & 310908 & 1119179 & 316225 & 460213 \\
763917 & 337540 & 876372 & 1016103 & 574666
\end{pmatrix}
$$

$$
= \begin{pmatrix}
0 & 0 & 396 & 108 \\
0 & 0 & 4 & 555 \\
0 & 470 & 0 & 812 \\
455 & 0 & 0 & 926 \\
194 & 761 & 550 & 0
\end{pmatrix}
\begin{pmatrix}
0 & 260 & 681 & 695 & 985 \\
847 & 0 & 978 & 543 & 366 \\
217 & 522 & 0 & 851 & 191 \\
169 & 208 & 874 & 0 & 13
\end{pmatrix}.
$$

Note that the construction relies on a rather different approach; constructing such examples using a geometric approach would be more difficult as three-dimensional NPP are more difficult to visualize and solve (Theorem 2.19).

---

[36]Actually, the authors only prove that these Exact NMFs are locally unique. However, there is strong numerical evidence that they are also globally unique; namely, any numerical solution generated by the algorithm proposed in [460] to compute Exact NMFs always coincides with the given Exact NMF, up to permutation and scaling.

4.2.6 ▪ Further readings

The two main identifiability results of this section (Theorems 4.12 and 4.21) and the necessary condition (Theorem 4.34) directly apply to the symNMF problem where $W^\top = H$. This is straightforward: if the Exact NMF (W, H) of X is unique and $W^\top = H$, the symmetric Exact NMF also has a unique solution (since the solution space of symmetric Exact NMF is smaller than that of Exact NMF because of the additional constraint $W^\top = H$). We refer the reader to [251] for more details. We also refer the interested reader to the recent paper [409] for other results on the identifiability of symNMF.

For a more in-depth look at this topic, we recommend the recent paper "Uniqueness of Nonnegative Matrix Factorizations by Rigidity Theory" by Krone and Kubjas [288]. The authors provide necessary conditions for the uniqueness of Exact NMF of size $r = \text{rank}(X)$ using rigidity theory.

4.3 ▪ Regularized Exact NMF

Among Exact NMF solutions, we may look for one that satisfies some additional constraints or that minimizes some criterion. This allows us to reduce the number of solutions and leads to identifiability under milder assumptions than for Exact NMF. Let us formally define this problem.

Problem 4.1 (Regularized Exact (N)MF). *Given a matrix $X \in \mathbb{R}^{m \times n}$, the sets $\mathcal{W} \subseteq \mathbb{R}^{m \times r}$ and $\mathcal{H} \subseteq \mathbb{R}^{r \times n}$, and the objective function $f : \mathcal{W} \times \mathcal{H} \mapsto \mathbb{R}$, define the following optimization problem:*

$$\min_{W, H} \quad f(W, H)$$

$$such\ that \quad X = WH, \tag{4.20}$$

$$W \in \mathcal{W}\ and\ H \in \mathcal{H}.$$

Any optimal solution to (4.20) is referred to as a regularized Exact (N)MF of X of size r.

We put the N in parentheses in (N)MF because for some of the models described in this section, nonnegativity is not a necessary condition for identifiability. If there is no objective function in a regularized Exact (N)MF problem, we write $f = 0$. As for the Exact NMF problem (Definition 4.1), we say that a regularized Exact (N)MF is unique if the optimal solution to (4.20) is unique, up to permutation and scaling of the rank-one factors.

In this section, we present four key models of the form (4.20) that lead to unique solutions under appropriate conditions, namely

- separable NMF that requires H to be a separable matrix (Section 4.3.1),

- ONMF that requires H to have orthogonal rows (Section 4.3.2),

- minimum-volume NMF that requires the convex hull of the columns of W to have the smallest possible volume (Section 4.3.3), and

- sparse NMF that requires H to be sparse (Section 4.3.4).

In Section 4.3.5, we summarize the identifiability results obtained for regularized Exact (N)MF; see in particular Table 4.2 (page 155).

4.3.1 ▪ Separable NMF

Let us define separable NMF which assumes that H is separable (see Definition 4.10 on page 107).

Definition 4.35 (Separable NMF). *Given a matrix $X \in \mathbb{R}^{m \times n}$ and a factorization rank r, separable NMF is the regularized Exact (N)MF problem with $f = 0$,*

$$\mathcal{W} = \mathbb{R}^{m \times r} \quad and \quad \mathcal{H} = \{H \in \mathbb{R}_+^{r \times n} \mid H \text{ is separable}\}. \tag{4.21}$$

Remark 4.1 (Terminology). *The use of NMF in "separable NMF" is arguably not suitable since W is not required to be nonnegative. However, we follow the terminology mostly used in the literature. The same choice will be made for ONMF and min-vol NMF in the next two sections.*

The matrix H in separable NMF contains the identity matrix as a submatrix (up to permutation and scaling); see Lemma 4.11. Given $X = WH$ where $H \in \mathbb{R}^{r \times n}$ is separable therefore implies that there exists an index set $\mathcal{K}$ of size r such that $W = X(:, \mathcal{K})$, up to scaling of the columns of W. Geometrically, this means that there exists a subset of r columns of X whose conical hull contains all the columns of X; see Figure 4.9 (left) for an illustration. If $X \geq 0$ and the columns of X are scaled to have unit ℓ_1 norm, then separability means that the columns of W are the vertices of $\mathrm{conv}(X)$ since the columns of H have unit ℓ_1 norm as well (Lemma 2.1); see Figure 4.9 (right) for an illustration.

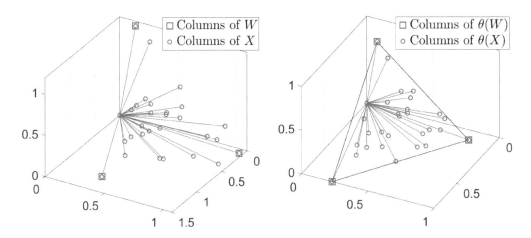

Figure 4.9. *Geometric illustration of separable NMF for $m = r = 3$ and $n = 25$, with $W = X(:, \mathcal{K})$ for some index set $\mathcal{K}$ of size $r = 3$. Both figures represent the same data set. The figure on the right is the normalization to unit ℓ_1 norm of the columns of X and W from the figure on the left.*

We have the following lemma.

Lemma 4.36. *Let $X = WH$ where $H \in \mathbb{R}_+^{r \times n}$ is separable. Then*

$$\mathrm{cone}(W) = \mathrm{cone}(X).$$

Proof. Since $X = WH$ for $H \geq 0$, $\text{cone}(X) \subseteq \text{cone}(W)$. Since H is separable, there exists an index set $\mathcal{K}$ such that $H(:, \mathcal{K})$ is a diagonal matrix with positive diagonal elements (Lemma 4.11). W.l.o.g. let us assume this diagonal matrix is the identity matrix (after scaling the columns of W accordingly). This implies that $W = X(:, \mathcal{K})$ and hence $\text{cone}(W) \subseteq \text{cone}(X)$.   □

We can now prove identifiability of separable NMF.

Theorem 4.37. *Let $X = WH$ where $H \in \mathbb{R}_+^{r \times n}$ is separable and $r = \text{rank}(X)$. Then (W, H) is the unique separable NMF of X.*

Proof. Let $(\bar{W}, \bar{H})$ be a solution of separable NMF. By Lemma 4.36,

$$\text{cone}(W) = \text{cone}(\bar{W}) = \text{cone}(X).$$

Since $r = \text{rank}(X) = \text{rank}(W)$, $\text{cone}(X)$ has r extreme rays which are the columns of W. Similarly, since $r = \text{rank}(X) = \text{rank}(\bar{W})$, the columns of $\text{rank}(\bar{W})$ are the r extreme rays of $\text{cone}(X)$. Therefore, the columns of W and $\bar{W}$ must coincide, up to scaling and permutation.

Given W, H is unique and equal to $W^\dagger X$ where $W^\dagger$ is the left inverse of W.   □

Note that separable NMF can be cast as another equivalent regularized Exact (N)MF, where the constraints (4.21) are replaced with

$$\mathcal{W} = \{W \mid W = X(:, \mathcal{K}), |\mathcal{K}| = r\} \quad \text{and} \quad \mathcal{H} = \mathbb{R}_+^{r \times n}.$$

In fact, $X = WH$, $W = X(:, \mathcal{K})$, and $r = \text{rank}(X)$ implies that H is separable since it implies that $H(\mathcal{K}, :) = I_r$. Hence separable NMF is closely related to the column subset selection problem; see Chapter 7 for more details.

The condition $r = \text{rank}(X)$ in Theorem 4.12 can be relaxed to the condition that $\text{cone}(X)$ has r distinct extreme directions while preserving the uniqueness of W. However, the uniqueness of H is not preserved. For example, if $r > \text{rank}(X)$, a point corresponding to a column of H with positive entries is in the interior of $\text{cone}(X)$ and can be constructed with multiple linear combinations of extreme rays of X. To preserve uniqueness of H, there is an additional condition: all columns of X are located on $(d-1)$-dimensional faces of $\text{cone}(X)$ generated by exactly $d - 1$ extreme directions. In that case, the convex combinations given by H are unique [442]. For example, consider a cone in three dimensions with more than three extreme directions. For H to be unique, the data points can only belong to the border of that cone; hence the columns of H can have at most two positive entries corresponding to two adjacent extreme directions.

Remark 4.2 (Link with semi-NMF). *Since W does not need to be nonnegative, separable NMF is related to semi-NMF, which is the problem where $X = WH$ with $H \geq 0$ while W is not required to be nonnegative; see Section 5.4.3. However, like unconstrained LRMA, semi-NMF does not have a unique solution for most matrices without further assumptions like separability or sparsity [203].*

Is the separability a reasonable condition in practice?

In Chapter 7, we will analyze in detail algorithms for separable NMF. In particular, we will see that separable NMF can be solved in polynomial time. Moreover, in the presence of noise, separable NMF algorithms can recover the columns of W up to some error bounds that depend on the noise level. We will also discuss several applications where separability makes sense such as hyperspectral unmixing, audio source separation, facial feature extraction, and document classification. Separable NMF is therefore a particularly interesting NMF model because it resolves the two main issues of NMF:

NP-hardness and identifiability. This is the reason why Chapter 7 is dedicated to this particular NMF variant. Note that separability is a rather strong condition that does not hold in many practical scenarios. However, separable NMF algorithms can be used as effective initializations strategies for more sophisticated NMF models; see Section 8.6.

4.3.2 ▪ Orthogonal NMF

ONMF is an NMF variant where H is required to have orthogonal rows, that is, $HH^\top = I_r$; see for example [130, 132, 91, 495, 489, 312, 17, 476] and the references therein. The exact variant is defined as follows.

Definition 4.38 (Exact ONMF). *Given a matrix $X \in \mathbb{R}^{m \times n}$ and a factorization rank r, Exact ONMF is the regularized Exact (N)MF problem with $f = 0$,*

$$\mathcal{W} = \mathbb{R}^{m \times r} \quad and \quad \mathcal{H} = \{H \in \mathbb{R}_+^{r \times n} \mid HH^\top = I_r\}.$$

Together with nonnegativity of H, orthogonality implies that each column of H has a single positive entry.

Lemma 4.39. *Let $H \in \mathbb{R}_+^{r \times n}$ satisfy $HH^\top = I_r$. Then each column of H has at most a single positive entry.*

Proof. The proof follows from the observation that two nonnegative vectors x and y are orthogonal if and only if they have disjoint supports, that is, for $x \geq 0$ and $y \geq 0$, $x^\top y = 0$ if and only if $x_i y_i = 0$ for all i. The constraint $HH^\top = I_r$ implies that every two rows of H are orthogonal and hence have disjoint supports. This implies that each column of H can have at most a single positive entry.   □

Lemma 4.39 implies that ONMF is a clustering problem: each column of X is approximated using a single properly scaled column of W. It can be shown that ONMF is a particular variant of spherical k-means; see Section 5.5.3. This also means that Exact ONMF corresponds to separable NMF where all columns of X are multiples of columns of W. Hence the uniqueness result of separable NMF applies to ONMF. However, due to the particular nature of Exact ONMF, the condition $r = \operatorname{rank}(X)$ can be relaxed.

Theorem 4.40. *Let $X = WH$ where $H \in \mathbb{R}_+^{r \times n}$ satisfies $HH^\top = I_r$ and the columns of $W \in \mathbb{R}^{m \times r}$ are not multiples of one another. Then the Exact ONMF (W, H) of X of size r is unique.*

Proof. By Lemma 4.39, under the constraints $H \geq 0$ and $HH^\top = I_r$, each column of X is a multiple of a column of W, that is, for all j, there exists k such that $X(:, j) = W(:, k)H(k, j)$. Moreover, $HH^\top = I_r$ implies that each row of H must have at least one positive entry, and hence that every column of W appears at least once as a column of X (up to a proper scaling; see below). Since the columns of W are not multiples of one another, there is only one way to construct W in an Exact ONMF: pick r columns of X that are not multiples of one another, up to a proper scaling (see below). Given W, H is unique and assigns each column of X to its corresponding column of W that it is multiple with. For columns of X equal to zero, the corresponding column of H is equal to zero since no column of W is equal to zero (because columns of W are not multiples of one another). Note that, as opposed to separable NMF, there is no scaling ambiguity in Exact ONMF because the normalization $\|H(k, :)\|_2 = 1$ for all k imposes a particular scaling for the rows of H, and hence for the columns of W.   □

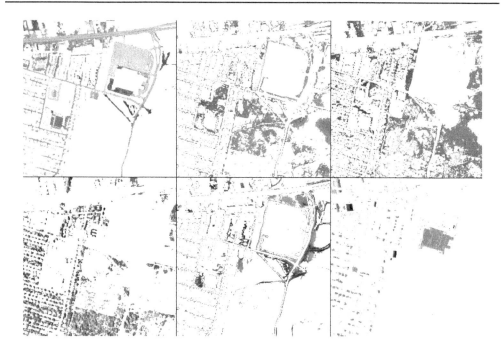

Figure 4.10. *Illustration of the clustering performed by ONMF with $r = 6$ on the Urban hyperspectral image (Figure 1.6). Each cluster shown above corresponds to a row of H, reshaped as an image. The first cluster corresponds to the road, the second and third to two different types of grass, the fourth to the trees, the fifth to dirt, and the last to roof tops. [Matlab file: ONMF_Urban.m].*

Is orthogonality a reasonable condition in practice? The ONMF model is rather strong and applies only in clustering settings. For example, in hyperspectral imaging (Section 1.3.2), ONMF requires that each pixel contains a single material; this assumption is violated in most cases. However, it may be useful to apply ONMF on such data sets in order to cluster the pixels according to the material they contain in the largest proportion. Figure 4.10 displays the ONMF solution for the Urban hyperspectral image obtained with the EM-ONMF algorithm [384] based alternating optimization (see Section 8.3.1 for more details on this strategy) and initialized with a separable NMF algorithm, namely, the successive projection algorithm (SPA) (see Section 7.4.1). ONMF allows us to extract different meaningful clusters for this relatively simple hyperspectral image.

In facial feature extraction where X is a pixel-by-subject matrix (Section 1.3.1), ONMF approximates each input facial image with a single face: this would be useful if one wants to cluster the faces according to the subject they display (assuming the different images of the same subject are approximately multiples of one another). Applying ONMF on $X^\top$ instead generates facial features that are disjoint: pixels are clustered into regions corresponding to different facial features, that is, sets of pixels behaving similarly among the subjects; see Figure 4.11 for an illustration of ONMF on the CBCL facial images using the same algorithmic approach as for the Urban hyperspectral image.

Remark 4.3 (Spatial coherence). *On the two ONMF examples shown in Figures 4.10 and 4.11, the spatial coherence of the clusters (that is, the rows of H) is automatically obtained while it was not imposed explicitly. ONMF was applied on a matrix X obtained by reshaping the input images as its rows; hence the spatial information is lost and not used by ONMF. The reason*

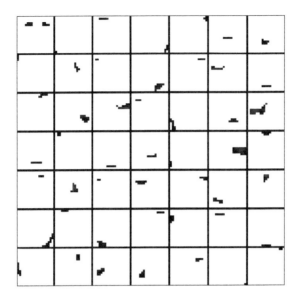

Figure 4.11. *Illustration of the clustering performed by ONMF with* $r = 49$ *on the CBCL facial images (Figure 1.2). Each image on the above 7-by-7 grid is obtained by reshaping a row of H as an image.* [`Matlab file: ONMF_CBCL.m`].

for this spatial coherence is that neighboring pixels contain similar information and hence are likely to have the same cluster assignment. For example, in the hyperspectral image, neighboring pixels are likely to contain the same materials. Spatially more coherent solutions can be obtained by using this prior; see Section 5.3.

4.3.3 ▪ Minimum-volume NMF

Let us consider a first min-vol NMF model, which is a regularized Exact (N)MF model $X = WH$ where the columns of H belong to the unit simplex, that is, $H \geq 0$ and $H^\top e = e$. The assumption $H^\top e = e$ can be made w.l.o.g. for Exact NMF by scaling the columns of X to unit ℓ_1 norm (this follows from Lemma 2.1); see Section 7.2.2 for further discussion on this issue when X is not nonnegative. This normalization implies

$$\mathrm{conv}(X) \subseteq \mathrm{conv}(W),$$

that is, each column of X is a convex combination of the columns of W; see Section 2.1.2. Among the m-by-r matrices W such that $\mathrm{conv}(X) \subseteq \mathrm{conv}(W)$, it makes sense to look for a matrix W whose convex hull has minimum volume (see the next subsection for a formal definition). Intuitively, we look for basis vectors as close as possible to the data points. If the data points are sufficiently spread within $\mathrm{conv}(W)$, then such a W is unique; see Theorem 4.43, which requires H to satisfy the SSC. As we will see later, there are other min-vol NMF models, based on other normalizations of W and H.

The main idea behind min-vol NMF dates back to Full, Erlich, and Klovan [178] (1981) and from Craig's belief [109] (1994) that a minimum-volume solution would correspond to the true materials within a hyperspectral image, given that the data points (that is, the pixels) are sufficiently spread within $\mathrm{conv}(W)$.

A first NMF-based formulation and algorithm was proposed in [343] (2007), and many works followed, especially in the area of hyperspectral unmixing; see for example [79, 236, 8, 516] and the references therein. It was also later used for blind source separation [407, 370], chemometrics [327], and machine learning applications such as facial feature extraction [513], topic modeling [171, 259], and community detection [246]. The first identifiability result of min-vol NMF (see Theorem 4.43 below) was proved by Fu et al. [176] (2015), and a similar result was proved by Lin et al. [320] (2015). These papers were worked out independently, as explained in [176, footnote 3], and the proofs are rather different.

This section about min-vol NMF is organized as follows. We first explain how to measure the volume of $\text{conv}(W)$ (Section 4.3.3.1). Then we present three different min-vol NMF models using different normalizations (Sections 4.3.3.2–4.3.3.4). Although these models essentially lead to the same identifiability result in noiseless settings, they behave rather differently in the presence of noise; this is discussed in Section 4.3.3.5. Finally, we discuss the issue of solving min-vol NMF (Section 4.3.3.6) and whether the SSC on H which is required for the identifiability in min-vol NMF is reasonable in practice (Section 4.3.3.7).

4.3.3.1 ▪ How to measure the volume of $\text{conv}(W)$

Let $a_1, a_2, \ldots, a_r \in \mathbb{R}^{r-1}$ be r points in dimension $r - 1$ (for example, the vertices of a triangle in two dimensions). The volume of the convex hull of these points is given by the determinant of a particular matrix [437]:

$$\text{volume}\big(\text{conv}(\{a_1, a_2, \ldots, a_r\})\big) = \frac{1}{(r-1)!} \left| \det \left(\begin{array}{cccc} a_2 - a_1 & a_3 - a_1 & \ldots & a_r - a_1 \end{array} \right) \right|.$$

In fact, recall that the volume of the parallelotope[37] generated by the columns of a matrix A is given by $|\det(A)|$. The volume of $\text{conv}(\{a_1, a_2, \ldots, a_r\})$ is equal to the volume of $\text{conv}(\{0, a_2 - a_1, \ldots, a_r - a_1\})$ (it is just a translation by $-a_1$) which is proportional to the volume of the parallelotope generated by $a_2 - a_1, \ldots, a_r - a_1$.

For an m-by-r matrix W with $m \geq r$, the volume of $\text{conv}(W)$ is zero (for example, the volume of a two-dimensional triangle in three dimensions is zero). What we are interested in is the volume of $\text{conv}(W)$ within the affine hull of the columns of W, which is an $(r - 1)$-dimensional subspace, given that $r = \text{rank}(W)$. Hence computing the volume of $\text{conv}(W)$ for an m-by-r matrix W with $m \geq r$ requires us to perform a linear dimensionality reduction to represent the columns of W in an $(r - 1)$-dimensional subspace. This can be achieved for example using PCA as done in [343].

However, there is another measure related to the volume of $\text{conv}(W)$ that is more convenient for proving the identifiability results of min-vol NMF. We focus on this measure in this book. Add the vector 0 to the set of points $a_i \in \mathbb{R}^r$ $(i = 1, 2, \ldots, r)$ above to obtain

$$\text{volume}\big(\text{conv}(\{0, a_1, a_2, \ldots, a_r\})\big) = \frac{1}{r!} \left| \det \left(\begin{array}{cccc} a_1 & a_2 & \ldots & a_r \end{array} \right) \right| = \frac{1}{r!} |\det(A)|, \quad (4.22)$$

where $A = [a_1, a_2, \ldots, a_r]$.

Lemma 4.41. *Let $W \in \mathbb{R}^{m \times r}$ with $m \geq r$ and $r = \text{rank}(W)$. Then*

$$\frac{1}{r!} \sqrt{\det(W^\top W)}$$

is the volume of the convex hull of the columns of W and the origin in the linear subspace spanned by the columns of W.

---

[37]The parallelotope generalizes to higher dimensions the parallelogram in two dimensions and parallelepiped in three dimensions.

Proof. This follows directly from the observations above. First note that the affine hull of $[0, W]$ is the column space of W. Let $W = U\Sigma V^\top$ be the compact SVD of W where $U \in \mathbb{R}^{m \times r}$ and $V \in \mathbb{R}^{r \times r}$ have orthogonal columns, and Σ is a diagonal matrix whose diagonal entries are the nonnegative singular values of W (see Section 6.1.1 for more information on the SVD). The matrix U is an orthogonal basis of $\operatorname{col}(W)$. Using the formula (4.22), the volume of the convex hull of the columns of W and the origin in the subspace $\operatorname{col}(W)$ is given by $\frac{1}{r!}$ times

$$| \det(\Sigma V^\top)| = \det(\Sigma) = \sqrt{\det(W^\top W)},$$

since the singular values of $W^\top W = V \Sigma^2 V^\top$ are equal to the square of the singular values of $W = U\Sigma V^\top$. □

In the following three sections, we discuss three min-vol NMF models of the form

$$\min_{W,H} \det \left(W^\top W\right) \quad \text{such that} \quad X = WH, H \geq 0,$$

under different normalizations of H or W, namely $H^\top e = e$ (Section 4.3.3.2), $He = e$ (Section 4.3.3.3), and $W^\top e = e$ (Section 4.3.3.4). Note that some normalization is necessary; otherwise, W will go to zero because of the scaling degree of freedom in NMF, that is, $WH = (\alpha W)(H/\alpha)$ for any $\alpha > 0$.

4.3.3.2 ▪ Min-vol NMF (1): $H^\top e = e$

One min-vol NMF model that leads to unique solutions imposes the columns of H to have unit ℓ_1 norm, that is, $H^\top e = e$.

Definition 4.42 (Min-vol NMF (1)). *Given a matrix $X \in \mathbb{R}^{m \times n}$ and a factorization rank r, min-vol NMF (1) is the regularized Exact (N)MF problem with*

$$\mathcal{W} = \mathbb{R}^{m \times r}, \ \mathcal{H} = \{H \in \mathbb{R}_+^{r \times n} \mid H^\top e = e\}, \ \text{and} \ f(W, H) = \det(W^\top W). \tag{4.23}$$

We have the following identifiability result for min-vol NMF (1).

Theorem 4.43. *[176, Theorem 1] Let $X = WH$ where $H \in \mathbb{R}_+^{r \times n}$ satisfies the SSC, $H^\top e = e$, and $r = \operatorname{rank}(X)$. Then (W, H) is a unique solution to the min-vol NMF (1) of X of size r.*

Proof. By assumption, (W, H) is a feasible solution of min-vol NMF (1), that is, of (4.20) with $\mathcal{W}, \mathcal{H}$, and f defined in (4.23). Let $(\bar{W}, \bar{H})$ be another feasible solution of min-vol NMF (1). Since $r = \operatorname{rank}(X)$, $\operatorname{rank}(\bar{W}) = \operatorname{rank}(\bar{H}) = \operatorname{rank}(W) = \operatorname{rank}(H) = r$ so that there exists an r-by-r invertible matrix Q such that

$$\bar{W} = WQ^{-1} \ \text{and} \ \bar{H} = QH;$$

see Lemma 4.4. We also have

$$\bar{H}H^\dagger = Q,$$

where $H^\dagger$ is the right inverse of H, that is, $HH^\dagger = I_r$, which exists since $\operatorname{rank}(H) = r$. By assumption, $e^\top H = e^\top \bar{H} = e^\top$ so that

$$e^\top Q = e^\top \bar{H} H^\dagger = e^\top H^\dagger = (e^\top H)H^\dagger = e^\top. \tag{4.24}$$

Moreover, $\bar{H} = QH \geq 0$ since $\bar{H} \in \mathcal{H}$, hence, by Corollary 4.7,

$$\text{cone}(Q^\top) \subseteq \text{cone}^*(H). \tag{4.25}$$

Since H satisfies SSC1, $\mathcal{C} \subseteq \text{cone}(H)$, hence $\text{cone}^*(H) \subseteq \mathcal{C}^*$ by duality (Lemma 4.17). Together with (4.25), this implies that $\text{cone}(Q^\top) \subseteq \mathcal{C}^*$, that is,

$$Q(j,:)e \geq \|Q(j,:)\|_2 \quad \text{for} \quad j = 1, 2, \ldots, r. \tag{4.26}$$

Therefore,

$$|\det(Q)| \leq \prod_{j=1}^r \|Q(j,:)\|_2 \leq \prod_{j=1}^r Q(j,:)e \leq \left(\frac{\sum_{j=1}^r Q(j,:)e}{r} \right)^r = \left(\frac{e^\top Q e}{r} \right)^r = 1,$$

where

- the first inequality is the Hadamard's inequality,

- the second follows from (4.26),

- the third follows from the arithmetic-geometric mean inequality, that is, $\prod_{i=1}^r x_i \leq \left(\frac{1}{r} \sum_{i=1}^r x_i \right)^r$ for $x \in \mathbb{R}_+^r$,

- the last equality follows from (4.24).

If $|\det(Q)| = 1$, all inequalities above are equalities, hence, for all j,

$$Q(j,:)e = \|Q(j,:)\|_2 = 1$$

and $|\det(Q)| = \prod_{j=1}^r \|Q(j,:)\|_2$, implying that $Q^\top$ is orthogonal.[38] Using $\text{cone}(Q^\top) \subseteq \text{cone}^*(H)$, we obtain by duality that $\text{cone}(H) \subseteq \text{cone}(Q^\top)$ since, by Lemma 4.19, $\text{cone}^*(Q^\top) = \text{cone}(Q^\top)$. Finally, since H satisfies SSC2, $Q^\top$ is orthogonal, and $\text{cone}(H) \subseteq \text{cone}(Q^\top)$, $Q^\top$ can only be a permutation matrix.

Suppose now $(\bar{W}, \bar{H})$ is an optimal solution to (4.20), that is, it is a regularized Exact (N)MF of X, which is not obtained by permutation of the rank-one factors of WH (note that the scaling degree of freedom is absent due to the constraint $H^\top e = e$). Then Q (as defined above) cannot be a permutation matrix (Theorem 4.5) implying that $|\det(Q)| < 1$. We have

$$\det\left(\bar{W}^\top \bar{W} \right) = \det\left(Q^{-\top} W^\top W Q^{-1} \right)$$
$$= \det\left(W^\top W \right) |\det(Q)|^{-2}$$
$$> \det\left(W^\top W \right).$$

This contradicts the optimality of $(\bar{W}, \bar{H})$ since (W, H) is a feasible solution. □

Remark 4.4. *The proof of Theorem 4.43 follows the proof from [176] to show the identifiability of min-vol NMF. In [320], the authors use a slightly more restrictive condition on H; see (4.5) and the discussion that follows (page 111).*

As stated above, the condition $H^\top e = e$ is not restrictive when $X \geq 0$ and in the absence of noise, as it can be assumed w.l.o.g. by normalizing the columns of the input matrix X to have unit ℓ_1 norm; see Lemma 2.1. (The condition $H^\top e = e$ can also be assumed w.l.o.g. even when

---

[38]This is a standard linear algebra result and can be proved for example using Lemma 7.11 (page 227).

$X \not\geq 0$ using a proper scaling; see Section 7.2.2.) However, in the presence of noise, such a scaling might not be desirable. For example, the columns of X with small norm that typically contain less information and are more easily affected by noise (such as background pixels in a hyperspectral image, or documents containing only a few words) are given the same importance as columns with large norms [294].

In [169] (2018), Fu, Huang, and Sidiropoulos were able to relax the condition $H^\top e = e$ to the condition $He = e$ (see Theorem 4.45 below). The condition $He = e$ can be assumed w.l.o.g., even in the presence of noise: normalizing the r rows of H simply removes the scaling degree of freedom in NMF. At this point, it is interesting to observe the following: although these two results are equivalent in noiseless settings (after normalizing the input matrix), they might behave rather differently in noisy scenarios; this will be illustrated in Example 4.48.

4.3.3.3 ▪ Min-vol NMF (2): $He = e$

It is also possible to obtain identifiability results for min-vol NMF using a model that normalizes the rows of H instead of the columns.

Definition 4.44 (Min-vol NMF (2)). *Given a matrix $X \in \mathbb{R}^{m \times n}$ and a factorization rank r, min-vol NMF (2) is the regularized Exact (N)MF problem with*

$$\mathcal{W} = \mathbb{R}^{m \times r}, \quad \mathcal{H} = \{H \in \mathbb{R}_+^{r \times n} \mid He = e\}, \quad and \quad f(W, H) = \det(W^\top W). \qquad (4.27)$$

Min-vol NMF (2) is more relaxed than min-vol NMF (1) because row normalization can be assumed w.l.o.g. We have the following identifiability result.

Theorem 4.45. *[169, Theorem 1] Let $X = WH$ where $H \in \mathbb{R}_+^{r \times n}$ satisfies the SSC, $He = e$, and $r = \mathrm{rank}(X)$. Then (W, H) is the unique solution to min-vol NMF (2) of X of size r.*

Proof. The proof follows exactly the same steps as the proof of Theorem 4.43; there are only a few differences due to the different normalizations.

By assumption, (W, H) is a feasible solution of min-vol NMF (2), that is, of (4.20) with $\mathcal{W}$, $\mathcal{H}$ and f as defined in (4.27). Let $(\bar{W}, \bar{H})$ be another feasible solution of min-vol NMF (2). Since $r = \mathrm{rank}(X)$, there exists an r-by-r invertible matrix Q such that

$$\bar{W} = WQ^{-1} \quad \text{and} \quad \bar{H} = QH;$$

see Lemma 4.4. Since $He = \bar{H}e = e$,

$$e = \bar{H}e = QHe = Qe. \qquad (4.28)$$

For the same reasons as in Theorem 4.43, $\mathrm{cone}(Q^\top) \subseteq \mathcal{C}^*$, that is, Q satisfies (4.26). Therefore,

$$|\det(Q)| \leq \prod_{j=1}^{r} \|Q(j,:)\|_2 \leq \prod_{j=1}^{r} Q(j,:)e = 1,$$

where the first inequality is the Hadamard's inequality, the second follows from (4.26), and the equality follows from the inequalities (4.28).

The remainder of the proof is exactly the same as that of in Theorem 4.43. □

Although the constraint $He = e$ relaxes the constraint $H^\top e = e$, we have observed that min-vol NMF (2) does not perform well *in the presence* of noise when the rank-one factors are unbalanced. Assume there exists k and ℓ such that

$$\|W(:,k)H(k,:)\|_1 \gg \|W(:,\ell)H(\ell,:)\|_1.$$

The normalization $He = e$ implies $\|H(k,:)\|_1 = 1$ for all k since $H \geq 0$ and hence implies that

$$\|W(:,k)H(k,:)\|_1 = \|W(:,k)\|_1 \gg \|W(:,\ell)\|_1 = \|W(:,\ell)H(\ell,:)\|_1.$$

This leads to two practical problems (see Example 4.48 for a numerical experiment). First, this makes the influence of these two columns of W in the objective $\det(W^\top W)$ unbalanced. The model favors $W(:,k)$, that is, the objective will be decreased more by making $W(:,k)$ closer to the data points than $W(:,\ell)$.

Second, W is ill-conditioned, that is, it has a large condition number, which leads to slower convergence of first-order methods, and possibly numerical issues. The conditioning of a matrix W is defined as $\kappa(W) = \frac{\sigma_{\max}(W)}{\sigma_{\min}(W)}$ where $\sigma_{\min}(W)$ and $\sigma_{\max}(W)$ are the smallest and largest singular values of W, respectively. Using the inequalities $\sigma_{\max}(W) \geq \max_j \|W(:,j)\|_2$ and $\sigma_{\min}(W) \leq \min_j \|W(:,j)\|_2$, we obtain

$$\kappa(W) \geq \frac{\max_j \|W(:,j)\|_2}{\min_j \|W(:,j)\|_2} \geq \frac{\|W(:,k)\|_2}{\|W(:,\ell)\|_2} \gg 1.$$

4.3.3.4 ▪ Min-vol NMF (3): $W^\top e = e$

The authors in [310] (2019) formulate a min-vol NMF model using the condition $W^\top e = e$ that normalizes the columns of W instead of the rows of H.

Definition 4.46 (Min-vol NMF (3)). *Given a matrix $X \in \mathbb{R}^{m \times n}$ and a factorization rank r, min-vol NMF (3) is the regularized Exact (N)MF problem with*

$$\mathcal{W} = \{W \in \mathbb{R}^{m \times r} \mid W^\top e = e\}, \quad \mathcal{H} = \mathbb{R}_+^{r \times n}, \quad \text{and} \quad f(W,H) = \det(W^\top W). \tag{4.29}$$

Min-vol NMF (3) does not seem like a significant modification compared to min-vol NMF (2), scaling W instead of H. However, given that the volume of the convex hull of W and the origin is minimized, this prevents the two drawbacks of min-vol NMF (2) explained above: this normalization balances the importance of the columns of W and leads to more well-conditioned W. Our experience has shown that the normalization $W^\top e = e$ performs much better and is more numerically stable in noisy settings. This min-vol model with the constraint $W^\top e = e$ was considered earlier in [513] but no identifiability guarantees were provided.

Theorem 4.47. *[310, Theorem 1] Let $X = WH$ where $H \in \mathbb{R}_+^{r \times n}$ satisfies the SSC, $W^\top e = e$, and $r = \text{rank}(X)$. Then (W,H) is the unique solution to min-vol NMF (3) of X of size r.*

Proof. The proof follows the same steps as the proof of Theorem 4.43; there are only a few differences due to the different normalizations.

By assumption, (W,H) is a feasible solution of min-vol NMF (3), that is, of (4.20) with $\mathcal{W}$, $\mathcal{H}$, and f as defined in (4.29). Let $(\bar{W},\bar{H})$ be another feasible solution of min-vol NMF (3). Since $r = \text{rank}(X)$, there exists an r-by-r invertible matrix Q such that

$$\bar{W} = WQ^{-1} \quad \text{and} \quad \bar{H} = QH;$$

see Lemma 4.4. Since $W^\top e = \bar{W}^\top e = e$,

$$e = \bar{W}^\top e = Q^{-\top} W^\top e = Q^{-\top} e. \tag{4.30}$$

Multiplying $Q^{-\top} e = e$ by $Q^\top$ leads to $e = Q^\top e$.

For the same reasons as in Theorem 4.43, $\text{cone}(Q^\top) \subseteq \mathcal{C}^*$, that is, Q satisfies (4.26). Therefore,

$$
|\det(Q)| \le \prod_{j=1}^{r} \|Q(j,:)\|_2 \le \prod_{j=1}^{r} Q(j,:)e \le \left(\frac{\sum_{j=1}^{r} Q(j,:)e}{r} \right)^r = \left(\frac{e^\top Q e}{r} \right)^r = 1,
$$

where the first inequality follows from the Hadamard's inequality, the second from (4.26), and the third from the arithmetic-geometric mean inequality, and the equality follows from (4.30).

The remainder of the proof is exactly the same as in Theorem 4.43.   □

The normalization $W^\top e = e$ slightly weakens the generality of the identifiability results from min-vol NMF (1)–(2), even in noiseless conditions. In Theorems 4.43 and 4.45, the only constraint on W is the implicit constraint that $r = \text{rank}(W)$. Therefore, we could have $e^\top W(:,j) \le 0$ for some j, in which case W cannot be scaled to satisfy $W^\top e = e$ while H remains nonnegative. However, if $W \ge 0$ and $r = \text{rank}(W)$, the scaling $W^\top e = e$ can be assumed w.l.o.g., and hence the identifiability result is equivalent in noiseless conditions to the ones using $H^\top e = e$ and $He = e$. Additionally, the condition $W \ge 0$ can actually be relaxed to $X \ge 0$. If $X = WH \ge 0$, then $\text{conv}(W^\top) \subseteq \text{cone}^*(H) \subseteq \mathcal{C}^*$ (Corollary 4.7 combined with SSC1 and duality), hence $e^\top W(:,j) \ge \|W(:,j)\|_2 > 0$ for all j as $r = \text{rank}(W)$.

4.3.3.5 ▪ Comparison of min-vol NMF models in the presence of noise

Min-vol NMF (1)–(2) lead to the same identifiability result in the absence of noise (after scaling of the input matrix), while min-vol NMF (3) is slightly less general as it requires $W^\top e = e$ which can be assumed w.l.o.g. when $X = WH \ge 0$; see the discussion after Theorem 4.47.

In noisy settings, we are looking for (W, H) such that $WH \approx X$ and W has a small volume, and hence one needs to balance the data fitting term $D(X, WH)$ and the volume regularization. For example, a standard min-vol NMF formulation that has been shown to be very successful in practice is the following:

$$
\min_{W \in \mathcal{W}, H \in \mathcal{H}} D(X, WH) + \lambda \, \text{logdet}(W^\top W + \delta I_r), \tag{4.31}
$$

where $\lambda > 0$ is a parameter balancing the data fitting and minimum-volume terms, and $\delta > 0$ is a small parameter preventing $\text{logdet}(W^\top W + \delta I_r)$ from going to $-\infty$ when $\text{rank}(W) < r$ (see Example 8.11 for the description of an algorithm for tackling this problem). Using the logarithm of $\det(W^\top W)$ has been shown to produce better results compared to $\det(W^\top W)$, being less sensitive to very small and very large singular values of W [172, 12]. In fact,

$$
\det(W^\top W + \delta I) = \prod_{i=1}^{r} \left(\sigma_i^2(W) + \delta \right),
$$

while

$$
\text{logdet}(W^\top W + \delta I) = \sum_{i=1}^{r} \log \left(\sigma_i^2(W) + \delta \right).
$$

Moreover, it has been shown that the logarithm of the determinant is closely related to a generative model where the columns of H are generated following a Dirichlet distribution [360].

In this context, min-vol NMF (1) is rather restrictive because it imposes that the columns of H have unit ℓ_1 norm, or requires the input matrix to be preprocessed if this assumption is not satisfied; see the discussion in Section 4.3.3.2 and in [169]. In many applications, this constraint is not satisfied, for example in audio source separation or in text mining. However, in practice, most algorithms developed so far have focused on min-vol NMF (1), most likely due to the following reasons:

- In hyperspectral unmixing where most of the literature on min-vol NMF can be found, the so-called sum-to-one constraint $H^\top e = e$ agrees with the linear mixing model (see Figure 1.5, page 8) since the entries of H correspond to the abundances of the materials in the pixels.

- The intuition behind min-vol NMF came about with the interpretation of NMF in terms of nested convex hulls. The constraints $He = e$ and $W^\top e = e$ combined with the min-vol NMF are less intuitive. In both cases, the columns of W form a cone containing $\text{cone}(X)$, while the volume of their convex hull with the origin is minimized. With the constraint $He = e$, it is difficult to interpret $\text{conv}([W, 0])$ since the norms of the columns of W can be arbitrarily large or small. On the contrary, with $W^\top e = e$, the columns of W are on the affine set $\{x \mid e^\top x = 1\}$, and hence minimizing the volume of $\text{conv}([W, 0])$ makes more sense.

- Min-vol NMF (2-3) are much more recent models. Note, however, that the model with the constraints $W^\top e = e$ was considered relatively early in [513] (2011) but has not attracted much attention, possibly due to the reasons above and because no identifiability guarantees were provided.

Compared to min-vol NMF (2), min-vol NMF (3) has two important practical advantages, as already pointed out above:

- normalizing the columns of W better balances their importance in the volume regularization term, and

- it corresponds to W matrices which are better conditioned and hence leads to more stable numerical algorithms.

For example, in hyperspectral imaging, assume an endmember (say, the kth) is present in a much higher proportion than the others (for example the grass in the Urban image; see Figure 1.6). In that case, the rank-one factor $W(:, k)H(k, :)$ corresponding to this endmember has a much higher norm than the others. Using the normalization[39] $He = e$ of min-vol NMF (2) makes the corresponding column of W have a much larger norm than the other columns and hence makes W ill-conditioned; see Example 4.48 below. The authors of [310] observed empirically on audio source separation problems that the condition number of W was growing to values larger than 10^{16}; this is what motivated them to consider the constraint $W^\top e = e$ instead. The constraint $W^\top e = e$ balances the importance of the columns of W while bounding the condition number of W as follows [310]:

$$
\begin{aligned}
\kappa(W^\top W + \delta I) &= \frac{\sigma_{\max}(W^\top W + \delta I)}{\sigma_{\min}(W^\top W + \delta I)} \\
&= \frac{\sigma_{\max}(W)^2 + \delta}{\sigma_{\min}(W)^2 + \delta} \\
&\leq \frac{\left(\sqrt{r} \max_k \|W(:, k)\|_2\right)^2 + \delta}{\delta} \leq 1 + \frac{r}{\delta},
\end{aligned}
$$

---

[39] In the paper [169], it is shown that the normalization $He = e$ can be replaced with $He = \rho e$ for any $\rho > 0$, but this degree of freedom does not resolve the issue of W being ill-conditioned (it simply allows the entries of W to be scaled by a constant which does not modify its condition number).

where we used the facts that

- $\sigma_{\max}(W) = \max_{\|x\|_2 \le 1} \|Wx\|_2 \le \sqrt{r} \max_k \|W(:,k)\|_2$, since

$$\sigma_{\max}^2(W) \le \|W\|_F^2 = \sum_k \|W(:,k)\|_2^2 \le r \max_k \|W(:,k)\|_2^2;$$

- $\|W(:,k)\|_2 \le \|W(:,k)\|_1 = 1$ for $W \ge 0$.

In the following example, we observe the behavior of the different min-vol NMF models on the Urban hyperspectral image.

Example 4.48 (Urban hyperspectral image). Let us apply the three min-vol NMF models on the Urban hyperspectral image (see Figure 1.6). We performed 1000 iterations of the optimization scheme that optimizes W and H alternately as described in Example 8.11 to solve (4.31) with 10 inner iterations for the updates of W and H, and with $\delta = 0.1$. We tune the penalty parameter λ in (4.31) so that all models achieved a relative error of about 5%, that is, so that $\frac{\|X-WH\|_F}{\|X\|_F} \approx 0.05$, while a separable NMF algorithm (namely SNPA) is used as an initialization (see Section 7.4.4). Figure 4.12 reports the spectral signatures extracted (columns of W) and Figure 4.13 shows the corresponding abundance maps (reshaped rows of H).

We observe the following:

- The solution computed by min-vol NMF (2) has one of the columns of W close to zero (the fourth one), while the first and third columns have very large norms. As explained in the previous paragraph and in the discussion after Theorem 4.45, the reason is that min-vol NMF (2) gives more importance to endmembers present in large proportions (here, the grass and trees) because of the constraint $He = e$. For this reason, the matrix $W^\top W + \delta I$ is ill-conditioned: MATLAB issued a warning while running the algorithm as this matrix needs to be inverted in the algorithm used; see Example 8.11.

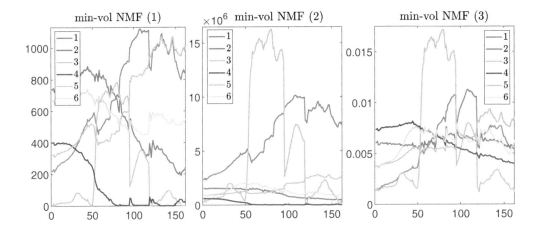

Figure 4.12. *Spectral signatures (that is, columns of W) extracted by min-vol NMF. From left to right: min-vol NMF (1), min-vol NMF (2), and min-vol NMF (3). The abundance maps corresponding to these spectral signatures are shown in Figure 4.13, respecting the same ordering.*

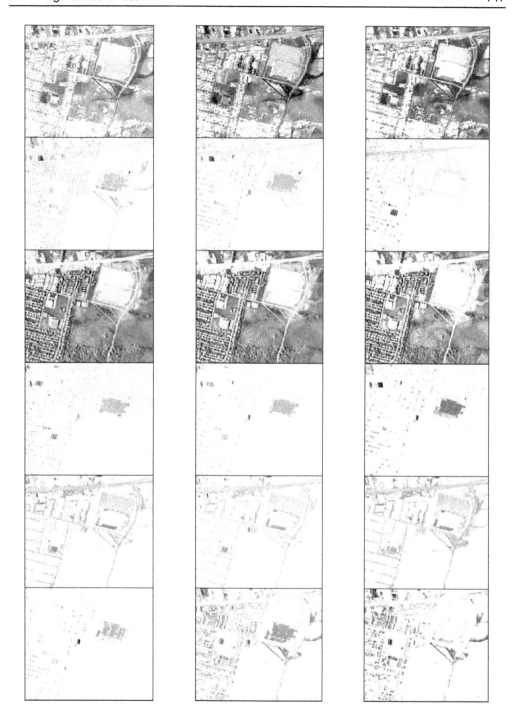

Figure 4.13. *Abundance maps (that is, reshaped rows of H) extracted by min-vol NMF. From left to right: min-vol NMF (1), (2), and (3). The spectral signatures corresponding to these abundance maps are shown in Figure 4.12, respecting the same ordering.* [`Matlab file: minvolNMF_Urban.m`].

- For min-vol NMF (1), the third and fourth extracted spectral signatures are not realistic as many entries are close or equal to zero, similarly as for min-vol NMF (2). Moreover, the abundance maps are denser and noisier than that of min-vol NMF (3); in particular the second and fourth abundance maps. This is an indication that the extracted spectral signatures do not correspond to well-defined materials.

- Min-vol NMF (3) provides very good results. All spectral signatures are meaningful and correspond well to results found in the literature; see for example [514]. Although the abundance maps do not separate all materials perfectly (see Figure 1.6 for a comparison), they are spatially coherent and rather sparse. The first one corresponds to grass, the second to road, the third to trees and grass, the fourth to roof tops 2, the fifth to road and dirt, and the sixth to roof tops 1 and dirt.

Hence, on this hyperspectral image, min-vol NMF (3) provides the best results among the three min-vol NMF models. This is the typical behavior we have observed on several data sets.   ∎

Remark 4.5 (Link with the nuclear norm). *The nuclear norm of a matrix is the sum of its singular values (see also Section 3.4.8). In particular,*

$$\|W\|_* = \sum_{i=1}^{r} \sigma_i(W)$$

for an r-by-m matrix W with $m \geq r$. It is a widely used surrogate for the rank function in order to obtain low-rank solutions; see [394] and the references therein. The rationale is that the nuclear norm of a matrix is the ℓ_1 norm of the vector of singular values, while the rank is its ℓ_0 norm. Considering the minimum-volume regularizer

$$g(W) = \operatorname{logdet}(W^\top W + \delta I) = \sum_{i=1}^{r} \log\left(\sigma_i^2(W) + \delta\right),$$

one can check that the function $g(W)$ is a sharper (but nonconvex) surrogate (after a proper scaling and translation) for the rank function than the nuclear norm for δ sufficiently small; see Figure 4.14. Hence the function $g(W)$ can also be used when looking for an m-by-r matrix W whose rank is smaller than r as done in [156, 350]. This would be useful for example for multispectral images for which the number of materials r can be larger than the number of spectral bands m [309]. Note that using the nuclear norm for a min-vol NMF model provides reasonable results but does not perform as well as $g(W)$ [12].

4.3.3.6 ▪ Can min-vol NMF be solved efficiently?

If there are no constraints on the input matrix X, finding the convex hull $\operatorname{conv}(W)$ with r vertices of minimum volume containing $\operatorname{conv}(X)$ is NP-hard [372]. Therefore, although the SSC relaxes the separability condition, it comes at a cost: the corresponding optimization problem is harder to solve. Moreover, as far as we know, the behavior of min-vol NMF in the presence of noise is not well-understood, while there are quite a few theoretical results in this direction for separable NMF (Chapter 7). Understanding min-vol NMF in the presence of noise is an important direction of research.

Another important direction of research is to identify when min-vol NMF is solvable in polynomial time. In particular, it was conjectured that it is the case when H is sufficiently scattered [170]. Proving this conjecture, and providing an efficient algorithm to solve min-vol NMF

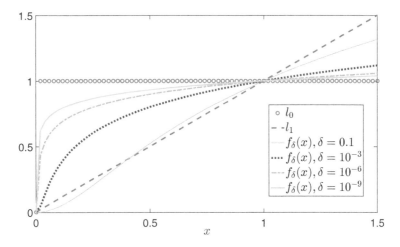

Figure 4.14. *Function $f_\delta(x) = \frac{\ln(x^2+\delta)-\ln(\delta)}{\ln(1+\delta)-\ln(\delta)}$ for different values of δ, ℓ_1 norm (= $|x|$), and ℓ_0 norm (= 0 for $x = 0$, = 1 otherwise). This show that $\ln(x^2 + \delta)$ becomes a sharper surrogate to the ℓ_0 norm as δ decreases (up to a constant term, and scaling). Figure adapted from [309].*

under this assumption, would be a key result in the NMF literature. Note that the former does not necessarily imply the latter (for example, Exact NMF can be solved in polynomial time for r fixed, but it remains to provide a practical algorithm to do so even for 4-by-4 matrices; see the discussion around Theorem 2.21, page 52). Also, adapting such algorithms in the presence of noise would be necessary for real-world applications.

An important effort in this direction is the algorithm proposed in [321]: the authors provide a provably correct algorithm for min-vol NMF (1). The idea is to first compute the maximum-volume ellipsoid contained in $\text{conv}(X)$, which can be formulated as a semidefinite program. If H is sufficiently scattered, they can show that this ellipsoid touches each facet of $\text{conv}(W)$, which allows them to compute its vertices. However, this approach requires all facets of $\text{conv}(X)$ to be computed (each of them corresponds to a constraint in the semidefinite program), and there might be exponentially many facets for a polytope with n vertices in dimension m; see Equation (3.9), page 74. Hence their algorithm does not run in polynomial time. However, they are able to solve middle-scale problems, with $m = 224$, $n = 1000$, and $r = 8$, in a few minutes. This idea has been recently revisited, and a new algorithm has been proposed leading to better performances [319].

For large-scale real-world problems, practitioners rely on formulations of the form (4.31) and use standard nonlinear optimizations strategies. A typical approach is to optimize alternatively over the variables W for H fixed and vice versa. The problem in H is convex, while the problem in W is nonconvex, because of the term $\text{logdet}(W^\top W)$, but a convex quadratic auxiliary function (that is, an upper bound which is tight at the given iterate) can be easily constructed [172, 309]; see Example 8.11 in Section 8.1.3.

4.3.3.7 ▪ Is the SSC on $W^\top$ or H reasonable in practice?

In [170], the authors conjecture that the SSC is also necessary for min-vol NMF to be unique. This is an important open question. Note, however, that this conjecture does not hold if the model incorporates the constraint $W \geq 0$; see the matrix in (4.18) for a counterexample.

In any case, the SSC on one of the two factors in NMF is arguably rather mild. It drastically relaxes the separability condition and was shown to be very powerful in many applications; see the introduction of this section. It only requires that one of the two factors in NMF is sufficiently sparse; see Section 4.2.3.4. This sparsity assumption on one of the two factors holds in most applications; see the discussion in the beginning of the following section.

4.3.4 ▪ Sparse NMF

Sparse approximations have attracted a lot of attention in the last two decades, including sparse low-rank approximations [505], sparse PCA [115], and dictionary learning [4], to cite a few.

Although NMF naturally leads to sparse factors W and H (see Section 1.2), one may want to enhance this property of NMF. For example, let us look back at the four applications from Section 1.3 to see how sparsity arises:

- Facial features extraction: each facial image contains a small subset of all possible facial features (H is sparse). Moreover, facial features are located within a small region of the image, hence activating only a few pixels (W is sparse); see Figure 1.2 (page 7).

- Hyperspectral unmixing: each pixel contains a small number of materials (typically, at most 5) making H sparse; see Figure 1.6.

- Text mining: most documents typically discuss only a few topics (H sparse) while most topics use only a small subset of words within the dictionary (W sparse).

- Audio source separation: most sources are not active during all time windows (H sparse) while the signature of most sources does not cover the full frequency domain (W sparse); see Figure 1.8 (page 11).

Variants of NMF requiring sparse W and/or H are referred to as sparse NMF. Sparse NMF was introduced early on because it enhances the ability of NMF to learn a parts-based representation and produces more easily interpretable factors. For example, in facial feature extraction, sparsity leads to more localized features, while fewer features are used to reconstruct each input image. Moreover, as we will show in this subsection, sparsity leads to identifiable solutions. The first landmark paper on sparse NMF was by Hoyer [243] (2004). Hoyer introduced the following measure of sparsity based on the ratio between the ℓ_1 and ℓ_2 norms: for $x \in \mathbb{R}^n$ and $x \neq 0$,

$$\text{spar}(x) = \frac{\sqrt{n} - \frac{\|x\|_1}{\|x\|_2}}{\sqrt{n} - 1} \in [0, 1]. \tag{4.32}$$

We have that $\text{spar}(x) = 0$ if and only if $\|x\|_1 = \sqrt{n}\|x\|_2$ so that all entries of x are equal to one another. Also, $\text{spar}(x) = 1$ if and only if $\|x\|_1 = \|x\|_2$ and hence $\|x\|_0 = 1$, where $\|x\|_0$ counts the number of nonzero entries of x. Hoyer then proposed an efficient algorithm to project a vector onto a set of given sparsity, used this projection within an NMF algorithm, and applied it for facial feature extraction. Many other algorithms and applications were later considered, including audio source separation [471], bioinformatics [277], and hyperspectral unmixing [387].

Rather surprisingly, identifiability results for sparse NMF are very scarce. We are only aware of the result by Theis, Stadlthanner, and Tanaka [448], which has rather strong conditions. However, in the dictionary learning literature, many such results were obtained. Most works focus on models where the positions of the zero entries and the values of the nonzero entries of H are randomly generated. Also, most works focus on the overcomplete case, that is, $r \gg n$. It is beyond the scope of this book to present these results; we refer the interested reader to [220] and the references therein. Instead, we focus on a relatively simple recent result that applies in a deterministic scenario in the undercomplete case, that is, $m \geq r$ and $r = \text{rank}(W)$, which has been our focus so far. As we will see, this result allows us to shed some light on min-vol NMF in comparison with sparse NMF.

This section is organized as follows. We first define a particular sparse NMF model, referred to as k-sparse matrix factorization (MF) (Section 4.3.4.1). Then we provide the identifiability result for this model (Section 4.3.4.2), which we compare to min-vol NMF in Section 4.3.4.3.

4.3.4.1 ▪ k-sparse MF

Let us first define the sparse NMF variant for which we present an identifiability result. Recall that a vector is k-sparse if it has at most k nonzero entries.

Definition 4.49 (k-sparse MF). *Given a matrix $X \in \mathbb{R}^{m \times n}$ and the integers $k \leq r$, the k-sparse MF problem is the regularized Exact (N)MF problem with $f = 0$,*

$$\mathcal{W} = \mathbb{R}^{m \times r}, \quad \text{and} \quad \mathcal{H} = \{H \in \mathbb{R}^{r \times n} \mid H(:, j) \text{ is } k\text{-sparse for all } j\}.$$

As for separable, orthogonal, and min-vol NMF, W does not need to be nonnegative. Moreover, H does not need to be nonnegative either.

4.3.4.2 ▪ Identifiability of k-sparse MF

Before providing the identifiability result, let us introduce some notation and two useful lemmas. The Kruskal rank of a matrix X, denoted k-rank(X), is the maximum r such that any subset of r columns of X are linearly independent [289]. We have k-rank$(X) \leq \text{rank}(X)$, but k-rank(X) can be arbitrarily smaller than rank(X) (in particular, k-rank$(X) = 0$ if X contains a zero column). The Kruskal rank is closely related to the notion of *spark*, which is the smallest p such that p columns of X are linearly dependent, widely used in the dictionary learning literature, so that the spark of X is equal to k-rank$(X) + 1$.

Given $W \in \mathbb{R}^{m \times r}$, let us denote, for $j = 1, 2, \ldots, r$,

$$\mathcal{F}_j(W) = \text{col}(W(:, \mathcal{J})) \quad \text{where} \quad \mathcal{J} = \{1, 2, \ldots, r\} \backslash \{j\}.$$

We now state a simple lemma showing that identifying the r columns of W is equivalent to identifying the r subspaces $\{\mathcal{F}_j(W)\}_{j=1}^r$ spanned by $r - 1$ columns of W.

Lemma 4.50. *Let $W \in \mathbb{R}^{m \times r}$ and $W' \in \mathbb{R}^{m \times r}$ be full column rank. If there exists a permutation π such that*

$$\mathcal{F}_j(W) = \mathcal{F}_{\pi(j)}(W') \text{ for } j = 1, 2, \ldots, r,$$

then W and W' are equal to each other, up to permutation and scaling of their columns.

Proof. This follows directly from a simple linear algebra argument since, for all j,

$$\text{col}(W(:, j)) = \cap_{i \neq j} \mathcal{F}_i(W)$$

for any m-by-r matrix W of rank r.   □

Let us now prove a key lemma. It states that if there are sufficiently many data points on one of the subspaces $\{\mathcal{F}_j(W)\}_{j=1}^r$, then it must be identified in any sparse factorization.

Lemma 4.51. *[102, Lemma 3.5] Let $A \in \mathbb{R}^{m \times p}$ be such that* $\mathrm{rank}(A) = \mathrm{k\text{-}rank}(A) = r - 1$. *Let also $A = WH$ where $W \in \mathbb{R}^{m \times r}$ has rank r, and the columns of H are k-sparse. Then*

$$p \geq \left\lfloor \frac{r(r-2)}{r-k} \right\rfloor + 1 \quad \Rightarrow \quad \text{there exists } j \text{ such that } \mathcal{F}_j(W) = \mathrm{col}(A).$$

Proof. For simplicity, let us denote $\mathcal{F}_j = \mathcal{F}_j(W)$. For $j = 1, 2, \ldots, r$, let us define

$$S_j = \{A(:,i) \mid A(:,i) \in \mathcal{F}_j\}$$

so that S_j is the set of columns of A contained in the jth hyperplane $\mathcal{F}_j$ generated by W. If there exists some j such that $|S_j| \geq r-1$, then $\mathrm{col}(S_j) = \mathcal{F}_j$ since, by assumption, $\mathrm{k\text{-}rank}(A) = r-1$ and the dimension of $\mathcal{F}_j$ is $r - 1$. In other words, any hyperplane $\mathcal{F}_j$ containing strictly more than $r - 2$ data points of A satisfies $\mathcal{F}_j = \mathrm{col}(A)$. Moreover, every column of A lies on at least $r - k$ subspaces $\{\mathcal{F}_j\}_{j=1}^r$ since $\|H(:,i)\|_0 \leq k$ for all i. Hence one can check using the pigeonhole principle that the maximum number of columns that A can contain such that each of the r hyperplanes generated by W contains at most $r - 2$ columns of A is given by

$$p_{\max} = \left\lfloor \frac{r(r-2)}{r-k} \right\rfloor.$$

Hence $p \geq p_{\max} + 1 > p_{\max}$ implies $\mathcal{F}_j = \mathrm{col}(A)$ for some j, which completes the proof. □

We can now state an identifiability theorem for k-sparse MF.

Theorem 4.52. *Let $X = WH$ where the columns of $H \in \mathbb{R}^{r \times n}$ are k-sparse and $\mathrm{rank}(W) = \mathrm{rank}(X) = r$, and assume there exist subsets $\{\mathcal{I}_j\}_{j=1}^r$ such that*

(i) $\mathcal{I}_j \subseteq \{i \mid H(j,i) = 0\}$, *so that the columns of $X(:,\mathcal{I}_j)$ belong to $\mathcal{F}_j(W)$,*

(ii) $\mathrm{k\text{-}rank}(X(:,\mathcal{I}_j)) = r - 1$ *for $j = 1, 2, \ldots, r$, and*

(iii) $|\mathcal{I}_j| \geq \left\lfloor \frac{r(r-2)}{r-k} \right\rfloor + 1$ *for $j = 1, 2, \ldots, r$.*

Then (W, H) is the unique solution to the k-sparse MF of X of size r.

Proof. By (i) and (ii), $\mathcal{F}_j(W) = \mathrm{col}(X(:,\mathcal{I}_j))$ for all j. Uniqueness of W follows from Lemma 4.51 using $A = X(:,\mathcal{I}_j) = WH(:,\mathcal{I}_j)$. The conditions (ii) and (iii) imposed on the $\mathcal{I}_j$'s imply that $\mathrm{rank}(A) = \mathrm{k\text{-}rank}(A) = r - 1$ so that for any k-sparse MF (W', H') of $X = W'H'$, $\mathcal{F}_j(W') = \mathrm{col}(X(:,\mathcal{I}_j)) = \mathcal{F}_j(W)$ for all j (up to permutation of the columns of W'). Lemma 4.50 allows us to conclude that identifying the column of W or its subspaces $\mathcal{F}_j(W)$ is equivalent.

Uniqueness of H follows from the assumption that $\mathrm{rank}(W) = r$ implying $r = \mathrm{rank}(W)$, and hence there is a unique H such that $X = WH$. □

Theorem 4.52 tells us that a k-sparse MF $X = WH$ is unique if on each subspace spanned by all but one column of W, there are $\left\lfloor \frac{r(r-2)}{r-k} \right\rfloor + 1$ columns of X with Kruskal rank[40] $r - 1$. For example, for $r = 3$ and $k = 2$, Theorem 4.52 guarantees that having four distinct points

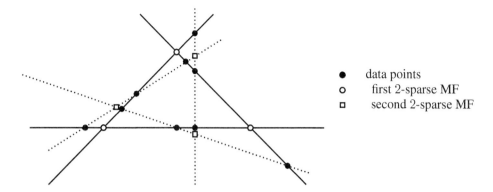

Figure 4.15. *A scenario where k-sparse MF is not unique for $r = 3$ and $k = 2$: each data point is a linear combination of two out of the three columns of W. The data points and the subspaces are projected onto the affine space $\{x \mid e^\top x = 1\}$ so that we may visualize them in two dimensions. The k-sparse MF of these data points would be unique by adding a single point on any of the subspaces [102, Theorem 3.8]. Figure adapted from [102, Figure 1].*

on each subspace spanned by two columns of W makes k-sparse MF unique. However, having three points on each subspace is not always enough; see Figure 4.15 for an illustration.

Theorem 4.52 can be used to compute a minimum value of the number n of columns of a matrix X to have a unique k-sparse MF. Since each column of X belongs to $r - k$ subspaces among $\{\mathcal{F}_j(W)\}_{j=1}^r$, it may belong to $r - k$ subsets $\mathcal{I}_j$'s, and hence the condition (4.52) implies that

$$n \geq \frac{\sum_{j=1}^r |\mathcal{I}_j|}{r - k} \geq \frac{r}{r - k} \left(\left\lfloor \frac{r(r - 2)}{r - k} \right\rfloor + 1 \right).$$

For example, for $r = 3$ and $k = 2$, we need $n = 12$ points.

The conditions of Theorem 4.52 can be relaxed; see the discussion in [102]. Also, it could be improved by taking nonnegativity of W and/or H into account; see the discussion in [103]. This is a topic for future research.

4.3.4.3 • Sparse vs. minimum-volume NMF

The SSC that makes min-vol NMF identifiable requires some degree of sparsity; see Theorem 4.28, and see Section 4.3.4.3 for a discussion. Moreover, minimizing the volume of W in min-vol NMF leads to Exact NMFs with a certain degree of sparsity for H. For example, it can be easily checked that the support of the rows of H in a min-vol NMF solution form an antichain (otherwise, a solution with smaller volume can be constructed using the same construction as in Theorem 4.34).

When the columns of H are k-sparse, the SSC guaranteeing min-vol NMF to succeed requires H to have at least $\frac{r(r-1)}{r-k}$ columns (see the discussion after Theorem 4.28). Sparse NMF requires more columns, namely $\mathcal{O}(r^3/(r - k)^2)$ columns (Theorem 4.52). However, the condition on the nonzero entries is milder for sparse NMF; in particular, the Kruskal rank condition is satisfied with probability one if the nonzero entries of H are generated randomly. This is not the case for the SSC that requires the columns of H to be sufficiently spread. This means that neither min-vol NMF nor sparse NMF is superior to the other. One should choose the right model

---

[40]We make a slight abuse of language here: the Kruskal rank of a set of columns is equal to the Kruskal rank of the matrix obtained by concatenating these columns.

depending on the application at hand. In some cases, min-vol NMF will lead to identifiability while sparse NMF will not, and vice versa. Let us illustrate this with a simple example.

Example 4.53. Let us consider the matrix from Example 4.16:

$$H(\omega) = \begin{pmatrix} \omega & 1 & 1 & \omega & 0 & 0 \\ 1 & \omega & 0 & 0 & \omega & 1 \\ 0 & 0 & \omega & 1 & 1 & \omega \end{pmatrix}.$$

For $\omega < 0.5$, H satisfies the SSC; see Figure 4.4 (page 112). Hence $X = WH(\omega)$ is identifiable using any of the min-vol NMF models and for any W of rank three as long as $\omega < 0.5$. This is not the case for 2-sparse MF: there is another solution for this particular problem; see Figure 4.4, where one can check that there is another triangle going through the six columns of H. Hence there exists another 2-sparse MF (W', H') where the columns of H' are 2-sparse (but the volume of W' is larger):

$$H = QH' = \begin{pmatrix} \omega-1 & 1 & 1 \\ 1 & \omega-1 & 1 \\ 1 & 1 & \omega-1 \end{pmatrix} \begin{pmatrix} \frac{1-\omega}{2-\omega} & 0 & 0 & \frac{1-\omega}{2-\omega} & \frac{1}{2-\omega} & \frac{1}{2-\omega} \\ 0 & \frac{1-\omega}{2-\omega} & \frac{1}{2-\omega} & \frac{1}{2-\omega} & \frac{1-\omega}{2-\omega} & 0 \\ \frac{1}{2-\omega} & \frac{1}{2-\omega} & \frac{1-\omega}{2-\omega} & 0 & 0 & \frac{1-\omega}{2-\omega} \end{pmatrix},$$

so that $W'H' = (WQ)H'$ is another 2-sparse MF, where $\det(Q) = \omega^3 - 3\omega^2 + 4 > 1$ for $\omega \in [0,1)$.

Now, let us consider the matrix

$$H(\omega_1, \omega_2) = \begin{pmatrix} H(\omega_1) & H(\omega_2) \end{pmatrix}.$$

For any choice of $\omega_1 \neq \omega_2$ such that $\min(\omega_1, \omega_2) \geq 0.5$, $H(\omega_1, \omega_2)$ does not satisfy the SSC, and min-vol NMF is not identifiable (see Example 4.16, from which one can construct other min-vol NMFs). Looking at Figure 4.4, the intuition is that adding 2-sparse columns to H is useless to make min-vol NMF identifiable as long as these columns are not sufficiently spread within $\mathbb{R}^3_+$. On the other hand, sparse NMF is identifiable for any $\omega_1 \neq \omega_2$: each two-dimensional subspace $\mathcal{F}_j(W)$ contains four points that have Kruskal rank two. ∎

In summary, sparse NMF does not require the data points to be sufficiently scattered within $\text{cone}(W)$ but requires that there are enough of them on the different facets generated by $r - 1$ columns of W.

From a practical point of view, our experience has shown that sparse NMF is a more difficult problem to handle than min-vol NMF. As mentioned above, under the SSC, it appears that min-vol NMF algorithms provide good estimates of the ground truth solutions, even in the presence of noise. As far as we know, no such observations have been made for sparse NMF. Imposing the columns of H to be k-sparse is a problem of a combinatorial nature, while min-vol NMF is a continuous optimization problem; see for example (4.31). Also, in most applications, the levels of sparsity of the columns of H are different; hence choosing a single value of k and imposing all columns of H to be k-sparse is not often reasonable. In practice, most algorithms rely on strategies that add a sparsity-promoting penalty term in the objective such as the ℓ_1 norm [277] or try to minimize the data fitting error $D(X, WH)$ while achieving a certain degree of sparsity [243].

4.3.4.4 ▪ Sparse NMF via facet identification

More recently, several models and algorithms with identifiability guarantees have been proposed for sparse NMF that rely on identifying the facets of $\text{conv}(W)$:

- In [183], Ge and Zou identify facets of $\mathrm{conv}(W)$ using the fact that a data point can be reconstructed using convex combinations of points within the same facet. For identifiability, they require that, among the facets of $\mathrm{conv}(X)$, only the facets of $\mathrm{conv}(W)$ contain points in their relative interior.

- In [2], authors identify the facets of $\mathrm{conv}(W)$ by selecting the facets of $\mathrm{conv}(X)$ containing the largest number of columns of X. To obtain identifiability, it is only required that $d \geq \mathrm{rank}(X)$ data points are present on each facet of $\mathrm{conv}(W)$, while no facet of $\mathrm{conv}(X)$ which is not a facet of $\mathrm{conv}(W)$ contains more than d points.

We refer the interested reader to [2] and the references therein for more information on such approaches.

4.3.5 ▪ Summary of identifiability for regularized Exact (N)MF

In terms of generality, we have the following inclusions:

$$\text{Exact ONMF} \subseteq \text{separable NMF} \subseteq \text{min-vol NMF} \quad \text{(1-3)}.$$

Sparse NMF generalizes Exact ONMF where all columns of H are 1-sparse, but not separable NMF, which only requires r columns of H to be 1-sparse while the others are unconstrained. As we have shown, neither sparse NMF nor min-vol NMF dominates the other.

However, min-vol NMF and sparse NMF are difficult optimization problems to be solved, as opposed to Exact ONMF and separable NMF. Table 4.2 summarizes the identifiability results discussed in this chapter.

Table 4.2. *Summary of identifiability/uniqueness results for various regularized Exact (N)MF models of the form (4.20) for $r = \mathrm{rank}(X)$, that is, $X = WH$ where $W \in \mathbb{R}^{m \times r}$ and $H \in \mathbb{R}^{r \times n}$, implying $\mathrm{rank}(W) = \mathrm{rank}(H) = r$. That is, solutions to (4.20) are unique when each model satisfies the corresponding conditions. When no further conditions are imposed on W or H, we indicate /. When no objective function is used, we also indicate /. Some of these models do not need W and/or H to be nonnegative, hence these identifiability results apply to a broader class than NMF problems; this explains the notation (N)MF. Note that separability (Definition 4.10) and SSC (Definition 4.15) require nonnegativity.*

Model	$\mathcal{W}$	$\mathcal{H}$	$f(X, W, H)$	Theorem
Exact NMF	SSC†	SSC†	/	4.21
separable NMF	$W = X(:, \mathcal{K})$	H separable	/	4.37
Exact ONMF	/	$HH^\top = I_r, H \geq 0$	/	4.40
min-vol NMF (1)	/	SSC, $H^\top e = e$	$\det(W^\top W)$	4.43
min-vol NMF (2)	/	SSC, $He = e$	$\det(W^\top W)$	4.45
min-vol NMF (3)	$W^\top e = e$	SSC	$\det(W^\top W)$	4.47
k-sparse MF	/	k-sparse columns	/	4.52

†One of these two conditions can be relaxed to SSC1 (Corollary 4.22).

4.3.6 ▪ Other models and further research

As we will see in Chapter 5, there exist numerous other regularized NMF models. Such models are typically designed to take into account the prior information of the application at hand, which leads to better practical performance. This allows the search space to be reduced and hence leads, in many cases, to the sought decomposition. However, for many of them, no identifiability results are available. Therefore, a topic for further research is to provide identifiability guarantees for other regularized Exact (N)MF models. Another important topic for further research is to analyze the sensitivity of these models in the presence of noise (Chapter 7 discusses the case of separable NMF). Also, it is important to keep in mind that uniqueness does not necessarily lead to tractability. For example, solving sparse NMF is hard; we "only" have the guarantee that, under some conditions, the globally optimal solution is unique. An interesting recent work in these directions is a recovery guarantee for NMF when using alternating optimization (that is, optimizing alternatively for W with H fixed and vice versa; see Chapter 8) under some generative model and assuming the initial solution is close enough to the sought solution [316].

4.4 ▪ Take-home messages

The two main take-home messages from this chapter are as follows:

1. In most cases, you should not expect Exact NMF to have a unique solution.

   The conditions for uniqueness are rather strong and do not apply for most real-world scenarios. An exception is for *simple* mixtures of audio signals where it can be expected that the sources and the activations are sufficiently scattered.

2. If in your application of interest, identifiability is crucial (such as in hyperspectral imaging or audio source separation), you should pay particular attention to your NMF formulation.

   In practice, you should therefore look for NMF solutions with additional properties as it leads to much weaker conditions for identifiability, often met in practical situations. Identifiability in this context was discussed in Section 4.3. Two notable examples are (1) the SSC on one of the two factors $W^\top$ or H, and (2) the sparsity of W and/or H. Both conditions are rather mild in the sense that they are reasonable in most applications while leading to the identifiability via min-vol and sparse NMF models, respectively.

Part II

Approximate factorizations

In the three previous chapters, we have focused on exact decompositions. In the next five chapters, we focus on approximate decompositions, and in particular on the NMF problem defined in Problem 1.1 (page 4).

In Chapter 5, we discuss NMF models. This includes the choice of the error measure and of the factorization rank. Because NMF suffers from identifiability issues (see Chapter 4), practitioners often use additional constraints and regularizers within the NMF model. These are designed depending on the particular application in order to recover the sought unique solution. We review some of these models. We also link some of these models with well-known data analysis techniques such as k-means, and PLSA/PLSI.

In Chapter 6, we address the computational complexity issue of NMF, building up on the results from Chapter 2. We discuss the influence of the error measure as well as of the use of regularizations and constraints within the model. Not surprisingly, NMF is a difficult problem, as it allows us to tackle Exact NMF as a particular case. However, as we will see, there are some differences. For example, Exact NMF for $r = 1$ is trivial (all columns of the input matrix are multiples of one another), while some NMF models are NP-hard even for $r = 1$.

A possible way to get around the two main difficulties of NMF (complexity and identifiability) is to make additional assumptions on the input matrix. This is the topic of Chapter 7, where separable NMF is discussed (recall that separable NMF requires $W = X(:, \mathcal{K})$ for some index set $\mathcal{K}$). It is an NMF model that is identifiable (see Chapter 4) and can be solved in polynomial time, even in the presence of noise.

However, because the input matrix does not necessarily admit a separable decomposition, and NMF is a hard nonconvex optimization problem in general (Chapter 6), practitioners usually rely on standard nonlinear optimization schemes to tackle NMF. These algorithms are presented in Chapter 8; they include in particular the MU popularized by Lee and Seung.

Finally, in Chapter 9, we discuss some practical issues when using NMF and present three more applications of NMF, namely, SMCR, microarray data analysis, and recommender systems.

Chapter 5

NMF models

In this book, what we call an NMF model is an optimization model that requires the choice of

- the variables (in the standard NMF model, the factors W and H),

- the objective function (such as the standard least squares error $\|X - WH\|_F^2$) with or without regularizers (such as $\|H\|_1$ to induce sparse solutions), and

- constraints on the variables (such as nonnegativity of W and H in the standard NMF model, and orthogonality with $HH^\top = I$ in the ONMF model).

Over the years, many NMF models have been introduced. The goal of this chapter is to review the most important ones and explain which model is the most appropriate in which situation. This is a key challenge when using NMF in practice.

The main motivation for designing various NMF models is identifiability: introducing prior knowledge within a model increases the chances of obtaining the sought solution. Let us quote Vincent et al. [470]:

> Guiding separation improves the accuracy of the parameter estimates, which in turn improves separation.

In the paper of Paatero and Tapper [371] (1994), it is mentioned that

> With nontrivial regularizers, the degeneracy of an NMF problem usually disappears. The solution of NMF is unique.

For example, in multispectral imaging, the number of sources, r, is sometimes larger than the ambient dimension, m. Since the nonnegative rank is bounded by the ambient dimension (Theorem 3.1), we have $\text{rank}_+(X) \leq m < r$. Hence the NMF cannot be unique without additional constraints because some columns of W and/or rows of H can be set to zero; for example one feasible solution is $X = [I_m, 0_{m \times r-m}][X; 0_{r-m \times n}]$. Therefore, one has to impose additional constraints to obtain meaningful solutions. In this case, possible options are to look for a sparse factor H (Section 4.3.4) or a factor W with minimum volume (Section 4.3.3).

Organization of the chapter This chapter is divided into five parts. The first part discusses the choice of the error measure $D(X, WH)$ in NMF which is related to the statistical properties of the noise present in the data set (Section 5.1). The second part briefly discusses the choice of the factorization rank r, also know as the order of the model (Section 5.2). Selecting the order of an NMF model is a difficult task. The third part (Section 5.3) discusses regularized NMF models, where regularization terms are added to the objective function to promote some structures on the factors. The fourth part (Section 5.4) discusses NMF variants that add constraints into the NMF model; it extends Section 4.3, where several such models were introduced, namely orthogonal, separable, min-vol, and sparse NMF. Finally, in Section 5.5, NMF is put into perspective by providing links with other data analysis models, namely (nonnegative) PCA, (spherical) k-means clustering, PLSA/PLSI, and neural networks. We conclude with take-home messages.

5.1 ▪ Error measures

As already mentioned in the introduction of this book (Chapter 1), one defining characteristic of an LRMA model is the choice of the error measure, $D(X, WH)$, used to evaluate the quality of the approximation, WH, of X. Ideally, this choice should be dictated by the statistical properties of the noise added to the low-rank matrix. However, in practice, the distribution of the noise is usually unknown, while the linear model, $X = WH$, is only a model ("all models are wrong"; see the discussion in the introduction of this book and in Section 9.2). Therefore, in most cases, one needs to rely on some strategy to decide which error measure to use. The choice of the error measure is of course not specific to NMF, nor LRMA, and is faced when designing any regression model. Note that the error measure is also referred to as the objective/loss/cost function, or the data fitting term.

This section is organized as follows. We first explain how error measures are obtained as maximum likelihood estimators of statistical distributions of the noise (Section 5.1.1). Then we focus on the choice of the error measure in practice (Section 5.1.2). Finally, we discuss in more detail beta-divergences, which are the most widely used in the NMF literature (Section 5.1.3), and for which we will present algorithms in Chapter 8.

5.1.1 ▪ Statistical model and maximum likelihood

Let us briefly recall the principle of maximum likelihood estimators. Suppose that the entry at position (i, j) of matrix X contains the observations of a random variable, $\tilde{X}_{ij}$, defined by the parameter $(\hat{W}\hat{H})_{ij}$. For example, these random variables may follow a linear mixing model with additive noise, $\tilde{X} = \hat{W}\hat{H} + \tilde{N}$, where the factors $\hat{W} \geq 0$ and $\hat{H} \geq 0$ are deterministic, but unknown, and the noise is i.i.d. Gaussian with mean 0 and standard deviation σ. In this case it is equivalent to say that the entries of $\tilde{X}$ follow a Gaussian distribution,

$$\tilde{X}_{ij} \sim \mathcal{N}\left((\hat{W}\hat{H})_{ij}, \sigma\right) \quad \text{for all } i, j \text{ and some } \sigma > 0.$$

Thus the probability density function of $\tilde{X}_{ij}$ is

$$p\left(\tilde{X}_{ij}; (\hat{W}\hat{H})_{ij}, \sigma\right) = \frac{1}{\sqrt{2\pi}\sigma} e^{-\frac{1}{2\sigma^2}(\tilde{X}_{ij} - (\hat{W}\hat{H})_{ij})^2}.$$

Since the noise is assumed to be i.i.d., the likelihood of the sample X with respect to $(\hat{W}\hat{H})_{ij}$ and σ is

$$\ell(X; \hat{W}\hat{H}, \sigma) = \prod_{i,j} p(X_{ij}; (\hat{W}\hat{H})_{ij}, \sigma). \tag{5.1}$$

Given a sample X, the unknown parameters, $\hat{W}$, $\hat{H}$, and σ, can be estimated by solving the optimization problem

$$\max_{W \geq 0, H \geq 0, \sigma} \ell(X; WH, \sigma).$$

The solution[41] to this optimization problem is known as the maximum likelihood estimator for $(\hat{W}, \hat{H})$. Most NMF models are written in a simpler form, minimizing the logarithm of the likelihood multiplied by -1 (up to some multiplicative and additive constant factors that do not influence the optimization problem) and can be written as

$$\min_{W \geq 0, H \geq 0} D(X, WH).$$

For the example above with i.i.d. Gaussian noise,

$$D(X, WH) = \sum_{i,j} (X - WH)_{ij}^2 = \|X - WH\|_F^2,$$

which is obtained by taking the logarithm of (5.1), and simplifying the expression (in particular, getting rid of the parameter $\sigma > 0$ which does not influence the estimation of $\hat{W}$ and $\hat{H}$). This is the well-known least squares objective function.

Let us list a few other important examples where we assume i.i.d. noise, unless specified otherwise. We do not explicitly derive the likelihood functions and the corresponding NMF objective functions; these can be found in standard textbooks.

- *Poisson.* The entries of $\tilde{X}$ are distributed as a Poisson distribution of parameter $(\hat{W}\hat{H})_{ij}$, that is, for all i, j,

$$\mathbb{P}(\tilde{X}_{ij} = k) = \frac{(\hat{W}\hat{H})_{ij}^k}{k!} e^{-(\hat{W}\hat{H})_{ij}} \quad \text{for } k = 0, 1, 2, \ldots.$$

This is not an additive noise. The Poisson distribution makes particular sense for count data, such as vector of word counts used in text mining (see Section 1.3.3) and this model is closely related to PLSI/PLSA (see Section 5.5.4). It is also used in imaging, motivated by the fact that capturing an image is a photon-counting process with Poisson noise; see [232] and the references therein. Note that $(\hat{W}\hat{H})_{ij} = 0$ implies $\tilde{X}_{ij} = 0$ since we have $\mathbb{P}(\tilde{X}_{ij} = 0) = 1$ for $(\hat{W}\hat{H})_{ij} = 0$. This implies that positive entries of the sample X cannot be approximated by zeros, that is, $X(i,j) > 0$ implies $(\hat{W}\hat{H})_{ij} > 0$. The maximum likelihood estimator corresponding to the Poisson distribution is obtained by minimizing the β-divergence for $\beta = 1$ between the matrices X and WH. It is defined as

$$D(X, WH) = \sum_{i,j} X_{ij} \log\left(\frac{X_{ij}}{(WH)_{ij}}\right) - X_{ij} + (WH)_{ij}$$

and is known as the Kullback–Leibler (KL) divergence between X and WH. The β-divergences are discussed in Section 5.1.3.

- *Uniform.* The entries of $\tilde{X}$ follow the distribution

$$\tilde{X}_{ij} \sim \mathcal{U}((\hat{W}\hat{H})_{ij} - u, (\hat{W}\hat{H})_{ij} + u) \text{ for all } i, j,$$

where $\mathcal{U}(a, b)$ is the uniform distribution in the interval $[a, b]$. The corresponding maximum likelihood estimator is obtained by minimizing the componentwise infinity norm[42] between X and WH, that is,

$$D(X, WH) = \|X - WH\|_\infty = \max_{i,j} |X - WH|_{ij}.$$

---

[41]Of course, computing this solution might be difficult (see Chapter 6), and the solution might not be unique (see Chapter 4), and these are two main issues in practice.

[42]In the linear algebra community, the infinity norm of matrix A is defined as $\max_{\|x\|_\infty \leq 1} \|Ax\|_\infty$. We use in this book the terminology for the componentwise infinity norm of a matrix, as is often done in the machine learning community. We make the same choice for the ℓ_1 norm.

This is an additive noise, $\tilde{X} = \hat{W}\hat{H} + \tilde{N}$ with $\tilde{N}_{ij} \sim \mathcal{U}(-u, u)$ for all i, j. Such a noise model makes sense, for example, for quantized low-rank matrices, that is, low-rank matrices whose entries are rounded up to some accuracy. For example, rounding entries of a matrix to the nearest integer is closely related to adding uniform noise in the interval $[-0.5, 0.5]$; see for example [209].

- *Laplace.* The probability density function of the Laplace distribution is

$$p\left(\tilde{X}_{ij}; (\hat{W}\hat{H})_{ij}, \sigma\right) = \frac{1}{2\sigma}e^{-\frac{1}{\sigma}|\tilde{X}_{ij} - (\hat{W}\hat{H})_{ij}|} \quad \text{for all } i, j,$$

and the corresponding maximum likelihood estimator is obtained by minimizing the componentwise ℓ_1 norm of the residual, that is,

$$D(X, WH) = \|X - WH\|_1 = \sum_{i,j} |X - WH|_{ij}.$$

This is an additive noise with $\hat{X} = \hat{W}\hat{H} + \tilde{N}$ where $\tilde{N}_{ij} \sim \mathcal{L}(0, \sigma)$ for all i, j, where $\mathcal{L}(a, b)$ is the Laplace distribution of mean a and diversity b. In practice, the componentwise ℓ_1 norm is often used in the presence of outliers as it is less sensitive to large deviations than the Frobenius norm[43] and more generally than β-divergences [273]. This problem is closely related to robust PCA; see [212] and the discussion therein.

- *Multiplicative gamma.* All models above, except for the Poisson distribution, correspond to additive noise. Let us consider the multiplicative noise model $\hat{X} = (\hat{W}\hat{H}) \circ N$ where N is the noise. If the noise N follows a gamma distribution of mean 1, the corresponding maximum likelihood estimator is obtained by minimizing the β-divergence for $\beta = 0$ between X and WH. It is defined as

$$D(X, WH) = \sum_{i,j} \frac{X_{ij}}{(WH)_{ij}} - \log\left(\frac{X_{ij}}{(WH)_{ij}}\right) - 1$$

and is known as the Itakura–Saito (IS) divergence; see Section 5.1.3 for more details. As we will see, this model is invariant to scaling, that is, $D(\gamma X, \gamma WH) = D(X, WH)$ for any $\gamma > 0$, and is particularly meaningful for audio data sets [158].

- If the noise is additive Gaussian, independently but not identically distributed, that is, $\tilde{X}_{ij} \sim \mathcal{N}((\hat{W}\hat{H})_{ij}, \sigma_{ij})$ for all i, j and for some $\sigma_{ij} \geq 0$, then the maximum likelihood estimator is obtained by minimizing

$$D(X, WH) = \sum_{i,j} P_{ij}(X - WH)_{ij}^2, \quad \text{where} \quad P_{ij} = \frac{1}{\sigma_{ij}^2}. \tag{5.2}$$

This is a WLRA problem. Intuitively, the larger the variance, the less importance is given to the corresponding entry in the objective function. If the entry at position (i, j) is not observed, then $\sigma_{ij} = \infty$ and a weight of zero is associated with that entry. This corresponds to the low-rank matrix completion problem with noise: given a low-rank matrix contaminated with i.i.d. Gaussian noise and with missing entries, recover these missing entries (approximately); see Section 9.5 for a numerical example. If the variance is zero, there is no noise and the corresponding entry of X should be approximated perfectly. For other types of noise such as the ones mentioned above, weighted variants can be derived to take into account independently but not identically distributed noise.

---

[43]Recall that, in one dimension, $\operatorname{argmin}_{x \in \mathbb{R}} \sum_i |a_i - x|$ is the median of the a_i's, while $\operatorname{argmin}_{x \in \mathbb{R}} \sum_i (a_i - x)^2$ is their average.

Table 5.1 summarizes some important objective functions for NMF depending on the distribution of the data as discussed above. The table includes Tweedie distributions that correspond to the maximum likelihood estimators obtained by minimizing β-divergences between X and WH; we refer the reader to [98, 329, 421] for more details on these distributions.

Table 5.1. *Several error measures for NMF and the corresponding distribution.*

Acronym	$D(X, WH)$	Distribution[†]
ℓ_2-NMF [303]	$\|X - WH\|_F^2 = \sum_{i,j}(X - WH)_{ij}^2$	Gaussian
Weighted NMF [179]	$\sum_{i,j} P_{ij}(X - WH)_{ij}^2$	independently distributed entries, Gaussian
ℓ_1-NMF [273]	$\|X - WH\|_1 = \sum_{i,j}|X - WH|_{ij}$	Laplace
ℓ_∞-NMF [209]	$\|X - WH\|_\infty = \max_{i,j}|X - WH|_{ij}$	Uniform
KL-NMF [303]	$D_1(X, WH)$	Poisson
IS-NMF [158]	$D_0(X, WH)$	multiplicative Gamma
β-NMF [160]	$D_\beta(X, WH)$	Tweedie distributions

[†]If not specified, the noise is i.i.d.

Discussion For nonnegative input data, additive noise of mean 0 (such as Gaussian or Laplace; see above) is not always reasonable since this implies a positive probability of observing negative entries in the input matrix [86]. This includes all NMF models using componentwise ℓ_p norms, which we refer to as ℓ_p-NMF, with $D(X, WH) = \min_{W \geq 0, H \geq 0} \sum_{i,j}|X - WH|_{i,j}^p$. In particular, such models do not make much sense for sparse data sets (such as document data sets) or data sets with many small entries (such as audio data sets). For document and audio data sets, the β-divergences with $\beta \in [0, 1]$ have been shown to perform much better; see for example [158, 201] and the references therein. However, for data sets with mostly positive entries (such as images), ℓ_p-NMF are reasonable and have been used successfully in many scenarios.

The ℓ_p norms and β-divergences are the most popular error measures for NMF. However, many other measures have been used, including Bregman divergences [433], α-divergences [98], the earth mover's distance [405], and γ-divergences [335], to cite a few. We do not cover them in this book and refer the reader to the corresponding papers for more information. We also refer the reader to [241] for a nice introduction to the topic in the context of low-rank tensor approximation, with examples of maximum likelihood estimators for the Bernoulli distribution for binary data and the Rayleigh distribution for nonnegative continuous data.

5.1.2 ▪ Choice of the error measure

Choosing the right objective function for your NMF model can be crucial. Let us quote Lu, Yang, and Oja [329]:

> The performance of NMF can be improved by using the most suitable β-divergence, not restricting to the squared Euclidean distance or KL divergence, as in the paper of Lee and Seung [303].

To the best of our knowledge, there are currently mainly four ways to handle this situation. Let us briefly discuss them.

Empirical choice The user chooses empirically the objective function she/he believes is the most suitable for the application at hand. This is, as far as we know, the simplest and most widely used approach. For example, as mentioned above, for count data such as document data sets, the KL divergence is particularly meaningful and often used [86]. For image data sets, the Frobenius norm is reasonable and easier to handle from a computational point of view (see Chapter 8) and hence is the most widely used error measure.

Cross validation The objective function is automatically selected using cross validation. If the ground truth factors (W, H) are known, then they can be used to assess the quality of solutions computed with different error measures. Otherwise, part of the data set can be hidden; typically a subset of the entries of X is randomly selected. Then NMF models are learned on this data set for different error measures, using a weighted error measure that takes into account the missing data similarly as in (5.2). Then the model is tested on the remaining entries: how well can it predict the missing entries? The model that best fits the hidden entries is the one selected; see [352, 90, 159]. For example,

- for music transcription based on NMF, Vincent, Bertin, and Badeau [469] argued that the β-divergence with $\beta = 0.5$ performs best;

- for hyperspectral images (Section 1.3.2), Févotte and Dobigeon [159] observed that using β-divergences with $\beta \approx 1.5$ performs best.

Statistical approaches The most suitable objective function is chosen using some statistically motivated criteria. For example, score matching minimizes the expected squared Euclidean distance between the scores of the true distribution and the model [329]. A maximum likelihood approach can also be used to assess whether the observed data is more likely to follow a given distribution [129]. Using this strategy, the authors in [129] showed for example that for piano and jazz signals, the β-divergence for $\beta \approx 0$ was the most suitable. This is expected since $\beta = 0$ corresponds to the IS divergence and is known to be suitable for audio signals.

Distributional robustness More recently, a distributionally robust NMF (DR-NMF) model was proposed in [201]. DR-NMF computes an NMF solution that is robust to different types of noise distributions. For example, using several β-divergences, DR-NMF looks for a solution (W, H) that minimizes the maximum value among these β-divergences,[44] that is,

$$\min_{W \geq 0, H \geq 0} \max_{\beta \in \Omega} D_\beta(X, WH),$$

where $D_\beta(X, WH)$ is the β-divergence between X and WH (see the next section for more details), and Ω is a subset of β's of interest. This problem can be tackled by minimizing a weighted sum of the different objective functions where the weights assigned to the different objective functions are automatically tuned within the iterative process. For example, for audio signals where both KL and IS divergences are often used, using DR-NMF with $\Omega = \{0, 1\}$ leads to a low reconstruction error for both IS and KL divergences.

5.1.3 ▪ β-divergences

An important class of estimators is based on the β-divergences; see Section 5.1.1 and in particular Table 5.1. We will mostly focus on this class of error measures in the remainder of this book.

---

[44]For DR-NMF to make sense, the objective functions $D_\beta(X, WH)$ should be scaled appropriately.

Given two nonnegative scalars z and y, the β-divergence between z and y is defined as follows:

$$d_\beta(z,y) = \begin{cases} \frac{z}{y} - \log\frac{z}{y} - 1 & \text{for } \beta = 0, \\ z\log\frac{z}{y} - z + y & \text{for } \beta = 1, \\ \frac{1}{\beta(\beta-1)}\left(z^\beta + (\beta-1)y^\beta - \beta z y^{\beta-1}\right) & \text{for } \beta \neq 0,1. \end{cases} \quad (5.3)$$

Figure 5.1 displays the function $d_\beta(z,y)$ for $z = 1$ and $\beta = -1, 0, 1, 2, 3$.

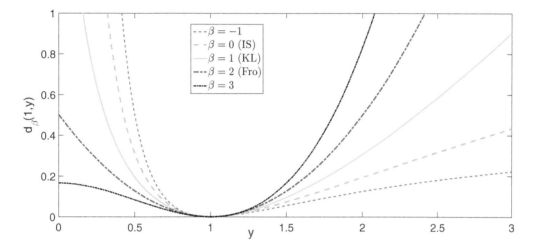

Figure 5.1. *Illustration of the β-divergences $d_\beta(1,y)$ for $\beta = -1, 0, 1, 2, 3$.*

Let us mention two important properties of the β-divergences:

- *Convexity.* The function $d_\beta(z,y)$ is convex in the second argument, y, for $\beta \in [1,2]$. This implies that $D_\beta(X, WH)$ is convex in H for W fixed and vice versa. This makes it easier to design the alternating optimization strategy, where W and H are optimized alternatively, for $\beta \in [1,2]$; see Chapter 8.

- *Scaling.* For any $\gamma > 0$, $z, y \geq 0$,

$$d_\beta(\gamma z, \gamma y) = \gamma^\beta d_\beta(z,y).$$

This implies that the larger the β, the more sensitive is the β-divergence to large values of z, and vice versa. In particular, the β-divergence for $\beta = 0$, which is the IS divergence, is invariant to scaling. This means that small and large entries in X are given the same importance in terms of relative error; for example, approximating 10^{-3} with 10^{-6} leads to the same error as approximating 10^6 with 10^3 since only the ratio $\frac{z}{y}$ plays a role in $d_0(z,y)$. This is because the IS divergence corresponds to an underlying noise model which is multiplicative; see Section 5.1.1. This property is interesting in audio source separation as low-power frequency bands can perceptually contribute as much as high-power frequency bands. Moreover, it can be motivated using a meaningful statistical model generating the data; see [158] for the details.

The β-divergence between two matrices A and B is

$$D_\beta(A,B) = \sum_{i,j} d_\beta(A_{ij}, B_{ij}).$$

The NMF problem using the β-divergence, which we refer to as β-NMF, is the following: Given $X \in \mathbb{R}_+^{m \times n}$ and r, solve

$$\min_{W \in \mathbb{R}_+^{m \times r}, H \in \mathbb{R}_+^{r \times n}} D_\beta(X, WH). \qquad (5.4)$$

The three best-known and most widely used β-divergences are the Frobenius norm ($\beta = 2$), the KL divergence ($\beta = 1$), and the IS divergence ($\beta = 0$).

From Figure 5.1, we observe that, for $\beta \leq 1$, $d_\beta(z, y)$ goes to infinity as y goes to zero (because of the term $y^{\beta-1}$). This is related to our observation in Section 5.1.1: in the presence of Poisson noise ($\beta = 1$), a positive entry (here $z = 1$) cannot be approximated by zero. On the other hand, as β increases, the β-divergences for $y \leq z$ decreases. This means that β-NMF for $\beta \leq 1$ tends to overapproximate the entries of the input matrix, that is, the entries of the solution WH will be on average larger than those of X, because approximating X by smaller entries has a larger cost. For $\beta \geq 1$, the opposite behavior is observed and β-NMF will tend to underapproximate the input matrix. This is illustrated in the following example.

Example 5.1 (Over-/underapproximations). Let us examine how the value of the parameter β used in the β-divergence to measure dissimilarity affects the entries of the estimated solution, WH. For each of 100 independent trials with $\beta = 0$ and with $\beta = 2$, we generate a new X using the function `sprand(100,100,0.5)` of MATLAB and compute a β-NMF (W, H) for $r = 10$ via 100 iterations of the MU technique that will be described in Section 8.2; see [`Matlab file: Example51.m`]. For $\beta = 0$ (IS-NMF), we observe that in all cases

$$\frac{\|\max(0, WH - X)\|_F}{\|X - WH\|_F} \geq 99.86\% \quad \text{while} \quad \frac{\|\max(0, X - WH)\|_F}{\|X - WH\|_F} \leq 2.03\%,$$

so that WH overapproximates X in all cases as most entries of WH are larger than X. For $\beta = 2$ (ℓ_2-NMF), we observe that in all cases

$$\frac{\|\max(0, WH - X)\|_F}{\|X - WH\|_F} \leq 60.21\% \quad \text{while} \quad \frac{\|\max(0, X - WH)\|_F}{\|X - WH\|_F} \geq 79.83\%,$$

so that WH is more balanced around X although it tends to underapproximate it.

In fact, it can be proved that any stationary point (W, H) of ℓ_2-NMF satisfies the inequality $\|WH\|_F \leq \|X\|_F$ [194, Theorem 11]; see Section 8.1.2 for the proof. Note that, for $\beta = 1$ (KL-NMF), any stationary point (W, H) preserves the row sum and the column sum of X, that is, $WHe = Xe$ and $e^\top WH = e^\top X$; see Theorem 6.9 in the next chapter. This implies that stationary points (W, H) have entries balanced around X since $(X - WH)e = 0$ and $e^\top(X - WH) = 0$. ∎

Domain of $d_\beta(z, \cdot)$ and its derivative

It is important to note that $d_\beta(z, \cdot)$ for $z = 0$ is not defined for all values of β:

$$d_\beta(0, y) = \begin{cases} \text{not defined} & \text{for } \beta \leq 0, \\ \frac{1}{\beta} y^\beta & \text{for } \beta > 0. \end{cases}$$

This means that, for NMF, one should use β-divergences for $\beta \leq 0$ only when the input matrix is positive. Table 5.2 provides the domain of $d_\beta(z, y)$, that is, the values of y such that $d_\beta(z, y)$ is defined, depending on the values of β and z. In the context of NMF, Table 5.2 implies, for example, that for $\beta \leq 1$ all positive entries of X must be approximated by positive entries: $X_{i,j} > 0$ implies $(WH)_{i,j} > 0$, because otherwise $d_\beta(X_{i,j}, (WH)_{i,j})$ is not defined as $d_\beta(X_{i,j}, (WH)_{i,j})$ goes to infinity as $(WH)_{i,j}$ goes to zero when $X_{i,j} > 0$; see also the discussion in Section 5.1.1 about the Poisson distribution and KL divergence (page 161), and see Figure 5.1.

Table 5.2. *Domain of $d_\beta(z, \cdot)$ depending on the values of β and z.*

	$\beta \leq 0$	$\beta \in (0, 1]$	$\beta > 1$
$z = 0$	$\emptyset$	$\mathbb{R}_+$	$\mathbb{R}_+$
$z > 0$	$\mathbb{R}_{++}$	$\mathbb{R}_{++}$	$\mathbb{R}_+$

Let us now consider the derivative of $d_\beta(z, y)$ with respect to y, denoted $d'_\beta(z, y)$; it will play an instrumental role in designing algorithms and deriving optimality conditions for NMF in Chapter 8. We have

$$d'_\beta(z, y) = \frac{\partial d_\beta(z, y)}{\partial y} = y^{\beta-1} - z y^{\beta-2} \text{ for } z > 0$$

and

$$d'_\beta(0, y) = \begin{cases} \text{not defined} & \text{for } \beta \leq 0, \\ y^{\beta-1} & \text{for } \beta > 0. \end{cases}$$

Table 5.3 provides the domain of $d'_\beta(z, \cdot)$ depending on the values of β and z. The domain of $d'_\beta(z, \cdot)$ is contained in the domain of $d_\beta(z, \cdot)$; however, they differ when $\beta \in (0, 1)$ and $z = 0$, or when $\beta \in [1, 2)$ and $z > 0$: in these cases, y has to be positive for $d'_\beta(z, y)$ to be defined, which is not the case for $d_\beta(z, y)$.

Table 5.3. *Domain of $d'_\beta(z, \cdot)$ depending on the values of β and z.*

	$\beta \leq 0$	$\beta \in (0, 1)$	$\beta \in [1, 2)$	$\beta \geq 2$
$z = 0$	$\emptyset$	$\mathbb{R}_{++}$	$\mathbb{R}_+$	$\mathbb{R}_+$
$z > 0$	$\mathbb{R}_{++}$	$\mathbb{R}_{++}$	$\mathbb{R}_{++}$	$\mathbb{R}_+$

5.2 ▪ Model-order selection

The choice of the factorization rank, also known as model-order selection, is a long-standing problem, studied well before NMF was introduced. This is not a problem specific to NMF. Most data models face this issue; see for example [435] and the references therein. It is beyond the scope of this book to review this topic. We refer the interested reader to [445, 432, 325] and the references therein for model-order selection specifically designed for NMF.

Let us briefly mention some model-order selection strategies:

- *Expert insight.* In some applications, the domain experts might have prior information on the value of r. This is typically the case when the factors W and H have a physical meaning. For example, in analytical chemistry (Section 1.4.1), the chemists usually know the number of chemical components present in the chemical reaction (or at least have a small range of possible values). Similarly, in hyperspectral unmixing (Section 1.3.2), the number of materials present in a scene can be estimated. For airborne hyperspectral images, the materials present in large proportion can sometimes be identified with the naked eye. In audio source separation (Section 1.3.4), the model order can also sometimes be known; for example, in a piano recording, it corresponds to the number of notes used in the piece.

- *Cross validation.* For machine learning–related applications, such as facial feature extraction, text mining (see Section 1.3), or recommender systems (see Section 9.5), the model order, like the choice of objective function, is typically estimated using cross validation. Part of the data set is hidden, and the models with different values of r are compared in terms of how well they predict the hidden entries. In the same spirit, if NMF is used as a preprocessing step for another data analysis task (for example, classification of faces or documents), then one can choose the model order that leads to the best performance for that particular task.

- *Statistical approaches.* For PCA, numerous approaches exist to select the factorization rank, for example by looking at the decay of the singular values of the input data matrix. Again, as for the choice of the objective function, different scores can be used to compare different models, such as the maximum likelihood; see [435] and the references therein.

5.3 ▪ Regularizations

Over the years, numerous regularizations have been proposed in the NMF literature. It is not possible to list them all and we present here some of the most widely used ones. We refer the interested reader to [98, 478] and the references therein for more examples. As explained in the introduction of this chapter, the main motivation behind these regularizations is identifiability: the goal is to use prior information to refine the NMF model and hence obtain better estimates of the factors. By design, these regularizers are application dependent, should be carefully thought of by the practitioners, and should be validated on synthetic and real data sets. As for the choice of the error measure and the factorization rank, this issue is not specific to NMF; regularizations for inverse problems date back to Tikhonov [453] (1963).

A class of NMF models that is frequently used is the following:

$$\min_{W \geq 0, H \geq 0} D(X, WH) + \alpha_W f_W(W) + \alpha_H f_H(H),$$

where $f_W(W)$ and $f_H(H)$ are regularizers that promote solutions with a specific structure, and α_W and α_H are positive penalty parameters. If only one of the two factors is regularized (that is, $\alpha_W = 0$ or $\alpha_H = 0$), these models should be used in combination with some normalization of W or H; otherwise, because of the scaling degree of freedom, one of the two factors will tend to zero for most regularizers. For example, for min-vol NMF in Section 4.3.3, we used $H^\top e = e$, $He = e$, or $W^\top e = e$.

Let us mention a few important examples:

- *Minimum-volume.* As discussed in length in Section 4.3.3, it makes sense to look for a factor W that is as close as possible to the data points. This can be achieved by promoting solutions W with small volume. The most natural regularizer is $f_W(W) = \det(W^\top W)$, but others are also used, for example $f_W(W) = \mathrm{logdet}(W^\top W + \delta I_r)$, $f_W(W) = \sum_{i<j} \|W(:,i) - W(:,j)\|_2^2$ [452], or $f_W(W) = \mathrm{tr}(W^\top(I_r - \frac{1}{r}ee^T)W)$ [367]; the last two have the advantage of being convex.

- *Sparsity.* The importance of sparse factors W and H was discussed in Section 4.3.4. The most widely used regularizer for sparsity is the ℓ_1 norm, that is, $f_W(W) = \|W\|_1$ and/or $f_H(H) = \|H\|_1$ [277]. However, one has to be careful when using such penalty terms. Consider the case where we want to promote sparsity in H. This would be, for example, particularly meaningful for the four applications described in Section 1.3; see the discussion in Section 4.3.4. As mentioned above, we need to normalize W to avoid pathological solutions where the entries of H tend to zero and the entries of W tend to infinity. For example, we could enforce the constraint[45] $\|W(:,k)\|_2 \leq 1$ for all k. Note that we prefer not to impose an equality constraint (namely, $\|W(:,k)\|_2 = 1$ for all k) in order to keep the feasible set convex. Moreover, the sparsity promoting ℓ_1 penalty on H combined with the scaling degree of freedom of the rank-one factors, $W(:,k)H(k,:)$ for $1 \leq k \leq r$, implies that these inequality constraints will be active at optimality. It is rather common in practice to have rank-one factors that have rather different norms (see for example the discussion on page 142). For example, if

$$\|W(:,p)H(p,:)\|_2 \gg \|W(:,\ell)H(\ell,:)\|_2 \quad \text{for some } p, \ell,$$

then $\|H(p,:)\|_2 \gg \|H(\ell,:)\|_2$. Therefore, one should not use the same penalty parameter for all rows of H, as is done implicitly when using $\|H\|_1 = \sum_{k=1}^{r} \|H(k,:)\|_1$, especially if one is looking for rows of H with similar sparsity levels. Hence, one should rather use a penalty term on the norm of the rows of H, such as $\sum_{k=1}^{r} \lambda_k \|H(k,:)\|_1$ for some penalty parameters $\lambda_k > 0$ $(1 \leq k \leq r)$; see for example [186], where such parameters are tuned automatically to achieve a desired level of row sparsity. One could also be interested in having columns of H with a given sparsity level (as is standard in dictionary learning; see also Section 4.3.4), in which case a penalty of the form $\sum_{j=1}^{n} \lambda_j \|H(:,j)\|_1$ would be meaningful.

- *Orthogonality.* To promote the features W and/or the activations H to overlap as little as possible, orthogonality can be used. More precisely, one can impose the columns of W or the rows of H to be orthogonal via the penalty terms $f_W(W) = \|W^\top W - I_r\|_F^2$ and $f_H(H) = \|HH^\top - I_r\|_F^2$, respectively; see also Section 4.3.2. This regularization encourages the columns of W and/or the rows of H to be as different as possible, which is the opposite goal of minimum-volume regularizers. When the constraint $HH^\top = I_r$ is enforced, NMF becomes a clustering problem; see Sections 4.3.2 and 5.5.3.

- *Smoothness.* In many applications, the columns of W and/or rows of H are discretizations of (piecewise) smooth functions; for example, in hyperspectral images, the columns of W are the spectral signatures of the pure materials [262], and, in audio source separation, the rows of H provide the activations of the sources over time [157]. In that case, one could, for example, use a regularizer such as $f_W(W) = \sum_{k=1}^{r} \sum_{i=1}^{m-1} \left(W(i,k) - W(i+1,k)\right)^2$, and similarly for the rows of H.

  More specific shapes can also be considered, for example unimodal functions [65, 9] or polynomials [122, 234]. This can also be achieved using dictionaries $W = DY$, where D is a dictionary of functions; see Section 5.4.2.

- *Spatial information.* When NMF is used for feature extraction among a set of images, these images are typically first vectorized and stacked next to each other as the columns or rows of X; see for example Section 1.3 for facial and hyperspectral images. Hence, the spatial information is lost; see also Remark 4.3 (page 137). Regularizers can be used to incorporate this information: since the columns of W (or rows of H) should represent basis images, they should have some spatial coherence. In order to keep the sharp edges

---

[45]We could also normalize H, for example using $\|H(k,:)\|_\infty = 1$ for all k, but this leads to a nonconvex problem in H even when W is fixed.

present in images, ℓ_1-based regularizers are more suitable; for example, total variation is often used and has the form

$$f_W(W) = \sum_{j=1}^{r} \sum_{(i,\ell) \in E} |W(i,j) - W(\ell,j)|,$$

where E is the set of neighboring pixels in the image [255]. This regularization can be interpreted as a type of smoothness regularizer.

- *Graph-based regularization.* Assume we want to respect the geometry of the input data points when they are mapped in the low-dimensional subspace spanned by the columns of W. In other words, we would like the distances between the columns of H to match, as best as possible, the distances between the columns of X, that is, if $\|X(:,i) - X(:,j)\|_2$ is small, we expect $\|H(:,i) - H(:,j)\|_2$ to be small as well, and vice versa. This can be achieved using the regularizer

$$f_H(H) = \sum_{i,j} P(i,j)\|H(:,i) - H(:,j)\|_2^2,$$

where $P(i,j)$ is inversely proportional to the distance between the data points i and j. The matrix P can be computed in different ways, a standard choice being $P(i,j) = e^{-\alpha\|X(:,i)-X(:,j)\|_2}$ for some parameter $\alpha > 0$. The entries in the matrix P can be interpreted as the weights in a graph that connects the data points, which explains the name of this regularized NMF model: graph-regularized NMF [71].

Another setting where this regularizer is useful is when some label information is available (for example, for a set of facial images, we know subsets of images that contain the same person), and one can, for example, use $P \in \{0,1\}^{n \times n}$ where $P(i,j) = 1$ if and only if the data points i and j have the same label. Such approaches taking into account partial label information are referred to as semisupervised NMF [305], in contrast with NMF which is unsupervised.

On top of designing proper regularizers, one also has to tune the values of the parameters α_W and α_H. Possible approaches are similar as for the choice of the objective function and factorization rank (see Sections 5.1.2 and 5.2), for example cross validation, the use of expert insight, or statistically motivated approaches. Other strategies have also been designed to automatically tune such parameters; see for example the work of Bobin and collaborators for sparse decompositions [50, 391, 274] and see [516] for min-vol NMF.

5.4 ▪ NMF variants

As discussed in this chapter, the standard NMF model can be modified by the choice of error measure, or the inclusion of constraints or regularizers on the factors. NMF variants are models that adapt the standard NMF formulation to be more suitable for specific scenarios or applications. Numerous such variants have been proposed in the literature, and we review here only some of the most influential ones. Moreover, we do not investigate these models in detail; we simply present them and explain in which situations they are meaningful. For example, we do not delve into the important issues of identifiability or complexity; note, however, that some of these results can be derived from the identifiability and complexity results for NMF, for example, identifiability results for symNMF (see Section 4.2.6).

In this section, we present several NMF variants that are summarized in Table 5.4 (page 186). We illustrate their usefulness on several applications, including feature extraction in images, blind HU, community detection, and topic modeling.

5.4.1 ▪ Orthogonal and projective NMF

As already presented in Section 4.3.2, ONMF requires $H^\top H = I_r$. This implies that each column of H has at most a single positive entry, hence each data point is approximated using a single column of W: ONMF is a clustering problem; see Section 5.5.3 for the details.

Interestingly, if H is fixed in ONMF, the optimal W can be easily computed.

Lemma 5.2. *If $HH^\top = I_r$, then $\operatorname{argmin}_W \|X - WH\|_F^2 = XH^\top$.*

Proof. The necessary and sufficient first-order optimality conditions of the convex quadratic problem $\min_W \|X - WH\|_F^2$ are given by

$$\nabla_W \|X - WH\|_F^2 = 2(X - WH)H^\top = 0,$$

and hence $W = XH^\top$ at optimality since $HH^\top = I_r$. $\qquad\square$

Note that Lemma 5.2 does not require W, H, or X to be nonnegative. Of course, if $H \geq 0$ and $X \geq 0$, then $W = XH^\top \geq 0$. Lemma 5.2 shows that, in ONMF, the columns of W are linear combinations of the data points. Since H has orthogonal rows, each data point is used in the linear combination of a single column of W, and the columns of W can be interpreted as cluster centroids.

Lemma 5.2 implies that ONMF using the Frobenius norm can be equivalently formulated as

$$\min_{H \geq 0, HH^\top = I} \|X - XH^\top H\|_F.$$

This motivated Yuan and Oja [497] and Yang and Oja [490] to introduce projective NMF by relaxing the orthogonality constraint on this reformulation:

$$\min_{H \geq 0} \|X - XH^\top H\|_F.$$

Projective NMF is an NMF variant with the constraint $W = XH^\top$. Projective NMF allows data points to be assigned to more than one cluster and hence can be interpreted as a soft clustering model. The entries of H in projective NMF provide dual information. On the one hand, $H(k, j)$ tells us the importance of the kth cluster or label to the jth data point, that is, how well the jth column of X is represented by the kth column of $W = XH^\top$. On the other hand, $H(k, j)$ is also the weight of the jth data point in the linear combination used to construct the kth centroid. Projective NMF was shown to provide much sparser solutions than NMF and was used successfully for clustering tasks [490].

Figure 5.2 provides an illustration of projective NMF applied on the Urban hyperspectral image. As opposed to ONMF (see Figure 4.10, page 137), projective NMF allows data points to belong to several clusters. In particular, the last three clusters (second row of Figure 5.2) correspond to the dirt and two types of grass, and we observe that many pixels contain more than one of these three endmembers (some pixels contain dirt and grass, many contain the two types of grass).

Figure 5.3 provides an illustration on the CBCL data set. We observe that projective NMF obtains a solution that is much sparser than NMF (see Figure 1.2, page 7) but denser than ONMF as it allows some overlap between features (see Figure 4.11, page 138).

Projective NMF can also be interpreted as a projection operator applied on the rows of X, namely, $H^\top H$ projects X onto $XH^\top H$ (hence the name). The rows of X are projected onto an r-dimensional subspace; this is closely related to nonnegative PCA (see Section 5.5.2). If the data is streaming or not all samples are observed simultaneously, incoming samples can be

Figure 5.2. *Illustration of projective NMF with $r = 6$ on the Urban hyperspectral image (see Figure 1.6). We used the MU of Yang and Oja [490] initialized with SPA; see* [`Matlab file: projectiveNMF.m`]. *Each cluster shown above corresponds to a row of H, reshaped as an image. The first cluster corresponds to the trees, the second to road and dirt, the third to roof tops, the fourth to dirt, and the last two to two different types of grass.* [`Matlab file: ProjectiveNMF_Urban.m`].

projected directly onto this r-dimensional subspace via the projector computed from the previously observed samples, $H^\top H$, since $X(j,:) \approx X(j,:)H^\top H$ for all j.

5.4.2 ▪ Dictionary-based, convex and separable NMF

Dictionary-based NMF requires $W = DC \geq 0$ where D is a given dictionary. For example, in hyperspectral imaging, there exist libraries of spectral signatures from which the columns of W can be constructed from [256, 101]. Another example is the use of Gaussian radial basis functions to construct Raman spectra [500].

Convex NMF [133] is a self-dictionary NMF model with $D = X$ and requires the basis elements, W, to be conic combinations[46] of the data points, that is, $W = XC$ for some $C \geq 0$. Convex NMF is equivalent to *archetypal analysis* [110] with the additional constraint that the linear combinations are convex, that is, $C^\top e = e$. Convex NMF is a rather restrictive model. However, it has the advantage of being more easily interpretable as each basis vector is a linear combination of data points. This model can be relaxed, allowing C to contain small negative entries so that W can extend beyond cone(X) [118]. Convex NMF is closely related to concept factorization where the input matrix is not required to be nonnegative; see [506] and the references therein.

---

[46]The name convex NMF may be confusing since the linear combinations are not convex but conic. However, we keep the name introduced in [133].

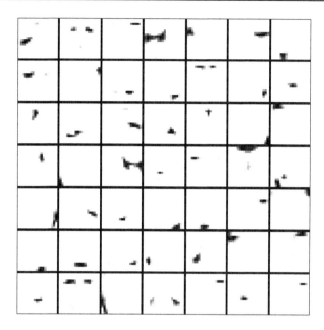

Figure 5.3. *Illustration of projective NMF with* $r = 49$ *applied on the transpose of the pixel-by-face CBCL data set with* $r = 49$ *(see Figure 1.2, page 7). We used the MU of Yang and Oja [490] initialized with SPA; see* [Matlab file: projectiveNMF.m]. *The basis vectors shown on the image were extracted by projective NMF applied on* $X^\top$; *hence the computed basis vectors* W *satisfy* $W \geq 0$ *and* $X \approx WW^\top X$. [Matlab file: CBCL_projectiveNMF.m].

Separable NMF, which we have discussed previously in Section 4.3.1, is a special case of convex NMF where the conic combinations of data points have only one nonzero weight. That is, $W = X(:, \mathcal{K})$ for some index set $\mathcal{K}$, so that the columns of W are extreme rays of the conical hull of the columns of X; see Chapter 7 for more details.

5.4.3 ▪ Semi-nonnegative matrix factorization

Semi-nonnegative matrix factorization (semi-NMF) is a variant of NMF where the nonnegativity constraints on the factor W are relaxed, that is, semi-NMF only requires the factor H to be nonnegative in the decomposition $X \approx WH$ [133]. Hence semi-NMF can be meaningfully applied to an input matrix X with negative entries. Like NMF, semi-NMF approximates the columns of X with conic combinations of the columns of W since $H \geq 0$; see Section 2.1. From a complexity point of view, this problem is rather different from NMF for at least two reasons. First, the semi-nonnegative rank of a matrix can be bounded by a function of its rank. This is not true for the nonnegative rank of a matrix; see Chapter 3. The semi-nonnegative rank is defined as the smallest r such that an exact semi-NMF of X exists, that is, $X = WH$ for some W with r columns and $H \geq 0$ with r rows. It turns out that the semi-nonnegative rank is equal to the rank or the rank plus one and can be computed in polynomial time [92, 203]. Moreover, the semi-nonnegative rank of a nonnegative matrix X is always equal to its rank.[47] Second, for a nonnegative matrix X, the semi-NMF problem using the Frobenius norm as the error measure can be solved in polynomial time using a simple transformation of the truncated SVD. Recall that a nonnegative square matrix A is irreducible if and only if the directed graph induced by A

---

[47]Simply construct a factorization $X = WH$ where H is made of $\operatorname{rank}(X)$ linearly independent rows of X.

(meaning that A is the adjacency matrix of that graph) is strongly connected, that is, there is a path between every pair of vertices.

Theorem 5.3. *[203, Corollary 3.5] Let $X \geq 0$ and $X^\top X$ be irreducible. Then an optimal solution to*

$$\min_{W,H} \|X - WH\|_F^2 \quad such\ that \quad H \geq 0 \tag{5.5}$$

can be computed in polynomial time.

Proof. Let $(U_r, V_r) \in \mathbb{R}^{m \times r} \times \mathbb{R}^{n \times r}$ be such that $U_r V_r^\top$ is an optimal rank-r approximation of X. The pair (U_r, V_r) can be computed in polynomial time via the truncated SVD; see the Eckart–Young theorem in the next chapter (Theorem 6.3). Moreover, recall that in the truncated SVD the columns of U_r (resp. V_r) are the eigenvectors of $XX^\top$ (resp. $X^\top X$). Since $X^\top X$ is irreducible, we can assume w.l.o.g. that $V_r(:,1) > 0$ using the Perron–Frobenius theorem [38]: the eigenvector of an irreducible matrix $X^\top X$ corresponding to the largest eigenvalue has only positive entries (or only negative ones).

Then, using a simple transformation of (U_r, V_r), let us construct (W, H) such that $H \geq 0$ and $WH = U_r V_r^\top$, which will conclude the proof. We let $H(1,:) = V_r(:,1)^\top > 0$ and, for $2 \leq k \leq r$, we let α_k be sufficiently large so that

$$H(k,:) = V_r(:,k)^\top + \alpha_k V_r(:,1)^\top \geq 0.$$

The simplest choice is to take the smallest α_k's such that $H(k,:) \geq 0$, that is, for $2 \leq k \leq r$, take

$$\alpha_k = \max_{\{i|V_r(i,k)<0\}} \frac{-V_r(i,k)}{V_r(i,1)}.$$

Then let $W(:,k) = U_r(:,k)$ for $2 \leq k \leq r$, and

$$W(:,1) = U_r(:,1) - \sum_{k=1}^{r} \alpha_k U_r(:,k),$$

which gives $WH = U_r V_r^\top$. $\square$

In contrast to the above result, when the input matrix X is allowed to have negative entries, semi-NMF (5.5) is NP-hard, even for $r = 1$ [203, Theorem 4.1]. This is also the case for NMF; see Theorem 6.5 in the next chapter.

Theorem 5.3 shows that, for a nonnegative input matrix, semi-NMF is essentially equivalent to the truncated SVD. Moreover, the solution obtained following the construction of Theorem 5.3 will in general have a single zero entry in each row of H (except for the first row which is positive), meaning that almost all data points are reconstructed using all basis vectors. Hence, like the SVD, semi-NMF does not bring a sparse or an easily interpretable decomposition. For this reason, it makes more sense to consider regularized or constrained semi-NMF models [203]. In Section 4.3, we discussed four such important models: separable, orthogonal, min-vol, and sparse NMF. In these four cases, we do not require the matrix W to be nonnegative to obtain identifiability results but require additional structure on W and/or H (see Table 4.2, page 155).

Figure 5.4 displays the basis vectors of semi-NMF computed from the transpose of the CBCL data set as described in Section 1.3.1 (where the columns are vectorized gray-level facial images), leading to a decomposition $X \approx WH$ where $W \geq 0$. Since $X \geq 0$, an optimal solution can be computed via the construction of Theorem 5.3, and we can see that it is dense, like the SVD.

Figure 5.4. *Basis vectors extracted by semi-NMF applied on the transpose of the pixel-by-faces CBCL data set with $r = 49$ (see Figure 1.2). These basis vectors $W \geq 0$ are obtained via a simple translation of the truncated SVD basis vectors; see Theorem 5.3.* [Matlab file: CBCL_semiNMF.m].

5.4.4 ▪ Sparse NMF

Sparse NMF is one of the NMF models that has attracted the most attention. Numerous models exist, and regularization is the most common approach to obtain sparser NMF solutions, typically using the ℓ_1 norm; see Section 5.3. In his seminal paper, Hoyer [243] (2004) introduced a sparse NMF model with the following additional constraints:

$$\text{spar}(H(k,:)) \geq s_H \ \text{ and } \ \text{spar}(W(:,k)) \geq s_W \text{ for } k = 1, 2, \ldots, r,$$

where $\text{spar}(.) \in [0, 1]$ is a measure of sparsity defined in (4.32) (see page 150) that uses the ratio of the ℓ_1 norm to the ℓ_2 norm, and s_H and s_W are constants in the interval $[0, 1]$ that impose a minimal sparsity level to the rows of H and the columns of W, respectively. Compared to the model based on the ℓ_0 norm presented in Section 4.3.4, the sparsity measure $\text{spar}(.)$ has the advantage of being continuous; see [449] for a discussion. For example, $\text{spar}([1, 10^{-6}, 10^{-6}]) > \text{spar}([1, 1, 0])$, which makes sense numerically as $[1, 10^{-6}, 10^{-6}]$ is very close to the 1-sparse vector $[1, 0, 0]$. Also, $\text{spar}(x)$ is invariant to scaling, that is, $\text{spar}(x) = \text{spar}(\alpha x)$ for any $\alpha \neq 0$. Moreover, it is easy to project a vector onto the closest vector with a given sparsity, that is, to solve

$$\min_x \|x - y\|_2 \text{ such that } \text{spar}(y) \geq s,$$

for a given vector y and sparsity level s.

A possible drawback of Hoyer's sparse NMF model is that all of the rows of H must have the same sparsity level, and all of the columns of W must have the same sparsity level. Of course, the sparsity level can be chosen to be different for each of these vectors. However, this leads to a large number of parameters to be tuned. Instead, one can also impose an average sparsity

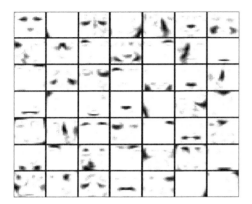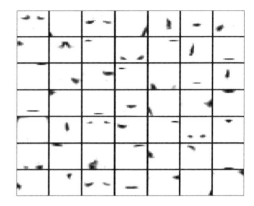

Figure 5.5. *Illustration of the basis vectors extracted by NMF versus sparse NMF on the CBCL facial image data set with $r = 49$. We used the algorithm from [207] (see [`Matlab file: sparseNMF.m`]) with SPA as an initialization. For sparse NMF, the columns of W are imposed to have an average Hoyer sparsity of 85%. The solution of NMF has average Hoyer sparsity of 69%. The relative error of NMF is $\|X - WH\|_F/\|X\|_F = 8.07\%$ and of sparse NMF is 9.33%. Hence for an increase of 1.26% relative error, the sparsity increases by 16%. [`Matlab file: CBCL_sparseNMF.m`].*

level on the rows of H and columns of W. It turns out one can also efficiently project onto this set [207]. Figure 5.5 illustrates sparse NMF on the CBCL facial images. Note that the sparsity measure of Hoyer has been used in other contexts, for example in dictionary learning [449] and deep neural networks [487].

5.4.5 ▪ Affine NMF and nonnegative matrix underapproximation

As we have seen in Section 4.2.5, sparsity on the input matrix may lead to identifiability. Affine NMF [299], also known as NMF with offset, tries to leverage this observation by considering the model

$$X \approx WH + we^\top,$$

where w is an offset vector that is used to reconstruct all columns of X (because of the vector e of all ones). If $we^\top$ is computed before the NMF (W, H) is computed, it makes sense to look for w such that $X - we^\top \geq 0$, which leads to

$$w(i) = \min_i X(i,j) \text{ for all } i.$$

This implies that $Y = X - we^\top \geq 0$ has at least one zero entry per row, hence the NMF of Y is more likely to admit a unique NMF; see Section 4.2.5. A similar preprocessing can be performed on the rows. A different form of preprocessing was proposed in [186]. The data matrix X is multiplied by a matrix Q such that $XQ \geq 0$ is sparse while the inverse of Q is nonnegative, that is, $Q^{-1} \geq 0$. Hence an NMF of the preprocessed matrix $XQ = WH$ directly provides an NMF for $X = W(HQ^{-1})$.

Affine NMF is related to NMU [195], which is an NMF model with the constraint $WH \leq X$. There are two main reasons to consider NMU:

1. NMU provably leads to sparser decompositions since $X(i,j) = 0$ implies $(WH)_{i,j} = 0$.

2. NMU allows NMF to be solved sequentially, like PCA, by computing a single rank-one matrix each step. The recursive method sets $R_1 = X$ and computes

$$R_{k+1} = R_k - W(:,k)H(k,:) \geq 0 \text{ for } k = 1, 2, \ldots, r,$$

where $W(:,k)H(k,:)$ is a rank-one NMU of R_k, that is, $W(:,k)H(k,:) \leq R_k$. We refer the interested reader to Section 8.7.1 for more details; see also [Matlab file: recursiveNMU.m].

For example, Figure 5.6 shows the basis elements extracted sequentially by rank-one NMU on the CBCL facial images. We observe that NMU extracts more localized factors as the factorization unfolds, because the residual matrices R_k are getting sparser and sparser and so are the rank-one matrices $W(:,k)H(k,:)$.

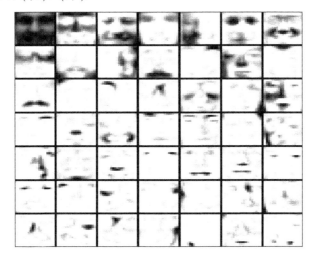

Figure 5.6. *Basis vectors extracted sequentially by NMU from the CBCL facial image data set with $r = 49$; see Figure 5.5 for a comparison with NMF.* [Matlab file: CBCL_NMU.m].

NMU has also been shown to be useful for blind HU [284, 76], medical imaging [286], and climate data and clustering tasks [447].

5.4.6 ▪ Convolutive NMF

One property that is not taken into account by the standard NMF model is the possible correlation between data points. For example, in audio source separation, adjacent columns of X correspond to the frequency response of an audio signal for adjacent time windows and hence are highly correlated. This correlation can be accounted for using regularization; see the previous section. Moreover, correlation can also exist between nonadjacent time windows in audio signals, because sources typically activate in similar ways over time (possibly with different intensities); for example, a piano note in a piece; see Figure 1.8 (page 11). Figure 5.7 shows the spectrogram of a bird song containing several repeated patterns in the spectrogram.

Convolutive NMF [423, 424] requires that the sources are activated in the same way over time. More precisely, it attempts to find r matrices $W_\ell \in \mathbb{R}_+^{m \times p}$ ($1 \leq \ell \leq r$) and a matrix $H \in \mathbb{R}_+^{r \times n}$ such that

$$X \approx \sum_{\ell=1}^{r} \sum_{k=1}^{p} W_\ell(:,k) \left[0_{1 \times (k-1)} \; H(\ell, 1 : n - k + 1) \right].$$

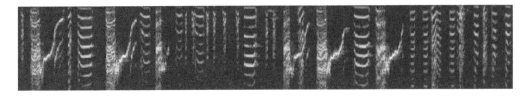

Figure 5.7. *Spectrogram of a bird song [336]; x-axis is time, y-axis is frequency.*

The factors $H(\ell,:)$ $(1 \leq \ell \leq r)$ move toward the right, putting zeros on their left-hand side. This means that the patterns given in the matrices W_ℓ identify the way each source activates over time and always appear as a block in the data matrix; see Figure 5.8 for an illustration. The parameter p indicates the maximum time duration of a source when it is activated ($p = 3$ in Figure 5.8). Note that for $p = 1$, NMF and convolutive NMF coincide, hence convolutive NMF generalizes NMF.

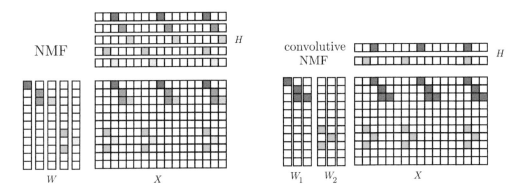

Figure 5.8. *Illustration of NMF ($r = 5$) versus convolutive NMF ($r = 2, p = 3$).*

5.4.7 ▪ Symmetric NMF

SymNMF requires $W = H^\top$, that is, $X \approx WW^\top$. SymNMF makes sense only when the input matrix is symmetric since $WW^\top$ is symmetric. Let us try to interpret this decomposition. When a nonnegative matrix $X \in \mathbb{R}_+^{n \times n}$ is symmetric, it can be thought of as the adjacency matrix of a graph where $X(i,j) = X(j,i)$ indicates the strength of the link between nodes i and j. The simplest case is when X is binary which corresponds to an unweighted graph. A key task in graph theory is to find subsets of nodes that are highly connected. In particular, finding subsets of nodes that are all connected to one another is the celebrated clique problem. More generally, subsets of nodes that are densely connected are referred to as communities. Finding these communities is a central problem in large graphs and networks; see for example the survey [167].

SymNMF allows us to perform such a task. SymNMF decomposes X as follows:

$$X \approx WW^\top = \sum_{k=1}^{r} W(:,k)W(:,k)^\top;$$

hence X is decomposed as the sum of r symmetric and nonnegative rank-one matrices $W(:,k)W(:,k)^\top$ for $k = 1, 2, \ldots, r$. The nonzero entries of these rank-one matrices correspond to square submatrices that connect subsets of nodes in the graph. This follows from the additive nature of symNMF. In the exact case, when $X = WW^\top$, X is decomposed into r cliques. This is similar to the interpretation that Exact NMF decomposes a biadjacency matrix, X, into bi-cliques; see Section 3.4.4. In summary, each rank-one matrix $W(:,k)W(:,k)^\top$ in a symNMF of X corresponds to a subset of nodes that are highly connected. Hence symNMF allows a matrix to be decomposed into (possibly overlapping) communities. For example, if $X(i,j)$ indicates the similarity between pixels in an image, performing a symNMF of X provides a soft clustering of the pixels into homogeneous regions. If $X(i,j)$ indicates the similarity between documents in a corpus, symNMF classifies these documents into subsets of documents discussing similar topics; see also [235, 290, 271, 356] and the references therein.

Example 5.4 (Zachary's karate club). Let us illustrate the capacity of symNMF to split the nodes of a graph into different communities on a simple example using the Zachary's karate club data set [498]. Zachary is a researcher who studied the relationships between the members of a karate club. Each edge in the graph represents the friendship between two members of the club. There are 34 members and 78 friendship links. During his study, Zachary observed a dispute between the administrator and the instructor of the club, which resulted in the instructor leaving the club to start a new one, taking about half of the original club's members with him. Applying symNMF with $r = 2$ to the symmetric adjacency matrix of this graph, $X \in \{0,1\}^{34 \times 34}$, allows two communities to be identified, where each column of W represents a community. Note that $X(i,j)$ represents the affinity between i and j, and hence we set $X(i,i) = 1$ for $i = 1, 2, \ldots, n$. Figure 5.9 illustrates this decomposition. Using the W computed with [Matlab file: symNMF_karate], one can easily split the graph into two disjoint communities by assigning node j to cluster 1 if $W(j,1) > W(j,2)$ and to cluster 2 otherwise. We obtain

$$\mathcal{K}_1 = \{1, 2, 3, 4, 5, 6, 7, 8, 11, 12, 13, 14, 17, 18, 20, 22\},$$

which is the instructor's faction (node 1), and

$$\mathcal{K}_2 = \{9, 10, 15, 16, 19, 21, 23, 24, 25, 26, 27, 28, 29, 30, 31, 32, 33, 34\},$$

which is the administrator's faction (node 34). This clustering is consistent with the two groups within the club; see also [213].

Recall that symNMF leads to a soft clustering: some vertices belong to the two communities with different intensities. For example, node 9 is rather central in the graph and is shared among the two communities, with $W(9,1) = 0.32$ and $W(9,2) = 0.54$. This node is actually the only one "misclassified" by symNMF in the sense that the person represented by node 9 left the club with the instructor (node 1), not with the administrator (node 34). Note that Zachary's analysis based on network flows [498] led to the same classification as symNMF, namely $\mathcal{K}_1$ and $\mathcal{K}_2$ as shown above. As expected, $W(1,1) = 1.37 > W(1,2) = 0.04$, indicating that the instructor (node 1) belongs mostly to the first community, and $W(34,1) = 0.08 < W(34,2) = 1.38$, which confirms that the administrator (node 34) belongs much more to the second community. ∎

Completely positive matrices As already pointed out in Section 3.4.8, symNMF is closely related to the set of completely positive matrices

$$\mathcal{C}_+^n = \{A \in \mathbb{R}^{n \times n} \mid A = BB^\top \text{ for some } B \geq 0\}.$$

It is beyond the scope of this book to discuss this very rich topic. We refer the interested reader to [39, 36] and the references therein.

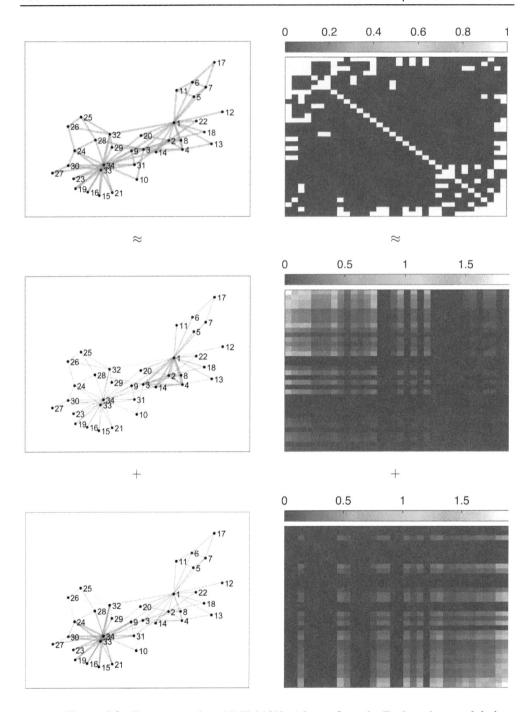

Figure 5.9. *Illustration of symNMF [461] with $r = 2$ on the Zachary karate club data set. On the left: representation of the decomposition in terms of graphs. Edges are represented using the weights of X (top), $W(:, 1)W(:, 1)^\top$ (middle), and $W(:, 2)W(:, 2)^\top$ (bottom). The width of the edges corresponds to their values in the corresponding matrices, and values smaller than 0.1 are represented by red edges. On the right: representation of the decomposition in terms of matrices: X (top), $W(:, 1)W(:, 1)^\top$ (middle), and $W(:, 2)W(:, 2)^\top$ (bottom). [Matlab file: symNMF_karate].*

5.4.8 ▪ Nonnegative matrix trifactorization

The NMF model with three factor matrices, referred to as nonnegative matrix trifactorization (tri-NMF), is the following: Given $X \in \mathbb{R}_+^{m \times n}$, r_1, and r_2, find $W \in \mathbb{R}_+^{m \times r_1}$, $S \in \mathbb{R}_+^{r_1 \times r_2}$, and $H \in \mathbb{R}_+^{r_2 \times n}$ such that

$$X \approx WSH.$$

To make things more concrete, let us interpret this model on a user-by-movie matrix X. Let us assume that the entries of the matrix X are binary and defined as follows: each row corresponds to a user and each column to a movie, and $X(i, j) = 1$ if user i has watched movie j. Applying NMF on this matrix provides r rank-one matrices $W(:, k)H(k, :)$ ($1 \leq k \leq r$). Each rank-one matrix identifies a subset of users and a subset of movies that highly interact with each other, that is, its positive entries correspond to a rectangular submatrix of X with many ones. Tri-NMF instead finds separately r_1 subsets of movies that are watched together (the columns of W) and r_2 subsets of users that behave similarly (the rows of H), while the matrix S tells us how these subsets interact together. More precisely, $S(k, j) > 0$ indicates that the jth subset of users, corresponding to the positive entries of $H(j, :)$, watches the movies from the kth subset of movies, corresponding to the positive entries of $W(:, k)$. In fact, in tri-NMF, X is approximated as follows:

$$X \approx \sum_{k=1}^{r_1} \sum_{j=1}^{r_2} W(:, k)S(k, j)H(j, :).$$

In other words, tri-NMF identifies communities of users that have a similar behavior (in the sense that they watch the same movies) and communities of movies that are similar (in the sense that they are watched by the same people), while connecting these communities through the nonnegative matrix S. Another application of tri-NMF is text mining (see Section 1.3.3), where it finds subsets of documents containing the same words (the columns of W) and subset of words appearing in the same documents (the rows of H) and links these subsets via the matrix S.

Tri-NMF does not in general have a unique solution. For example, letting $r_1 \geq r_2$, we have

$$WSH = \underbrace{\begin{pmatrix} WS & 0_{m \times (r_1 - r_2)} \end{pmatrix}}_{W'} \underbrace{\begin{pmatrix} I_{r_2} \\ 0_{(r_1 - r_2) \times r_2} \end{pmatrix}}_{S'} \underbrace{H}_{H'}$$

for any tri-NMF (W, S, H). Hence tri-NMF needs additional constraints or regularizations such as sparsity. In Section 5.4.10, we briefly discuss deep NMF, which is a generalization of tri-NMF.

Tri-ONMF   A particular tri-NMF variant that has attracted attention is orthogonal nonnegative matrix trifactorization (tri-ONMF) [132], where W is required to have orthogonal columns and H is required to have orthogonal rows. In other words, the factors are constrained so that $W^\top W = I_{r_1}$ and $HH^\top = I_{r_2}$. Recall that orthogonality together with nonnegativity implies that the columns of W and the rows of H have disjoint supports; see Lemma 4.39. In the context of the user-by-movie matrix, this means that tri-ONMF imposes that each user and each movie belongs to a single subset: tri-ONMF finds disjoint communities of users and movies, and connects them through the matrix S; see for example [82].

5.4.9 ▪ Symmetric nonnegative matrix trifactorization

Symmetric nonnegative matrix trifactorization (tri-symNMF) enriches the symNMF model with a third factor matrix S: Given a symmetric nonnegative matrix $X \in \mathbb{R}_+^{m \times m}$ and a factorization rank r, it looks for a nonnegative matrix $W \in \mathbb{R}_+^{m \times r}$ and a symmetric nonnegative matrix $S \in \mathbb{R}_+^{r \times r}$ such that

$$X \approx WSW^\top.$$

The interpretation of this model is a combination of the interpretations of symNMF and tri-NMF. As for symNMF, the columns of W identify communities, that is, highly correlated items in the data set. As for tri-NMF, tri-symNMF allows these communities to interact via the factor S. The entry $W(j,k)$ can be interpreted as the membership indicator of item j for community k. The entry $S(k,\ell)$ is the strength of the connection between communities k and ℓ. We have

$$X(i,j) \approx W(i,:)SW(j,:)^\top = \sum_{k=1}^{r}\sum_{k=1}^{r} W(i,k)S(k,\ell)W(j,\ell)$$

so that the value $X(i,j)$ reflects the memberships of items i and j in the different communities and how these communities interact together [475, 503].

5.4.9.1 ▪ Topic modeling

Let us discuss further the tri-symNMF model in the context of topic modeling. In this application area, the word-by-document matrix X is typically far from a low-rank matrix, and it does not follow the NMF model $X \approx WH$ very closely. Let us briefly explain why. In the NMF model, an observed document $X(:,j) \in \mathbb{R}^m$ for $j \in \{1,2,\ldots,n\}$ is a vector of word counts, that is, $X(i,j)$ is the number of times word i appears in this document. The vector $X(:,j)$ is a sample of a random variable $\tilde{x}_j \in \mathbb{R}^m$. The distribution of $\tilde{x}_j$ is such that $\mathbb{E}(\tilde{x}_j) = \hat{W}\hat{H}(:,j)$ where $(\hat{W}, \hat{H})$ are deterministic but unknown parameters to be estimated; see Section 5.1.1.

In the context of topic modeling, these parameters can be interpreted as follows:

- The columns of $\hat{W}$ correspond to topics. Assuming w.l.o.g. that the entries in each column of $\hat{W}$ sum to one, $\hat{W}(i,k)$ is the probability of picking the word i when discussing the topic k.

- The vector $\frac{\hat{H}(:,j)}{\|\hat{H}(:,j)\|_1}$ indicates the proportion of each topic discussed in the jth document, while $\|\hat{H}(:,j)\|_1$ equals the number of words present in the document.

This model is closely related to PLSA/PLSI. In fact, we show in Section 5.5.4 the equivalence between PLSA/PLSI and NMF based on the KL divergence. Such models make sense for long documents, so that, for all j, the standard deviation of $\tilde{x}_j$ is not too large and $\hat{h}_j$ can be estimated accurately. However, most documents do not use many of the words associated with a topic they discuss, and hence $X(:,j) \approx \hat{W}\hat{H}(:,j)$ is not satisfied for most documents. In particular, short documents (for example, tweets) cannot be well-approximated in this way, even when they discuss only one topic.

However, even short documents provide an important piece of information, namely, they indicate *the words that occur together*. This information can be found in the word co-occurrence matrix $XX^\top$. The entry $(XX^\top)_{i,j}$ is equal to the number of different combinations of the words i and j appearing in the same document. The symmetric matrix $XX^\top$ can be interpreted as the weighted adjacency matrix of a graph connecting nodes corresponding to the words in the dictionary. The communities in the graph should correspond to topics, that is, sets of words that are highly connected.

Let us briefly explain why applying tri-symNMF on $XX^\top$ makes sense for a certain class of probabilistic topic models. Let the matrix $\hat{W} \in \mathbb{R}_+^{n \times r}$ be a deterministic, but unknown, word-by-topic matrix whose entry at position (i,k) contains the probability for word i to be used in topic k. Observe that $\hat{W}$ plays the same role as for NMF. Let the vector $\tilde{h}$ be a random variable corresponding to the proportions of the topics discussed within a document. Then the columns of X are assumed to be generated as follows: for $j = 1, 2, \ldots, n$,

1. let the vector $H(:, j) \in \Delta^r$ be a sample of the random variable $\tilde{h}$;

2. $X(:, j)$ is the sample of a multinomial distribution of parameters $\hat{W} H(:, j)$, that is, the probability to pick the ith word in the dictionary is given by $(\hat{W} H(:, j))_i$. The number of words sampled per document is not important, as long as there are least two words picked within each document.

There are two key differences of the above model with NMF:

- The columns of X are sampled from the same distribution with the same parameters; for example, $\tilde{h}$ could follow a Dirichlet distribution, as in the latent Dirichlet allocation model [47].

  In NMF, the columns of X are sampled from the same distributions but with different parameters, namely with parameters $\hat{W} \hat{H}(:, j)$ for the jth column of X.

- When the number of words sampled in the j document, $e^\top X(:, j)$, is not sufficiently large, we will not have

$$\frac{X(:, j)}{e^\top X(:, j)} \approx \hat{W} H(:, j). \tag{5.6}$$

  This means that, as opposed to NMF, the above model does not need (5.6) to be satisfied, because it does not require the number of words in each document to be sufficiently large to make sense.

Finally, as the number n of sampled documents goes to infinity, $XX^\top$ gets closer to $\mathbb{E}(\hat{W} \tilde{h} \tilde{h}^\top \hat{W}^\top)$, up to a constant factor: we have

$$\lim_{n \to \infty} \frac{XX^\top}{e^\top XX^\top e} = \mathbb{E}\left(\hat{W} \tilde{h} \tilde{h}^\top \hat{W}^\top\right) = \hat{W} \underbrace{\mathbb{E}\left(\tilde{h} \tilde{h}^\top\right)}_{=S} \hat{W}^\top,$$

where $S \in \mathbb{R}^{r \times r}$ is the topic-by-topic matrix, which is the second-order moment of $\tilde{h}$. If the number of documents observed is sufficiently large, the use of the tri-symNMF,

$$\frac{XX^\top}{e^\top XX^\top e} \approx \hat{W} S \hat{W}^\top,$$

is justified by the probabilistic topic models as described above. We refer the interested reader to [14] for more details.

We will see in Section 7.8 how the matrices $\hat{W}$ and S can be recovered from X in polynomial time under the separability assumption. Separability requires that, for each topic, there exists a so-called anchor word associated only to that topic. An anchor word has a positive probability to be observed for only one topic. Mathematically, it requires $\hat{W}^\top$ to be separable (Definition 4.10, page 107), that is, for each topic k, there exists a word i such that $\hat{W}(i, k) > 0$ while $\hat{W}(i, p) = 0$ for all $p \neq k$.

The idea of working with $XX^\top$ instead of X can be extended to higher-order statistics by considering the tensors of the co-occurrences of more than two words. We refer the interested reader to [260] for more on this topic.

Other applications   Tri-symNMF is also particularly meaningful for the identification of hidden Markov models [170] and for detecting overlapping and correlated communities in graphs [246]. Tri-symNMF is closely related to the mixed membership stochastic block model [5] where the observation X is binary, and $X(i, j)$ is a Bernoulli distribution of parameter $(WSW^\top)_{ij}$ [340, 374, 1, 246]. We also refer the interested reader to the discussion in the survey paper [170], and the references therein.

5.4.10 ▪ Multilayer/deep NMF

Let us provide another interpretation of tri-NMF. A tri-NMF of $X \approx WSH$ can be interpreted as two NMFs: $X \approx W'H = WH'$ with $W' = WS$ and $H' = SH$. Each NMF has a different basis, namely W versus WS, and different activations, namely, SH versus H. The basis elements $W' = WS$ are constructed as nonnegative linear combinations of the columns of W. Hence the columns of W' are denser than that of W (except for columns of W' corresponding to columns of S with a single nonzero entry); see Figure 5.10 for an illustration on the CBCL face data set. The columns of W can be interpreted as the atoms of a dictionary, and W' is constructed with this dictionary. Hence tri-NMF can also be interpreted as a dictionary-based NMF model (see Section 5.4.2) where the dictionary is learned from the data.

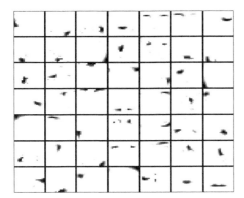
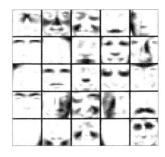

Figure 5.10.   *Illustration of tri-NMF (deep NMF with three factor matrices) $X \approx WSH$ with $r_1 = 49$ and $r_2 = 25$ on the CBCL face data set: on the left, $W \in \mathbb{R}_+^{361 \times 49}$; on the right, $W' = WS \in \mathbb{R}_+^{361 \times 25}$. Since $W' = WS$, features on the right (W') are obtained via nonnegative linear combinations of features from the left (W). A similar figure can be found in [496, Figure 2], where three layers are used. [*`Matlab file: CBCL_triNMF.m`*].*

Deep NMF, also know as multilayer NMF, generalizes tri-NMF by considering more than three factors matrices:

$$X \approx WH_1H_2 \ldots H_t.$$

This allows several layers of features to be extracted, namely W, WH_1, $WH_1H_2, \ldots$, that become denser and denser as the factorization unfolds. Hence W contains rather localized features which are combined progressively within the layers of deep NMF to construct the final high-level features $WH_1 \ldots H_{t-1}$ that linearly reconstruct the data set with the weight factor H_t. As noted above for tri-NMF, such models are typically highly nonunique and require additional constraints or regularizations such as sparsity [228] (sparse NMF was used to construct the solution in Figure 5.10). We refer the interested reader to the recent survey [117] for more details on deep NMF.

5.4.11 ▪ Other variants

Other NMF variants include the following:

- *Binary and Boolean NMF.* For a binary input matrix, binary and Boolean NMF impose that W and H have binary entries [504]. Moreover, Boolean NMF, also known as the discrete basis problem [344], uses a Boolean sum, that is, $a \oplus b = \max(a, b)$, so that Boolean NMF requires one to approximate X with $\min(WH, 1)$ where the entries of W and H are binary. This problem is closely related to the rectangle covering problem and the Boolean rank; see Sections 3.4.3 and 3.4.4.

- *Interval-valued NMF.* The entries of X are intervals rather than scalars [314, 282], and the goal is to compute WH whose entries fall into these intervals.

- *Kernel NMF.* Inspired by kernel methods, NMF is applied to a nonlinear transformation of the data points, $\Phi(X(:, j))$ for all j [501, 67].

- *Bilinear NMF.* In blind HU (see Section 1.3.2), if the light hits two materials before being measured, its spectral signature is equal to $W(:, k) \circ W(:, \ell)$, where $W(:, k)$ and $W(:, \ell)$ are the spectral signatures of these two materials. Taking this nonlinear effect into account leads to the NMF model

$$X(:, j) \approx \sum_{k=1}^{r} W(:, k)H(k, j) + \sum_{k<\ell} \big(W(:, k) \circ W(:, \ell)\big)H^\circ(k, \ell, j) \ \text{ for all } j,$$

where $W, H \geq 0$, and $H^\circ(k, \ell, j) \geq 0$ is the intensity of the interaction between materials k and ℓ in pixel j. This model is referred to as the bilinear mixing model and is closely related to the Nascimento model which was introduced in [362]. In blind HU, additional sum-to-one constraints are typically imposed on H and H°; see [136] for a survey on nonlinear unmixing models. We refer the interested reader to [126] for a discussion on the identifiability of this model and to [275] for a provably correct algorithm under a separability-like assumption.

- *Online NMF.* It updates the factors of NMF taking into account continuously arriving data samples, while not storing past data samples; see [507] and the references therein.

Moreover, there are many variants that are combinations of the above models; for example sparse symNMF [30], sparse separable NMF [359], separable convolutive NMF [123], kernel NMU [285], or separable tri-symNMF (Section 7.8), and regularizations can also be incorporated. There is a close link between regularizations and NMF variants. In particular, many NMF variants are solved using penalty approaches which in turn amount to solving regularized NMF models. For example, symNMF can be solved via NMF by adding the penalty term $\|W - H^\top\|_F^2$ [290]. Designing NMF models tailored to particular applications is still an active area of research; see for example [463] for a recent model where a temporal NMF problem with Wasserstein metric loss is proposed to tackle motion segmentation in biological imaging scenarios, [166] where NMF is generalized for input matrices where the entries are quaternions in order to deal with spectropolarimetric images, and [218] for designing an NMF model to deal with ordinal data, that is, the entries of X belong to ordered categories such as movie ratings that belong to $\{0, 1, \ldots, 5\}$ (0 means that the user did not watch the movie).

Table 5.4 summarizes the NMF variants we have discussed in this chapter.

5.5 ▪ Models related to NMF

In this section, we present several well-known data analysis models and explain how they are related to NMF variants, namely PCA (Section 5.5.1), nonnegative PCA (Section 5.5.2), k-means

Table 5.4. *NMF variants for a given data matrix $X \in \mathbb{R}^{m \times n}$ and a factorization rank r. Unless specified otherwise, $X \approx WH$, $W \in \mathbb{R}^{m \times r}$, and $H \in \mathbb{R}^{r \times n}$.*

Name	Model
NMF	$W \geq 0, H \geq 0$
ONMF	$W \geq 0, H \geq 0, HH^\top = I_r$
projective NMF	$W = XH^\top, H \geq 0$
convex NMF	$W = XC, C \geq 0, H \geq 0$
separable NMF	$W = X(:, \mathcal{K})$ with $|\mathcal{K}| = r, H \geq 0$
dictionary NMF	$W = DC \geq 0, D$ dictionary, $H \geq 0$
semi-NMF	$H \geq 0$
sparse NMF	$W \geq 0, H \geq 0, W$ and/or H sparse
affine NMF	$X \approx WH + we^\top, W \geq 0, H \geq 0, w \geq 0$
NMU	$WH \leq X, W \geq 0, H \geq 0$
convolutive NMF	$X \approx \sum_{\ell=1}^{r} \sum_{k=1}^{p} W_\ell(:, k) [0_{1 \times (k-1)} \, H(\ell, 1 : n - k + 1)]$, $W_\ell \in \mathbb{R}_+^{m \times p} (1 \leq \ell \leq r), H \in \mathbb{R}_+^{r \times n}$
symNMF	$W = H^\top \geq 0$
tri-NMF	$X \approx WSH, W \in \mathbb{R}_+^{m \times r_1}, S \in \mathbb{R}_+^{r_1 \times r_2}, H \in \mathbb{R}_+^{r_2 \times n}$
tri-ONMF	tri-NMF & $W^\top W = I_{r_1}, HH^\top = I_{r_2}$
tri-symNMF	tri-NMF & $W = H^\top, S = S^\top$
deep NMF	$X \approx WH_1 H_2 \ldots H_t, W \geq 0, H_i \geq 0$ for all i
binary NMF	$W \in \{0, 1\}^{m \times r}, H \in \{0, 1\}^{r \times n}$
Boolean NMF	$X \approx \min(WH, 1), W \in \{0, 1\}^{m \times r}, H \in \{0, 1\}^{r \times n}$
interval-valued NMF	$(WH)_{i,j} \in X(i, j) = [a(i, j), b(i, j)]$
kernel NMF	$\Phi(X) \approx WH, W \geq 0, H \geq 0$
bilinear NMF	$W \geq 0, H \geq 0, H^\circ \geq 0$
	$X(:, j) \approx WH(:, j) + \sum_{k < \ell} (W(:, k) \circ W(:, \ell)) H^\circ(k, \ell, j)$

and spherical k-means (Section 5.5.3), PLSA and PLSI (Section 5.5.4), and neural network and autoencoders (Section 5.5.5).

5.5.1 ▪ Unconstrained factorizations and PCA

For an Exact NMF of X of size $r = \text{rank}(X)$, that is, $X = WH$ with $W \in \mathbb{R}_+^{m \times r}$ and $H \in \mathbb{R}_+^{r \times n}$, we have that $\text{col}(W) = \text{col}(X)$ and $\text{col}(H^\top) = \text{col}(X^\top)$; see Lemma 2.7. This implies that for any exact unconstrained factorization of $X = AB$ of size r with $A \in \mathbb{R}^{m \times r}$ and $H \in \mathbb{R}^{r \times n}$, we have $\text{col}(W) = \text{col}(A)$ and $\text{col}(H^\top) = \text{col}(B^\top)$. Such unconstrained factorizations can be computed via the SVD (see Section 6.1.1) and are closely related to PCA (see the introduction of the book). This observation implies that there exists an invertible matrix Q such that

$$W = AQ \quad \text{and} \quad H = Q^{-1}B; \tag{5.7}$$

see in particular Theorems 2.21 and 4.4. When $\text{rank}(X) = \text{rank}_+(X)$, Exact NMF therefore provides an invertible transformation (that is, the matrix Q) of unconstrained factorizations such as PCA to make the factors nonnegative. In the presence of noise, the assumption (5.7) is typically violated, and hence NMF provides solutions which are not invertible transformations of PCA solutions. However, some NMF algorithms are constructed explicitly to generate solutions which are invertible transformations of unconstrained factorizations; see Section 8.7.3.

5.5.2 ▪ Nonnegative PCA

Let us reformulate the unconstrained LRMA problem

$$\min_{W,H} \|X - WH\|_F^2. \tag{5.8}$$

In an unconstrained decomposition $X \approx WH$, we can assume w.l.o.g. that

- W and $H^\top$ have full column rank. If some columns of W and/or rows of H are linearly dependent, they can be discarded to reduce the size of the factorization (since there are no nonnegativity constraints).

- H has orthogonal rows, that is, $HH^\top = I_r$. Since H has rank r, there exists an invertible r-by-r matrix Q satisfying $QQ^\top = HH^\top$ (use for example the Cholesky factorization). Replacing W with WQ and H with $Q^{-1}H$ gives the results since $Q^{-1}HH^\top Q^{-\top} = I_r$ since $HH^\top = QQ^\top$.

Under these assumptions, for H fixed, the optimal W is given by $XH^\top$: this follows from the first-order optimality condition $\nabla_W \|X - WH\|_F^2 = 2(X - WH)H^\top = 0$ and the convexity of $\|X - WH\|_F^2$ with respect to W (for H fixed); see Lemma 5.2. Hence, (5.8) is equivalent to

$$\min_{H, HH^\top = I_r} \|X - XH^\top H\|_F^2.$$

This problem has the same objective function as projective NMF (Section 5.4.1) but has an orthogonality constraint instead of a nonnegativity constraint for H. Let us further reformulate this problem as follows:

$$\begin{aligned}
\|X - XH^\top H\|_F^2 &= \langle X, X \rangle - 2 \langle X, XH^\top H \rangle + \langle XH^\top H, XH^\top H \rangle \\
&= \|X\|_F^2 - 2 \langle XH^\top, XH^\top \rangle + \langle XH^\top, XH^\top \rangle \\
&= \|X\|_F^2 - \|XH^\top\|_F^2,
\end{aligned}$$

where we used the facts that $HH^\top = I_r$, $\|A\|_F^2 = \langle A, A \rangle$, and $\langle A, BC \rangle = \langle AC^\top, B \rangle$. Hence (5.8) can be formulated as

$$\max_{H, HH^\top = I_r} \|XH^\top\|_F^2 = \text{tr}\left(HX^\top XH^\top\right).$$

For $r = 1$, we obtain

$$\max_{h \in \mathbb{R}^n, \|h\|_2 = 1} h^\top (X^\top X) h,$$

which is the standard formulation for obtaining the first principal component of $X^\top X$ (although PCA usually first centers the data by removing the mean).

Nonnegative PCA solves the same problem with nonnegativity constraints on h:

$$\max_{h \in \mathbb{R}^n_+, \|h\|_2=1} h^\top (XX^\top)h; \tag{5.9}$$

see for example [354] and the references therein. If $X \geq 0$, the two problems are equivalent because we can assume w.l.o.g. that $h \geq 0$. Moreover, (5.9) is equivalent to rank-one semi-NMF; this follows from the derivations above. This implies that (5.9) is NP-hard when $X \not\geq 0$, and can be solved in polynomial-time when $X \geq 0$; see Section 5.4.3.

Adding the assumption that the rows of H are mutually independent leads to nonnegative independent component analysis; we refer the interested reader to [383] and the references therein.

5.5.3 ▪ k-means and spherical k-means

Let $\{X(:,j)\}_{j=1}^n$ be a set of n data points in dimension m. The goal of k-means is to find k centroids $\{W(:,\ell)\}_{\ell=1}^k$ such that the sum of the squared Euclidean distances between the data points and their closest centroid is minimized. To be consistent with the terminology of k-means, we make an exception in this section and use k instead of r to denote the number of columns of W. The k-means problem can be formulated as follows: Given $X \in \mathbb{R}^{m \times n}$ and k,

$$\min_{W \in \mathbb{R}^{m \times k}} \sum_{j=1}^n \left(\min_{1 \leq \ell \leq k} \|X(:,j) - W(:,\ell)\|_2^2 \right). \tag{5.10}$$

Note that if $X \geq 0$, then w.l.o.g. W can be assumed to be nonnegative. Let us reformulate k-means in a form that resembles NMF:

$$\min_{\substack{W \in \mathbb{R}^{m \times k}_+ \\ H \in \{0,1\}^{k \times n}}} \sum_{j=1}^n \|X(:,j) - WH(:,j)\|_2^2 \text{ such that } HH^\top \text{ is diagonal}, H^\top e = e. \tag{5.11}$$

By Theorem 4.39, $HH^\top$ is diagonal if and only if its rows have disjoint supports if and only if each column has at most a single positive entry. The constraint $H^\top e = e$ ensures that each data point is associated with a centroid, with a weight of one. Without this constraint, there would be a subtle difference: data points would be allowed to be approximated by zero. In other words, the vector of zeros would be a "hidden" centroid in (5.11), and $H(:,j) = 0$ would mean that the jth data point, $X(:,j)$, was attached to the hidden centroid. Interestingly, the diagonal entries of $HH^\top$ indicate the number of data points associated with each cluster, that is, $(HH^\top)_{\ell\ell}$ is the number of data points in the ℓth cluster ($1 \leq \ell \leq k$).

If we relax the constraint $H \in \{0,1\}^{k \times n}$ to $H \geq 0$, and remove the constraint $H^\top e = e$, we obtain ONMF:

$$\min_{W \in \mathbb{R}^{m \times k}_+, H \in \mathbb{R}^{k \times n}_+} \|X - WH\|_F^2 \quad \text{such that } HH^\top = I_k, \tag{5.12}$$

In (5.12), we assume w.l.o.g. that $HH^\top = I_k$ because of the scaling degree of freedom: For any pair W and H such that $HH^\top$ is diagonal, we have

$$\underbrace{W \operatorname{diag}(t)}_{W'} \underbrace{\operatorname{diag}(t)^{-1} H}_{H'}, \text{ with } H'H'^\top = I_k$$

by taking $t_\ell = \sqrt{H(\ell,:)H(\ell,:)^\top} = \sqrt{(HH^\top)_{\ell,\ell}}$ for $\ell = 1, 2, \ldots, k$. These derivations show that ONMF is a relaxation of k-means, where data points are approximated by multiples of the columns of W.

It turns out that ONMF is more closely related to spherical k-means. Spherical k-means aims at finding k directions $\{W(:,\ell)\}_{\ell=1}^k$ such that the sum of the cosines of the angles between the data points and their closest direction is maximized; see (5.14) below. To show this, let us observe the following simple fact. For two vectors a and b,

$$\mathrm{argmin}_{\alpha \in \mathbb{R}} \|a - \alpha b\|_2^2 = \frac{a^\top b}{\|b\|_2^2},$$

so that

$$\min_{\alpha \in \mathbb{R}} \|a - \alpha b\|_2^2 = \|a\|_2^2 - 2\alpha a^\top b + \alpha^2 \|b\|_2^2 = \|a\|_2^2 - \frac{(a^\top b)^2}{\|b\|_2^2}.$$

Let us use this observation for the jth column of X, with $a = X(:,j)$, $b = W(:,\ell)$, and $\alpha = H(\ell,j)$. Let $\ell^* = \mathrm{argmax}_{1 \le \ell \le k} \frac{X(:,j)^\top W(:,\ell)}{\|W(:,k)\|_2^2}$. Since $H(:,j)$ minimizes $\|X(:,j) - WH(:,j)\|_2^2$ in ONMF, the nonzero entry of $H(:,j)$ is $H(\ell^*,j) = \frac{X(:,j)^\top W(:,\ell^*)}{\|W(:,\ell^*)\|_2^2}$. Note that $H(:,j) = 0$ if and only if $X(:,j)^\top W(:,\ell) = 0$ for all ℓ, which would require all columns of W to have their support disjoint to that of $X(:,j)$, since W and X are nonnegative. We can remove the variable H from (5.12) to obtain the following equivalent problem (since $\|X(:,j)\|_2^2$ are constant terms):

$$\max_{W \in \mathbb{R}_+^{m \times k}} \sum_{j=1}^n \left(\max_{1 \le \ell \le k} \frac{X(:,j)^\top W(:,\ell)}{\|W(:,\ell)\|_2^2} \right). \tag{5.13}$$

Let $\tilde{X}(:,j) = \frac{X(:,j)}{\|X(:,j)\|_2}$ be the normalization of the jth data points ($1 \le j \le n$). Using the scaling degree of freedom, we can assume w.l.o.g. that $\|W(:,\ell)\|_2 = 1$ for $\ell = 1, 2, \ldots, k$, and ONMF is equivalent to

$$\max_{W \in \mathbb{R}_+^{m \times k}} \sum_{j=1}^n \|X(:,j)\|_2^2 \left(\max_{1 \le \ell \le k} \tilde{X}(:,j)^\top W(:,\ell) \right)$$

$$\text{such that } \|W(:,\ell)\|_2 = 1 \text{ for } \ell = 1, 2, \ldots, k.$$

Without the terms $\|X(:,j)\|_2^2$, this is exactly the spherical k-means problem, which has the following formulation:

$$\max_{W \in \mathbb{R}^{m \times k}} \sum_{j=1}^n \max_{1 \le \ell \le k} \left(W(:,\ell)^\top \tilde{X}(:,j) \right) \text{ such that } \|W(:,\ell)\|_2 = 1 \text{ for } \ell = 1, 2, \ldots, k. \tag{5.14}$$

Hence ONMF is a weighted variant of spherical k-means; see [384] for more details and discussions. ONMF gives more importance to data points with large norm, which is reasonable in many applications as data points with small norm typically contain less information and are more easily affected by noise. This would be the case, for example, in text mining where long documents should be given more importance than short ones; see for example the discussion in [294].

5.5.4 ▪ Probabilistic latent semantic analysis and indexing

Probabilistic latent semantic analysis (PLSA), also known as probabilistic latent semantic analysis indexing (PLSI), is a probabilistic topic model. Let us briefly describe it here. In PLSA, the number of documents, n, is assumed to be fixed, while the dictionary contains m words. The observation is a matrix of word counts, $X \in \mathbb{Z}_+^{m \times n}$, where $X(i, j)$ is the number of times word i appears in document j. The matrix X is also referred to as the co-occurrence matrix of the words and documents. The total number of words observed in all of the documents is referred to as the *length* of a set of documents and is equal to $\ell = e^\top X e$. Let us explain how an observation X is generated according to PLSA. First, let us define

- the vector $\hat{s} \in \mathbb{R}_+^r$ where $\hat{s}(k)$ is the probability of a word sampled randomly to be associated to with the kth topic for $k = 1, 2, \ldots, r$, with $\hat{s}^\top e = 1$,

- the matrix $\hat{A} \in \mathbb{R}_+^{m \times r}$ where $\hat{A}(i, k)$ is the probability of using the ith word in the dictionary assuming we are discussing the kth topic, for $i = 1, 2, \ldots, m$ and $k = 1, 2, \ldots, r$, with $\hat{A}^\top e = e$, and

- the matrix $\hat{B} \in \mathbb{R}^{r \times n}$ where $\hat{B}(k, j)$ is the probability of being within the jth document assuming we are discussing the kth topic, for $k = 1, 2, \ldots, r$ and $j = 1, 2, \ldots, n$, with $\hat{B}e = e$.

Then, PLSA assumes the word co-occurrence matrix X of length ℓ is a sample of a random variable $\tilde{X}$ and is generated by sampling ℓ words as follows:

0. Set $X(i, j) = 0$ for $i = 1, 2, \ldots, m$ and $j = 1, 2, \ldots, n$.

1. For $p = 1, 2, \ldots, \ell$

   1.1 Pick a topic $k \in \{1, 2, \ldots, r\}$ with probability given by $\hat{s}$.

   1.2 Pick a word $i \in \{1, 2, \ldots, n\}$ with probability given by $\hat{A}(:, k)$.

   1.3 Pick a document $j \in \{1, 2, \ldots, m\}$ with probability given by $\hat{B}(k, :)$.

   1.4 $X(i, j) = X(i, j) + 1$.

PLSA assumes that each word sampled in the data set is generated so that *the words and documents are conditionally independent given the hidden topic*. The above model implies that

$$\frac{1}{\ell} \mathbb{E} \left(\tilde{X} \right) = \hat{A} \operatorname{diag}(\hat{s}) \hat{B}$$

since $\frac{1}{\ell} \mathbb{E}(\tilde{X}_{ij}) = \sum_{k=1}^r \hat{s}(k) \hat{A}(i, k) \hat{B}(k, j)$. As the number of words in the data set, ℓ, increases, $\frac{1}{\ell} X$ gets closer to $\frac{1}{\ell} \mathbb{E}(\tilde{X})$, and hence the factorization model

$$X \approx \ell \hat{A} \operatorname{diag}(\hat{s}) \hat{B}$$

makes sense, given that ℓ is sufficiently large. More precisely, this requires that the number of words in each document is relatively large, since we need $X(:, j) \approx \ell \hat{A} \operatorname{diag}(\hat{s}) \hat{B}(:, j)$ for all j. This is a relatively strong assumption in practice; see also the discussion in Section 5.4.9.1.

   Given X and r, the goal of PLSA is to estimate $\hat{s}$, $\hat{A}$, and $\hat{B}$. To do so, PLSA assumes $\tilde{X}(i, j)$ follows a Poisson distribution of parameter $(\hat{A} \operatorname{diag}(\hat{s}) \hat{B})_{ij}$, similar to KL-NMF; see Section 5.1.1. It then uses the maximum likelihood estimator for $(\hat{A}, \hat{s}, \hat{B})$ which is obtained by solving

$$\max_{(A, s, B) \geq 0} \sum_{i, j, k} X_{ij} \log(A \operatorname{diag}(s) B)_{ij} \text{ such that } s^\top e = 1, A^\top e = e \text{ and } Be = e. \quad (5.15)$$

A solution (A, s, B) of (5.15) can be directly used to construct an NMF (W, H) of X, by choosing $W = A$ and $H = \ell \operatorname{diag}(s)B$ so that

$$X \quad \approx \quad \ell A \operatorname{diag}(s)B \quad = \quad WH.$$

The question is, To what NMF model does this decomposition correspond? In other words, is there a particular NMF model for which we can construct a solution (A, s, B) of (5.15) from a solution (W, H)? The main issue is that NMF solutions do not necessarily satisfy the sum-to-one constraints. Let (W, H) be an NMF of X of size r. Let $s_w = W^\top e$ and $s_h = He$. Assume w.l.o.g. that $s_w > 0$ and $s_h > 0$; otherwise remove the factors equal to zero. Then, we construct

- $A = W \operatorname{diag}(s_w^{-1})$ with $A^\top e = \operatorname{diag}(s_w^{-1})W^\top e = \operatorname{diag}(s_w^{-1})s_w = e$,

- $B = \operatorname{diag}(s_h^{-1})H$ with $Be = \operatorname{diag}(s_h^{-1})He = \operatorname{diag}(s_h^{-1})s_h = e$, and

- $s = \frac{1}{\ell}s_w \circ s_h$ where $\circ$ is the componentwise multiplication so

$$\operatorname{diag}(s) = \frac{1}{\ell} \operatorname{diag}(s_w) \operatorname{diag}(s_h).$$

This construction gives $WH = \ell A \operatorname{diag}(s)B$ with $A^\top e = e$ and $Be = e$. However, there is no guarantee in general that $s^\top e = 1$. Nevertheless, it turns out that PLSA is equivalent to KL-NMF [182, 131]. Let us explain this result.

Recall that KL-NMF is the NMF problem minimizing the KL divergence (that is, the β-divergence for $\beta = 1$) between X and WH:

$$\min_{W \geq 0, H \geq 0} D_1(X, WH), \tag{5.16}$$

where

$$
\begin{aligned}
D_1(X, WH) &= \sum_{i,j} X_{ij} \log \frac{X_{ij}}{(WH)_{ij}} - X_{ij} + (WH)_{ij} \\
&= \sum_{i,j} (WH)_{ij} - X_{ij} \log(WH)_{ij} + c,
\end{aligned}
$$

where c is a constant. It can be proved that any stationary point (W, H) of KL-NMF (5.16) preserves the row and column sum of X, that is, $WHe = Xe$ and $e^\top WH = e^\top X$; see Theorem 6.9 in the next chapter. This means that at a stationary point, $e^\top WHe = e^\top Xe = \ell$. Therefore, constructing (A, s, B) as described above using a stationary point (W, H) of KL-NMF, we obtain

$$\ell = e^\top WHe = e^\top \ell A \operatorname{diag}(s)Be = \ell e^\top \operatorname{diag}(s)e = \ell e^\top s,$$

and hence s satisfies the sum-to-one constraint. This implies that (A, s, B) is a feasible solution to PLSA (5.15). Moreover, for any stationary point (W, H) of KL-NMF, we have

$$D_1(X, WH) = \ell - X_{ij} \log(WH)_{ij} + c,$$

which shows that the objective functions of PLSA (5.15) and KL-NMF (5.16) coincide, up to constant terms, given that $WH = \ell A \operatorname{diag}(s)B$. In particular, this holds for optimal solutions of (5.16) since they must be stationary points, and hence PLSA and KL-NMF are equivalent in the sense that the optimal solutions of (5.15) and (5.16) coincide using the transformations defined above.

Remark 5.1 (Algorithms for PLSA and KL-NMF). *For both PLSA and KL-NMF, the algorithm based on MU is the most popular; see Section 8.2. However, the MU for PLSA are not equivalent to the MU for KL-NMF because they take the normalization constraints into account explicitly in the updates, while KL-NMF achieves normalization only at the limit (assuming the MU converge to stationary points; see Section 8.2.4). This may lead the MU of PLSA and of KL-NMF to different solutions; see the discussion in [131] for more details.*

Other topic models   There exist numerous topic models. PLSA is an early and simple model. It is a reasonable one in scenarios where the documents are long enough, because the total number of words observed, ℓ, must be sufficiently large to be able to estimate $\hat{B}$. This implies that KL-NMF is also only well-suited in this scenario (as they are equivalent; see above). For example, tri-symNMF applied on the word co-occurrence matrix $XX^\top$ is arguably more reasonable. It does not require the data set to contain only long documents. Intuitively, with the data set X, PLSA needs to estimate $mr + r + nr$ parameters[48] (namely, $\hat{A}$, $\hat{s}$, and $\hat{B}$), while tri-symNMF only needs to estimate $mr + r^2$ parameters (namely, $\hat{W}$ and S). Tri-symNMF is closely related to the latent Dirichlet allocation model [47]. Such models only require that sufficiently many documents are available; see Section 5.4.9. We refer the interested reader to [46] and the references therein for more information on probabilistic topic models.

5.5.5 ▪ Neural networks and autoencoders

Let us consider a neural network with one hidden layer to classify data points according to their labels; see Figure 5.11 for an illustration. Let $z_i \in \mathbb{R}^p$ be a p-dimensional input vector and let $y_i \in \mathbb{R}^m$ be the corresponding m-dimensional output labels, for $i = 1, 2, \ldots, n$. Construct an input data matrix by concatenating the n vectors, $Z = [z_1, z_2, \ldots, z_n] \in \mathbb{R}^{p \times n}$, and similarly an output matrix $Y = [y_1, y_2, \ldots, y_n] \in \mathbb{R}^{m \times n}$. Note that if $Z = Y$ then we obtain a so-called autoencoder neural network which is useful to extract features automatically within the hidden layer.

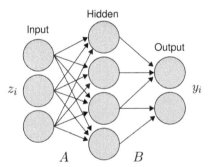

Figure 5.11. *Illustration of a one-hidden-layer neural network, with $p = 3$, $r = 4$, and $m = 2$.*

Defining the weights between the input and hidden layers as $A \in \mathbb{R}^{r \times p}$, the hidden variables are given by Az_i for $i = 1, 2, \ldots, n$. In the hidden neurons, a nonlinear function is applied to the hidden variables that we denote $h(\cdot)$. We also define $B \in \mathbb{R}^{m \times r}$ as the weights between the hidden and output layers so that, in an ideal situation (no noise, correct labels),

$$y_i = Bh(Az_i) \text{ for } i = 1, 2, \ldots, n, \text{ or, in matrix form, } Y = Bh(AZ),$$

where $h(\cdot)$ applied on a matrix is defined as applying $h(\cdot)$ columnwise to that matrix. Note that,

---

[48]There are actually fewer parameters, because of the sum-to-one constraints; however, for simplicity, they are not taken into account here. The same comment applies to tri-symNMF.

for simplicity, we have not considered in this model

- bias which can be taken into account using $y_i = Bh(Az_i + c_1) + c_2$ for all i for some vectors c_1 and c_2,

- nonlinear activations on the output layer, in which case the model would be $Y = h(Bh(AZ))$.

Let us consider the rectified linear unit (ReLU) as the nonlinear function h, that is, $h(x) = \max(0, x) = [x]_+$ (negative entries are set to zero). In practice, one has to compute the matrices A and B in order to minimize some loss function. Using least squares, we obtain the following optimization problem:

$$\min_{A,B} \|Y - B[AZ]_+\|_F^2. \tag{5.17}$$

To sum up, (5.17) optimizes the weights of a single-hidden-layer neural network with no bias that uses ReLU as the nonlinear activation in the hidden layer and the Euclidean distance as the error measure. There is a clear link between (5.17) and NMF, taking $X = Y$, $W = B$, and $H = [AZ]_+$. To have equivalence between the two models, we need $B \geq 0$ in (5.17) while we need H to have the form $H = [AZ]_+$ in NMF. Interestingly, when considering an autoencoder with $Z = Y = X$, we have $H = [AX]_+$ which is closely related to convex NMF (Section 5.4.2). Convex NMF applied on the rows of X is the model

$$X \approx W(AX), \quad \text{where } A \geq 0,$$

so that the rows of H are nonnegative linear combinations of the rows of X. Hence (5.17) is a generalization of convex NMF that could be very useful for several applications, including hyperspectral unmixing, document analysis, and audio source separation [426]. The geometric interpretation of NMF with $H = [AX]_+$ is particularly interesting. While convex NMF and archetypal analysis require A to be nonnegative, having negative entries in A in (5.17) allows the rows of AX to be outside the convex cone generated by the rows of X. If we do not want the rows of H to go too far away from this convex cone, we might want to change the constraint $A \geq 0$ to $A \geq -\epsilon$ for some $\epsilon > 0$. This idea was used in [118] and the corresponding problem was referred to as near-convex archetypal analysis, closely related to min-vol NMF.

Extending the observations above from one to many hidden layers allows us to link deep neural networks with deep NMF (Section 5.4.10).

5.6 ▪ Take-home messages

Choosing the right model for your data is a crucial aspect when using NMF in practice, as already pointed out in the previous chapter when discussing the identifiability of NMF. In this chapter, we have described various NMF models. As we have seen, there exist numerous NMF variants and we have not reviewed them all; we refer the interested reader to [98, 382, 478] and the references therein for more on this topic.

Interestingly, many NMF models are linked with existing problems in the literature. For example, separable NMF is closely related to the column subset selection problem (see Chapter 7), and tri-symNMF to the mixed membership stochastic block model (see Section 5.4.9). In practice, knowledge of these connections is important in order to use the most appropriate model for the data at hand. For example, in text mining, it is well-known that latent Dirichlet allocation is arguably a more appropriate model than PLSA (hence KL-NMF) and should be preferably used [47]. Also, as discussed in Section 5.4.9, it is more meaningful to apply NMF on the word co-occurrence matrix $XX^\top$ via tri-symNMF, for which it provides state-of-the-art results in topic modeling [14, 171]; see also Section 7.8, where it is shown how the factors of tri-symNMF can be computed in polynomial time under the separability assumption.

Chapter 6

Computational complexity of NMF

In Chapter 2, we have seen that Exact NMF is easily solvable when $\text{rank}(X) \leq 2$ (Theorem 2.6), and that it is NP-hard to check whether $\text{rank}(X) = \text{rank}_+(X)$ (Theorem 2.20). However, Exact NMF can be solved in time $\mathcal{O}((mn)^{r^2})$, implying it can be solved in polynomial time when the factorization rank r is not part of the input, that is, when r is assumed to be a fixed constant. However, this has not lead so far to practical algorithms for Exact NMF when r is small; see the discussion that follows Theorem 2.21.

When there exists an Exact NMF of size r, solving NMF, that is, Problem 1.1 (page 4), allows us to recover it because the objective function in NMF must satisfy the following property: $D(X, WH) = 0$ if and only if $X = WH$; see the discussion after Problem 1.1. Therefore, any hardness result that applies to Exact NMF applies to NMF.

Theorem 6.1. *[465] NMF is NP-hard.*

Proof. This follows from the NP-hardness result of Vavasis [465] for Exact NMF when r is part of the input. □

However, since Exact NMF can be solved in polynomial time when r is not part of the input, it is worth discussing the complexity of NMF when r is small and not part of the input.

Organization of the chapter We first focus on NMF based on the Frobenius norm that minimizes $\|X - WH\|_F^2$ (Section 6.1). It corresponds to the maximum likelihood estimator if the noise added to WH is i.i.d. Gaussian noise; see Section 5.1.1. It is one of the most widely used NMF objective functions. As we will see, it is possible to compute an optimal solution in the case $r = 1$. We then discuss the KL divergence (Section 6.2) and the infinity norm (Section 6.3) for which it is also possible to compute an optimal solution in the case $r = 1$. We then discuss the weighted Frobenius norm which can be used in the context of missing data (Section 6.4), and the componentwise ℓ_1 norm which is more tolerant to outliers (Section 6.5). For these two cases, it turns out that the problem is NP-hard even when $r = 1$. Finally, we briefly discuss the complexity of NMF variants in Section 6.6. We conclude with some take-home messages.

6.1 ▪ Frobenius norm

In this section, we discuss the complexity of the NMF problem using the Frobenius norm (ℓ_2-NMF): Given $X \in \mathbb{R}_+^{m \times n}$ and r, solve

$$\min_{W \in \mathbb{R}_+^{m \times r}, H \in \mathbb{R}_+^{r \times n}} \|X - WH\|_F^2. \tag{6.1}$$

This section is organized as follows. We first review the well-known result of Eckart and Young in the unconstrained case, that is, when W and H are not required to be componentwise nonnegative (Section 6.1.1). Then, we consider the special cases when $r = 1$ (Section 6.1.2), $r = 2$ (Section 6.1.3), and $r \geq 3$ (Section 6.1.4).

6.1.1 ▪ The singular value decomposition

Let us state one of the most important theorems in numerical linear algebra.

Theorem 6.2 (SVD). *[216] For $X \in \mathbb{R}^{m \times n}$ with $m \leq n$, there exist orthogonal matrices*

$$U = [u_1, u_2, \ldots, u_m] \in \mathbb{R}^{m \times m} \quad and \quad V = [v_1, v_2, \ldots, v_n] \in \mathbb{R}^{n \times n}$$

such that
$$U^\top X V = [\Sigma, 0_{m \times n - m}],$$

where Σ is a diagonal m-by-m matrix whose diagonal entries $\Sigma(i, i) = \sigma_i$ for $1 \leq i \leq m$ are such that $\sigma_1 \geq \sigma_2 \geq \cdots \geq \sigma_m \geq 0$ and are called the singular values of X. Since $UU^\top = I_m$ and $VV^\top = I_n$, we also have that $X = U\Sigma V^\top$, $XV = U\Sigma$, and $X^\top U = V\Sigma^\top$. The columns of U (resp. V) are the left (resp. right) singular vectors of X.

Using the SVD, the best rank-r approximation of a given matrix X can be computed as follows.

Theorem 6.3 (Eckart–Young theorem). *[141] Let $X \in \mathbb{R}^{m \times n}$ and let (U, Σ, V) be its SVD. Let also $r \leq \min(m, n)$ and $X_r = \sum_{j=1}^r \sigma_j U(:, j) V(:, j)^\top$ be the rank-r truncated SVD of X. Then*

$$\min_{\mathrm{rank}(B) \leq r} \|X - B\|_F^2 \quad = \quad \|X - X_r\|_F^2 \quad = \quad \sum_{j=r+1}^{\min(m,n)} \sigma_j^2.$$

The matrix X_r is also optimal for any unitarily invariant norms, that is, for any norm invariant under orthogonal transformations, such as the matrix ℓ_2 norm.

The SVD can be computed up to any precision ϵ in time polynomial in m, n, and $\mathcal{O}(\log(1/\epsilon))$; see [464, 456] and the references therein.

Why is the unconstrained low-rank matrix approximation in the Frobenius norm tractable?
Similar to the NMF formulation (6.1), the unconstrained best rank-r approximation can be formulated as follows: Given $X \in \mathbb{R}^{m \times n}$ and r, solve

$$\min_{W \in \mathbb{R}^{m \times r}, H \in \mathbb{R}^{r \times n}} \|X - WH\|_F^2. \tag{6.2}$$

This is a nonconvex optimization problem. However, surprisingly, it has a very nice property: every local minimum is global, and all other stationary points are saddle points. The set of stationary points can be characterized using the SVD. Let (U, Σ, V) be the SVD of X. The solution (W, H) is a stationary point of (6.2) if and only if

$$WH = \sum_{i \in \mathcal{S}} \sigma_i U(:, i) V(:, i)^\top,$$

where $\mathcal{S} \subseteq \{1, 2, \ldots, \min(m, n)\}$ and $|S| = r$; see for example [239, Chapter 1.3] and also [143]. Moreover, (6.2) can be formulated as a convex semidefinite optimization problem using the Ky

Fan 2-k norm [134]. Interestingly, (6.2) retains his nice properties as long as the optimization problem is not modified significantly; see [515, 229] and the references therein. For example, in the presence of missing entries, the objective function is modified to $\sum_{(i,j)\in\Omega}(X - WH)^2_{ij}$ where Ω is the set of observed entries. If the missing entries are not too numerous and their position is picked randomly, the problem remains tractable [515].

However, the nonnegativity constraints completely destroy the nice geometric properties of (6.2) and make the problem NP-hard in general (Theorem 6.1). The global minimum (W, H) is in general not located within the nonnegative orthant (orthogonality and nonnegativity of the columns of a matrix imply they have disjoint supports; see Chapter 4.3.2), so that the nonnegativity constraints generate in general numerous spurious local minima and saddle points on the boundary of the nonnegative orthant.

6.1.2 ▪ Case $r = 1$

Let us make the following simple observation. Let $X \in \mathbb{R}^{m\times n}_+$, $w \in \mathbb{R}^m$, and $h \in \mathbb{R}^n$. We have

$$\left\|X - wh^\top\right\|_F \quad \geq \quad \left\||X - |wh^\top|\right\|_F = \left\|X - |w||h^\top|\right\|_F \,,$$

where $|A|$ denotes the componentwise absolute value of matrix A. In fact, for a nonnegative real number x and any a, $(x - a)^2 \geq (x - |a|)^2$. This implies the following.

Theorem 6.4. *NMF* (6.1) *with $r = 1$ can be solved in polynomial time.*

Proof. Let $u_1\sigma_1 v^T_1$ be the rank-one truncated SVD of X, which can be computed in polynomial time (Theorem 6.3). Then $w = \sqrt{\sigma_1}|u_1|$ and $h = \sqrt{\sigma_1}|v_1|$ is an optimal solution of NMF (6.1) with $r = 1$, by Theorem 6.3. □

Let us make a few comments.

- The first rank-one terms (u_1, σ_1, v_1) computed by the SVD are not necessarily nonnegative. When using MATLAB, we often observe that $u_1 \leq 0$ and $v_1 \leq 0$, and hence it is important to take their absolute value to obtain an NMF solution; see also the discussion in [64].

- If the matrix $X^\top X$ or $XX^\top$ is not irreducible (see page 173 for the definition) and $\sigma_1(X) = \sigma_2(X)$, there might exist optimal solutions of the rank-one unconstrained matrix approximation problem with negative entries. This is related to the Perron–Frobenius theorem, which states that the largest eigenvalue of an irreducible matrix is simple and that the entries of the corresponding eigenvector are nonzero and have the same sign [38]. The simplest example is the 2-by-2 identity matrix $I_2 = \begin{pmatrix} 1 & 0 \\ 0 & 1 \end{pmatrix}$ with $\sigma_1 = \sigma_2 = 1$. The optimal rank-one approximations have the form

$$\begin{pmatrix} \lambda^2_1 & \lambda_1\lambda_2 \\ \lambda_1\lambda_2 & \lambda^2_2 \end{pmatrix} \quad \text{for any } \lambda^2_1 + \lambda^2_2 = 1.$$

For example, with $\lambda_1 = -\lambda_2 = \frac{\sqrt{2}}{2}$, we obtain

$$\begin{pmatrix} 1/2 & -1/2 \\ -1/2 & 1/2 \end{pmatrix}$$

with negative entries. However, taking its absolute value leads to a nonnegative optimal rank-one approximation (corresponding to $\lambda_1 = \lambda_2 = \frac{\sqrt{2}}{2}$).

A somewhat surprising result is that if the input matrix X is not nonnegative, then the rank-one NMF problem is NP-hard.

Theorem 6.5. *[194, Corollary 1], [200, Corollary 1] Given $X \in \mathbb{R}^{m \times n}$, solving*

$$\min_{w \in \mathbb{R}_+^m, h \in \mathbb{R}_+^n} \|X - wh^\top\|_F^2$$

is NP-hard.

Proof. [Sketch of the proof] Let us provide a sketch of the proof. We believe this might give some more insight into NMF and its relationship with an important graph-theoretic problem, namely the biclique problem. Let us describe this problem; see also Section 3.4.4, where its link with the nonnegative rank was discussed.

A bipartite graph G_b is a graph whose vertices can be divided into two disjoint sets, V_1 and V_2:

$$G_b = (V, E) = \Big(V_1 \cup V_2, E \subseteq (V_1 \times V_2)\Big).$$

A *biclique* K_b is a complete bipartite graph, that is, a bipartite graph for which all the vertices of V_1 are connected to all the vertices of V_2, that is, $E = V_1 \times V_2$. The so-called maximum-edge biclique problem in a bipartite graph $G_b = (V, E)$ is the problem of finding a biclique $K_b = (V', E')$ in G_b (that is, $V' \subseteq V$ and $E' \subseteq E$) maximizing its number of edges $|E'| = |V_1'||V_2'|$. The decision problem, *Given B, does G_b contain a biclique with at least B edges?*, has been shown to be NP-complete [380].

Let $M_b \in \{0,1\}^{m \times n}$ be the biadjacency matrix of the unweighted bipartite graph $G_b = (V_1 \cup V_2, E)$ with $V_1 = \{s_1, \ldots s_m\}$ and $V_2 = \{t_1, \ldots t_n\}$, that is, $M_b(i,j) = 1$ if and only if $(s_i, t_j) \in E$. With this notation, the maximum-edge biclique problem in G_b can be formulated as

$$\min_{w \in \{0,1\}^m, h \in \{0,1\}^n} \left\|M_b - wh^\top\right\|_F^2 \quad \text{such that} \quad wh^\top \leq M_b,$$

where w (resp. h) is the indicator vector for the vertices in V_1 (resp. V_2), that is, for all i, $w(i) = 1$ if and only if s_i is in the biclique, and similarly for h and t. The objective minimizes the number of edges outside the biclique, which is equivalent to maximizing the number of edges contained in the biclique. The constraint $wh^\top \leq M_b$ ensures that (w, h) corresponds to a biclique since $M_b(i,j) = 0 \Rightarrow w(i) = 0$ or $h(j) = 0$, that is, both of the two nonadjacent vertices cannot belong to the biclique. The binary constraints on the variables w and h can be relaxed because the constraints $w_i h_j \leq M_b(i,j)$ for all i, j along with the objective imply that if $w_i h_j > 0$ then $w_i h_j = 1$: if $w_i h_j$ is positive, then the best possible choice to minimize the objective is to take $w_i h_j = 1$ (recall that M_b is binary). This implies that any optimal solution (w^*, h^*) of

$$\min_{w \in \mathbb{R}_+^m, h \in \mathbb{R}_+^n} \left\|M_b - wh^\top\right\|_F^2 \quad \text{such that} \quad wh^\top \leq M_b$$

is such that $w^* {h^*}^\top$ is binary. Then, the constraint $wh^\top \leq M_b$ can be removed while replacing the zero entries in M_b by a sufficiently large negative value $-d$ for $d > 0$. Intuitively, replacing the zero entries in M_b by large negative values penalizes $wh^\top$ to approximate zero entries in M_b by positive ones. We end up with the following rank-one nonnegative factorization problem:

$$\min_{w \in \mathbb{R}_+^m, h \in \mathbb{R}_+^n} \left\|M_d - wh^\top\right\|_F^2, \quad \text{where } M_d = M_b - (1 - M_b)d. \tag{6.3}$$

It can be shown that for $d \geq \max(m, n)$, optimal solutions of (6.3) are binary [200, Theorem 2], implying NP-hardness of (6.3).   □

Theorem 6.5 is not very encouraging. For example, given all the optimal rank-one factors in an NMF decomposition but one, that is, $W(:,k)H(k,:)$ for $1 \leq k \leq r - 1$, there is no guarantee that computing that last factor can be done efficiently as the residual $X - \sum_{k=1}^{r-1} W(:,k)H(k,:)$ might have negative entries. This is in contrast with rank-one NMF that can be solved in polynomial time (Theorem 6.4).

6.1.3 ▪ Case $r = 2$

For $r = 2$, we have the following result.

Theorem 6.6. *Let $X \in \mathbb{R}^{m \times n}$ and let X_2 be the rank-two truncated SVD of X. If $X_2 \geq 0$, then NMF (6.1) with $r = 2$ can be solved in polynomial time.*

Proof. By Theorem 6.3, X_2 is an optimal rank-two approximation of X. If $X_2 \geq 0$, its Exact NMF with $r = 2$ can be computed in polynomial time (Theorem 2.6); see in particular Algorithm 4.1. ☐

Of course, there is no guarantee that $X_2 \geq 0$. In particular, if X contains many entries equal to zero, X_2 typically contains negative entries. However, since X_2 approximates a nonnegative matrix, most of its entries are nonnegative. Hence one can easily adapt Algorithm 4.1 when X_2 has negative entries: simply use $\max(X_2, 0)$ (which is not necessarily a rank-two matrix) as the input of Algorithm 4.1 to identify two extreme columns of X_2 and construct W. Then compute $H^* = \operatorname{argmin}_{H \geq 0} \|X - WH\|_F$, which is a convex nonnegative least squares (NNLS) problem (see Section 8.3). This strategy was proposed in [202] and performs well in practice; see [Matlab file: Rank2NMF.m]. However, it comes with no optimality guarantee.

When the entries of X are all strictly positive, it is more likely for X_2 to be nonnegative, which can be formalized as follows.

Lemma 6.7. *[202, Corollary 2] Let $X \in \mathbb{R}_+^{m \times n}$ and let X_2 be the rank-two truncated SVD of X. If $x_{\min} = \min_{i,j} X(i,j) \geq \sigma_3(X)$, then $X_2 \geq 0$.*

Proof. By Theorem 6.3, $\|X - X_2\|_2 = \sigma_3(X)$. If $X_2(i,j) < 0$ for some (i,j), then

$$\|X - X_2\|_2^2 \geq (X_2 - X)_{i,j}^2 > X(i,j)^2 \geq x_{\min}^2 \geq \sigma_3^2(X),$$

a contradiction. ☐

Lemma 6.7 implies that if X is close to a rank-two matrix (that is, $\sigma_3(X)$ is small) and X has large positive entries, then its best rank-two NMF can be computed in polynomial time (Theorem 6.6).

However, as far as we know, there does not exist a polynomial-time algorithm that solves NMF (6.1) with $r = 2$ in all cases.

6.1.4 ▪ Case $r \geq 3$

As mentioned in the introduction of this section, since NMF allows us to solve Exact NMF, any complexity result for Exact NMF applies to NMF. Hence, NMF is NP-hard when r is part of the input.

For $r \geq 3$, even if the rank-r truncated SVD is nonnegative, its nonnegative rank might be large; see the paragraph "Rank-one perturbations can modify the nonnegative rank by more than one" (page 60) for a discussion on matrices whose nonnegative rank is much larger than their rank. Hence, we cannot invoke the result of Arora et al. [15, 16], which states that Exact NMF with r fixed can be solved in polynomial time to conclude that, in that case, one can solve NMF of rank r in polynomial time. This is in contrast with the case $r = 2$.

Nevertheless, Arora et al. [15, 16] provided an algorithm polynomial in m and n for the approximate problem but only up to some precision.

Theorem 6.8. *[16, Theorem 6.1] Let X be an m-by-n nonnegative matrix such that there exists an NMF of size r with $\|X - WH\|_F \leq \epsilon \|X\|_F$. Then there is an algorithm that produces a rank-r NMF solution (W', H') such that $\|X - W'H'\| \leq \mathcal{O}\big(\epsilon^{1/2} r^{1/4}\big)\|X\|_F$ in time $2^{\mathrm{poly}(r\log(1/\epsilon))} \mathrm{poly}(m,n)$.*

Hence, as opposed to Exact NMF, it is an open problem to know whether NMF (6.1) can be solved up to global optimality in polynomial time in m and n, while r is fixed and not part of the input.

6.2 ▪ Kullback–Leibler divergence

Another widely used measure to assess the quality of NMF solutions is the KL divergence. It is the maximum likelihood estimator when assuming i.i.d. Poisson noise and is the β-divergence for $\beta = 1$; see Section 5.1. Given two nonnegative scalars x and y, the KL divergence between x and y is defined as follows:

$$d_1(z,y) = \begin{cases} x\log\frac{z}{y} - z + y & \text{for } z > 0, \\ y & \text{for } z = 0. \end{cases}$$

The NMF problem using the KL divergence (KL-NMF) is the following: Given $X \in \mathbb{R}_+^{m\times n}$ and r,

$$\min_{W\in\mathbb{R}_+^{m\times r}, H\in\mathbb{R}_+^{r\times n}} D_1(X, WH), \tag{6.4}$$

where $D_1(X, WH) = \sum_{i,j} d_1\big(X_{i,j}, (WH)_{ij}\big)$. An interesting property of KL-NMF is that any stationary point preserves the row sum and the column sum of X. In the case $r = 1$, it allows us to prove that KL-NMF can be solved in polynomial time (Theorem 6.10).

Theorem 6.9. *[240, Theorem 1] Given $X \in \mathbb{R}_+^{m\times n}$ and r, any stationary point (W, H) of KL-NMF (6.4) satisfies*

$$Xe = WHe \quad \text{and} \quad e^\top X = e^\top WH,$$

that is, WH preserves the row sum and the column sum of X.

Proof. Let us focus on the first-order optimality conditions for W. If (W, H) is a stationary point of (6.4), then

$$W \circ \nabla_W D_1(X, WH) = W \circ \left(\frac{[WH - X]}{[WH]}\right) H^\top = 0;$$

see page 263 in Chapter 8.1.2. Rearranging the terms, we obtain

$$W \circ \big(ee^\top H^\top\big) = W \circ \left(\frac{[X]}{[WH]} H^\top\right). \tag{6.5}$$

Let us multiply both sides by e from the right. Observe that, given two matrices $A, B \in \mathbb{R}^{m\times n}$, we have

$$(A \circ B)e = \mathrm{diag}(AB^\top).$$

For the left-hand side of (6.5), we obtain using this equivalence twice

$$\left(W \circ \left(ee^\top H^\top\right)\right) e = \mathrm{diag}(WHee^\top) = \left((WH) \circ (ee^\top)\right) e = (WH)e.$$

Similarly, for the right-hand side, we obtain

$$\left(W \circ \left(\frac{[X]}{[WH]} H^\top\right)\right) e = \mathrm{diag}\left(WH\left(\frac{[X]}{[WH]}\right)^\top\right) = \left((WH) \circ \frac{[X]}{[WH]}\right) e = Xe.$$

This implies $(WH)e = Xe$, that is, the column sum of WH and X coincide. By symmetry, the same observation holds for the row sum of X, that is, $e^\top(WH) = e^\top X$, using the optimality conditions for H. □

6.2.1 ▪ Case $r = 1$

It turns out that, for $r = 1$, an optimal solution of (6.4) can be computed in polynomial time, as for the Frobenius norm, although this result is not as well-known.

Theorem 6.10. *[240, Theorem 1] Given $X \in \mathbb{R}_+^{m \times n}$ with $X \neq 0$, the unique optimal solution of (6.4) for $r = 1$ is, up to scaling,*

$$w^* = \frac{Xe}{\sqrt{e^\top Xe}} \quad and \quad h^* = \frac{X^\top e}{\sqrt{e^\top Xe}}.$$

Proof. Let (w, h) be a stationary point of (6.4) for $r = 1$. By Theorem 6.9, $Xe = wh^\top e$ and $e^\top X = e^\top wh^\top$. Since $X \neq 0$, $w \neq 0$ and $h \neq 0$. Assume w.l.o.g. that $h^\top e = 1$. Then $w = Xe$ and $h = \frac{X^\top e}{e^\top w} = \frac{X^\top e}{e^\top Xe}$, which coincides with the solution (w^*, h^*), up to scaling. Since all the stationary points (which include the optimal solutions) have this form and have the same objective function value, they are all optimal solutions. □

Note that, for $X = 0$, the solution is not unique since any solution of the form $(w, 0)$ or $(0, h)$ is optimal.

We refer the reader to [249], where the above results are extended to tensors, that is, the authors show that the sums along all modes of the input tensor are preserved by stationary points. Some numerical experiments are also presented.

6.2.2 ▪ Case $r \geq 2$

For r fixed and $r \geq 2$, the complexity of KL-NMF (6.4) is unknown. Recall that KL-NMF (6.4) would solve Exact NMF and hence is NP-hard when r is part of the input; see Theorem 6.1. As far as we know, even the unconstrained problem, in which W and H are not required to be nonnegative, has not been studied much and its complexity is unknown. Studying the complexity of KL-NMF (6.4) for $r \geq 2$ is a topic of future research.

As opposed to ℓ_2-NMF (6.1), X and its approximation WH have to be nonnegative because the KL divergence used in KL-NMF (6.4) is defined only for nonnegative inputs. In other words, KL-NMF (6.4) where the nonnegativity constraint on W and H is discarded has a hidden nonnegativity constraint, namely $WH \geq 0$.

6.2.3 ▪ Other β-divergences

For other β-divergences, complexity issues have not been studied much. For example, $\beta = 0$ corresponds to the IS divergence, associated to multiplicative Gamma noise particularly meaningful in audio source separation; see Section 5.1.3. The complexity is unknown even for

$r = 1$, as far as we know. Moreover, the optimization problem in W for H fixed, that is, $\min_{W \geq 0} D_0(X, WH)$, is not convex for the IS divergence, as opposed to β-divergences for $\beta \in [1, 2]$. Hence it is even unknown whether the subproblem in W for H fixed can be solved in polynomial time, and similarly for H when W is fixed by symmetry.

6.3 ▪ Infinity norm

Let us now consider NMF in the infinity norm (ℓ_∞-NMF), that is,

$$\min_{W \in \mathbb{R}_+^{m \times r}, H \in \mathbb{R}_+^{r \times n}} \|X - WH\|_\infty, \tag{6.6}$$

where $\|X - WH\|_\infty = \max_{i,j} |X - WH|_{ij}$. ℓ_∞-NMF is the maximum likelihood estimator when assuming i.i.d. uniform noise added to WH; see Section 5.1.1.

As for the Frobenius norm and the KL divergence, the problem can be solved efficiently when $X \geq 0$ and $r = 1$.

Theorem 6.11. *[209, Corollary 1] For $r = 1$, $X \geq 0$ and given the scalar $c \geq 0$, checking whether the optimal value of (6.6) is smaller than c can be done in polynomial time.*

Proof. Checking whether the optimal value of (6.6) for $r = 1$ is smaller than c is equivalent to checking whether there exist $w \in \mathbb{R}_+^m$ and $h \in \mathbb{R}_+^n$ such that

$$-c \leq w_i h_j - X_{i,j} \leq c \quad \text{for all } i, j. \tag{6.7}$$

First, notice that if the ith row of X (resp. jth column of X) has all its entries smaller than c, we can take $w_i = 0$ (resp. $h_j = 0$). This means that we can assume w.l.o.g. that every column and row of X have at least one entry larger than c, by discarding rows and columns for which all entries are smaller than c. Then, all rows and columns of X must be approximated by a positive entry: for $X_{i,j} > c$, we must have $w_i h_j > 0$, otherwise (6.7) cannot be satisfied. Hence we can assume w.l.o.g. that $w > 0$ and $h > 0$. Moreover, by the scaling degree of freedom, we can assume w.l.o.g. that $w \geq e$. Rearranging the terms of (6.7), we obtain

$$X_{i,j} - c \leq w_i h_j \leq X_{i,j} + c \quad \text{for all } i, j.$$

Let us denote $s_i = \frac{1}{w_i} \leq 1$ for $i = 1, 2, \ldots, m$ and multiply both sides of the above inequalities by s_i to obtain

$$s_i(X_{i,j} - c) \leq h_j \leq s_i(X_{i,j} + c) \quad \text{for all } i, j.$$

Together with the constraints $0 \leq s \leq 1$ and $h \geq 0$, this is a linear system of inequalities that can be solved in polynomial time [342]. □

Surprisingly, the unconstrained low-rank matrix approximation problem in the infinity norm is NP-hard, even for $r = 1$ [209, Theorem 3]. This is in contrast with the Frobenius norm for which the unconstrained low-rank matrix approximation problem can be solved in polynomial time for any input matrix and any value of r (Theorem 6.3). The difference with (6.6) is that when $X \not\geq 0$, we cannot assume w.l.o.g. that $w \geq 0$ and $h \geq 0$, which is key in the proof of Theorem 6.11.

As for the KL divergence, the case $r \geq 2$ has not been studied much and the same comments apply; see the first part of Section 6.2.2.

6.4 ▪ Weighted Frobenius norm

Another important norm that is often used in practice is the weighted Frobenius norm; see for example [179, 434, 239]. The corresponding NMF problem, referred to as weighted NMF (which is a variant of WLRA) is the following: Given $X \in \mathbb{R}_+^{m \times n}$, $P \in \mathbb{R}_+^{m \times n}$, and r, solve

$$\min_{W \in \mathbb{R}_+^{m \times r}, H \in \mathbb{R}_+^{r \times n}} \sum_{i,j} P_{ij}(X - WH)_{ij}^2. \tag{6.8}$$

The two main cases for which weighted NMF is particularly useful is when data is missing ($P(i,j) = 0$ if $X(i,j)$ is missing) and when the variance of the Gaussian noise is different among the entries of X (the noise is not identically distributed); see Section 5.1.1.

Interestingly, if $\operatorname{rank}(P) = 1$, the problem can be reduced to standard NMF. In that case, there exists u and v such that $P(i,j) = u_i^2 v_j^2$ for all i,j so that

$$\sum_{i,j} P(i,j)(X - WH)_{ij}^2 = \sum_{i,j} u_i^2 v_j^2 (X - WH)_{ij}^2$$

$$= \sum_{i,j} (X_{ij} u_i v_j - (WH)_{i,j} u_i v_j)^2$$

$$= \sum_{i,j} (X' - W'H')_{ij}^2,$$

where $X'(i,j) = X(i,j)u_i v_j$, $W'(i,k) = W(i,k)u_i$, and $H'(k,j) = H(k,j)v_j$ for all i,j,k. Otherwise, the problem is hard in general.

Theorem 6.12. *[196, Theorems 1.1 and 1.2] For $r = 1$, weighted NMF (6.8) is NP-hard. This holds whether P has only positive entries or if P is binary (missing data) and whether X is nonnegative or not.*

Proof. [Sketch of the proof] Let us focus on the case when P has positive entries (the case P binary is more difficult to analyze). In this case, the proof relies on the reduction of weighted NMF from the maximum-edge biclique problem, as in Theorem 6.5. Taking $X = M_b$ where M_b is the biadjacency matrix of a bipartite graph, and $P = M_b + d(1 - M_b)$ where d is sufficiently large, weighted NMF approximates X by giving much more importance to the zero entries of M_b since $P_{ij} = d$ when $M_b(i,j) = 0$; otherwise $P_{ij} = 1$. Hence, for d sufficiently large, the optimal solution corresponds to the largest biclique in M_b which is NP-hard to compute [380]. □

In the unconstrained case, there exist approximation algorithms; we refer the reader to [393, 20] and the references therein.

6.5 ▪ Componentwise ℓ_1 norm

In the presence of outliers, the Frobenius norm typically performs poorly. In that case, the componentwise ℓ_1 norm is known to perform better [273]. It is the maximum likelihood estimator when assuming i.i.d. Laplacian noise; see Section 5.1.1. The NMF problem using the componentwise ℓ_1 norm (ℓ_1-NMF) is defined as

$$\min_{W \in \mathbb{R}_+^{m \times r}, H \in \mathbb{R}_+^{r \times n}} \|X - WH\|_1, \tag{6.9}$$

where $\|X - WH\|_1 = \sum_{i,j} |X - WH|_{ij}$. This problem is closely related to robust PCA; see [212] and the discussion therein. As for the weighted norm, the corresponding NMF problem is NP-hard.

Theorem 6.13. *[212, Theorem 3] For $r = 1$, ℓ_1-NMF (6.9) is NP-hard.*

Proof. [Sketch of the proof] The proof of this result contains two main steps. The first and most difficult step is to prove that if the input matrix X is binary, then there always exists an optimal binary solution to ℓ_1-NMF (6.9). This result is rather intriguing because it proves that, for $r = 1$ and $X \in \{0, 1\}^{m \times n}$, the optimal solution of ℓ_1-NMF (6.9) without any constraint[49] can be assumed w.l.o.g. to be binary.

The second step is to prove that finding the best rank-one binary approximation of a binary input matrix is NP-hard. This is a combinatorial problem, related to finding the densest subgraph in a bipartite graph.   □

NP-hardness also holds in the unconstrained case. As for the weighted Frobenius norm, approximation algorithms have been developed [428] and are a direction of future research in the context of NMF. Similar observations hold for componentwise ℓ_p norms; we refer the interested reader to [19].

6.6 ▪ Other NMF models

So far, we have not discussed the complexity of NMF models that were presented in Sections 4.3 and 5.4. Not surprisingly, most of these models are NP-hard. Let us discuss three of the most important models.

- *Separable NMF* in the presence of noise can be formulated as follows: find an index set $\mathcal{K}$ of cardinality r and a nonnegative matrix H such that the quantity $D\big(X, X(:, \mathcal{K})H\big)$ is minimized. Without any assumption on the input matrix X and no nonnegativity constraint on H, the problem is NP-hard [77, 78]; it is a particular column subset selection problem. Note that the number of possible solutions is $\binom{n}{r}$: once the index set $\mathcal{K}$ is selected, H can be computed efficiently in most cases. For example, for the Frobenius norm, this is a NNLS problem; see Chapter 8.3.1. Hence, the problem can be solved in polynomial time when r is not part of the input since $\binom{n}{r} = \mathcal{O}(n^r)$ by using a brute-force approach (that is, enumerating all solutions).

  However, if the input matrix X is close to a matrix of the form WH where H is separable (see Chapter 4.3.1), then separable NMF can be solved efficiently. Designing provably correct algorithms for separable NMF, even in the presence of noise, is the topic of the next chapter.

- *Min-vol NMF* is NP-hard for general input matrices X; see the discussion in Chapter 4.3.3. This is not surprising as it generalizes separable NMF. However, an important open question is to determine whether min-vol NMF can be solved in polynomial time under some additional assumptions on X, in particular when $X = WH$ where H is sufficiently scattered. (If the input matrix X is close to a matrix of the form WH where H is separable, then it can be solved efficiently via separable NMF as both problems coincide.)

---

[49]Even the nonnegativity constraints can be removed since nonnegativity of W and H can be assumed w.l.o.g. when $X \geq 0$, as for the Frobenius norm; see Section 6.1.2.

- *Sparse NMF* is also hard. For most NMF models, if one is given one of the two factors W or H, the corresponding subproblem is convex and efficiently solvable. This does not hold for sparse NMF: estimating H with a given sparsity level (say, k-sparse columns) for W fixed is NP-hard [363].

Other NMF models that have been shown to be NP-hard include NMU (Sections 5.4.5 and 8.7.1) and semi-NMF (Section 5.4.3), even when $r = 1$.

6.7 ▪ Take-home messages

NMF is at least as hard as Exact NMF, which is NP-hard when r is part of the input. Moreover, we have seen that for some error measures, even the problem for $r = 1$ is NP-hard (namely, for the weighted Frobenius norm and the componentwise ℓ_1 norm). Therefore, unless the input matrix satisfies additional assumptions (as done in Chapter 7, where we analyze separable NMF), one typically relies on heuristic algorithms to tackle NMF; these are presented in Chapter 8.

The same observations hold true for most constrained LRMA models, or LRMA models with other objective functions: as soon as the error measure $D(X, WH)$ is not the Frobenius norm or the feasible domain has constraints, LRMA problems become difficult in most cases. An active direction of research is developing approximation algorithms for such problems; see for example [19] and the references therein. Deriving such algorithms dedicated to NMF is a direction of future research.

NMF for $r = 1$ can be solved in some cases in polynomial time (namely, for the Frobenius norm, the KL divergence, and the ℓ_∞ norm). Hence it is tempting to build higher rank NMFs sequentially, one rank-one factor at a time. Another motivation is that, in the unconstrained case, this approach using deflation leads to an optimal solution (this follows directly from Theorem 6.3). However, for NMF, implementing this idea is not straightforward because after the first rank-one factor $wh^\top$ is computed, the residual $X - wh^\top$ typically contains roughly half negative and half positive entries and hence can no longer be easily and meaningfully approximated by nonnegative rank-one factors. We refer the reader to Chapter 8.7.1, presenting such approaches.

Chapter 7

Near-separable NMF

As we have discussed in detail in the previous chapters, two key issues when facing NMF problems are the following:

1. NP-hardness: Exact NMF and NMF are NP-hard, and being able to compute efficiently globally optimal decompositions for these problems in general is unlikely, even when r is relatively small; see Section 2.3 and Chapter 6.

2. Identifiability: Most matrices X do not admit a unique NMF unless some additional constraints are imposed on the NMF factors (W, H). In other words, NMF is in general ill-posed, and the factors (W, H) are in most cases not identifiable; see Chapter 4.

These two issues can be overcome by assuming that the input matrix X has the form $X = WH$ where H is separable. Separability of H requires that all unit vectors are hidden among the columns of H, up to scaling (Definition 4.10, page 107). Because of the scaling degree of freedom in the decomposition $X = WH$, separability of H is equivalent to assuming that there exists an index set $\mathcal{K}$ such that $W = X(:, \mathcal{K})$. This means that the columns of W appear as columns of X; see Figure 7.1 for an illustration.

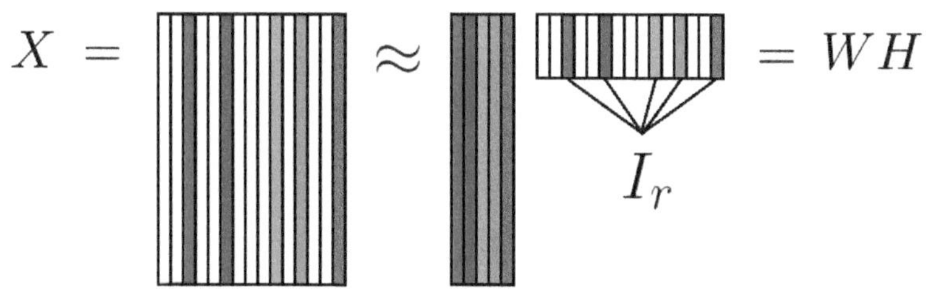

$$X = \quad \approx \quad \underbrace{\qquad}_{I_r} \quad = WH$$

Figure 7.1. *Illustration of $X \approx WH$ where H satisfies the separability assumption or, equivalently, where $W = X(:, \mathcal{K})$ for some index set $\mathcal{K}$.*

In practice, the input matrix X does not in general admit an exact separable NMF decomposition (because of noise and model misfit), and we focus in this chapter on the separable NMF problem in the presence of noise, referred to as near-separable NMF and defined as follows.

Problem 7.1 (Near-separable NMF). *Given a matrix $X \in \mathbb{R}^{m \times n}$, a factorization rank r, and a distance measure $D(.,.)$, solve*

$$\min_{\mathcal{K} \subseteq \{1,2,\ldots,n\}, H \in \mathbb{R}_+^{r \times n}} D\big(X, X(:,\mathcal{K})H\big) \quad \text{such that} \quad |\mathcal{K}| = r.$$

Let us make a few comments about near-separable NMF:

- The input matrix X is not required to be nonnegative, as for separable NMF, as discussed already in Section 4.3.1.

- Separable NMF is identifiable under the assumption that the input matrix has the form $X = X(:,\mathcal{K})H$ where $H \geq 0$ and $\text{rank}(X) = r$; see Theorem 4.37 and the discussion that follows. The identifiability of the factor $W = X(:,\mathcal{K})$ only requires that the columns of $X(:,\mathcal{K})$ are distinct extreme rays of $\text{cone}(X)$.

- Near-separable NMF is NP-hard in general; see Section 6.6. However, under the assumption that the input matrix X can be well-approximated with a decomposition of the form $X(:,\mathcal{K})H$ with $H \geq 0$ (see Assumption 7.1 for a formal description), then polynomial-time and provably efficient algorithms can be designed, even in the presence of noise. This is the main topic of this chapter.

Organization of the chapter   In Section 7.1, we review the context of near-separable NMF, several applications, and problems closely related to separable NMF. In Section 7.2, we present the model assumptions, define the notion of robustness to noise in the context of near-separable NMF in which the conditioning of W plays a central role, and discuss preprocessing of the input data. The next four sections present near-separable NMF algorithms. We start by analyzing an idealized algorithm that computes an optimal solution of near-separable NMF (Section 7.3). It allows us to provide the best possible achievable error bounds for any near-separable NMF algorithm; in other words, it provides the performance limits in terms of robustness to noise of any near-separable NMF algorithm. Then, we present the three most important classes of near-separable NMF algorithms: greedy algorithms (Section 7.4), heuristic algorithms (Section 7.5), and convex-optimization-based algorithms (Section 7.6). We summarize the computational cost, robustness to noise, and practical performances of near-separable NMF algorithms in Section 7.7. In Section 7.8, we explain how separability can be used to tackle another NMF model, namely tri-symNMF (see Section 5.4.9), which led to an important breakthrough in topic modeling by Arora et al. [14] (2013). We conclude this section with pointers to further readings (Section 7.9) and take-home messages (Section 7.10).

7.1 ▪ Context and applications

The terminology *separability* was first introduced by Donoho and Stodden [138] (2004). In their paper, Donoho and Stodden also make other strong sparsity assumptions to obtain an identifiability theorem for the standard NMF model. The same concept was used by Laurberg et al. [298] (2008) but was referred to as the sufficiently spread condition. It was later used and popularized in the seminal paper by Arora et al. [15] (2012) on the complexity of NMF and is now the standard terminology in the machine learning community.

However, separability has a long history in blind HU. In this context, each column of the matrix X is the spectral signature of a pixel, each column of the matrix W is the spectral signature of an endmember (a "pure" material), and each column of H gathers the abundances of the endmembers in a pixel; see Section 1.3.2 for more details. In the literature on blind HU, the separability assumption is referred to as the *pure-pixel assumption*: for each endmember, there must exist a pixel that contains only that endmember. Mathematically, for each $1 \leq k \leq r$, there exists i such that $W(:,k) = X(:,i)$. Equivalently, for each $1 \leq k \leq r$, there exists i such that $H(:,i) = e_k$, which means that H is separable, and we have $WH(:,i) = We_k = W(:,k) = X(:,i)$. The notion of pure pixels was introduced by Boardman, Kruse, and Green [49] (1995), who proposed an algorithm referred to as the pure-pixel index (PPI); see Section 7.5.1. If the spatial resolution of the input hyperspectral image is sufficiently high, this assumption is reasonable and has been used successfully and extensively for blind HU. Even if separability is not satisfied, near-separable NMF algorithms, referred to as *pure-pixel search algorithms* in the blind HU literature, are used extensively to initialize more sophisticated approaches such as min-vol NMF and nonlinear mixture models [136]. This is a very rich topic in blind HU with many algorithms, and we present in this chapter only a few of them, with a focus on algorithms that have theoretical guarantees to be robust to noise. We refer the interested reader to the surveys [45, 334] for comprehensive historical backgrounds and detailed discussions on blind HU under the pure-pixel assumption and beyond.

Applications Let us review the three applications, besides blind HU, described in the introduction (Section 1.3), and discuss whether the separability assumption makes sense in their context.

1. Feature extraction in a set of images. The input matrix X is a pixel-by-image matrix, the columns of W represent common features found across all images, and H indicates which feature belongs to which image. Separability of H requires that there are images in the data set containing a single feature: this is not a reasonable assumption. For example, for facial feature extraction (see Figure 1.2, page 7, for an illustration of NMF on the CBCL data set), we would need an image in the data set for each feature, which are nose, eyes, and lips. By construction, facial images contain more than one feature. Another example is the swimmer data set (see Figure 1.3) where this assumption is not satisfied as all images are combinations of five features. Separability of $W^\top$ is more realistic and requires that *each feature contains a pixel that does not belong to any other feature*. In other words, no feature is fully overlapped by the other features. Mathematically, for each feature $W(:,k)$ $(1 \leq k \leq r)$, there exists i such that $W(i,k) > 0$ while $W(i,\ell) = 0$ for $\ell \neq k$, that is, $W(i,:) = \alpha e_k^\top$ for some $\alpha > 0$. This is a realistic assumption in many image data sets, and it has been used successfully, for example, in [81] to unmix images.

   Let us in particular consider the two image data sets presented in Section 1.3.1, namely the CBCL data set and the swimmer data set. Figure 7.2 shows the basis elements extracted for the CBCL facial images by SPA, a workhorse near-separable NMF algorithm (see Section 7.4.1). Recall that each column of X is a vectorized facial image, and near-separable NMF applied on $X^\top$ provides a decomposition $X \approx WH$ where $H = X(\mathcal{K},:)$ and $W \geq 0$. We observe that separable NMF provides basis images similar to that of NMF (see Figure 1.2). Note that this explains why we used SPA applied on the transpose of X to initialize the NMF models on the CBCL face data set in Chapter 5.

   The transpose of the swimmer data set satisfies the separability assumption perfectly: each body part contains at least one pixel that does not belong to any other body parts. Therefore separable NMF applied on the transpose of the swimmer data set identifies the true underlying factors shown in Figure 1.4, namely the body and each limb in four positions [Matlab file: Swimmer.m].

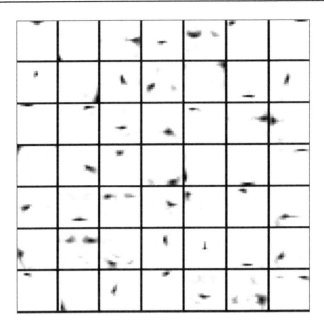

Figure 7.2. *Basis vectors extracted by SPA (a near-separable NMF algorithm; see Section 7.4.1) on the transpose of the CBCL facial image data set with* $r = 49$. *[Matlab file: CBCL_sepNMF.m]*.

2. Text mining. The input matrix X is a word-by-document matrix, W is a word-by-topic matrix, and H is a topic-by-document matrix. Separability of H requires that for each topic there is a document in the data set that discussed only that topic. Separability of $W^\top$ requires that for each topic there is a word associated only to that topic; Arora et al. [14] referred to these words as *anchor words*. It appears that the anchor word assumption is more reasonable [14]. Moreover, because of probabilistic considerations, it is preferable to use the word co-occurrence matrix (see Section 5.4.9), in which case separability of $W^\top$ can be used to recover it; see Section 7.8, where we describe this particular scenario.

3. Audio source separation. The matrix X is a frequency-by-time matrix, the factor W is a frequency-by-source matrix whose columns are the frequency content of the sources, and H is a source-by-time matrix indicating which source is active at which time window. Separability of H requires that for each source there is a time window where only that source is active. This assumption is not often satisfied in practice for complicated audio signals, especially when some sources (for example, a human voice) are represented by more than one rank-one factor. Separability of $W^\top$ requires that for each source there is a frequency where only that source is active. Again, for relatively complicated signals, this is not reasonable. Note, however, that the simple example from Figure 1.8 satisfies both assumptions: the piano notes are active at different time windows while their main peaks in the frequency response do not overlap. However, as far as we know, the separability assumption has not been considered much for audio source separation.

Other applications of separable NMF include community detection [374, 246, 339], crowd-sourcing that aims at producing accurate labels via integrating noisy and nonexpert labeling from annotators [253], time-resolved Raman spectra analysis [331] (see Section 9.3), blind source separation in nuclear magnetic resonance [148], video summarization, classification and outlier rejection [145], foreground-background separation in a video sequence [293], and the recovery of the joint probability of discrete random variables [252]. Moreover, as we will discuss in Section 8.6, near-separable NMF algorithms are particularly well-suited to initialize more sophisticated NMF algorithms.

Link with other problems Near-separable NMF is closely related to the column subset selection problem in numerical linear algebra; see for example [55] and the references therein. The only difference is the constraint on H: the column subset selection problem does not require H to be nonnegative in the decomposition $X \approx X(:, \mathcal{K})H$. Nonnegativity makes these two problems rather different. For example, in the exact case, that is, when $X = X(:, \mathcal{K})H$, any subset of $r = \text{rank}(X)$ linearly independent columns of X provides a solution to the column subset selection problem, and hence the solution is trivial to compute and highly nonunique (in general, most subsets of r columns provide an exact decomposition). This does not hold true for separable NMF for which the solution is unique (Section 4.3.1) and not as easy to compute.

Near-separable NMF is a special case of convex NMF where $W = XC$ for some $C \geq 0$; see Section 5.4.2. Hence, near-separable NMF algorithms are particularly suitable to initialize convex NMF algorithms [118].

7.2 ▪ Preliminaries

In this section, we discuss several important aspects of the near-separable NMF problem that will be used throughout the chapter, namely the model and assumptions (Section 7.2.1), the normalization of the input matrix (Section 7.2.2), the robustness to noise of near-separable NMF algorithms (Section 7.2.3), the conditioning of the factor W (Section 7.2.4), and pre- and post-processing (Section 7.2.5).

7.2.1 ▪ Model and assumptions

Throughout this chapter, we denote the input data matrix $\tilde{X}$, instead of X, which denotes the noiseless input matrix, and assume it satisfies the following assumption.

Assumption 7.1 (Near-separable factorization). *The matrix* $\tilde{X} \in \mathbb{R}^{m \times n}$ *admits a near-separable factorization if it has the form*

$$\tilde{X} = X + N \quad with \quad X = \underbrace{X(:, \mathcal{K}^*)}_{=W} H,$$

where

(i) $|\mathcal{K}^*| = r$ *equals the number of nonzero vertices of* $\text{conv}([X, 0])$, *that is, the columns of* W *are the nonzero vertices of* $\text{conv}([X, 0])$,

(ii) $\max_k \|W(:, k)\|_p = 1$ *for some* $p \geq 1$,

(iii) $H \in \mathbb{R}_+^{r \times n}$ *and* $H^\top e \leq e$, *and*

(iv) $\|N(:, j)\|_p \leq \epsilon$ *for all* $1 \leq j \leq n$ *for some* $\epsilon \geq 0$ *and* $p \geq 1$.

Let us observe the following:

- Item (i) implies that $\mathcal{K}^*$ has the smallest possible cardinality among the sets satisfying $X = X(:, \mathcal{K}^*)H$ for some H satisfying (iii). The columns of $X(:, \mathcal{K}^*)$ are the nonzero vertices of $\mathrm{conv}([X, 0])$, hence, by definition, they cannot be represented as convex combinations of other data points within $\mathrm{conv}([X, 0])$.

- Item (ii) can be assumed w.l.o.g. by multiplying X by a positive constant. It is made for the simplicity of the presentation as it allows us to get rid of the parameter $\max_k \|W(:, k)\|_p$ that appears in most robustness analysis of near-separable NMF algorithms.

- The value of p in items (ii) and (iv) will be either $p = 1$ or $p = 2$. Current near-separable algorithms have been proven to be robust to noise in these two cases.

Of course, any matrix $\tilde{X}$ admits a near-separable factorization (taking ϵ sufficiently large). Also, any randomly generated matrix X with $m \geq n$ admits a near-separable factorization with $r = n$ and $\epsilon = 0$ with probability one, since $X = XI_n$ and $\mathrm{rank}(X) = \min(m, n) = n$ with probability one. We focus on the case when r and ϵ are small; otherwise the problem is either trivial (when $r = n$) or NP-hard (for ϵ large; see the introduction of this section).

7.2.2 • Normalization/scaling of the input matrix

The assumption $H^\top e \leq e$ in item (iii) of Assumption 7.1 requires that the entries in each column of H sum to at most one, that is, each column of H belongs to the convex hull of the unit simplex Δ^r and the origin which we denote as

$$\mathcal{S}^r = \left\{ h \in \mathbb{R}^r \mid h \geq 0, \sum_{i=1}^r h_i \leq 1 \right\} = \mathrm{conv}\left([\Delta^r, 0]\right).$$

This assumption can be made w.l.o.g., which we explain below. A reason to make this assumption a priori is that it simplifies the analysis. In fact, the noise added to each column of X needs to be proportional to the norm of that column because columns with smaller norm deviate more easily from the data distribution. In other words, if no scaling is assumed, we would need to replace the the assumption $\|N(:, j)\|_p \leq \epsilon$ for all j by $\|N(:, j)\|_p \leq \epsilon \alpha_j$ where α_j is a quantity related to the norm of the jth column of X.

Let us explain why this assumption can be made w.l.o.g. Let v be a vector such that $v^\top X > 0$. Such a vector exists if the columns of X belong to the interior of the half space $\{x | v^\top x \geq 0\}$, where v is the normal vector of that half space. For example,

- for $X \geq 0$, the nonzero columns of X are within the interior of the half space defined by $\{x | e^\top x \geq 0\}$ that contains the nonnegative orthant; this normalization was used in Section 2.1.2 to transform the interpretation of Exact NMF in terms of nested cones to the interpretation in terms of nested convex hulls (see in particular Lemma 2.1);

- for $X = WH$, $H \geq 0$ with no zero column and $\mathrm{rank}(W) = r$, such a vector v always exists: take a solution of the system $W^\top v = e$; for example $v = (W^\dagger)^\top e$ where $W^\dagger$ is the left-inverse of W (that is, $W^\dagger W = I_r$) for which

$$v^\top X = v^\top WH = e^\top H > 0,$$

since $H \geq 0$ and H does not have zero columns. Because $\mathrm{rank}(W) = r$ is a typical assumption in practice, normalizing X is typically not an issue.

Since we do not know W, a possible way to compute v is, for example, to solve the convex quadratic optimization problem

$$\min_{v} \|v\|_2^2 \quad \text{such that} \quad v^\top X \geq e,$$

which is always feasible under the conditions $\text{rank}(W) = r$, $H \geq 0$ and X does not have zero columns (since $v = \alpha(W^\dagger)^\top e$ is feasible for α sufficiently large).

Note that, in some degenerate cases, such a v might not exist, for example taking three vectors in two dimensions that are not within the same half space, such as

$$X = \begin{pmatrix} 1 & 0 & -1 \\ 0 & 1 & -1 \end{pmatrix}.$$

When such a v has been identified, each column of X can be projected onto the affine space $\{x | v^\top x = 1\}$ by dividing each column $X(:,j)$ by the constant $\alpha_j = v^\top X(:,j) > 0$ for $j = 1, 2, \ldots, n$. After this transformation,

$$e^\top = v^\top X = v^\top (X(:,\mathcal{K})H) = e^\top H;$$

hence $H^\top e = e$ since all columns of X belong to the same affine subspace, $\{x | v^\top x = 1\}$. Note that zero columns of X need to be removed (or not scaled) as in Lemma 2.1.

Let us make a few observations:

- Since the normalization described above leads to $H^\top e = e$, one may wonder why we only assume $H^\top e \leq e$ in Assumption 7.1. The reason is that it makes the model more general. In particular, some real data sets satisfy naturally this condition (hence there is no need for normalization) while they could violate the condition $H^\top e = e$. This is, for example, the case for hyperspectral images for which the illumination across the pixels in the image is not uniform. Avoiding column normalization, if possible, is usually preferable as it does not intensify the noise present in columns with small norms (for example, background pixels in hyperspectral images).

- Under this normalization, near-separable NMF is closely related to the so-called simplex-structured matrix factorization problem, which looks for a factorization of the type WH where the columns of H lie on the unit simplex [321, 192, 2].

- Since the normalization described above simply scales the columns of the input matrix, any column of W that is an extreme ray of $\text{cone}(X)$ remains an extreme ray, by definition.

  However, a column of W that is a vertex of $\text{conv}([X, 0])$ might not remain a vertex of $\text{conv}([X, 0])$ after the normalization. The reason is that being a vertex of $\text{conv}([X, 0])$ does not guarantee being an extreme ray of $\text{cone}(X)$. A simple example is the matrix

$$W = \begin{pmatrix} 1 & 0 & 1 \\ 0 & 1 & 1 \end{pmatrix}.$$

The point $(1,1)$ is not in the convex hull of $(1,0)$, $(0,1)$ and the origin but is not an extreme ray of $\text{cone}(W)$. Hence, in such cases, the matrix X from Assumption 7.1 would be impacted by the normalization: any column of W that is not an extreme ray of $\text{cone}(X)$ will be lost in this procedure (in the sense that it will be projected onto the convex hull of other columns of W). However, in practice, we typically do not want to recover these columns: they do not allow us to reduce the error in the original near-separable NMF formulation as they are nonnegative linear combinations of other columns of W; see Problem 7.1. Moreover, this situation can happen only when $\text{rank}(W) < r$, which is usually not the case in practice.

7.2.3 ▪ Robustness to noise

Let us first briefly recall the geometric interpretation behind separable NMF; see Figure 4.9 for an illustration (page 134). In the noiseless case, separable NMF is equivalent to identifying the extreme directions of $\mathrm{cone}(X)$. After normalization of the columns of X, separable NMF is equivalent to identifying the vertices of $\mathrm{conv}(X)$. This is a rather simple geometric problem for which many algorithms exist. Most near-separable NMF algorithms (in particular in the blind HU literature) are based on concepts from convex geometry.

In the presence of noise, the problem becomes more challenging. The question can be asked as follows:

Given a certain noise level ϵ, can we quantify the error between the recovered factor $\tilde{X}(:,\mathcal{K})$ and the ground truth $W = X(:,\mathcal{K}^*)$?

To quantify the error, let us define the following error measure:

$$q_p(\mathcal{K}) \ = \ \max_{1 \le k \le r} \min_{j \in \mathcal{K}} \|W(:,k) - \tilde{X}(:,j)\|_p, \tag{7.1}$$

which quantifies the quality of a solution. We only focus on the recovery of W because, when W is rank deficient, the corresponding matrix H is not necessarily unique. This is the standard in the literature on near-separable NMF. In particular, robustness theorems will have the following form: given that $\tilde{X}$ satisfies Assumption 7.1 and that ϵ is sufficiently small, the algorithm outputs a solution $\mathcal{K}$ such that $q(\mathcal{K}) \le \delta$ for some δ. The difficulty is to quantify ϵ and δ. This allows us to compare the behavior of different algorithms in the presence of noise.

The noise is assumed to be adversarial: we do not make any assumption on the noise except that it is bounded; see item (iv) in Assumption 7.1. Most analysis of near-separable NMF algorithms uses this model, and a direction for future research is to provide noise robustness results under specific noise distributions.

7.2.4 ▪ Conditioning of W

A key parameter in the analysis of near-separable NMF algorithms is the conditioning of $W = X(:,\mathcal{K}^*)$. It will influence the bounds on the noise level ϵ allowed by near-separable NMF algorithms along with the error $q_p(\mathcal{K})$ they can achieve. In this section, we use two parameters: the rth singular value of W, that is, $\sigma_r(W)$, and $\gamma_p(W)$ defined as

$$\gamma_p(W) = \min_{1 \le k \le r} \min_{h \in \mathcal{S}^{r-1}} \left\| W(:,k) - W(:,\bar{k})h \right\|_p,$$

where $\bar{k} = \{1,2,\dots,r\}\backslash\{k\}$. The quantity $\gamma_p(W)$ is the smallest distance between a column of W and the convex hull of the other columns of W and the origin. For $p = 2$ and an m-by-r matrix W, the quantity $\gamma_2(W)$ is always larger than $\sigma_r(W)$.

Lemma 7.1. *For any matrix* $W \in \mathbb{R}^{m \times r}$,

$$\gamma_2(W) \ \ge \ \sigma_r(W).$$

Proof. For any $k \in \{1,2,\dots,r\}$,

$$\{x \mid x(k) = 1, \|x(\bar{k})\|_1 \le 1, x \ge 0\} \ \subset \ \{x \mid \|x\|_2 \ge 1\};$$

hence, by virtue of relaxation, we have for all $k \in \{1,2,\dots,r\}$

$$\sigma_r(W) = \min_{x, \|x\|_2 \ge 1} \|Wx\|_2 \ \le \ \min_{h \in \mathcal{S}^{r-1}} \|W(:,k) - W(:,\bar{k})h\|_2;$$

hence $\gamma_2(W) \ge \sigma_r(W)$. □

Under Assumption 7.1 which requires $\max_k \|W(:,k)\|_2 = 1$, $\gamma_2(W) \le 1$ (take $h = 0$ in the definition above) so that $\sigma_r(W) \le \gamma_2(W) \le 1$.

An upper bound on the noise ϵ   Let us start with a simple example. Assume that $r = 3$ and take

$$W = \begin{pmatrix} 1 & 1 & 0 \\ 0 & 1 & 1 \end{pmatrix}.$$

The columns of W are the vertices of a triangle in the plane. Moving the second column of W toward the middle of the segment joining the other two, it will become harder and harder to distinguish that column from the data points within $\mathrm{conv}([W, 0])$ in the presence of noise (the triangle becomes flatter). In particular, using $N(:, 2) = -1/2\, e$ makes the second columns of $\tilde{X}$ belong to the convex hull of $W(:, [1, 3])$. This means that the second column of W "disappeared" (in the sense that it cannot be distinguished from a point in the convex hull of the other columns). This shows that the noise level has to be smaller than the distance between each column of W and its projection onto the convex hull of the other columns and the origin, that is, $\epsilon < \gamma_p(W)$.

Another example is when W has the form

$$W = \begin{pmatrix} W' & 0 \\ 0 & \sigma \end{pmatrix}.$$

We have $\sigma_r(W) = \min\left(\sigma_{r-1}(W'), \sigma\right)$. Let us assume that $\sigma_{r-1}(W') > \sigma$ so that $\sigma_r(W) = \sigma$. In order to be able to recover the last column of W via near-separable NMF, the noise level ϵ has to satisfy $\epsilon < \sigma_r(W)$, otherwise the last column of W can be set to zero by properly choosing the noise matrix N (take $\left(\begin{smallmatrix} 0 \\ -\sigma \end{smallmatrix}\right)$ for the column of N corresponding to the last column of W) and hence cannot be recovered under Assumption 7.1. In other words, the matrix $\tilde{X}$ can be made near-separable with $|\mathcal{K}^*| = r - 1$.

Comparison of $\sigma_r(W)$ and $\gamma_2(W)$ As we will see in this chapter, some near-separable NMF algorithms rely on the condition that $\sigma_r(W) > 0$, while others rely on the weaker assumption that $\gamma_p(W) > 0$. For near-separable NMF, it is not necessary that $\sigma_r(W) > 0$ (that is, $\mathrm{rank}(W) = r$) to be able to recover W. In other words, we might have $\sigma_r(W) = 0$ while still being able to recover the columns of W in the presence of noise. For this to be possible, it suffices that each column of W does not belong to the convex hull of the other columns and the origin, that is, $\gamma_p(W) > 0$.

For well-conditioned matrices $W \in \mathbb{R}^{m \times r}$ with $m \ge r$, we have that $\gamma_2(W) \approx \sigma_r(W)$, and the distinction between $\sigma_r(W)$ and $\gamma_2(W)$ is not crucial. Under Assumption 7.1, σ_r is close to one since

$$1 \approx \kappa(W) = \frac{\sigma_1(W)}{\sigma_r(W)} \ge \frac{\max_j \|W(:, j)\|_2}{\sigma_r(W)} = \sigma_r^{-1}(W),$$

while $\sigma_r(W) \le \gamma_2(W) \le 1$ (Lemma 7.1). For example, we have randomly generated 1000 matrices W of dimension 100-by-10 using the uniform distribution (`rand(100,10)` in MATLAB) which generates well-conditioned matrices with high probability, and divided all entries by $\max_j \|W(:, j)\|_2$ to satisfy Assumption 7.1. The average value and standard deviation of $\sigma_r(W)$ is 0.35 ± 0.02 and of $\gamma_2(W)$ is 0.43 ± 0.02.

When W is ill-conditioned (resp. $\mathrm{rank}(W) < r$), $\gamma_2(W)$ can be significantly larger than zero while $\sigma_r(W)$ can be arbitrarily close to zero (resp. equal to zero). For example, we have randomly generated 1000 matrices of dimension 10-by-20 using the uniform distribution (`rand(10,20)` in MATLAB) and divided all entries by $\max_j \|W(:, j)\|_2$ to satisfy Assumption 7.1. While $\sigma_r(W) = 0$ in all cases since $m < r$, the average value and standard deviation of $\gamma_2(W)$ is 0.13 ± 0.03. We refer the reader to Section 7.4.6 (page 245) for a similar numerical example in the ill-conditioned case for which $0 < \sigma_r(W) \ll \gamma_2(W)$.

7.2.5 ▪ Pre- and postprocessing

It is standard in the blind HU literature to assume $H^\top e = e$. Therefore it makes sense to pre-process the data set by identifying the $(r-1)$-dimensional affine subspace that best approximates the columns of X; see for example [334]. Another similar approach is to use the truncated rank-r SVD of $\tilde{X}$ [361]. Both these approaches filter the noise and provide a cleaner data set. It is also possible to randomly project the data points within a smaller dimensional subspace. This is particularly useful when m is very large [35].

Another useful preprocessing step is to identify and remove outliers. This may be particularly crucial for near-separable NMF since outliers are outside $\mathrm{conv}(W)$ and hence are extracted by most near-separable NMF algorithms. Similarly, postprocessing can be used to identify outliers extracted by near-separable NMF algorithms.

We will not discuss this data processing further as it falls outside the range of near-separable NMF. However, it is important to keep in mind that using such approaches might be important in practice, and they are embedded in some near-separable algorithms.

In the next sections, we review three classes of important near-separable NMF algorithms: greedy, heuristic, and convex-optimization-based algorithms. We do not follow the chronological order in which these methods were introduced. Note that greedy algorithms can be seen as a special case of heuristic algorithms; however, because this subclass of heuristics is of particular interest, we present it separately. Before presenting and analyzing these three types of algorithms, we first present an idealized algorithm. It will provide bounds on the maximum allowed noise level ϵ and minimal achievable error $q_p(\mathcal{K})$ for any near-separable NMF algorithm.

7.3 ▪ Idealized algorithm

An optimal solution of near-separable NMF can be computed using brute force: for all possible index sets $\mathcal{K}$ of size r, compute

$$d(\mathcal{K}) \;=\; \min_{H \geq 0} D\left(\tilde{X}, \tilde{X}(:,\mathcal{K})H\right),$$

which is a convex optimization problem if $D(\cdot, \cdot)$ is convex in its second argument, and keep the best solution $\mathcal{K}$, that is, the solution that minimizes $d(\mathcal{K})$. This approach requires trying $\binom{n}{r} = \frac{(n-r)!}{n!r!}$ possible index sets $\mathcal{K}$, and hence it is impractical for large n, even when r is small. For example, even for $r = 2$, it requires testing $\mathcal{O}(n^2)$ index sets. For each index set, one needs to solve an optimization problem that requires $\Omega(mnr)$ operations since computing $\tilde{X}(:,\mathcal{K})H$ requires $\mathcal{O}(mnr)$ operations, for a total of $\Omega(mrn^3)$ operations. This would be too expensive, for example, for hyperspectral images for which n is typically in the order of millions.

However, it could be interesting to investigate clever ways to explore the $\binom{n}{r}$ possible index sets, using strategies such as branch and bound. However, as far as we know, no such techniques have been developed so far for near-separable NMF; this could be an interesting direction of research.

The goal of this section is to analyze an idealized algorithm that computes the best index set $\mathcal{K}$. More precisely, we consider the algorithm that outputs an optimal solution of

$$\min_{\mathcal{K},|\mathcal{K}|=r} f(\mathcal{K}), \quad \text{with } f(\mathcal{K}) = \max_{1 \leq j \leq n} \min_{h \in \mathcal{S}^r} \left\| \tilde{X}(:,j) - \tilde{X}(:,\mathcal{K})h \right\|_2. \tag{7.2}$$

Let us make two observations:

- For simplicity we chose the ℓ_2 norm, although the analysis we provide below can be adapted to other norms such as the ℓ_1 norm.

- The problem (7.2) minimizes the largest approximation error among all data points: this choice is rather natural if the goal is to approximate the true $W = \tilde{X}(:, \mathcal{K}^*)$, that is, to minimize $q_2(\mathcal{K})$ as defined in (7.1). Intuitively, to approximate well the columns of W (which are among the columns of X), $\mathcal{K}$ has to correspond to columns near the columns of W: the columns of W are the vertices of $\mathrm{conv}(X)$ and hence cannot be well-approximated with convex combinations of other columns (by definition). This is quantified via the parameter $\gamma_2(W) > 0$ which is the smallest distance between a column of W and convex combinations of other columns of W and the origin.

The reason we analyze this idealized algorithm is twofold:

- Analyzing the robustness to noise of this idealized algorithm is relatively simple, at least compared to the other algorithms presented in the following sections and for which we will not provide robustness proofs (such proofs are lengthy and rather technical). Hence this analysis allows the reader to get insights and intuitions on such robustness proofs at a higher level.

- It provides limits in terms of performance for any near-separable NMF algorithm. The intuition is that any polynomial-time algorithm has to perform worse than this brute-force idealized algorithm.

If you are not interested in these theoretical derivations, you can skip this part and go directly to "Take-home message from the idealized algorithm" (page 222).

The developments of this section follow closely the paper[50] [192].

7.3.1 ▪ Robustness to noise

In this section, we prove robustness to noise of an optimal solution of (7.2) to solve the near-separable NMF problem. The first lemma shows that the solution $\mathcal{K}^*$ (which contains the indices corresponding to the columns of W) achieves an error $f(\mathcal{K}^*)$ of at most 2ϵ.

Lemma 7.2. *[192, Lemma 1] Let $\tilde{X}$ satisfy Assumption 7.1 with $p = 2$. Then $f(\mathcal{K}^*) \leq 2\epsilon$ where f is defined in (7.2).*

Proof. By Assumption 7.1, we have for all j that

$$\tilde{X}(:, j) = X(:, j) + N(:, j) = X(:, \mathcal{K}^*)H(:, j) + N(:, j),$$

where $H(:, j) \in \mathcal{S}^r$, and we also have $\tilde{X}(:, \mathcal{K}^*) = W + N(:, \mathcal{K}^*)$. Therefore, using the triangle inequality, we have for all j that

$$\begin{aligned}
\min_{h \in \mathcal{S}^r} \|\tilde{X}(:, j) - \tilde{X}(:, \mathcal{K}^*)h\|_2 &= \min_{h \in \mathcal{S}^r} \|X(:, j) + N(:, j) - Wh - N(:, \mathcal{K}^*)h\|_2 \\
&\leq \min_{h \in \mathcal{S}^r} \|X(:, j) - Wh\|_2 + \|N(:, \mathcal{K}^*)h\|_2 + \|N(:, j)\|_2 \\
&\leq \underbrace{\|X(:, j) - WH(:, j)\|_2}_{=0} + \|NH(:, j)\|_2 + \epsilon \\
&\leq \max_{h \in \mathcal{S}^r} \|Nh\|_2 + \epsilon \leq 2\epsilon,
\end{aligned}$$

---

[50]We have clarified the presentation providing more details, improved some bounds (in particular for Theorem 7.6), and corrected some errors.

where we used

$$\max_{h \in \mathcal{S}^r} \|N(:,\mathcal{K}^*)h\|_2 = \max_{j \in \mathcal{K}^*} \|N(:,j)\|_2 \leq \max_j \|N(:,j)\|_2 \leq \epsilon,$$

which follows from Assumption 7.1.   □

The second lemma provides a lower bound on the error for any feasible solution $\mathcal{K}$.

Lemma 7.3. *[192, Lemma 2] Let $\tilde{X} = WH + N = X + N$ satisfy Assumption 7.1 with $p = 2$. Let $\mathcal{K}$ be an index set of size r and let $B = H(:,\mathcal{K}) \in \mathbb{R}^{r \times r}$ so that $X(:,\mathcal{K}) = WB$, and let $\alpha = \min_k \max_j B(k,j) \leq 1$. Then*

$$f(\mathcal{K}) \geq (1 - \alpha)\gamma_2(W) - 2\epsilon. \tag{7.3}$$

Proof. Before providing the proof, let us interpret the value of α. The matrix B contains the weights necessary to reconstruct $X(:,\mathcal{K})$ using the columns of W. The kth row of B indicates the importance of the kth column of W to reconstruct $X(:,\mathcal{K})$. The largest entry of $B(k,:)$ is the largest proportion of $W(:,k)$ in the columns of $X(:,\mathcal{K})$. In particular, if $\|B(k,:)\|_\infty = 1$, then one of the columns of $X(:,\mathcal{K})$ is equal to $W(:,k)$ since $B(:,j) \in \mathcal{S}^r$ (Assumption 7.1). Therefore $\alpha = \min_k \|B(k,:)\|_\infty$ is the smallest proportion of a column of W used to reconstruct the columns of $X(:,\mathcal{K})$.

Let us now prove the lemma. Clearly, (7.3) holds for $\alpha = 1$ since $f(\mathcal{K}) \geq 0$ for any $\mathcal{K}$ (in this case, B is the identity matrix, up to permutation, that is, $W = X(:,\mathcal{K})$, up to permutation). Otherwise, $\alpha < 1$ and let k be such that $\alpha = \max_j B(k,j)$. Let us denote $\tilde{W} = \tilde{X}(:,\mathcal{K}^*)$, and let us lower bound $f(\mathcal{K})$ by focusing on the error of the approximation of $\tilde{W}(:,k)$ which is one of the columns of $\tilde{X}$. We have, using a similar derivation as in Lemma 7.2, that

$$f(\mathcal{K}) \geq \min_{h \in \mathcal{S}^r} \left\| \tilde{W}(:,k) - \tilde{X}(:,\mathcal{K})h \right\|_2 \geq \min_{h \in \mathcal{S}^r} \|W(:,k) - X(:,\mathcal{K})h\|_2 - 2\epsilon.$$

Let us use the notation $\bar{k} = \{1, 2, \ldots, r\} \backslash \{k\}$. We have

$$\begin{aligned}
\min_{h \in \mathcal{S}^r} \|W(:,k) - X(:,\mathcal{K})h\|_2 &= \min_{h \in \mathcal{S}^r} \|W(:,k) - WBh\|_2 \\
&= \min_{h \in \mathcal{S}^r} \left\| (1 - (Bh)_k)W(:,k) - W(:,\bar{k})(Bh)_{\bar{k}} \right\|_2 \\
&\geq \min_{h \in \mathcal{S}^r} (1 - (Bh)_k) \left\| W(:,k) - W(:,\bar{k})\frac{(Bh)_{\bar{k}}}{(1 - (Bh)_k)} \right\|_2 \\
&\geq (1 - \alpha)\gamma_2(W).
\end{aligned}$$

The last inequality follows from the definition of $\gamma_2(W)$ and because $\frac{(Bh)_{\bar{k}}}{(1-(Bh)_k)} \in \mathcal{S}^{r-1}$ since $Bh \in \mathcal{S}^r$ and $(Bh)_k \leq \alpha < 1$. In fact, $Bh \in \mathcal{S}^r$ implies $\sum_j (Bh)_j = \sum_{j \neq k}(Bh)_j + (Bh)_k \leq 1$ so that

$$\sum_{j \neq k} \frac{(Bh)_j}{1 - (Bh)_k} \leq 1. \qquad \square$$

Robustness without duplicates and near-duplicates   Under Assumption 7.1, the matrix H in $X = WH$ can be written as

$$H = [I_r, H']\Pi \tag{7.4}$$

for some permutation matrix Π and some $H' \geq 0$ with $H'^\top e \leq e$. If the entries of H' are strictly smaller than one, then no column of H' is a unit vector and there are no duplicated columns of W in the data set. Under this condition and if the noise level ϵ is sufficiently small, then the optimal solution of (7.2) is $\mathcal{K}^*$ which recovers W up to the noise level ϵ.

Theorem 7.4. *[192, Theorem 1] Let $\tilde{X} = X + N = WH + N$ satisfy Assumption 7.1 with $p = 2$ where $H = [I_r, H']\Pi$ has the form (7.4). Let us assume that $\beta := \max_{i,j} H'(i,j) < 1$, and*

$$\epsilon < \frac{(1-\beta)\gamma_2(W)}{4}.$$

Then the optimal solution of (7.2) coincides with the true index set $\mathcal{K}^*$ for which

$$q_2(\mathcal{K}^*) = \max_{1 \leq k \leq r} \min_{j \in \mathcal{K}^*} \left\| W(:,k) - \tilde{X}(:,j) \right\|_2 \leq \epsilon.$$

Proof. Let $\mathcal{K}$ be a solution distinct from $\mathcal{K}^*$, $X(:,\mathcal{K}) = WH(:,\mathcal{K})$, $B = H(:,\mathcal{K})$, and $\alpha = \min_k \max_j B(k,j)$. Since $\mathcal{K} \neq \mathcal{K}^*$, at least one column of B corresponds to a column of H'. Since the entries in H' are smaller than β, this implies that α as defined in Lemma 7.3 is smaller than β: there is at least one row of B whose maximum entry is strictly smaller than one since there are at most $r-1$ columns of B being columns of the identity matrix. That maximum entry has to be smaller than β since, except for the unit vectors, all entries in H are smaller than β. Therefore, using Lemma 7.3, any solution $\mathcal{K} \neq \mathcal{K}^*$ has error $q_2(\mathcal{K})$ at least $(1-\beta)\gamma_2(W) - 2\epsilon$.

By Lemma 7.2, $\mathcal{K}^*$ has error at most 2ϵ. Therefore, $\mathcal{K}^*$ is the unique optimal solution of (7.2) as long as $(1-\beta)\gamma_2(W) - 2\epsilon > 2\epsilon$, that is, as long as $\epsilon < \frac{(1-\beta)\gamma_2(W)}{4}$, while $\mathcal{K}^*$ leads to an error on W smaller than ϵ since $\tilde{X}(:,\mathcal{K}^*) = W + N(:,\mathcal{K}^*)$ and $\max_j \|N(:,j)\|_2 \leq \epsilon$ (Assumption 7.1). □

The bound of Theorem 7.4 is tight since it is not possible to have $q_2(W)$ smaller than ϵ, because every column of N could potentially have a norm of ϵ; see Assumption 7.1.

Unfortunately, in most practical settings, there are near-duplicated columns of W in the data set, that is, nearby data points. For example, in hyperspectral images satisfying the pure-pixel assumption, there is usually more than one pure pixel per endmember, and many pixels contain mostly one material.

Robustness in the presence of duplicates and near-duplicates In the presence of near-duplicated columns of W, we have the following robustness result when solving (7.2).

Theorem 7.5. *[192, Theorem 2] Let $\tilde{X}$ satisfy Assumption 7.1 with $p = 2$, and let us assume that $\epsilon < \frac{\gamma_2(W)}{4}$. Then any optimal solution $\mathcal{K}$ of (7.2) satisfies*

$$q_2(\mathcal{K}) \leq \frac{8\epsilon}{\gamma_2(W)} + \epsilon.$$

Proof. Let $\mathcal{K}$ be an optimal solution of (7.2), $X(:,\mathcal{K}) = WB$ with $B = H(:,\mathcal{K})$, and $\alpha = \min_k \max_j B(k,j) \leq 1$ (as in Lemma 7.3). We have

$$2\epsilon \underset{\text{Lem. 7.2}}{\geq} f(\mathcal{K}^*) \underset{\mathcal{K} \text{ optimal}}{\geq} f(\mathcal{K}) \underset{\text{Lem. 7.3}}{\geq} \gamma_2(W)(1-\alpha) - 2\epsilon.$$

This implies that $\gamma_2(W)(1-\alpha) \leq 4\epsilon$, and hence $\alpha \geq 1 - \frac{4\epsilon}{\gamma_2(W)}$. By definition $\alpha \leq \max_j B(k,j)$ for all k, and hence for each row of B there is at least one entry with value at least $1 - \frac{4\epsilon}{\gamma_2(W)}$, that is, for each k there exists j_k such that $B(k, j_k) \geq 1 - \frac{4\epsilon}{\gamma_2(W)}$. Hence for all k there exists j_k such that

$$
\begin{aligned}
\|W(:,k) - WB(:,j_k)\|_2 &= \|W(:,k) - W(:,k)B(k,j_k) - W(:,\bar{k})B(\bar{k},j_k)\|_2 \\
&= \|(1 - B(k,j_k))W(:,k) - W(:,\bar{k})B(\bar{k},j_k)\|_2 \\
&\leq \frac{4\epsilon}{\gamma_2(W)}\|W(:,k)\|_2 + \frac{4\epsilon}{\gamma_2(W)} \max_{h \in \mathcal{S}^{r-1}} \|W(:,\bar{k})h\|_2 \\
&\leq \frac{8\epsilon}{\gamma_2(W)} \max_i \|W(:,i)\|_2 = \frac{8\epsilon}{\gamma_2(W)}.
\end{aligned}
$$

The first inequality follows from the triangle inequality and the facts that $B(:,j_k) \in \mathcal{S}^r$ and $B(k,j_k) \geq 1 - \frac{4\epsilon}{\gamma_2(W)}$, which implies $\|B(\bar{k},j_k)\|_1 \leq \frac{4\epsilon}{\gamma_2(W)}$; hence $\left(\frac{\gamma_2(W)}{4\epsilon}\right)B(\bar{k},j_k) \in \mathcal{S}^{r-1}$. The last equality follows from Assumption 7.1(ii), that is, $\max_i \|W(:,i)\|_2 = 1$. This concludes the proof since

$$
\tilde{X}(:,\mathcal{K}) = X(:,\mathcal{K}) + N(:,\mathcal{K}) = WB + N(:,\mathcal{K});
$$

hence, for all k the j_kth column of $\tilde{X}(:,\mathcal{K})$ is at distance at most $\frac{8\epsilon}{\gamma_2(W)} + \epsilon$ from $W(:,k)$, that is, $q_2(\mathcal{K}) \leq \frac{8\epsilon}{\gamma_2(W)} + \epsilon$.   □

A natural question is whether the bound of Theorem 7.5 is tight. In the following, we prove a lower bound on the best possible accuracy achievable by solving (7.2). This proves that the bounds in Theorem 7.5 are tight up to a multiplicative factor of two for the leading term in $\frac{\epsilon}{\gamma_2(W)}$.

Theorem 7.6. *[192, Theorem 3] There exists a class of near-separable matrices $\tilde{X}$ satisfying Assumption 7.1 with $p = 2$ and $\epsilon < \frac{\gamma_2(W)}{4}$ such that the optimal solution $\mathcal{K}$ of (7.2) satisfies*

$$
q_2(\mathcal{K}) > 4\frac{\epsilon}{\gamma_2(W)} + 4\epsilon.
$$

Proof. Let us construct a matrix $\tilde{X}$ satisfying Assumption 7.1. For that, we need to construct W, H, and N. The construction is illustrated in Figure 7.3.
Construction of W. Let

$$
W = \begin{pmatrix} 1 & 0 & \frac{1}{2} + \frac{\sqrt{2}}{2}\gamma \\ 0 & 1 & \frac{1}{2} + \frac{\sqrt{2}}{2}\gamma \end{pmatrix},
$$

where $0 < \gamma \leq 1 - \sqrt{2}/2$ so that $\max_k \|W(:,k)\|_2 = 1$. Let us show that $\gamma_2(W) = \gamma$. By convexity and symmetry of the problem $\min_y \|W(:,3) - W(:,1:2)y\|_2$, an optimal solution is $y^* = (0.5, 0.5)$ for which $\|W(:,3) - W(:,1:2)y^*\|_2 = \gamma$. In fact, if y_1 and y_2 are optimal solutions of a convex optimization problem, so is $\eta y_1 + (1 - \eta)y_2$ for any $\eta \in [0,1]$). The optimal solution of

$$
\min_y \|W(:,1) - W(:,2:3)y\|_2
$$

is achieved for $y(1) = 0$ as $W(:,1)$ and $W(:,2)$ are orthogonal, while $y(2) = W(1,3)^{-1}$ such that $W(:,3)y(2) = (0.5, 0.5)^\top$ with error $\sqrt{2}/2 \geq \gamma$. The same reasoning applies to $W(:,2)$ by symmetry.
Construction of H. Let $H = [I_3, h]$ where $h^\top = (0, 1 - \lambda, \lambda)$ for some $\lambda \in [0,1]$ to be defined later. This means that $\mathcal{K}^* = \{1, 2, 3\}$ and that the fourth column of $X = WH$ is a linear combination of the second and third columns of W, that is,

$$
X(:,4) = (1 - \lambda)W(:,2) + \lambda W(:,3) = (\lambda/2 + \gamma\lambda/\sqrt{2}, 1 - \lambda/2 + \gamma\lambda/\sqrt{2}).
$$

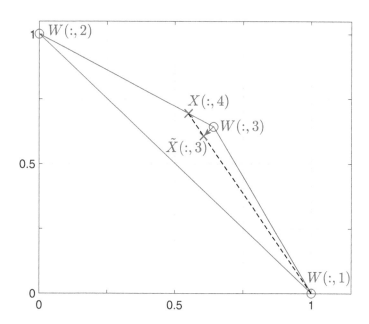

Figure 7.3. *Illustration of the construction from Theorem 7.6 with $\gamma = 0.2$ and $\lambda = 6/7$.*

Construction of N. We do not add noise to the first, second, and fourth columns of X, that is, $N(:, [1, 2, 4]) = 0$. We add noise to the third column of X as follows:

$$N(:, 3) = -\frac{\epsilon}{\sqrt{2}} \, e,$$

with $\|N(:, 3)\|_2 = \epsilon$, where e is the vector of all ones of appropriate dimension.

In the following, we show that, for a suitable choice of λ, the optimal solution of (7.2) is $\{1, 2, 4\}$ for some $\epsilon < \frac{\gamma}{4}$ and that $\|W(:, 3) - \tilde{X}(:, 4)\|_2 > \sqrt{2}\frac{\epsilon}{\gamma} + \sqrt{2}\epsilon$, which will conclude the proof.

The intuition is as follows. Let $\mathcal{K} = \{1, 2, 4\}$. We are going to choose λ such that $\tilde{X}(:, 3)$ belongs to the convex hull of $\tilde{X}(:, \mathcal{K})$, hence $f(\mathcal{K}) = 0$, since $N(:, \mathcal{K}) = 0$. On the other hand, $f(\mathcal{K}^*) > 0$ since $\tilde{X}(:, 4)$ does not belong to the convex hull of $\tilde{X}(:, [1, 2, 3])$ (see Figure 7.3).
Choice of λ such that $f(\mathcal{K}) = 0$. Let us construct the vector

$$v = \mu X(:, 4) + (1 - \mu) W(:, 1)$$

on the segment joining $X(:, 4)$ and $W(:, 1)$ (see Figure 7.3) with $\mu = \frac{1}{2 - \lambda}$ so that $v = (v_1, v_2)$ with $v_1 = v_2 = \frac{1}{2} + \frac{\gamma\lambda}{\sqrt{2}(2 - \lambda)}$. The vector v approximates $X(:, 3) = W(:, 3)$ using a convex combination of $X(:, 2)$ and $X(:, 4)$. We have

$$W(:, 3) - v = \frac{\gamma}{\sqrt{2}} \left(1 - \frac{\lambda}{2 - \lambda} \right) e = \frac{\gamma}{\sqrt{2}} \frac{1 - \lambda}{2 - \lambda} e.$$

Hence, for $\epsilon = \gamma \frac{1 - \lambda}{2 - \lambda}$,

$$\tilde{W}(:, 3) = W(:, 3) + N(:, 3) = W(:, 3) - \frac{\epsilon}{\sqrt{2}} e = v,$$

implying $f(\mathcal{K}) = 0$.

Value of $q_2(\mathcal{K})$. We have

$$
\begin{aligned}
q_2(\mathcal{K}) = \|W(:,3) - \tilde{X}(:,4)\|_2 &= \|W(:,3) - X(:,4)\|_2 \\
&= \|W(:,3) - (1-\lambda)W(:,2) - \lambda W(:,3)\|_2 \\
&= (1-\lambda)\|W(:,3) - W(:,2)\|_2 \\
&> \frac{1}{\sqrt{2}}(1-\lambda),
\end{aligned}
$$

since

$$
\|W(:,3) - W(:,2)\|_2 = \left\| \left(\frac{1}{2} + \frac{\gamma}{\sqrt{2}}, \frac{1}{2} - \frac{\gamma}{\sqrt{2}} \right) \right\|_2 = \sqrt{\frac{1+\gamma^2}{2}}.
$$

This implies that $q_2(\mathcal{K}) > \sqrt{\frac{1+\gamma^2}{2}}(1-\lambda)$.

Finally, let us choose for example λ such that $\epsilon = \gamma\frac{1-\lambda}{2-\lambda} = \frac{\gamma}{8} < \frac{\gamma}{4}$, that is, $\lambda = 6/7$. We have

$$
\frac{\epsilon}{\gamma} = \frac{1}{8},
$$

so that

$$
q_2(\mathcal{K}) > \sqrt{\frac{1+\gamma^2}{2}}(1-\lambda) \geq \frac{1}{2} + \frac{\gamma}{2} = 4\frac{\epsilon}{\gamma} + 4\epsilon,
$$

where we used[51] $\sqrt{\frac{1+\gamma^2}{2}} > \frac{1}{2} + \frac{\gamma}{2}$ for $0 < \gamma < 1$. Note that since $\gamma < 1$, this does not contradict Theorem 7.5 since $4\frac{\epsilon}{\gamma} + 4\epsilon < 8\frac{\epsilon}{\gamma}$.   □

Optimality of the idealized algorithm   Let us explain why the idealized algorithm achieves optimal bounds, up to constant multiplicative factors. This essentially follows from the construction of Theorem 7.6 where the "wrong" solution $\mathcal{K} = \{1,2,4\}$ leads to a perfect reconstruction of $\tilde{X}$ while $q_2(\mathcal{K}^*) > \mathcal{O}(\epsilon/\gamma)$.

First, the condition $\epsilon < \gamma_2(W)$ is a necessary condition; otherwise a column of $\tilde{X}(:,\mathcal{K}^*)$ could belong to the convex hull of the other columns and hence is not identifiable; see Section 7.2.4.

Second, using the construction of Theorem 7.6 but with a smaller noise level, we can construct $\tilde{X}$ such that

$$
\tilde{X} = X(:,[1,2,3])H + N = X(:,[1,2,4])H' + N',
$$

where $\max_j \|N(:,j)\|_2 = \max_j \|N'(:,j)\|_2$, while the distance between $X(:,4)$ and $W(:,3)$ is in $\mathcal{O}(\epsilon/\gamma)$. This means that the optimal solution of the near-separable NMF algorithm is non-unique (both solutions lead to the same reconstruction error). This is an important observation: for the solution of near-separable NMF to be unique, ϵ must be sufficiently small. Moreover, the distance between these solutions is in $\mathcal{O}(\epsilon/\gamma)$. Hence, whatever solution is returned by an algorithm, it has an error in $\mathcal{O}(\epsilon/\gamma)$ compared to the other one. We refer the interested reader to [187, Theorem 3.1] for further discussions on such an example.

Take-home message from the idealized algorithm   The main take-home message from this section is that an idealized near-separable NMF algorithm achieves the optimal bounds

$$
\epsilon \leq \mathcal{O}(\gamma_2(W)) \quad \Rightarrow \quad q_2(\mathcal{K}) \leq \mathcal{O}(\epsilon/\gamma_2(W)).
$$

---

[51] The function $\sqrt{\frac{1+\gamma^2}{2}}$ is convex, and $\frac{1+\gamma}{2}$ is its first-order Taylor approximation at $\gamma = 1$.

This result holds when the ℓ_2 norm is replaced with any ℓ_p norms with $p \geq 1$, replacing $\gamma_2(W)$ by $\gamma_p(W)$ in the bounds above. In fact, the proofs of the lemmas and theorems can be adapted accordingly as most of the steps only rely on the triangle inequality.

It is important to keep these bounds in mind to compare them with the bounds derived in this section for polynomial-time near-separable NMF algorithms.

7.4 ▪ Greedy/sequential algorithms

Greedy algorithms construct the solution $\mathcal{K}$ of near-separable NMF by identifying one index at a time. Most of them can be put in the following general framework. After initializing the index set to $\mathcal{K} = \emptyset$ and the residual to $R = \tilde{X}$, they consist of two main steps:

1. Selection step: select an index to add to $\mathcal{K}$ based on some criterion using the residual matrix R.

2. Projection step: update the residual matrix R by projecting its columns on a subspace taking into account the newly extracted column of $\tilde{X}$.

Both steps can be implemented in different ways leading to different algorithms.

Greedy algorithms are the most popular near-separable NMF algorithms: they are fast and scale well (typically running in $\mathcal{O}(mnr)$ operations), they are easy to understand and implement, and some of them come with provable guarantees even in the presence of noise. In this section, we review the most important near-separable NMF greedy algorithms with provable guarantees, namely

- SPA, which is based on orthogonal projections (Section 7.4.1),

- fast anchor words (FAW) that refine SPA with a second optimization phase (Section 7.4.2),

- vertex component analysis (VCA), which is closely related to SPA and is very popular in the blind HU literature (Section 7.4.3), and

- SNPA (Section 7.4.4), which takes into account nonnegativity in the projection step.

Then we discuss three different preconditionings that allow us to make greedy algorithms more robust to noise (Section 7.4.5). Finally, we perform some numerical experiments comparing greedy near-separable NMF algorithms and provide some take-home messages (Section 7.4.6).

7.4.1 ▪ The successive projection algorithm

SPA is the workhorse greedy near-separable NMF algorithm. It is arguably the simplest while having robustness guarantees (see Theorem 7.10 below). For the selection step, it picks the columns of the residual with the largest[52] ℓ_2 norm. The rationale behind this choice is that the ℓ_2 norm is always maximized at a vertex of the convex hull of a set of points. Moreover, this property holds under linear projections (see Theorem 7.9 for a formal proof). Once a column is selected, the residual is updated by performing a projection onto the orthogonal complement of the extracted column. This implies that the extracted column is projected onto zero and will not be extracted in the following steps. Algorithm 7.1 provides the pseudocode for SPA. It requires r as an input, or it can be terminated when the norm of the residual is below a certain threshold.

---

[52]In case of a tie, any column with maximum ℓ_2 norm can be picked. To break the ties, [210] for example uses the column that maximizes the ℓ_2 norm of the unprojected data set $\tilde{X}$ (in case of another tie, one such column is picked randomly).

Algorithm 7.1 Successive projection algorithm (SPA) [13] [`Matlab file: SPA.m`]

Input: The matrix $\tilde{X}$ that admits a near-separable factorization (Assumption 7.1), the number r of columns to extract, and/or a relative error δ to achieve (by default, $\delta = 0$).
Output: Set of r indices $\mathcal{K}$ such that $\tilde{X}(:, \mathcal{K}) \approx W$ (up to permutation).

1: Let $R = \tilde{X}, \mathcal{K} = \{\}, k = 1$.

2: **while** $k \leq r$ and $\dfrac{\|R\|_F}{\|\tilde{X}\|_F} \geq \delta$ **do**

3:     $p = \operatorname{argmax}_j \|R_{:j}\|_2$.   *% Selection step*

4:     $\mathcal{K} = \mathcal{K} \cup \{p\}$.

5:     $R = \left(I - \dfrac{R_{:p} R_{:p}^\top}{\|R_{:p}\|_2^2} \right) R$.   *% Projection step*

6:     $k = k + 1$.

7: **end while**

This section on SPA is organized as follows. We first prove the ability of SPA to extract the correct solution in the absence of noise (Section 7.4.1.1) and state the robustness result in the presence of noise (Section 7.4.1.2). Then we provide a geometric interpretation of SPA: it is a greedy algorithm to find a subset of columns with maximum volume (Section 7.4.1.3). In Section 7.4.1.4, we discuss the computational cost of SPA, in particular providing an efficient implementation for sparse matrices which does not require the computation of the residual matrix. In Section 7.4.1.5, we list the pros and cons of SPA, while in Section 7.4.1.6, we review when and how SPA came about in different research fields.

7.4.1.1 ▪ Correctness of SPA in the absence of noise

Before proving the correctness of SPA in the absence of noise, let us show the following simple lemma, and a corollary.

Lemma 7.7. *For any two vectors $w_1 \neq w_2 \in \mathbb{R}^m$ and $0 < \alpha < 1$,*

$$\|\alpha w_1 + (1 - \alpha) w_2\|_2 < \max_{i \in \{1,2\}} \|w_i\|_2. \tag{7.5}$$

Proof. If w_1 and w_2 are not multiples of one another, (7.5) follows from the Cauchy–Schwarz inequality

$$w_1^\top w_2 < \|w_1\|_2 \|w_2\|_2 \leq \max_{i \in \{1,2\}} \|w_i\|_2^2$$

and

$$\|\alpha w_1 + (1 - \alpha) w_2\|_2^2 = \alpha^2 \|w_1\|_2^2 + 2\alpha(1 - \alpha) w_1^\top w_2 + (1 - \alpha)^2 \|w_2\|_2^2$$
$$< \max_{i \in \{1,2\}} \|w_i\|_2^2,$$

since $\alpha(1 - \alpha) \neq 0$.

If w_1 and w_2 are multiples of one another, that is, $w_2 = \lambda w_1$ for some $|\lambda| < 1$ (w.l.o.g., since the case $|\lambda| > 1$ follows by exchanging the role of w_1 and w_2),

$$\begin{aligned}
\|\alpha w_1 + (1-\alpha)\lambda w_1\|_2 &= |\alpha + \lambda(1-\alpha)| \, \|w_1\|_2 \\
&\le (\alpha + |\lambda|(1-\alpha)) \, \|w_1\|_2 \\
&< (\alpha + (1-\alpha)) \, \|w_1\|_2 = \|w_1\|_2. \qquad (7.6)
\end{aligned}$$

□

Corollary 7.8. *For any matrix* $W \in \mathbb{R}^{m \times r}$ *and* $h \in \mathcal{S}^r$,

$$\|Wh\|_2 \le \max_{1 \le k \le r} \|W(:,k)\|_2.$$

Moreover, if the columns of W *are distinct and* h *is not a unit vector (that is,* $h \ne e_k$ *for* $1 \le k \le r$*), then the inequality is strict.*

Proof. The first part of the lemma follows from the triangle inequality and $h \in \mathcal{S}^r$:

$$\begin{aligned}
\|Wh\|_2 &= \left\| \sum_{k=1}^r W(:,k)h(k) \right\|_2 \\
&\le \sum_{k=1}^r \|h(k)W(:,k)\|_2 \\
&= \sum_{k=1}^r h(k) \, \|W(:,k)\|_2 \\
&\le \max_{1 \le k \le r} \|W(:,k)\|_2. \qquad (7.7)
\end{aligned}$$

For the second part, let $h \in \mathcal{S}^r$ which is not a unit vector. Let us define the set $\mathcal{K} = \{k | h(k) > 0\}$ and let us use induction. If $|\mathcal{K}| = 2$, that is $\mathcal{K} = \{k,j\}$ for some k,j, the result follows from Lemma 7.7, taking $w_1 = W(:,k)$, $w_2 = W(:,j)$, and $\alpha = h(k)$.

Now assume the result holds when $|\mathcal{K}| = p \ge 2$ and show that it implies it holds when $|\mathcal{K}| = p + 1$. Let k be some index in $\mathcal{K}$, and let us denote $\mathcal{J} = \mathcal{K}\backslash\{k\}$ and $h' = h(\mathcal{J})/\|h(\mathcal{J})\|_1 \in \Delta^{|\mathcal{J}|}$ which is not a unit vector (since all entries of $h(\mathcal{K})$ are nonzero). By the induction step,

$$\|W(:,\mathcal{J})h(\mathcal{J})\|_2 = \|h(\mathcal{J})\|_1 \|W(:,\mathcal{J})h'\|_2 < \|h(\mathcal{J})\|_1 \max_{j \in \mathcal{J}} \|W(:,j)\|_2.$$

Finally,

$$\begin{aligned}
\|Wh\|_2 &\le h(k)\|W(:,k)\|_2 + \|\sum_{j \in \mathcal{J}} h(j)W(:,j)\|_2 \\
&< h(k)\|W(:,k)\|_2 + (1-h(k)) \max_{j \in \mathcal{K}\backslash\{k\}} \|W(:,j)\|_2 \\
&\le \max_{1 \le k \le r} \|W(:,k)\|_2,
\end{aligned}$$

where we used $\|h(\mathcal{J})\|_1 = 1 - h(k)$. □

We can now prove correctness of SPA in the absence of noise.

Theorem 7.9. *Let* $X = X(:,\mathcal{K}^*)H = WH$ *be a matrix satisfying Assumption 7.1 with* $p = 2$, $N = 0$, *and* $r = \mathrm{rank}(W)$. *Then SPA (Algorithm 7.1) returns a set of indices* $\mathcal{K}$ *such that* $X(:,\mathcal{K}) = W\Pi$ *where* Π *is a permutation matrix.*

Proof. Let us prove the result by induction.

First step. This follows from Corollary 7.8: the ℓ_2 norm can only be maximized at a column of X corresponding to a column of W, that is, for a column $X(:, j) = WH(:, j)$ where $H(:, j)$ is a unit vector.

Induction step. Assume that SPA has already extracted columns of X corresponding to the columns of W in the set $\mathcal{L}$ with $|\mathcal{L}| \geq 1$. Let us denote by $Y = W(:, \mathcal{L})$ these columns, and let $P_Y^\perp$ be the projection onto the orthogonal complement of the columns of Y so $P_Y^\perp W(:, \mathcal{L}) = 0$. We have, for all $1 \leq j \leq n$,

$$P_Y^\perp X(:, j) = P_Y^\perp WH(:, j) = \sum_{k=1}^r P_Y^\perp W(:, k)H(k, j) = \sum_{k \notin \mathcal{L}} P_Y^\perp W(:, k)H(k, j).$$

Since W is full column rank, the columns $\{P_Y^\perp W(:, k)\}_{k \notin \mathcal{L}}$ are different from zero and distinct. Therefore, by Assumption 7.1 and Corollary 7.8, SPA identifies at the next step a column of W not extracted yet, which concludes the proof. $\qquad\qquad\qquad\qquad\square$

7.4.1.2 ▪ Robustness to noise

A main feature of SPA is that it is robust in the presence of noise.

Theorem 7.10. *[210, Theorem 3] Let $\tilde{X} = X + N = X(:, \mathcal{K}^*)H + N = WH + N$ be a matrix satisfying Assumption 7.1 with $p = 2$ and*

$$\epsilon = \max_j \|N(:, j)\|_2 \leq \mathcal{O}\left(\frac{\sigma_r^3(W)}{\sqrt{r}}\right).$$

Then SPA (Algorithm 7.1) returns a set of indices $\mathcal{K}$ such that

$$q_2(\mathcal{K}) \leq \mathcal{O}\left(\frac{\epsilon}{\sigma_r^2(W)}\right),$$

where $q_2(\mathcal{K})$ measures the distance between $\tilde{X}(:, \mathcal{K})$ and the ground truth W; see (7.1) page 214.

We do not provide the proof, which is rather long and technical, and refer the interested reader to [210]. The bound can also be obtained from the results of Arora et al. [14]; see the discussion in Section 7.4.2.

Let us discuss briefly the bounds of Theorem 7.10. In Assumption 7.1, $\max_k \|W(:, k)\|_2 = 1$; hence $\sigma_r(W) \leq 1$. If this assumption is removed, and denoting $K(W) = \max_k \|W(:, k)\|_2$, the bounds in Theorem 7.10 are replaced by [210, Theorem 3]

$$\epsilon = \max_j \|N(:, j)\|_2 \leq \mathcal{O}\left(\frac{\sigma_r^3(W)}{K(W)^2\sqrt{r}}\right) \quad \text{and} \quad q_2(\mathcal{K}) \leq \mathcal{O}\left(\frac{\epsilon K(W)^2}{\sigma_r^2(W)}\right).$$

Since $\frac{K(W)}{\sigma_r(W)} \leq \kappa(W)$, the bounds above have sometimes been replaced by

$$\epsilon = \max_j \|N(:, j)\|_2 \leq \mathcal{O}\left(\frac{\sigma_r(W)}{\sqrt{r}\kappa(W)^2}\right) \quad \text{and} \quad q_2(\mathcal{K}) \leq \mathcal{O}\left(\epsilon\kappa(W)^2\right).$$

Hence the noise level ϵ must be rather small for SPA to be provably robust, being proportional to $\frac{1}{\kappa(W)^2}$. However, it is important to keep in mind that this is a *worst-case analysis* and, as we will see in the numerical experiments in Section 7.4.6, SPA can identify the correct set of indices on randomly generated data for higher noise levels.

7.4.1.3 ▪ Geometric interpretation: volume maximization

SPA can be interpreted as a greedy algorithm that maximizes the volume of $\mathrm{conv}([\tilde{X}(:,\mathcal{K}),0])$. Let us explain this result which was reported in [77, 80, 439]. The volume of $\mathrm{conv}([Y,0])$ within the subspace spanned by Y is equal to $\det(Y^\top Y)$ given that Y has full column rank; see Lemma 4.41. At the first step, the column of $\tilde{X}$ that maximizes the volume of $\mathrm{conv}([\tilde{X}(:,j),0])$ is the one that maximizes $\|\tilde{X}(:,j)\|_2$, because $[\tilde{X}(:,j),0]$ is a segment of length $\|\tilde{X}(:,j)\|_2$. Now, let $Y = \tilde{X}(:,\mathcal{L})$ be the columns of $\tilde{X}$ already extracted by SPA, and let $y = \tilde{X}(:,j)$ be the next column to be selected. What is the column y that maximizes the volume of $\mathrm{conv}([Y,y,0])$, that is, that maximizes $\det([y,Y]^\top[y,Y])$? It turns out it is the column that has maximal ℓ_2 norm after projection onto the orthogonal complement of $\mathrm{col}(Y)$. This follows from the following lemma.

Lemma 7.11. *Let $Y \in \mathbb{R}^{m \times r}$ with $\mathrm{rank}(Y) = r$ and let $y \in \mathbb{R}^m$. Then*

$$\det([y,Y]^\top[y,Y]) = \det\begin{pmatrix} y^\top y & y^\top Y \\ Y^\top y & Y^\top Y \end{pmatrix} = \det(Y^\top Y)\|y^\perp\|_2^2,$$

where

$$y^\perp = \underbrace{\left(I_m - Y(Y^\top Y)^{-1}Y^\top\right)}_{P_Y^\perp} y,$$

and $P_Y^\perp$ is the projector onto the orthogonal complement of $\mathrm{col}(Y)$.

Proof. This follows from the Schur formula: given four matrices A, B, C, and D with appropriate dimensions and with D invertible, it states that

$$\det\begin{pmatrix} A & B \\ C & D \end{pmatrix} = \det(D)\det(A - BD^{-1}C),$$

where $A - BD^{-1}C$ is the Schur complement of $[A\,B;C\,D]$. Since Y is full column rank, $Y^\top Y$ is positive definite, hence

$$\begin{aligned}
\det\begin{pmatrix} y^\top y & y^\top Y \\ Y^\top y & Y^\top Y \end{pmatrix} &= \det(Y^\top Y)\det\left(y^\top y - y^\top Y(Y^\top Y)^{-1}Y^\top y\right) \\
&= \det(Y^\top Y)\det\left(y^\top(I - Y(Y^\top Y)^{-1}Y^\top)y\right) \\
&= \det(Y^\top Y)\det\left(y^\top P_Y^\perp y\right) \\
&= \det(Y^\top Y)\|P_Y^\perp y\|_2^2 = \det(Y^\top Y)\|y^\perp\|_2^2,
\end{aligned}$$

since $(P_Y^\perp)^\top = P_Y^\perp$ and $P_Y^\perp P_Y^\perp = P_Y^\perp$. □

Lemma 7.11 implies that SPA is a greedy algorithms that selects the columns of $\tilde{X}$ so that the volume of the convex hull of the selected columns and the origin is maximized.

7.4.1.4 ▪ Computational cost

Let us explain how to implement the different steps of SPA in an efficient way. First, it is key not to compute the residual matrix R explicitly. At the first step, we only need to compute $\|\tilde{X}(:,j)\|_2^2$ for all j, which can be done in $2mn$ operations. Then, for the next steps, to avoid forming R explicitly, one should use the following formula sequentially: for any $u, x \in \mathbb{R}^m$ with $\|u\|_2 = 1$,

$$\begin{aligned}
\|(I - uu^\top)x\|_2^2 &= \|x - u(u^\top x)\|_2^2 \\
&= \|x\|_2^2 - 2(u^\top x)^2 + (u^\top x)^2\|u\|_2^2 \\
&= \|x\|_2^2 - (u^\top x)^2.
\end{aligned}$$

Denoting u_k the normalized column of $\tilde{X}$ extracted at the kth step of SPA and projected onto the orthogonal complement of $[u_1, u_2, \ldots, u_{k-1}]$, we obtain

$$\|(I - u_k u_k^\top) \ldots (I - u_1 u_1^\top)x\|_2^2 = \|(I - u_1 u_1^\top - \cdots - u_k u_k^\top)x\|_2^2$$
$$= \|x\|_2^2 - (u_1^\top x)^2 - \cdots - (u_k^\top x)^2,$$

since $u_i^\top u_j = 0$ for $i \neq j$. Applying this formula to the columns of $\tilde{X}$ to compute the norms of the columns of the residual R leads to a computational cost of $2mnr + \mathcal{O}(mr^2)$ operations. At the kth step, the main cost is to first project u_k onto the orthogonal complement of $[u_1, u_2, \ldots, u_{k-1}]$ which requires $\mathcal{O}(mk)$ operations, and then compute $u_j^\top \tilde{X}$ which requires $2mn$ operations.

This implementation also handles sparse matrices efficiently, as it requires $\mathcal{O}(r\,\mathrm{nnz}(\tilde{X}))$ operations where $\mathrm{nnz}(\tilde{X})$ is the number of nonzero entries in $\tilde{X}$. This is not the case of the naive implementation of Algorithm 7.1 that computes the residual explicitly and requires $2mnr$ operations; see Example 7.16 for a numerical experiment comparing the two implementations. This makes SPA extremely fast: its main computational cost is equivalent to one matrix-matrix product $U^\top \tilde{X}$ where U is an r-by-m matrix.

7.4.1.5 ▪ Pros and cons of SPA

The mains advantages of SPA are the following:

- It is very fast, running in $2mnr + \mathcal{O}(mr)$ operations for a dense input matrix $\tilde{X}$ and in $\mathcal{O}(r\,\mathrm{nnz}(\tilde{X}))$ operations for a sparse input matrix $\tilde{X}$.

- There is no parameter to tune: one only needs to choose the number of columns of $\tilde{X}$ to extract, or the residual error to achieve.

- Even if the input matrix $\tilde{X}$ does not satisfy Assumption 7.1, SPA will extract r columns that are well-spread within the data set.

However, SPA also has several drawbacks:

- It is not very robust to noise. The error bounds in Theorem 7.10 are rather poor; in particular $\epsilon \leq \mathcal{O}\left(\frac{\sigma_r(W)}{r\kappa(W)^2}\right)$ means that the columns of N must have a very small norm, especially when W is ill-conditioned. In real-world scenarios, this bound will be violated in most cases. However, this is a worst-case analysis and SPA typically leads to good practical performances.

- It is not applicable when $r > \mathrm{rank}(W)$.

- Like most near-separable NMF algorithms, it is sensitive to outliers. This can be leveraged via pre- and postprocessing; see Section 7.2.

Another weak point of SPA is that it focuses on the vertices of $\mathrm{conv}(\tilde{X})$: it does not leverage the knowledge of many data points within $\mathrm{conv}(\tilde{X})$. In other words, it does not take the data fitting term $\|\tilde{X} - \tilde{X}(:,\mathcal{K})H\|_F$ into account in the selection step. In order to overcome this drawback, a variant, referred to as robust SPA [193], identifies several columns of $\tilde{X}$ in the selection step (instead of just one in SPA) and picks the one that decreases the data fitting error the most.

7.4.1.6 ▪ History

SPA has been discovered many times and in different contexts. As far as we know, its first use in the context of NMF and blind source separation is in the paper [13] (2001), where it is referred to as the successive projections algorithm and is applied on spectrophotometric data. However, it did not draw much attention initially (the citation number increased significantly after 2010). SPA has been rediscovered many times in different application areas:

- Ren and Chang [396] (2003) called it the automatic target generation process. They used it for blind HU.

- Jiang, Liang, and Ozaki [264] (2003) called it SIMPLEX1 and used it in the context of SMCR (see Sections 1.4 and 9.3).

- Chan et al. [80] (2011) called it the successive volume maximization algorithm. They proved that SPA is a greedy algorithm to maximize the convex hull of the extracted columns of $\tilde{X}$ (see Lemma 7.11). The same observation was also made in [77] (2009) in a rather different context (see the second bullet point below).

- Sun et al. [439] (2013) focused on the volume interpretation and the use of the formula from Lemma 7.11. They referred to their algorithm as the fast Gram determinant based algorithm and applied it on hyperspectral images.

Moreover, SPA has been discovered and used in other contexts:

- *Numerical linear algebra—QR with column pivoting.* Golub and Businger [69] construct QR factorizations of matrices by performing, at each step of the algorithm, the Householder reflection with respect to the column of $\tilde{X}$ whose projection onto the orthogonal complement of the previously extracted columns has maximum ℓ_2 norm, exactly as in SPA. Hence SPA can be interpreted as a modified Gram–Schmidt algorithm with column pivoting.

- *Theoretical computer science—maximum-volume submatrix.* Çivril and Magdon-Ismail [77, 78] showed that SPA is a good greedy heuristic for the problem of identifying a subset of columns of a given matrix whose convex hull has maximum volume. More precisely, unless P=NP, they proved that the approximation ratio guaranteed by SPA is within a logarithmic factor of the best possible achievable ratio by any polynomial-time algorithm.

- *Compressive sensing—orthogonal matching pursuit.* Fu et al. [175] showed that SPA can be interpreted as a greedy algorithm to solve a convex-optimization-based near-separable model (see Section 7.6). This interpretation shares some similarity with orthogonal matching pursuit, which is a greedy algorithm widely used in compressive sensing to solve least squares problem under sparsity constraints.

Variants of SPA   Let us briefly mention a few variants of SPA:

1. TRI-P [7] generalizes SPA by replacing the ℓ_2 norm in the selection step by the ℓ_p norm for $p > 1$.

2. FastSepNMF [210] generalizes SPA by replacing the ℓ_2 norm in the selection step by any strongly convex function f such that $f(0) = 0$. In particular, Lemma 7.7 applies to any such function. SPA remains robust to noise under this generalization, and it is shown that the ℓ_2 norm leads to the best bounds (in the worst case) when $p = 2$ in Assumption 7.1.

3. FAW [14] sweeps one more times over the indices extracted by SPA in order to improve its solution; see the next section.

4. SNPA [188] uses a projection onto the convex cone spanned by the columns extracted so far and the origin, instead of an orthogonal projection; see Section 7.4.4.

7.4.2 ▪ Fast anchor words

Arora et al. [14] designed a SPA-like algorithm, FAW, under the assumption that $H^\top e = e$. They used it in the context of topic modeling; see Section 7.8. FAW has two main phases. The first phase is very similar to SPA, while the second one goes through the extracted indices once more in order to improve the solution. Hence FAW is not a fully greedy algorithm: it is a greedy algorithm with one round of local search.

First phase of FAW   The only difference between SPA and the first phase of FAW is that the orthogonal projections of FAW are performed with respect to the affine subspace spanned by the extracted vertices, instead of their linear subspace as in SPA. This means that FAW relies on the assumption $H^\top e = e$ (as many near-separable NMF algorithms). The first column extracted by FAW is the same as SPA, that is, the one with the largest ℓ_2 norm. After the first column is extracted, say $\tilde{X}(:,k)$, all data points are translated by $-\tilde{X}(:,k)$, so that $\tilde{X}(:,k)$ is translated to the origin. After this first step, FAW runs exactly the same steps as SPA. In particular, the second extracted column will be the one the farthest away from the origin, that is, the farthest away from $\tilde{X}(:,k)$. Under the assumption $H^\top e = e$, FAW performs slightly better than SPA as it uses one less orthogonal projection than SPA. Therefore, it can extract one more column of $\tilde{X}$, that is, it can extract up to $\mathrm{rank}(\tilde{X}) + 1$ columns, as opposed to SPA, which can only extract $\mathrm{rank}(\tilde{X})$ columns. In particular, if $r = m + 1$ (for example, the columns of W form a triangle in the plane; see Figure 7.3, page 221, for an illustration), FAW will be able extract the r columns of W, while SPA will only be able to extract $r - 1$ of them.

For FAW to handle the case $H^\top e \le e$, it suffices to perform the orthogonal projection starting at the first step: the first phase of FAW is SPA. Equivalently, one can add the zero vector within the data set, which is a vertex of $\mathrm{conv}([\tilde{X}, 0])$. If the zero vertex is extracted as one of the first two vertices in FAW (or if one modifies FAW so that the zero vertex is extracted in the first step), then SPA and the first phase of FAW are the same algorithm. Since the theoretical analysis of these algorithms does not depend on the order in which the vertices are extracted, any result for SPA applies to the first phase of FAW and vice versa.

As we focus in this chapter on the assumption $H^\top e \le e$, we consider FAW adapted to the case $H^\top e \le e$ using projections with respect to the linear subspace spanned by the extracted columns of $\tilde{X}$, instead of their affine subspace; see Algorithm 7.2, where SPA is used in the first phase of FAW.

Second phase of FAW   After the first phase has identified $\mathcal{K}$, FAW goes through each index k in $\mathcal{K}$ (in the same order in which they were extracted) and checks whether this index k corresponds to the column of $\tilde{X}$ which is the farthest away from the affine hull spanned by the other extracted columns $\tilde{X}(:, \mathcal{K}\backslash\{k\})$. Again, to deal with the relaxed assumption $H^\top e \le e$, we adapt this step by looking at the distance to the linear subspace spanned by these columns. The column extracted in the first phase is replaced with the one that maximizes that quantity which increases the volume of $\mathrm{conv}([\tilde{X}(:,\mathcal{K}), 0])$; see Lemma 7.11. Algorithm 7.2 described this procedure. The second phase could be applied multiple times to improve the solution further; this would be equivalent to a local search heuristic used in combinatorial optimization and this idea

---

Algorithm 7.2 Fast anchor words (FAW) [14] adapted to the assumption $H^\top e \leq e$ [Matlab file: FastAnchorWords.m]

Input: The matrix $\tilde{X}$ that admits a near-separable factorization (Assumption 7.1), the number r of columns to extract, and/or a relative error δ to achieve (by default, $\delta = 0$).

Output: Set of indices $\mathcal{K}$ such that $\tilde{X}(:,\mathcal{K}') \approx W$ (up to permutation).

1: *% First phase of FAW*
2: $\mathcal{K} = \text{SPA}(\tilde{X}, r, \delta)$. *% See Algorithm 7.1*
3: *% Second phase of FAW*
4: **for** $k \in \mathcal{K}$ **do**
5:   Compute R, the projection of $\tilde{X}$ onto the orthogonal complement of $\tilde{X}(:,\mathcal{K}\backslash\{k\})$.
6:   $\mathcal{K} \leftarrow \mathcal{K} \backslash \{k\} \cup \text{argmax}_j \|R(:,j)\|_2$.
7: **end for**

---

was already proposed in the blind HU literature: it is the spirit behind the NFIND-R algorithm; see Section 7.5.2. However, this does not improve the theoretical robustness to noise of FAW.

Robustness to noise   Because of its second phase, FAW is slightly more robust to noise than SPA: although the upper bound on the noise level ϵ is the same,[53] it provides a more accurate estimation of W, from an error of order $\mathcal{O}\big(\frac{\epsilon}{\sigma_r^2(W)}\big)$ to an error of order $\mathcal{O}\big(\frac{\epsilon}{\sigma_r(W)}\big)$.

Theorem 7.12. *[14, Theorem 4.3] Let $\tilde{X} = X + N = X(:,\mathcal{K}^*)H + N = WH + N$ be a matrix satisfying Assumption 7.1 with $p = 2$, and let*

$$\epsilon = \max_j \|N(:,j)\|_2 \leq \mathcal{O}\left(\frac{\sigma_r^3(W)}{\sqrt{r}}\right).$$

Then FAW (Algorithm 7.2) returns a set of indices $\mathcal{K}$ such that

$$q_2(\mathcal{K}) \leq \mathcal{O}\left(\frac{\epsilon}{\sigma_r(W)}\right).$$

Note that, in [14, Theorem 4.3], the bound on ϵ is proportional to $\frac{1}{r}$; however, it can be improved to $\frac{1}{\sqrt{r}}$ using the analysis of SPA [210, Theorem 3].

Tightness of the bounds   As for SPA, the bound on ϵ is weak and requires the noise level to be rather low for Theorem 7.12 to apply. However, for well-conditioned matrices, the bound on $q_2(\mathcal{K})$ is close to being on par with the idealized algorithm (Theorem 7.5). As discussed in Section 7.2.4, for well-conditioned matrices $W \in \mathbb{R}^{m \times r}$ with $m \geq r$, $\gamma_2(W) \approx \sigma_r(W)$, in which case the bound on $q_2(\mathcal{K})$ of FAW matches the bound of the idealized algorithm given that the noise is sufficiently small.

Computational cost   FAW essentially costs r times the cost of SPA because the second phase requires projecting all data points onto each subset of $r - 1$ extracted columns, for a total cost of $\mathcal{O}(r^2 \text{nnz}(\tilde{X}))$ operations.

---

[53]Although the analysis of SPA in [210, Theorem 3] and of the first phase of FAW in [14, Lemma 4.4] are rather different, they lead to similar bounds; see the discussion in [211].

7.4.3 ▪ Vertex component analysis

VCA was proposed by Nascimento and Bioucas-Dias [361] (2005); see [`Matlab file: VCA.m`]. It differs from SPA in only two aspects:

1. It has a built-in preprocessing which consists in using the truncated SVD to filter the noise (Theorem 6.3).

2. It uses the following selection step: First it generates a random vector c, then selects the column of R that maximizes the linear function $f(x) = c^\top x$. As for the ℓ_2 norm in SPA, a linear function is also maximized at a vertex of the convex hull of a set of points. However, this property is only true with probability one when c is generated randomly: if c is parallel to the normal vector of a facet, then any interior point of that facet also maximizes that function.

In terms on computational cost, the core of VCA has exactly the same cost as SPA, while the preprocessing step allows one to filter the noise but can be used prior to any near-separable NMF algorithm; see Section 7.2.

Because of its selection step based on linear functions, VCA is not robust to noise: for an arbitrarily small noise level (that is, any $\epsilon > 0$ in Assumption 7.1), it is easy to construct examples for which VCA fails. In particular, take a data point in the middle of a segment whose endpoints are two columns of W. Then add noise to this data point such that it goes toward the outside of $\mathrm{conv}(W)$. More precisely, letting $X(:,j)$ be this data point, it suffices to take $N(:,j) = \delta\left(X(:,j) - \bar{w}\right)$ where $\bar{w} = We/r$ is the average of the columns of W. For any $\delta > 0$,

$$\tilde{X}(:,j) = X(:,j) + N(:,j)$$

is outside $\mathrm{conv}(W)$ and there is a positive probability for this point, which is far from any vertex, to be extracted by VCA. This adversarial setting will be tested in the numerical experiments in Section 7.4.6.

Therefore SPA should be preferred to VCA. However, VCA has been extensively used in the blind HU literature and many researchers still use it. We believe the reasons are that (i) VCA is one of the first greedy algorithms proposed in this application area, (ii) the authors provided an efficient code to run VCA, and (iii) the concept (and proofs) of robustness to noise were only later brought to light by Arora et al. [15] (2012).

7.4.4 ▪ The successive nonnegative projection algorithm

A main drawback of SPA, FAW, VCA, and any algorithm relying on orthogonal projections is that they require $\mathrm{rank}(W) = r$. In some real settings, this assumption might be violated. This is, for example, the case for multispectral images which have only a few spectral bands m (from 5 to 30) so that it may happen that $m < r$, in which case $\mathrm{rank}(W) \leq m < r$.

To get rid of this assumption and only require $\gamma_p(W) > 0$ to successfully recover W, one should instead use a projection onto the convex hull of the extracted columns and the origin. This simple and natural idea leads to SNPA [188]; see Algorithm 7.3. For example, if the columns of W are the vertices of a triangle in the plane ($W \in \mathbb{R}^{2 \times 3}$; see Figure 7.3, page 221, for an illustration), SPA can only extract two columns (the residual is equal to zero after two steps because $\mathrm{rank}(W) = 2$), while, in most cases, SNPA is able to identify the three vertices. Even when $\mathrm{rank}(W) = r$, SNPA is more robust to noise because Theorem 7.10 applies to SNPA [188, Theorem 3.17]. Intuitively, performing orthogonal projections decreases the norm of the residual faster, and hence more information is lost within the projection steps of SPA than of SNPA; we will observe this behavior on ill-conditioned matrices W (Section 7.4.6).

Algorithm 7.3 Successive nonnegative projection algorithm (SNPA) [188] [Matlab file: SNPA.m]

Input: The matrix $\tilde{X}$ admits a near-separable factorization (Assumption 7.1), the number r of columns to extract, and/or a relative error δ to achieve (by default, $\delta = 0$).

Output: Set of r indices $\mathcal{K}$ such that $\tilde{X}(:, \mathcal{K}) \approx W$ (up to permutation).

1: Let $R = \tilde{X}$, $\mathcal{K} = \{\}$, $k = 1$.

2: **while** $k \leq r$ and $\dfrac{\|R\|_F}{\|\tilde{X}\|_F} \geq \delta$ **do**

3:     $p = \operatorname{argmax}_j \|R_{:,j}\|_2$.   % *Selection step*

4:     $\mathcal{K} = \mathcal{K} \cup \{p\}$.

5:     $H = \operatorname{argmin}_{Y \geq 0, Y^\top e \leq e} \|\tilde{X} - \tilde{X}(:, \mathcal{K})Y\|_F^2$.

6:     $R = \tilde{X} - \tilde{X}(:, \mathcal{K})H$.   % *Projection step*

7:     $k = k + 1$.

8: **end while**

This section on SNPA is organized as follows. In Section 7.4.4.1, we provide several definitions that will be used in the analysis in SNPA. The correctness of SNPA in the absence of noise is presented in Section 7.4.4.2 and in the presence of noise in Section 7.4.4.3, where it is compared with SPA. In Section 7.4.4.4 we discuss the computational cost of SNPA and in Section 7.4.4.5 its connection with another algorithm dubbed XRAY.

7.4.4.1 ▪ Preliminaries

Under which conditions does SNPA identify the correct set of columns of X? To answer this question, we need to define the quantity $\beta(W)$. To do so and simplify the notation, let us first define the projection operator $\mathcal{R}_B(.)$: For a vector $x \in \mathbb{R}^m$ and a matrix $B \in \mathbb{R}^{m \times r}$,

$$\mathcal{R}_B(x) = x - Bh^*, \quad \text{where} \quad h^* = \min_{h \in \mathcal{S}^r} \|x - Bh\|_2.$$

For a matrix $X \in \mathbb{R}^{m \times n}$, we let $\mathcal{R}_B(X)_{:,j} = \mathcal{R}_B\big(X(:, j)\big)$ for all j, that is, $\mathcal{R}_B(X)$ is obtained by applying $\mathcal{R}_B(.)$ columnwise on the matrix X. Hence, the residual matrix R in step 6 of SNPA is equal to $R = \mathcal{R}_{W(:, \mathcal{K})}(X)$.

Given $W \in \mathbb{R}^{m \times r}$, we define

$$\beta(W) = \min\left(\gamma_2(W), \frac{1}{\sqrt{2}}\nu(W)\right).$$

where

$$\nu(W) = \min_{\substack{1 \leq i, j \leq r, i \neq j \\ \mathcal{J} \subseteq \{1, 2, \ldots, r\} \setminus \{i, j\}}} \left\|\mathcal{R}_{W(:, \mathcal{J})}(w_i) - \mathcal{R}_{W(:, \mathcal{J})}(w_j)\right\|_2.$$

The quantity $\nu(W)$ is the minimum among the distances between the residuals of two columns of W when projected on any subset of the other columns of W. The quantity $\beta(W)$ is the minimum between $\nu(W)$ multiplied by the constant $\frac{1}{\sqrt{2}}$, and $\gamma_2(W)$. Note that, in most cases, $\nu(W) > 0$ because it is unlikely for the residual of two columns of W after projections onto other columns to be exactly equal to one another (this happens with probability zero if W is generated randomly). However, there exist some pathological cases. For example, the matrix

$$W = \begin{pmatrix} 4 & 1 & 3 \\ 0 & 1 & 1 \end{pmatrix}$$

satisfies $\beta(W) = 0$ while $\gamma_2(W) > 0$, and any data point on the segment $[W(:, 2), W(:, 3)]$ could be extracted at the second step of SNPA because

$$\mathcal{R}_{W(:,1)}\Big(W(:, 2)\Big) = \mathcal{R}_{W(:,1)}\Big(W(:, 3)\Big) = \begin{pmatrix} 0 \\ 1 \end{pmatrix}.$$

Note, however, that $\beta(W) \geq \sigma_r(W)$ [188, Lemma 3.19]. In particular, for matrices W such that $\text{rank}(W) < r$, $\sigma_r(W) = 0$, hence SPA will fail, while SNPA will succeed if $\beta(W) > 0$. For example, let the columns of W be located on the unit circle within the nonnegative orthant in two dimensions so that $\text{rank}(X) = m = 2 < r$. In this case, SPA can only identify two vertices. However, $\gamma_2(W) > 0$ since no point on the unit circle is within the convex hull of other points on the unit circle and the origin. Moreover, we will generically have that $\nu(W) > 0$ so that SNPA will be able to extract all r vertices.

7.4.4.2 ▪ SNPA in the absence of noise

Before proving SNPA works in the absence of noise, let us show the following lemma.

Lemma 7.13. *[188, Lemma 3.1] For any matrices $A \in \mathbb{R}^{m \times r}$ and $B \in \mathbb{R}^{m \times p}$, and any vector $h \in \mathcal{S}^r$,*

$$\|\mathcal{R}_B(Ah)\|_2 \leq \|\mathcal{R}_B(A)h\|_2.$$

Proof. Let us denote $Y(:, j) = \text{argmin}_{y \in \mathcal{S}^p} \|A(:, j) - By\|_2$ for all j, that is, $\mathcal{R}_B(A) = A - BY$. We have

$$\|\mathcal{R}_B(Ah)\| = \min_{y \in \mathcal{S}^p} \|Ah - By\|_2 \leq \|Ah - BYh\| = \|\mathcal{R}_B(A)h\|.$$

The inequality follows from $Yh \in \mathcal{S}^p$, since $Y(:, j) \in \mathcal{S}^p$ for all j and $h \in \mathcal{S}^r$. $\square$

We can now prove the correctness of SNPA in the absence of noise.

Theorem 7.14. *[188, Theorem 3.2] Let $X = X(:, \mathcal{K}^*)H = WH$ be a matrix satisfying Assumption 7.1 with $p = 2$, $N = 0$, and $\beta(W) > 0$. Then SNPA (Algorithm 7.3) returns a set of indices $\mathcal{K}$ such that $X(:, \mathcal{K}) = W\Pi$ where Π is a permutation matrix.*

Proof. We prove the result by induction.
First step. The first step is the same as SPA; see the proof of Theorem 7.9.
Induction step. Assume SNPA has extracted some indices $\mathcal{K}$ corresponding to columns of W, that is, $X(:, \mathcal{K}) = W(:, \mathcal{I})$ for some $\mathcal{I}$. Let us show that the next extracted index will be a column of W different from the extracted columns $W(:, \mathcal{I})$. For any $h \in \mathcal{S}^r$,

$$\|\mathcal{R}_{W(:,\mathcal{I})}(Wh)\|_2 \underset{\text{(Lemma 7.13)}}{\leq} \|\mathcal{R}_{W(:,\mathcal{I})}(W)h\|_2 = \left\| \sum_{k=1}^{r} h(k)\mathcal{R}_{W(:,\mathcal{I})}(W(:, k)) \right\|_2$$

$$\leq \sum_{k=1}^{r} h(k) \left\| \mathcal{R}_{W(:,\mathcal{I})}(W(:, k)) \right\|_2$$

$$\leq \max_{k} \left\| \mathcal{R}_{W(:,\mathcal{I})}(W(:, k)) \right\|_2.$$

Finally, the residual R after $|\mathcal{I}|$ steps of SNPA is equal to $\mathcal{R}_{W(:,\mathcal{I})}(X)$. Moreover,

- $\mathcal{R}_{W(:,\mathcal{I})}(W(:,k)) = 0$ for all $k \in \mathcal{I}$,

- $\mathcal{R}_{W(:,\mathcal{I})}(W(:,k)) \neq 0$ for all $k \notin \mathcal{I}$ since $\gamma_2(W) > 0$ as $\beta(W) > 0$,

- The nonzero columns of $\mathcal{R}_{W(:,\mathcal{I})}(W)$ are distinct since $\nu(W) > 0$ as $\beta(W) > 0$.

By Corollary 7.8, this implies that the second inequality above is strict unless $h = e_j$ for some j. Hence SNPA identifies a column of W not extracted yet, that is, it extracts $W(:,k)$ for some $k \notin \mathcal{I}$.   □

7.4.4.3 ▪ Robustness to noise

Let us state the robustness result for SNPA.

Theorem 7.15. *[188, Theorem 3.22] Let $\tilde{X} = X + N = X(:,\mathcal{K}^*)H + N = WH + N$ be a matrix satisfying Assumption 7.1 with $p = 2$ and $\beta(W) > 0$, and let*

$$\epsilon = \max_j \|N(:,j)\|_2 \leq \mathcal{O}\left(\beta(W)^4\right).$$

Then SNPA (Algorithm 7.3) returns a set of indices $\mathcal{K}$ such that

$$q_2(\mathcal{K}) \leq \mathcal{O}\left(\frac{\epsilon}{\beta(W)^3}\right).$$

As for SPA, the proof of robustness to noise is rather technical and we refer the interested reader to the corresponding paper. As for SPA, these error bounds are not known to be tight and could possibly be improved; this is a direction for future research.

Comparison with SPA   It has to be noted that the robustness proof of SPA applies to SNPA [188, Theorem 3.17]. For well-conditioned matrices $W \in \mathbb{R}^{m \times r}$ with $m \geq r$, we have $\gamma_2(W) \approx \sigma_r(W)$ (see the discussion after Theorem 7.12), in which case the bound of SPA and SNPA will be similar and both algorithms will perform similarly. In fact, if W is well-conditioned, $\sigma_r(W)$ is close to one so is $\beta(W)$ (since $\sigma_r(W) \leq \beta(W) \leq 1$), and $\beta(W)^4$ will not be significantly larger than $\frac{\sigma_r^3(W)}{\sqrt{r}}$ as long as r is not too large. However, if W is ill-conditioned or $\text{rank}(W) < r$, then $\beta(W)$ can be arbitrarily larger than $\sigma_r(W)$, in which case SNPA outperforms SPA. These observations will be confirmed in the numerical experiments reported in Section 7.4.6.

7.4.4.4 ▪ Computational cost

The main computational cost of SNPA is to compute H (step 5), which is a linearly constrained least squares problem. Using a first-order method, the main cost resides in the computation of the gradient, given by

$$\tilde{X}(:,\mathcal{K})^\top \left(\tilde{X}(:,\mathcal{K})Y - \tilde{X}\right) = \left(\tilde{X}(:,\mathcal{K})^\top \tilde{X}(:,\mathcal{K})\right)Y - \tilde{X}(:,\mathcal{K})^\top \tilde{X},$$

for a total of $2mr^2 + 2nr^2 + 2r\,\text{nnz}(\tilde{X})$ operations when $|\mathcal{K}| = r$; see Section 8.3 for more details on the NNLS problem. Hence SPA and SNPA have the same asymptotic computational cost, namely $\mathcal{O}(r\,\text{nnz}(\tilde{X}))$ operations, and SNPA is a constant factor slower than SPA. For example, using 100 iterations of a first-order method to compute H makes SNPA roughly 100 times slower than SPA (one gradient computation has roughly the same cost as running SPA when $|\mathcal{K}| = r$).

As in SPA, the residual R should not be computed explicitly. To compute H using a first-order method, one only needs the matrix-matrix product (see above). To compute the norm of the columns of the residual R, let us denote $A = \tilde{X}(:, \mathcal{K})$. We have

$$\|R(:, j)\|_2^2 = \|\tilde{X}(:, j) - AH(:, j)\|_2^2$$
$$= \|\tilde{X}(:, j)\|_2^2 - 2\tilde{X}(:, j)^\top AH(:, j) + \|AH(:, j)\|_2^2$$
$$= \|\tilde{X}(:, j)\|_2^2 - 2\tilde{X}(:, j)^\top AH(:, j) + H(:, j)^\top A^\top AH(:, j),$$

where $A^\top A$ can be computed once for all columns of R. Hence, to compute the norms of the columns of R, we need

- the norms of the columns of $\tilde{X}$ which require $\mathcal{O}(\mathrm{nnz}(\tilde{X}))$ operations,

- the products $\tilde{X}(:, j)^\top AH(:, j)$ which require $\mathcal{O}(r\,\mathrm{nnz}(\tilde{X}) + rn)$ operations (first compute $\tilde{X}^\top A$), and

- the products $H(:, j)^\top (A^\top A)H(:, j)$ which require $\mathcal{O}(mr^2)$ operations to compute $A^\top A$, and $\mathcal{O}(nr^2)$ operations to compute $H(:, j)^\top (A^\top A)H(:, j)$ for all j.

This way of computing the norm of the columns of R is also efficient when $\tilde{X}$ is sparse, as R is never formed explicitly.

Example 7.16 (Sparse data set). Let us apply SNPA on the "20 Newsgroups" document data set with $r = 20$. It is a 19949×43586 matrix with 99.82% of its entries equal to zero.

SNPA requires about 25 seconds to extract 20 indices, while its naive implementation that forms R explicitly requires about 6 minutes. For larger data sets, the naive implementation will go out of memory.[54]

As a comparison, SPA requires about 0.3 seconds (about 100 faster than SNPA, as expected since we used 100 iterations of a first-order method to compute H in the projection steps), while the naive implementation forming R explicitly requires about 3 minutes.

FAW requires about 6 seconds, which is also consistent as it should be about $r = 20$ times slower than SPA. ∎

7.4.4.5 ▪ XRAY, a closely related algorithm

XRAY [294] is an algorithm very similar to SNPA and was proposed before SNPA. It also uses the nonnegativity constraint in the projection step. However, it does not rely on the assumption that $H^\top e \le e$, and its goal is to extract the extreme directions of $\mathrm{cone}(\tilde{X})$. The projection step is a projection onto the conical hull of the columns extracted so far (it is the same as SNPA where the constraint $H^\top e \le e$ is removed), while the selection step is given by

$$\mathrm{argmax}_k \frac{R(:, i)^\top \tilde{X}(:, j)}{p^\top \tilde{X}(:, j)},$$

where i is chosen in different ways; for example $i = \mathrm{argmax}_\ell \|\tilde{X}(:, \ell)\|_2$. In some sense, the scaling is built in within the algorithm by the use of the denominator $p^\top \tilde{X}(:, j)$. In our experience, XRAY performs very similarly as SNPA although it does not perform as well in difficult scenarios when W is ill-conditioned or $\mathrm{rank}(W) < r$; see the comparison in [188]. Moreover, XRAY has not been proven to be robust to noise.

---

[54]The initial implementation of SNPA that I provided with the paper [188] is the naive one. The implementation provided in the MATLAB code of this book is the efficient one, not computing the residual explicitly.

7.4.5 ▪ Preconditioning

Given any matrix $\tilde{X} = WH + N$ admitting a near-separable factorization (Assumption 7.1), premultiplying $\tilde{X}$ by a matrix P generates another matrix

$$\tilde{Y} = P\tilde{X} = (PW)H + PN = W'H + N'$$

that admits a near-separable factorization. This operation simply premultiplies W and N by P.

In SPA, the key parameter that controls the robustness to noise is the smallest singular value of W (or the condition number of W if it is not normalized). In particular, if $\mathrm{rank}(W) = r$ and one is given its left inverse $P = W^\dagger$, we would have $W' = PW = I_r$ with $\sigma_r(W') = 1$. Of course, the left inverse of W is unknown (otherwise near-separable NMF would be solved), but it turns out it can be estimated, up to orthogonal transformations (see Section 7.4.5.2). Note that SPA is not affected by premultiplying the input matrix by an orthogonal matrix which amounts to rotating the data points (the ℓ_2 norm is invariant under such transformations). In the following, we present three preconditionings for near-separable NMF.

7.4.5.1 ▪ Preconditioning I: truncated SVD

Let (U_r, Σ_r, V_r) be the rank-r truncated SVD of $\tilde{X}$; see Section 6.1.1. Replacing $\tilde{X}$ with $U_r \Sigma_r V_r^\top \approx \tilde{X}$ allows us to filter the noise prior to using a near-separable NMF algorithm. Since U_r has orthogonal columns, it does not influence algorithms based on orthogonal projections such as SPA. In other words, applying SPA on the matrix $U_r \Sigma_r V_r^\top$ or on $\Sigma_r V_r^\top$ returns the same index set $\mathcal{K}$. The columns of $\Sigma_r V_r^\top$ are the coordinates of the columns of $\tilde{X}$ within the orthogonal basis U_r. This preprocessing is equivalent to using the preconditioning $P = U_r^\top$. Let $\tilde{X} = U_r \Sigma_r V_r^\top + U_r^\perp Y$ where $U_r^\perp$ is the orthogonal complement of U_r so that $U_r^\top U_r^\perp = 0$ and $(U_r^\perp)^\top U_r^\perp = I$ so that $Y = (U_r^\perp)^\top \tilde{X}$. We have

$$P\tilde{X} = U_r^\top \tilde{X} = U_r^\top \left(U_r \Sigma_r V_r^\top + U_r^\perp Y \right) = \Sigma_r V_r^\top;$$

see [Matlab file: lindimred.m].

This preprocessing/preconditioning does not much influence the conditioning of W (in the noiseless case, it does not affect the input matrix). In the next two subsections, we present two approaches that are able to reduce the conditioning of W.

Remark 7.1 (Prewhitening). *One may be tempted to use prewhitening, that is, replace $\tilde{X}$ with V_r^T using the preconditioning $P = \Sigma_r^{-1} U_r^\top$. However, this does not guarantee reducing the conditioning of W: it depends on how the data points are distributed within $\mathrm{conv}(W)$. In unbalanced cases (for example, most data points are located close to one of the columns of W so that the corresponding row of H has large norm), this preconditioning might increase significantly the conditioning of W; see the discussion in [206].*

7.4.5.2 ▪ Preconditioning II: minimum-volume ellipsoid

Assume $\mathrm{rank}(W) = r$ and $m = r$. Under these assumptions, $W^\dagger = W^{-1}$ and the question is how to estimate this matrix.

It turns out that this can be done by solving a minimum-volume ellipsoid (MVE) problem. Before providing a formal construction, let us give some geometric intuition. All columns of X are contained in $\mathrm{conv}\left([W, 0]\right)$, while we expect the MVE centered at the origin and containing $\mathrm{conv}\left([W, 0]\right)$ to have the columns of W on its border, while all the other columns of X will be strictly inside the MVE. In that case, we can identify the columns of W as the columns on the border of the MVE. This intuition was used in [347] to directly identify the columns of W.

However, in the presence of noise, more than r columns of $\tilde{X}$ might belong to the border of the MVE and it is unclear how to select among them—SPA was used in [347].

Let us now formally describe the preconditioning based on the MVE. An ellipsoid $\mathcal{E}$ centered at the origin in $\mathbb{R}^r$ is described via a PSD matrix $A \in \mathbb{S}^r_{++}$:

$$\mathcal{E} = \{\, x \in \mathbb{R}^r \mid x^\top A x \le 1 \,\}.$$

The axes of the ellipsoid are given by the eigenvectors of A, while their length is equal to the inverse of the square root of the corresponding eigenvalue. The volume of $\mathcal{E}$ is equal to $\det(A)^{-1/2}$ times the volume of the unit ball in dimension r. Given a matrix $\tilde{X} \in \mathbb{R}^{r \times n}$ of rank r, the MVE centered at the origin and containing the columns of matrix $\tilde{X}$ can be formulated as follows:

$$\min_{A \in \mathbb{S}^r_+} \ \log \det(A)^{-1} \quad \text{such that} \quad \tilde{X}(:,j)^\top A \tilde{X}(:,j) \le 1 \quad \text{for } j = 1, 2, \ldots, n. \qquad (7.8)$$

This problem is representable using semidefinite programming [56, p. 222]. We can now show that, under the separability assumption, the MVE allows us to compute W^{-1}, up to orthogonal transformations.

Theorem 7.17. *[211, Theorem 2.3] Let $\tilde{X} = X + N = X(:, \mathcal{K}^*)H + N = WH + N$ be a matrix satisfying Assumption 7.1 with $p = 2$, $m = r$, $N = 0$, and $\mathrm{rank}(W) = r$. Then the optimal solution of (7.8) is given by $A^* = (WW^\top)^{-1}$.*

Proof. The matrix $A^* = (WW^\top)^{-1}$ is a feasible solution of the primal (7.8): for all j, $X(:,j) = WH(:,j)$ so that

$$X(:,j)^\top A X(:,j) = H(:,j)^\top W^\top W^{-T} W^{-1} W H(:,j) = \|H(:,j)\|_2^2 \le \|H(:,j)\|_1^2 \le 1.$$

The dual of (7.8) is given by [56, p. 222]

$$\max_{y \in \mathbb{R}^n} \ \log \det \left(\sum_{j=1}^n y_j X(:,j) X(:,j)^\top \right) - e^\top y + r \quad \text{such that} \quad y \ge 0. \qquad (7.9)$$

One can check that $y^* \in \mathbb{R}^n_+$ defined as

$$y^*(j) = \begin{cases} 1 & \text{for } j \in \mathcal{K}^* \\ 0 & \text{otherwise} \end{cases} \quad \text{for } j = 1, 2, \ldots, n,$$

is a feasible solution of the dual. We have

$$\sum_{j=1}^n y_j X(:,j) X(:,j)^\top = \sum_{j=1}^r W(:,j) W(:,j)^\top = WW^\top \succ 0,$$

and $e^\top y = r$. Hence its objective function value coincides with that of the primal solution $A^* = (WW^\top)^{-1}$ which is therefore optimal by duality. $\qquad\Box$

Given the optimal solution $A^* = (WW^\top)^{-1} = W^{-\top} W^{-1}$ of (7.8), we can compute a symmetric factorization of the form $A^* = P^\top P$ (for example, using the Cholesky factorization). The matrix P will be equal to W^{-1}, up to an orthogonal transformation, so that $\kappa(PW^{-1}) = 1$ and SPA will perform significantly better in the presence of noise.

Algorithm 7.4 provides the pseudocode to compute the MVE preconditioning for near-separable NMF.

Algorithm 7.4 Minimum-volume ellipsoid (MVE) preconditioning for near-separable NMF [211] [Matlab file: `minvolell.m`]

Input: A matrix $\tilde{X} \in \mathbb{R}^{r \times n}$ that admits a near-separable factorization (Assumption 7.1).
Output: A preconditioning $P \approx W^{-1} \in \mathbb{R}^{r \times r}$ (up to orthogonal transformations).

1: Solve the MVE problem (7.8) for $\tilde{X}$ to obtain the optimal solution A^*.
2: Compute P such that $A^* = P^\top P$ using for example the Cholesky factorization.

Robustness to noise Using Algorithm 7.4 to precondition SPA provides the following robustness result which was first proved in [211, Theorem 2.9] with the bound $\epsilon \leq \mathcal{O}\left(\frac{\sigma_r(W)}{r\sqrt{r}}\right)$, and later improved to $\epsilon \leq \mathcal{O}\left(\frac{\sigma_r(W)}{\sqrt{r}}\right)$ in [348, Theorem 3].

Theorem 7.18. *[211, Theorem 2.9], [348, Theorem 3] Let*

$$\tilde{X} = X + N = X(:,\mathcal{K}^*)H + N = WH + N$$

be a matrix satisfying Assumption 7.1 with $p = 2$, $m = r$, $\mathrm{rank}(W) = r$, and

$$\epsilon = \max_j \|N(:,j)\|_2 \leq \mathcal{O}\left(\frac{\sigma_r(W)}{\sqrt{r}}\right).$$

Then MVE-SPA [Matlab file: `MVESPA.m`], that is, SPA (Algorithm 7.1) applied on $P\tilde{X}$ where P is the MVE preconditioning (Algorithm 7.4), returns an index set $\mathcal{K}$ such that

$$q_2(\mathcal{K}) \leq \mathcal{O}\left(\frac{\epsilon}{\sigma_r(W)}\right).$$

Tightness of the bounds The bounds of Theorem 7.18 are significantly better than for plain SPA (Theorem 7.10) and FAW (Theorem 7.12). Moreover, for well-conditioned matrices for which $\sigma_r(W) \approx \gamma_2(W)$, these bounds are close to being on par with the bounds of the idealized algorithm (neglecting the hidden constants). For the noise level ϵ, this is tight up to the factor $1/\sqrt{r}$; for the error $q_2(\mathcal{K})$, it is tight.

How to reduce the dimensionality for MVE-SPA Algorithm 7.4 requires $m = r$. If $m > r$, one can use any LDR technique to reduce the dimension from m to r. The truncated SVD seems to be a natural choice (this is, for example, the approach chosen by VCA). One might be tempted to believe that the truncated SVD is even the appropriate choice in the presence of Gaussian noise, and this is in fact the method used in the first papers considering the MVE preconditioning [347, 211, 348].

However, the truncated SVD and most other LDR techniques (such as random projections) are blind to the separable structure of the input matrix. It turns out that using a projection based on the basis matrix extracted by a near-separable NMF algorithm works better. For example, using the LDR that premultiplies $\tilde{X}$ by $\tilde{X}(:,\mathcal{K})^\dagger$ where $\mathcal{K}$ is extracted by SPA[55] works better in practice [349], especially in difficult scenarios where W is ill-conditioned. Figure 7.4 illustrates this observation, namely that truncated SVD and random projections[56] perform worse than the

---

[55]This preconditioning corresponds to the SPA preconditioning presented in the next section.
[56]We simply premultiply $\tilde{X}$ with $P \in \mathbb{R}^{r \times m}$ where each entry of P is generated using the Gaussian distribution of mean 0 and variance 1.

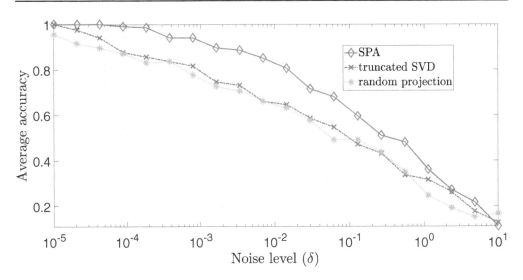

Figure 7.4. *Average accuracy of MVE-SPA (that is, SPA preconditioned with MVE) preprocessed with three different LDR techniques: truncated SVD, random projections, and SPA. The data set was generated using $m = 40$, $r = 20$, $n = 200$, ill-conditioned W ($\kappa(W) = 10^6$), Dirichlet distribution for H, and Gaussian noise for N; see Section 7.4.6. For each noise level, the figure reports the average accuracy (the percentage of correctly identified columns of W) over 10 randomly generated matrices. [Matlab file:* `compare_LDR_MVESPA.m`*].*

SPA-based LDR on a numerical example with ill-conditioned W, the Dirichlet distribution for H, and Gaussian noise for N (see Section 7.4.6 for more details on how these data sets are generated). Note that the truncated SVD performs slightly better than random projections (in particular, for $\delta = 10^{-5}$, truncated SVD has 100% average accuracy while random projections have 95.5% average accuracy). Note also that for well-conditioned W, the three approaches would provide similar results.

Computational cost   We first need to reduce the dimensionality which typically requires $\mathcal{O}(mnr)$ operations; this is the case for the truncated SVD, SPA, and random projections. The MVE preconditioning requires solving the semidefinite program (7.8) with r^2 variables (the matrix A) and n constraints. Using an interior-point method requires $\mathcal{O}(r^6 + n^3)$ operations per iteration (a linear system in $\mathcal{O}(r^2 + n)$ variables has to be solved, like in the Newton's method). The term n^3 is not reasonable in most applications; for example, in hyperspectral unmixing, n is the number of pixels in the image, and in text mining, n is the number of words (or documents). Luckily, only a few constraints will be active at optimality (a subset of at most $\frac{r(r+1)}{2}$ constraints is enough to solve the problem [265]) and active-set methods are very efficient to tackle such problems [440]. In practice, a very small number of iterations is required within the active-set method, and the algorithm rather behaves as $\mathcal{O}(r^6)$ (there are $\mathcal{O}(r^2)$ constraints active), which is typically negligible as $r \ll n$. For example, in [211], the authors apply it on a hyperspectral image with $n = 1.6\,10^5$ and $r = 16$ for which the semidefinite program (7.8) is solved in less than 10 seconds on a standard laptop.

7.4.5.3 ▪ Preconditioning III: SPA

Let us briefly discuss a third preconditioning based on SPA (Algorithm 7.5):

1. Run SPA on $\tilde{X}$ to obtain $\mathcal{K}$.

2. Compute the preconditioning $P = \tilde{X}(:,\mathcal{K})^\dagger$.

Algorithm 7.5 SPA preconditioning for near-separable NMF [206]

Input: A matrix $\tilde{X} \in \mathbb{R}^{m \times n}$ that admits a near-separable factorization (Assumption 7.1), and the number r of columns to extract.
Output: A preconditioning $P \approx W^\dagger \in \mathbb{R}^{r \times m}$.

1: Run SPA (Algorithm 7.1) on $\tilde{X}$ with r to obtain $\mathcal{K}$.
2: Let $P = \tilde{X}(:,\mathcal{K})^\dagger$.

Surprisingly, preconditioning SPA with itself allows us to improve its robustness to noise and obtain error bounds similar to FAW (recall that FAW is essentially SPA where a second phase is used to improve the solution; see Algorithm 7.2), by removing a factor $\sigma_r(W)$ in the error $q_2(\mathcal{K})$.

Theorem 7.19. *[206, Theorem 4.1] Let $\tilde{X} = X + N = X(:,\mathcal{K}^*)H + N = WH + N$ be a matrix satisfying Assumption 7.1 with $p = 2$, and let*

$$\epsilon = \max_j \|N(:,j)\|_2 \leq \mathcal{O}\left(\frac{\sigma_r^3(W)}{\sqrt{r}}\right).$$

Then SPA-SPA, that is, SPA (Algorithm 7.1) preconditioned with SPA (Algorithm 7.5), returns a set of indices $\mathcal{K}$ such that

$$q_2(\mathcal{K}) \leq \mathcal{O}\left(\frac{\epsilon}{\sigma_r(W)}\right).$$

Further discussions and readings   As far as we know, preconditionings have been analyzed only in the case $\text{rank}(W) = r$. It would be rather interesting to develop preconditioners and their analysis in the rank-deficient case. For example, we can use the MVE preconditioning with SNPA (computing the MVE in dimension $\text{rank}(W) < r$). However, there is no analysis in this case, and it is unclear how the robustness of SNPA is affected by such a preconditioning, and in particular how it affects the parameter $\beta(W)$.

There exist other preconditionings for near-separable NMF. This includes prewhitening (Remark 7.1) and approaches based on random projections and randomized low-rank approximations. We refer the interested reader to [206, 349] for more details.

7.4.6 ▪ Numerical comparison and discussion

In this section, we perform numerical experiments on synthetic data sets in order to illustrate some key differences between greedy near-separable NMF algorithms. The goal is not to perform an extensive comparison (these can be found in the corresponding papers) but rather to highlight the properties and compare the different algorithms. For example, we will see that VCA is not robust to noise in difficult scenarios (as pointed out in Section 7.4.3) and that preconditioned SPA is more robust than plain SPA.

7.4.6.1 ▪ Synthetic data sets

We generate $\tilde{X} \in \mathbb{R}^{m \times n}$ following Assumption 7.1 with

$$\tilde{X} = WH + N = W[I_r, H']\Pi + N,$$

where $W \in \mathbb{R}^{m \times r}$, $H'(:,j) \in \Delta^r$ for $j = 1, 2, \ldots, n' = n - r$, and $N \in \mathbb{R}^{m \times n}$ are generated in different ways. The matrix $\Pi \in \{0, 1\}^{n \times n}$ is a randomly generated permutation matrix.

We consider two types of distributions for H':

1. Dirichlet. Each column of H' is sampled following a Dirichlet distribution with its r parameters equal to θ. For $\theta = 1$, this distribution generates points uniformly within $\mathrm{conv}(W)$. As θ decreases, the columns of H' become sparser and sparser. We use $\theta = 0.5$ in our numerical experiments. Note that the Dirichlet distribution is a standard choice in the hyperspectral imaging literature [360].

2. Middle point. The columns of H' contain all possible combinations of two nonzero entries equal to $1/2$. This means that H' has $\binom{r}{2} = \frac{r(r-1)}{2}$ columns, and WH' contains all the points located between two columns of W. For example, for $r = 20$, which is the value we use, H' has $\binom{20}{2} = 190$ columns.

To generate N, we also consider two types, each of them associated with one of the two ways H' is generated:

1. Gaussian. This is used when H' is generated using the Dirichlet distribution. Each entry of N is first generated using the normal distribution $\mathcal{N}(0, 1)$. Then, given the parameter δ, N is scaled as follows:

$$N \leftarrow \delta N \frac{\|X\|_F}{\|N\|_F}.$$

This means that $\|\tilde{X} - X\|_F = \delta \|X\|_F$, and δ is the relative noise level.

2. Adversarial. This is used when H' is generated using the middle point setting. The columns of W are untouched, while the middle points are moved outward $\mathrm{conv}(X)$. More precisely, let $\bar{w} = We/r$ be the average of the columns of W. Given the parameter δ, we take

$$\tilde{X}(:,j) = X(:,j) + \delta \left(X(:,j) - \bar{w} \right).$$

Figure 7.5 illustrates this data generation for $r = 3$.

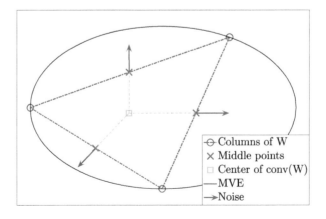

Figure 7.5. *Illustration of the middle point experiment in the case $r = 3$ (shown in a two-dimensional space, for simplicity). The noise moves each middle point, located between a pair of columns of W, outward $\mathrm{conv}(W)$.*

To generate W, we consider two cases:

1. **Well-conditioned.** Each entry of W is generated using the uniform distribution in the interval $[0,1]$, that is, $\mathcal{U}(0,1)$.

2. **Ill-conditioned.** The factor W is first generated using the uniform distribution as above. Then the compact SVD of $W = U\Sigma V^\top$ is computed and W is replaced with $W = U\Sigma'V^\top$ where Σ' is a diagonal matrix where the diagonal entries are between the parameter $\kappa < 1$ and 1 using a logarithmic scale, that is,

$$\log_{10}(\Sigma'(i,i)) = \log_{10}(\kappa) + \frac{i-1}{r-1}(1 - \log_{10}(\kappa)) \text{ for } i = 1,2,\dots,r;$$

in MATLAB notation, $\texttt{logspace}(\log_{10}(\kappa),\texttt{0},\texttt{r})$. This implies that $\kappa(W) = \kappa$. We use $\kappa = 10^{-6}$ in the experiments. Note that this transformation of W does not preserve nonnegativity, which is not an issue as this assumption is not required for near-separable NMF algorithms (nonnegativity of H is); see Assumption 7.1.

Finally, there are four ways to generate the data matrices: well-conditioned or ill-conditioned W, and Dirichlet associated with Gaussian noise or middle point associated with adversarial noise for H.

Remark 7.2 (Rank-deficient case). *We do not consider the case* $\operatorname{rank}(W) < r$ *in our experiments (for example, by taking* $m < r$*). In this scenario, only SNPA is able to extract a set of* r *indices (the residual in SPA and related methods is equal to zero after* m *steps). We refer the interested reader to the numerical experiments in [188], where SNPA is compared to XRAY.*

7.4.6.2 ▪ Comparison metrics

Since there are no duplicated columns of W in the synthetic data sets we generate,[57] it makes sense to count the number of correct indices (meaning indices corresponding to columns of W) extracted by an algorithm. Given the true location $\mathcal{K}^*$ of the columns of W and the set $\mathcal{K}$ returned by an algorithm with $|\mathcal{K}| = r$, we define to the accuracy of $\mathcal{K}$ as

$$\operatorname{accuracy}(\mathcal{K}) = \frac{|\mathcal{K} \cap \mathcal{K}^*|}{|\mathcal{K}^*|} \in [0,1].$$

It would also make sense to report $q_2(\mathcal{K})$ and/or the

$$\operatorname{relative\ error}(\mathcal{K}) = \frac{\min_{H \geq 0} \|\tilde{X} - \tilde{X}(:,\mathcal{K})H\|_F}{\|\tilde{X}\|_F},$$

which are standard choices in the literature. However, these metrics are highly correlated and, in most cases when there are no (near-)duplicated columns of W, if one algorithm performs better than another one for a metric, it also does for the other ones; see for example the numerical experiments in [204, 205]. Therefore, for the sake of simplicity, we show in this book only the accuracy.

7.4.6.3 ▪ Numerical experiments

We choose to compare the following algorithms:

- SPA: this is Algorithm 7.1.

---

[57]For the Dirichlet distribution of parameter $\theta = 0.5$ and $r = 20$, the average of the largest value in a column of H' is about $1/4$. Moreover, we have generated 10^6 such columns and all values of H' were smaller than 0.8. In other words, the probability for a column of W to have a proportion larger than 0.8 in a data point is very small.

- VCA: this is SPA where the selection step is replaced by selecting the column of $\tilde{X}$ that maximizes a randomly generated linear function, while the truncated SVD is used as a preprocessing step.

- FAW: this is SPA with a second phase refining the solution; see Algorithm 7.2.

- SNPA: this is Algorithm 7.3. We used 500 iterations of a fast gradient method to compute H for the projection step (for ill-conditioned W, it is important for these subproblems to be solved with relatively high precision).

- SPA-SPA: SPA preconditioned with SPA (Algorithm 7.5).

- MVE-SPA: SPA preconditioned with MVE (Algorithm 7.4), and where SPA is used as a preprocessing to reduce m to r. (This algorithm could be referred to as SPA-MVE-SPA, but we prefer MVE-SPA for simplicity.)

We use the values $m = 40$, $r = 20$, $n = 220$ (for Dirichlet), $\mathrm{cond}(W) = 10^6$ (for ill-conditioned). For each noise level δ, we generate 20 such matrices, and Figures 7.6 and 7.7 report the average accuracy for the Dirichlet and the middle point experiments, respectively.

Table 7.1 reports the computational times. Table 7.2 reports the largest values of δ so that an algorithm achieved an accuracy of 100% and 95%.

Table 7.1. *Total computational time in seconds to compute $\mathcal{K}$ for the 600 randomly generated data sets. The letter W corresponds to well-conditioned, I to ill-conditioned, D to Dirichlet, and M to middle point.*

Experiment	SPA	VCA	FAW	SNPA	MVE-SPA	SPA-SPA
W-D	0.77	3.59	10.4	110	1687	4.08
I-D	0.92	3.42	9.81	2069	1592	4.64
W-M	0.95	4.00	9.2	170	1462	4.11
I-M	0.72	3.41	8.86	2840	1392	3.83

Table 7.2. *Robustness defined as the maximum value of δ (the noise level) such that the algorithm has extracted on average 100% or 95% of the columns of W.*

Experiment		SPA	VCA	FAW	SNPA	MVE-SPA	SPA-SPA
W-D	100%	0.18	0.13	0.18	**0.31**	0.18	0.18
	95%	0.31	0.22	0.31	**0.40**	0.31	0.31
I-D	100%	$2.59 \ 10^{-5}$	0	$2.59 \ 10^{-5}$	**$4.89 \ 10^{-3}$**	$2.59 \ 10^{-5}$	$1.61 \ 10^{-5}$
	95%	$4.52 \ 10^{-4}$	0	$2.81 \ 10^{-4}$	**$8.53 \ 10^{-2}$**	$4.52 \ 10^{-4}$	$1.74 \ 10^{-4}$
W-M	100%	0.14	0	0.22	0.14	**0.43**	0.41
	95%	0.22	0	0.27	0.22	**0.43**	0.41
I-M	100%	$4.82 \ 10^{-3}$	0	$2.32 \ 10^{-2}$	$1.00 \ 10^{-3}$	**0.415**	0.145
	95%	$3.92 \ 10^{-2}$	0	$5.10 \ 10^{-2}$	$1.37 \ 10^{-2}$	**0.415**	**0.415**

We observe the following:

- For well-conditioned Dirichlet (Figure 7.6, top), all algorithms perform well. This is a relatively simple scenario. However, VCA performs worse, while SNPA performs best. All SPA variants perform similarly.

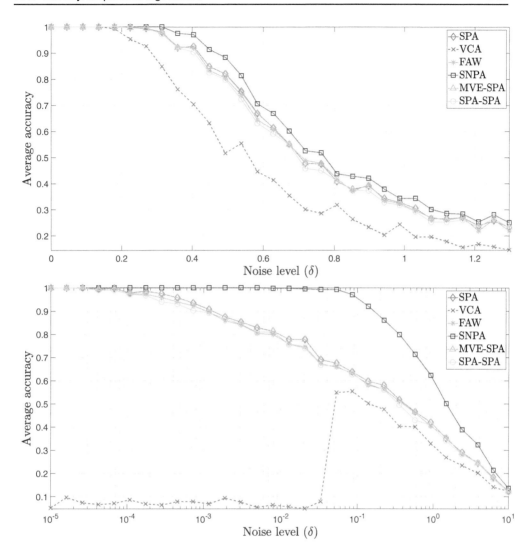

Figure 7.6. *Comparison of greedy separable NMF algorithms with the Dirichlet distribution for H and Gaussian noise for N. On the top: well-conditioned W. On the bottom: ill-conditioned W. [`Matlab file: compare_greedy_sepNMFalgo.m`].*

- For ill-conditioned Dirichlet (Figure 7.6, bottom), VCA fails completely. The reason is that it has a built-in preprocessing step that uses the truncated SVD that estimates the noise level automatically. For low noise levels and ill-conditioned cases, this preprocessing underestimates the dimension of the subspace spanned by X so that the truncated SVD thresholds the small singular value of X to zero, making it impossible for VCA to extract enough columns. Again, SPA variants perform similarly. This is somewhat unexpected.

SNPA outperforms all other approaches. The reason is that when W is ill-conditioned, $\sigma_r(W)$ is very close to zero, making SPA-based approaches very sensitive to noise (see the robustness theorems), while $\beta(W)$ can be significantly larger. For example, generating 1000 ill-conditioned matrices W (as described above) and dividing them by the quantity $\max_k \|W(:,k)\|_2$, the average value and standard deviation of $\sigma_r(W)$ is $2.1\,10^{-6} \pm 2.4\,10^{-7}$, while for $\beta(W)$ it is $0.01 \pm 4.5\,10^{-3}$.

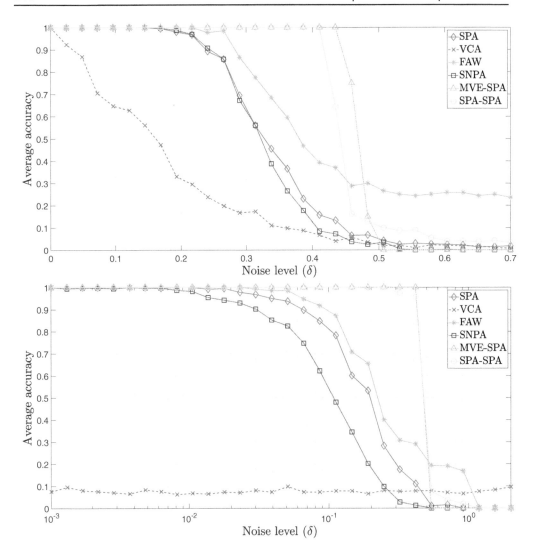

Figure 7.7. *Comparison of greedy near-separable NMF algorithms for the middle point selection for H and adversarial noise for N. On the top: well-conditioned W. On the bottom: ill-conditioned W. [Matlab file: `compare_greedy_sepNMFalgo.m`].*

- For the middle point experiments (Figure 7.7), the algorithms behave very differently than for the Dirichlet experiment. As expected, VCA fails as soon as the noise level is positive; see the discussion in Section 7.2. MVE-SPA outperforms all other approaches. The reason is that the generated matrices have a very particular form. The columns of $X(:,\mathcal{K})$ are not affected by the noise. Hence the MVE containing all columns of $\tilde{X}$ remains the same as long as the middle point remains inside the MVE; see Figure 7.5 for an illustration for $r = 3$. This explains the sharp phase transition of MVE-SPA: as long the columns of $\tilde{X}$ are within the MVE enclosing $\mathrm{conv}(W)$, MVE-SPA performs perfectly. Once the middle points get outside the MVE enclosing $\mathrm{conv}(W)$, the MVE changes drastically and MVE-SPA fails to identify any columns of W. Interestingly SPA-SPA performs almost as well as MVE-SPA (while being computationally much cheaper), while FAW performs better than SPA but worse than SPA-SPA. Finally, rather surprisingly, SNPA performs worse than

SPA, especially in the ill-conditioned cases. This is explained by the adversarial nature of this data set. For some reason, the orthogonal projections of SPA allow it to avoid middle points slightly more efficiently. This reminds us that the robustness proofs provided in this chapter are worst-case analysis and cannot explain the behavior of these algorithms on all data sets.

- In terms of computational time, SPA is the fastest, followed by VCA, SPA-SPA, and FAW, which are all rather fast (less than 0.02 seconds on average to output a solution). In well-conditioned cases, SNPA is also fast (less than 0.3 seconds on average to output a solution). However, in ill-conditioned cases, it is much slower (about 4 seconds to output a solution). The reason is that the iterative algorithm computing $H \geq 0$ in the projection step can stop early if it has converged. When W is well-conditioned, convergence is attained much faster, while in the ill-conditioned cases, 500 iterations are performed. MVE-SPA is relatively slow and takes on average about 2.5 seconds to output a solution.

7.4.6.4 ▪ Take-home messages from the numerical experiments

In real-world scenarios, like hyperspectral images or document data sets, the columns of H rather follow a Dirichlet distribution and the noise is not adversarial. In such cases, SNPA performs the best in terms of solution quality (although its computational cost is higher than the other greedy separable NMF algorithms) especially in ill-conditioned scenarios. In other cases, such as the adversarial middle point experiment, using preconditioning might improve the solution quality drastically. In practice, it is difficult to know which method will perform best, hence a standard approach is to run several algorithms and keep the solution $\mathcal{K}$ for which $\min_{H \geq 0} \|\tilde{X} - \tilde{X}(:, \mathcal{K})H\|_F$ is the smallest.

7.5 ▪ Heuristic algorithms

Many heuristics have been proposed for near-separable NMF, in particular in the hyperspectral unmixing literature. This includes approaches inspired by combinatorial optimization heuristics such as evolutionary algorithms [510] and particle-swarm optimization [333]. As mentioned in the introduction of this chapter, it is beyond the scope of this book to review this important literature; we refer the interested reader to [45, 334] and the references therein. In this section, we review only the two historically most important ones, namely PPI and N-FINDR.

7.5.1 ▪ Pure-pixel index

Boardman, Kruse, and Green [49] (1995) were the first to propose a pure-pixel search algorithm, PPI, within the hyperspectral unmixing literature. PPI uses the fact that, within a polytope, randomly generated linear functions (that is, functions $f(x) = c^\top x$ where $c \in \mathbb{R}^m$ is generated randomly) attain their minima and maxima on the vertices of that polytope with probability one. Recall that this idea is the one used in VCA to identify a vertex at each step (Section 7.4.3). However, as opposed to greedy near-separable NMF algorithms, PPI does not perform projections. It randomly generates a large number of linear functions and identifies the columns of $\tilde{X}$ maximizing and minimizing these functions. Under the separability assumption, these columns are, with probability one, the vertices of the convex hull of the columns of $\tilde{X}$, that is, the columns of W. (Note that PPI assumes $H^\top e = e$, as do many pure-pixel search algorithms.) By generating a large enough number of linear functions, PPI will identify, with high probability, the columns of W; see [111]. In the presence of noise, PPI assigns a score to each column of $\tilde{X}$ which is equal to the number of times that column minimized or maximized the randomly generated linear functions, and returns the r columns of $\tilde{X}$ with the largest score. In terms of computational cost, denoting K the number of generated linear functions, we have to evaluate K times linear functions over n columns in dimension m for a total computational cost of $\mathcal{O}(Kmn)$ operations. PPI has many drawbacks, including the following:

1. It is not guaranteed to work, even in the absence of noise, unless K goes to infinity. Even if K goes to infinity, it is not robust to noise for the same reason as VCA: linear functions can be maximized at any vertex of the convex hull of a set of points, and the noise can potentially make any column of $\tilde{X}$ a vertex of $\text{conv}(\tilde{X})$.

2. In the presence of (near-)duplicated columns of W, the score of these columns will be typically small (the score of these columns is split among them), while an isolated column which does not correspond to a column of W could potentially have a higher score and hence be extracted. Moreover, PPI might extract columns of $\tilde{X}$ corresponding to the same column of W.

For these reasons, PPI performs typically much worse than greedy algorithms such as SPA; we refer the interested reader to [210], for example, for some numerical experiments.

7.5.2 ▪ N-FINDR

In 1999, Winter [482] proposed N-FINDR. The goal of N-FINDR is to solve

$$\max_{\mathcal{K}, |\mathcal{K}|=r} \quad \text{volume}\left(\text{conv}\left(\tilde{X}(:,\mathcal{K})\right)\right), \tag{7.10}$$

that is, to find the index set $\mathcal{K}$ such that $\text{conv}\left(\tilde{X}(:,\mathcal{K})\right)$ has maximum volume (see Section 4.3.3 for a definition of this quantity). To find a solution to this problem, N-FINDR first initializes $\mathcal{K}$ randomly. Then, as long as the volume increases, it loops over each index in $\mathcal{K}$ and replaces it with the index $j \notin \mathcal{K}$ such that

$$\text{volume}\left(\text{conv}\left(\tilde{X}(:,\mathcal{K}\backslash\{k\} \cup \{j\})\right)\right)$$

is maximized. This procedure is not the most efficient way to identify the index j to replace k as it requires computing the volumes of $n-r$ matrices of size m-by-r. Identifying the best j to replace k can be performed by projecting $\tilde{X}$ onto the orthogonal complement of $\tilde{X}(:,\mathcal{K}\backslash\{k\})$, and then picking the column with the largest residual; see Lemma 7.11.

The main contribution of Winter is not so much the algorithm N-FINDR but rather the model (7.10) based on *volume maximization*. The majority of pure-pixel search algorithms rely on this insight [45, 334]. For example, SPA is a greedy algorithm to solve (7.10), while FAW is equivalent to performing one loop of N-FINDR using SPA as an initialization for $\mathcal{K}$.

7.6 ▪ Convex-optimization-based algorithms

In this section, we summarize another class of methods that are based on convex optimization algorithms and not relying on projections. As we will see, these algorithms have the advantage of being more robust to noise in ill-conditioned or rank-deficient cases. However, they are computationally more demanding, requiring $\Omega(mrn^2)$ operations. We first present the first near-separable NMF algorithm which was proved to be robust to noise (Section 7.6.1), and then present four closely related models using the notion of self-dictionary (Sections 7.6.2 to 7.6.5). In Section 7.6.6, we discuss the computational cost of these methods and discuss their practical performances.

7.6.1 ▪ Multiple linear programs

In their seminal paper on the complexity of NMF, Arora et al. [15] proposed the first near-separable NMF algorithm provably robust to noise. It is rather simple and natural and is closely related to ideas already present in the blind source separation literature [358]. However, they were the first to prove that their approach is robust in the presence of noise.

Let us present their algorithm, which we refer to as multiple linear programs (MLPs), in a high-level way focusing on the intuition behind it; see the paper [15] for the details. As before, let $\tilde{X} \in \mathbb{R}^{m \times n}$ satisfy Assumption 7.1. For $\delta \geq 0$ and $j \in \{1, 2, \ldots, n\}$, let us define

$$\mathcal{L}_j^\delta = \left\{ \ell \mid \min_\alpha \left\| \tilde{X}(:, \ell) - \alpha \tilde{X}(:, j) \right\|_1 > \delta \right\}.$$

The set $\mathcal{L}_j^\delta$ contains the indices of all columns of $\tilde{X}$ that are sufficiently far way from the cone generated by $\tilde{X}(:, j)$. In particular, the set $\mathcal{L}_j^0$ contains the indices of all columns of $\tilde{X}$ that are not multiples of $\tilde{X}(:, j)$. MLP solves one linear optimization problem for each data point: for $j = 1, 2, \ldots, n$, it computes

$$\zeta_j = \min_{h \geq 0} \left\| \tilde{X}(:, j) - \tilde{X}(:, \mathcal{L}_j^\delta) h \right\|_1,$$

which can be solved via LP. In the noiseless case, $\delta = 0$ is used and, by definition, $\tilde{X}(:, j)$ is an extreme ray of $\text{cone}(W)$ if and only if $\zeta_j > 0$, that is, if and only if $\tilde{X}(:, j)$ cannot be written as the conic combination of other columns of $\tilde{X}$. Hence, any columns of $\tilde{X}$ whose index is in $\mathcal{K} = \{j \mid \zeta_j > 0\}$ is a column of W, up to scaling.

In the presence of noise, one needs to use $\delta > 0$. The reason is that if a column of W is duplicated in the data set and/or there are nearby data points (which we refer to as near-duplicated columns), we might have ζ_j arbitrarily small because of the presence of noise, and hence that column could not be identified. The set of candidate indices is given by

$$\mathcal{K} = \{j \mid \zeta_j > \eta\}$$

for some η sufficiently large. Intuitively, the reason is that some data points could have $\zeta_j > 0$ while being far from a column of W. This is, for example, the case in our middle point experiment as the noise moves the middle points on the segment joining two columns of W outward $\text{conv}(W)$; see Figure 7.5. It can be shown that for well-chosen values of δ and η, any index in $\mathcal{K}$ corresponds to a columns of $\tilde{X}$ nearby a column of W, given that the noise is sufficiently small (see Theorem 7.20 below). Finally, because of possible near-duplicated columns of W, a post-processing step is necessary to cluster the columns corresponding to the indices in $\mathcal{K}$ to recover the r columns of W. Because the r clusters are well-separated and located around the columns of W, this step is not difficult and can be performed, for example, using spherical k-means (Section 5.5.3).

Let us provide the robustness theorem for MLP.

Theorem 7.20. *[15, Theorem 5.7] Let $\tilde{X} = X + N = X(:, \mathcal{K}^*)H + N = WH + N$ be a matrix satisfying Assumption 7.1 with $p = 1$ and $\gamma_1(W) > 0$, and let*

$$\epsilon = \max_j \|N(:, j)\|_1 \leq \mathcal{O}\left(\gamma_1(W)^2\right).$$

Then MLP returns a set of indices $\mathcal{K}$ such that

$$q_1(\mathcal{K}) \leq \mathcal{O}\left(\frac{\epsilon}{\gamma_1(W)}\right).$$

The condition on ϵ does not match, up to constant factors, the optimal bound $\epsilon \leq \mathcal{O}(\gamma_1(W))$; see Section 7.3.1. However, the error bound on $q_1(\mathcal{K})$ matches the optimal bound. For ill-conditioned or rank-deficient W, these bounds would outperform SPA variants that depend on the parameter $\sigma_r(W)$. Compared to SNPA, the bounds are also much stronger since SNPA depends on the parameter $\beta(W) \leq \gamma_2(W)$, where $\epsilon \leq \mathcal{O}(\beta(W)^4)$ for an error $q_2(\mathcal{K}) \leq \mathcal{O}(\epsilon/\beta(W)^3)$. However, for well-conditioned matrices for which $\sigma_r(W) \approx \gamma_2(W)$, MVE-SPA provides better bounds since it only requires $\epsilon \leq \mathcal{O}(\sigma_r(W)/\sqrt{r})$ for an error $q_2(\mathcal{K}) \leq \mathcal{O}(\epsilon/\sigma_r(W))$.

7.6.2 ▪ Self-dictionary I: exact model

We now present several closely related models that solve near-separable NMF in a global fashion, considering all data points simultaneously. They are based on the concept of self-dictionary, that is, using X as a dictionary to approximate the columns of X with itself.

Let us first consider the noiseless case, and let us introduce the set

$$\mathcal{Y} = \{Y \in \mathbb{R}_+^{n \times n} \mid X = XY\}.$$

The set $\mathcal{Y}$ contains all nonnegative matrices that can be used to reconstruct X with itself. Note that $I_n \in \mathcal{Y}$, which is not a very insightful decomposition. To recover a separable factorization, $Y \in \mathcal{Y}$ should have as few nonzero rows as possible. In fact, a row of zeros in the matrix Y means that the corresponding column in X is not used in the decomposition since

$$X = XY = \sum_{j=1}^{n} X(:,j)Y(j,:).$$

We have the following lemma.

Lemma 7.21. *Let $X = X(:, \mathcal{K}^*)H \in \mathbb{R}^{m \times n}$ satisfy Assumption 7.1 with $\epsilon = 0$. Let also*

$$Y^* \in \operatorname{argmin}_{Y \in \mathcal{Y}} \|Y\|_{row,0}, \tag{7.11}$$

where $\|Y\|_{row,0}$ counts the number of nonzero rows of Y. Then, $|\mathcal{K}^*| = \|Y^*\|_{row,0}$, and defining

$$\mathcal{K}_{Y^*} = \{k \in \{1, 2, \ldots, n\} \mid Y^*(k,:) \neq 0\},$$

we have $X(:, \mathcal{K}_Y) = X(:, \mathcal{K}^*)$ up to permutation and scaling, and

$$X = X(:, \mathcal{K}_{Y^*})Y^*(\mathcal{K}_{Y^*}, :)$$

provides a separable factorization of X.

Proof. By item (i) of Assumption 7.1, the columns of $X(:, \mathcal{K}^*)$ are distinct vertices of the set $\operatorname{conv}([X, 0])$. Hence, the only possible way to reconstruct $X(:, \mathcal{K}^*)$ using columns of X is to select $X(:, \mathcal{K}^*)$ themselves, or duplicated columns (up to scaling). This implies that for each column of $X(:, \mathcal{K}^*)$, there exists a nonzero row of Y corresponding to a column of X which is equal to $X(:, \mathcal{K}^*)$, up to scaling, since $X = XY$. This implies that $\|Y\|_{row,0} \geq r$ for any feasible solution of (7.11) and that r nonzero rows correspond to the columns of $X(:, \mathcal{K}^*)$. By Assumption 7.1, there exists a feasible solution Y with $\|Y\|_{row,0} = r$, namely take $Y(\mathcal{K}^*, :) = H$ while setting the other rows to zero (all columns of X are convex combinations of the columns of $X(:, \mathcal{K}^*)$ and the origin). The result then follows from the optimality of Y^* which uses as few nonzero rows as possible. $\square$

Let us make two observations that follow from Lemma 7.21. First, the model (7.11) does not need r as an input. As we will see, self-dictionary models can detect r automatically. Second, the near-separable NMF problem (Problem 7.1) can be formulated as

$$\min_{Y \in \mathcal{Y}} D(X, XY) \quad \text{such that} \quad \|Y\|_{\text{row},0} = r.$$

This motivates the self-dictionary models presented in the next sections.

7.6.3 ▪ Self-dictionary II: $\ell_{1,q}$ relaxations

As briefly discussed in Section 5.4.2, the self-dictionary model $X = XY$ is closely related to archetypal analysis and convex NMF. However, the first to use this model in the context of near-separable NMF were Esser et al. [148] (2012) and Elhamifar, Sapiro, and Vidal [145] (2012). More precisely, both papers develop the following model.

Since minimizing the number of nonzero rows of Y is a difficult problem of combinatorial nature, they instead minimize the convex surrogate

$$\|Y\|_{1,q} = \sum_{i=1}^{n} \|Y(i, :)\|_q,$$

which is the $\ell_{1,q}$ norm of Y. It is the ℓ_1 norm of the vector whose entries are the ℓ_q norms of the rows of Y. Note that this is a norm only for $q \geq 1$. For near-separable NMF, q should be chosen such that $q > 1$ because choosing $q = 1$ leads to a convex optimization problem but does not lead to row sparsity as it would be equivalent to minimizing $\|Y\|_1$, which would make Y sparse but not row sparse [145].

In practice, because of the presence of noise, the constraint $X = XY$ is replaced with $D(X, XY) \leq \delta$ for some appropriate distance measure $D(X, XY)$ and some parameter $\delta > 0$. Another standard approach is to add $D(X, XY)$ as a penalty in the objective function and solve the convex optimization problem

$$\min_{Y \geq 0} \|Y\|_{1,q} + \lambda D(X, XY) \tag{7.12}$$

for some well-chosen parameter $\lambda > 0$. Esser et al. [148] (resp. Elhamifar et al. [145]) designed an algorithm based on the alternating direction method of multipliers (ADMM) to tackle the model (7.12) using the squared Frobenius norm $D(X, XY) = \|X - XY\|_F^2$ and $q = \infty$ (resp. $q = 2$); see Section 8.3.5 for more details on ADMM. Esser et al. applied it on hyperspectral images and in nuclear magnetic resonance spectroscopy, with comparisons with VCA, NFIND-R, and SPA (which they refer to as QR), and Elhamifar et al. for video summarization (which consists in identifying key frames within a video sequence), classification using representatives, and outlier rejection.

Elhamifar et al. proved the correctness of the model for any $q > 1$ in the noiseless case (Esser et al. showed it for $q = \infty$), in the absence of duplicated columns of X, and in a normalized setting where $H^\top e = e$ is assumed. Let us prove the result for $q = \infty$ (which is slightly simpler).

Theorem 7.22. *[148, Lemma III.1], [145, Theorem 1] Let $\tilde{X} = X + N = WH + N$ satisfy Assumption 7.1 with $N = 0$ and where the columns of $X(:, \mathcal{K}^*)$ are not duplicated. Then any optimal solution Y^* of*

$$\min_{Y \in \mathcal{Y}} \|Y\|_{1,\infty} \tag{7.13}$$

satisfies

$$\mathcal{K}_{Y^*} = \{k \in \{1, 2, \ldots, n\} \mid Y^*(k, :) \neq 0\} = \mathcal{K}^*;$$

hence $X = X(:, \mathcal{K}_{Y^*})Y^*(\mathcal{K}_{Y^*}, :)$.

Proof. First, note that the solution Y^f defined as

$$Y^f(\mathcal{K}^*, :) = H$$

and $Y^f(j, :) = 0$ for $j \notin \mathcal{K}^*$ is feasible, and $\|Y^f\|_{1,\infty} = r$ since $H(\mathcal{K}^*, :) = I_r$ and $H \leq 1$ (Assumption 7.1). Second, as the columns of $X(:, \mathcal{K}^*)$ cannot be approximated with other columns and are not duplicated, we must have

$$Y(\mathcal{K}^*, \mathcal{K}^*) = I_r$$

for any feasible solution Y of (7.13), and hence $\|Y\|_{1,\infty} \geq r$ for any feasible solution Y. This implies that Y^f is optimal. Finally, since $Y^*(\mathcal{K}^*, \mathcal{K}^*) = I_r$ and having a nonzero entry in any other row of Y^* makes the objective function strictly larger than r, we must have $\mathcal{K}_{Y^*} = \mathcal{K}^*$ for any optimal solution. $\square$

In the papers [148, 145], their algorithm is used as a heuristic in the sense that no robustness result are provided.

In [174] robustness of the model with $q = \infty$ was proved but in the absence of near-duplicated columns of W. We will present in Section 7.6.5 a stronger robustness result allowing near-duplicated columns of W. As we will see, the solutions of these convex models are highly sensitive to duplicated and near-duplicated columns of W. In order to obtain robustness results without this strong and unrealistic assumption, the solution Y has to be postprocessed with care. The reason is that the optimal solution Y will in general not be row sparse, and some clustering of its rows are necessary, similarly as for the MLP approach presented in the previous section. Let us explain why Y is not row sparse in the presence of duplicates and near-duplicates. Assume the noiseless case, and assume each column of W is duplicated once. This implies that there exist 2^r separable factorizations: for each column of W, we have the choice between two different columns of X to represent it. Let Y_i ($1 \leq i \leq 2^r$) be the corresponding 2^r solutions of (7.13) with r nonzero rows (see the proof of Theorem 7.22). By convexity, the solutions $Y_\alpha = \sum_i \alpha_i Y_i$ for any $\alpha \in \Delta^r$ such that $\alpha_i \neq 0$ for all i are also optimal and have $2r$ nonzero rows. In the presence of noise, the behavior is worsened.

Link with multiple measurement vectors   It is worth noting that the self-dictionary model and the relaxations based on $\ell_{1,q}$ norms were already used in another context, where Y is not required to be nonnegative. This is the so-called problem of multiple measurement vectors, where usually the considered model is more general with $X = DY$ where D is any dictionary (not necessarily $D = X$). The theoretical results are rather different and closely related to the compressive sensing literature, where the assumptions on the dictionary D relate to incoherence conditions. We refer the interested reader to [107, 83] and the references therein for more details. Interestingly, one of the proposed greedy algorithms for the multiple measurement vectors problem, closely related to orthogonal matching pursuit [477], turns out to be equivalent to SPA [175].

7.6.4 ▪ Self-dictionary III: Hottopixx, an LP-based model

In [395] (2012), Recht et al. proposed the following self-dictionary model for near-separable NMF: given a matrix $\tilde{X}$ satisfying Assumption 7.1, a vector $v \in \mathbb{R}^n$, and a noise level $\delta > 0$, solve

$$\min_{Y \in \mathbb{R}^{n \times n}} \quad v^\top \operatorname{diag}(Y)$$

$$\text{such that} \quad \max_j \|\tilde{X}(:,j) - \tilde{X}Y(:,j)\|_1 \le \delta,$$

$$\operatorname{tr}(Y) = r, \tag{7.14}$$

$$0 \le Y(i,j) \le Y(i,i) \le 1 \text{ for all } i,j.$$

This model, referred to as Hottopixx, can be solved via LP. It can be interpreted as follows: we have to assign a value in the interval [0,1] to each diagonal entry of Y ($0 \le Y(i,i) \le 1$ for all i) for a total weight of r since $\operatorname{tr}(Y) = r$. Moreover, we cannot use a column of $\tilde{X}$ to reconstruct another column of $\tilde{X}$ with a weight larger than the corresponding diagonal entry of Y (because $Y(i,j) \le Y(i,i)$ for all i,j), while we have to guarantee that the approximation error is sufficiently small. For a matrix satisfying Assumption 7.1, in the noiseless case and in the absence of duplicated columns of W, it is easy to show that any feasible solution Y of (7.14) satisfies $Y(i,i) = 1$ if and only if $i \in \mathcal{K}^*$. The proof is essentially the same as that of Theorem 7.13. The main difference with the models based on the $\ell_{1,q}$ norms is that r is part of the input, while the objective function involves the vector v which allows breaking the symmetry of the near-separable NMF problem and hence allows us to deal with duplicated columns. In particular, in the noiseless case, if the entries of v are distinct (it was recommended in [395] to pick them at random), then the optimal solution of (7.14) is unique and has r nonzero rows.

Theorem 7.23. *[395, Theorem 3.1] Let $\tilde{X} = X = WH$ satisfy Assumption 7.1 with $N = 0$, and let the entries of v be distinct. Then the optimal solution Y^* of (7.13) with $\delta = 0$ satisfies*

$$X(:,\mathcal{K}_{Y^*}) = X(:,\mathcal{K}^*) = W,$$

up to permutation.

Proof. Let $k \in \mathcal{K}^*$ so that $X(:,k)$ is a column of W. Let also

$$\mathcal{J}_k = \{j \mid X(:,j) = X(:,k)\}.$$

Since $X(:,k)$ is a vertex of $\operatorname{conv}([W,0])$, we must have $\operatorname{tr}(Y(\mathcal{J}_k, \mathcal{J}_k)) \ge 1$ for all k to be able to reconstruct $X(:,k)$ exactly. Together with the constraint $\operatorname{tr}(Y) = r$ and the fact that the sets $\mathcal{J}_k$ do not intersect (since $W(:,i) \ne W(:,j)$ for $i \ne j$ by Assumption 7.1(i)), this implies $\operatorname{tr}(Y(\mathcal{J}_k, \mathcal{J}_k)) = 1$ for all k. For $k \in \mathcal{K}^*$, let $i = \operatorname{argmin}_{j \in \mathcal{J}_k} v(j)$. The optimal way to assign the weights to the diagonal entries of $Y(\mathcal{J}_k, \mathcal{J}_k)$ is with $Y(i,i) = 1$ and $Y(j,j) = 0$ for all $j \in \mathcal{J}_k \setminus \{i\}$, since the objective function is $v^\top \operatorname{diag}(Y) = \sum_{i=1}^n v(i)Y(i,i)$. □

In their paper, Recht et al. [395, Theorem 3.2] also provided a robustness proof in the absence of near-duplicated columns of W. It was later improved in [187, Theorem 3.2], getting rid of a factor $\gamma_1(W)^2$ in the bound for ϵ.

Theorem 7.24. *[187, Theorems 2.3 and 2.4] Let $\tilde{X} = X + N = WH + N$ satisfy Assumption 7.1 with $p = 1$ where $H = [I_r, H']\Pi$ as in (7.4). Let us assume that $\beta := \max_{i,j} H'(i,j) < 1$ and*

$$\epsilon < \frac{(1-\beta)\gamma_1(W)}{9(r+1)}.$$

Then picking the r largest entries of an optimal solution Y^* of (7.14) to form $\mathcal{K}$ guarantees $\mathcal{K} = \mathcal{K}^*$, and hence

$$q_1(\mathcal{K}) \leq \epsilon.$$

Moreover, the bound on ϵ in Theorem 7.24 is tight up to constant factors.

Unfortunately, in the presence of noise *and* near-duplicated columns of W, the analysis is more complicated and simply solving (7.14) and picking the r largest diagonal entries of Y does not allow us to obtain a robustness result. However, performing an appropriate postprocessing of the diagonal entries of Y^* allows us to resolve this issue. The idea is similar to that of MLP: the diagonal entries of Y^* are clustered together depending on the distances between the corresponding columns of $\tilde{X}$—this is a k-means problem with weights; see [187] for more details. This leads to the following robustness theorem, without the assumption on the absence of duplicates.

Theorem 7.25. *[187, Theorem 3.5] Let $\tilde{X}$ satisfy Assumption 7.1 with $p = 1$. If*

$$\epsilon < \frac{\omega(W)\gamma_1(W)}{99(r+1)},$$

where $\omega(W) = \min_{i \neq j} \|W(:,i) - W(:,j)\|_1$, then the index set $\mathcal{K}$ extracted by a proper postprocessing of the optimal solution of (7.14) satisfies

$$q_1(\mathcal{K}) \leq 49(r+1)\frac{\epsilon}{\gamma_1(W)} + 2\epsilon.$$

Compared to the idealized algorithm, the above bounds are tight up to the factor $\omega(W)/r$ for ϵ and up to the factor r for $q_1(\mathcal{K})$. Compared to MLP, the bound on ϵ is better when $\gamma_1(W) \leq \mathcal{O}(\omega(W)/r)$, while the error $q_1(\mathcal{K})$ is worse; this is not very satisfactory. However, in their paper, Recht et al. [395] showed that Hottopixx performs better than MLP on several synthetic data sets.

7.6.5 ▪ Self-dictionary IV: LP-based model, toward optimal bounds

Beyond having suboptimal error bounds, Hottopixx has another important drawback: the factorization rank r *and* the noise level δ have to be estimated. If these values are not well-chosen, (7.14) could be infeasible. In [204], these issues were resolved by proposing the following slightly modified LP-based model, which we refer to as self-dictionary via linear programming (SD-LP):

$$\min_{Y \in \mathbb{R}^{n \times n}} \quad v^\top \operatorname{diag}(Y)$$

$$\text{such that} \quad \max_j \|\tilde{X}(:,j) - \tilde{X}Y(:,j)\|_1 \leq \delta, \tag{7.15}$$

$$0 \leq Y(i,j) \leq Y(i,i) \leq 1 \text{ for all } i,j,$$

where $v \in \mathbb{R}^n$ has *positive* entries. In practice, it is recommended to choose these values of v distinct but close to one so that $v^\top \operatorname{diag}(Y) \approx \operatorname{tr}(Y)$ while allowing one to deal with duplicated columns. As for Hottopixx, the model is exact in the absence of noise (Theorem 7.23 and almost the same proof apply to this new model). The problem (7.15) is always feasible as $Y = I_n$ is a feasible solution. Moreover, it detects the factorization rank automatically; without going into the details, we have $r \approx \operatorname{tr}(Y)$ under Assumption 7.1 for ϵ sufficiently small and $\delta \approx \epsilon$.

In the setting without near-duplicated columns of W, we have the following robustness result.

Theorem 7.26. *[204, Theorem 3.2] Let $\tilde{X} = X + N = WH + N$ satisfy Assumption 7.1 where $H = [I_r, H']\Pi$ as in (7.4), and let $\delta = \rho\epsilon$ for some $\rho > 0$. Let us assume that $\beta := \max_{i,j} H'(i,j) < 1$ and*

$$\epsilon < \frac{(1-\beta)\gamma_1(W)\min(1,\rho)}{5(\rho+2)}.$$

Then picking the r largest entries of an optimal solution Y^* of (7.14) to form $\mathcal{K}$ guarantees $\mathcal{K} = \mathcal{K}^*$, and hence

$$q_1(\mathcal{K}) \leq \epsilon.$$

The bound of Theorem 7.26 allows δ to be different from ϵ, and hence it does not need to be estimated perfectly. The value of δ that leads to the best bound for ϵ is when $\rho = 1$. In that case, the bound on ϵ is $\frac{(1-\beta)\gamma_1(W)}{15}$ and it matches the bound of the idealized algorithm, namely $\epsilon < (1-\beta)\gamma_2(W)/4$, up to a small constant factor. This is rather nice: a polynomial-time algorithm can tackle near-separable NMF as well as the idealized algorithm, in the absence of near-duplicated columns of W.

Unfortunately, in the presence of near-duplicated columns of W, the robustness analysis cannot be improved compared to Hottopixx: Theorem 7.25 applies when solving (7.15) instead of (7.14). We believe that the bound can be improved by removing the factor r (which is not possible for Hottopixx [187]): this is a topic for future research.

Frobenius norm variant and fast gradient method   In the paper [205], the following model is considered:

$$\min_{Y \in \mathcal{Z}} \operatorname{tr}(Y) + \lambda\|\tilde{X} - \tilde{X}Y\|_F^2,$$

where

$$\mathcal{Z} = \{Y \in \mathbb{R}^{n \times n} \mid 0 \leq Y(i,j) \leq Y(i,i) \leq 1\}.$$

It is the same model as SD-LP except that the data fitting term is put in the objective and the Frobenius norm is used as an error measure. The advantage of this model is that the objective function is smooth, and the authors propose a projected fast gradient method (see Section 8.3.4 for more details on this optimization scheme) that runs in $\mathcal{O}(n^2 \max(m, \log n))$ operations. The main algorithmic contribution is the design of a projection on the set $\mathcal{Z}$ which requires $\mathcal{O}(n^2 \log n)$ operations.

Equivalence with the $\ell_{1,\infty}$ norm model   An interesting observation is that, under Assumption 7.1, SD-LP (7.15) is equivalent to the $\ell_{1,\infty}$ norm model [205, Theorem 2]. More precisely, let us define

$$\min_{Y \in \mathbb{R}_+^{n \times n}} \|Y\|_{1,\infty}$$

$$\text{such that } \max_{1 \leq j \leq n} \|\tilde{X}(:,j) - \tilde{X}Y(:,j)\|_2 \leq \epsilon, \tag{7.16}$$

and let us take $v = e$ and $\delta = \epsilon$ in (7.16). We have that

- at optimality, the objective function values of (7.15) and (7.16) coincide,

- any optimal solution of (7.15) is an optimal solution of (7.16), and

- any optimal solution of (7.16) can be easily transformed into an optimal solution of (7.15).

This result essentially follows from the observation that, for $v = e$,

$$e^\top Y = \operatorname{tr}(Y) = \|Y\|_{1,q}$$

for any feasible solution of (7.15) since $Y(i,j) \leq Y(i,i)$ for all i,j so that $\|Y(i,:)\|_\infty = Y(i,i)$ for all i.

7.6.6 ▪ Computational cost and practical performances

Convex-optimization-based methods are appealing in theory, being more robust to noise (in the worst-case setting). On synthetic data sets similar to the ones used in Section 7.4.6, they perform in general better than greedy algorithms; see the numerical experiments in [204, 205]. However, they have several important drawbacks:

- All the models presented in this section involve solving convex optimization problems in $\mathcal{O}(n^2)$ variables, or n linear programs in $\mathcal{O}(n)$ variables for MLP. Hence even first-order methods do not scale well. It is possible to use stochastic-based methods working on subsamples of the data set at each iteration, but such methods converge rather slowly; this is a strategy proposed for Hottopixx in [395]. In our experience, even when using first-order methods, the convergence can be slow; hence many iterations are needed to obtain high-quality solutions.

- In our experience, convex-optimization-based methods heavily rely on separability. If this assumption is highly violated (which often happens in practice), then the solution Y will typically not be row sparse and it will be difficult to identify the most important columns of $\tilde{X}$. On the contrary, greedy algorithms are better suited in these situations and will return subset of columns well-spread in the data set.

- Convex-optimization based methods rely on parameter tuning, namely λ for $\ell_{1,q}$ relaxations (7.12), δ and r for Hottopixx (7.14), and δ for SD-LP (7.15). This is a highly nontrivial task, and greedy algorithms do not have this issue. In practice, it may take some time to properly fine-tune these methods.

For these reasons, convex-optimization-based methods have not been been used much in practice so far.

A way to partially resolve the above drawbacks and obtain good performances in practice is to select a (small) subset of rows of Y to be nonzero, that is, to select initially important columns of $\tilde{X}$. This can be done, for example, using greedy near-separable NMF algorithms. In other words, convex-optimization-based methods can be used to aggregate solutions of simpler and faster methods. In the same spirit, the strategy proposed in [205] is to preselect columns of $\tilde{X}$ using a hierarchical clustering of the columns of $\tilde{X}$ and, for each cluster, choose a representative column of $\tilde{X}$. This was shown to outperform greedy algorithms for hyperspectral imaging. Another advantage of selecting a well-chosen subset of columns is that it resolves the issue of near-duplicated columns of W (since a well-chosen subset should typically not contain nearby columns), which deteriorates the performance of convex-optimization-based methods (see the discussions in the previous sections, in particular Theorem 7.26, and the discussion that follows).

7.7 ▪ Summary of provably robust near-separable NMF algorithms

Table 7.3 compares the different robustness recovery results for the different provably correct algorithms described in this chapter.

Table 7.3. *Comparison of robust near-separable NMF algorithms applied on a dense m-by-n matrix $\tilde{X}$ satisfying Assumption 7.1. Note that SNPA also achieves the bounds of SPA [188].*

Algorithm	Cost	Noise level (ϵ)	Error ($q_p(\mathcal{K})$)
Idealized [192]	$\binom{n}{r}\mathcal{O}(mnr)$	$\frac{\gamma_2(W)}{4}$	$8\frac{\epsilon}{\gamma_2(W)} + \epsilon$
SPA [210]	$2mnr + \mathcal{O}(mr^2)$	$\mathcal{O}\left(\frac{\sigma_r^3(W)}{\sqrt{r}}\right)$	$\mathcal{O}\left(\epsilon\sigma_r^2(W)\right)$
FAW [14]	$\mathcal{O}(mnr^2)$	$\mathcal{O}\left(\frac{\sigma_r^3(W)}{\sqrt{r}}\right)$	$\mathcal{O}\left(\epsilon\sigma_r(W)\right)$
SNPA [188]	$\mathcal{O}(mnr)$	$\mathcal{O}\left(\beta(W)^4\right)$	$\mathcal{O}\left(\frac{\epsilon}{\beta(W)^3}\right)$
MVE-SPA [211]	$\mathcal{O}(mnr + r^6)$	$\mathcal{O}\left(\frac{\sigma_r(W)}{\sqrt{r}}\right)$	$\mathcal{O}\left(\epsilon\sigma_r(W)\right)$
SPA-SPA [206]	$\mathcal{O}(mnr^2)$	$\mathcal{O}\left(\frac{\sigma_r^3(W)}{\sqrt{r}}\right)$	$\mathcal{O}\left(\epsilon\sigma_r(W)\right)$
MLP [15]	$\Omega(mn^2)$	$\mathcal{O}\left(\gamma_1(W)^2\right)$	$\mathcal{O}\left(\frac{\epsilon}{\gamma_1(W)}\right)$
Hottopix [395, 187] and SD-LP [204]	$\Omega(mn^2)$	$\mathcal{O}\left(\frac{\gamma_1(W)\omega(W)}{r}\right)$	$\mathcal{O}\left(\frac{r\epsilon}{\gamma_1(W)}\right)$

For well-conditioned matrices W for which $\sigma_r(W) \approx \gamma_2(W)$, MVE-SPA is the provably most robust algorithm. Note, however, that the analysis of SD-LP might not be tight, and the factor r could potentially be removed, in which case SD-LP would be provably more robust, performing almost as well as the idealized algorithm (this holds true in the absence of near-duplicated columns of W; see Theorem 7.26). For ill-conditioned ($\sigma_r \approx 0$) or rank-deficient matrices ($\sigma_r = 0$), SNPA or SD-LP should be preferred, as SPA variants either perform badly or are not even able to extract sufficiently many columns (for example, when $m < r$).

For large-scale data sets, where n is of the order of thousands or millions, greedy algorithms are the methods of choice. However, convex-optimization-based algorithms can be used as a post-processing to select columns among a smaller subset of candidates; see for example [205]. This is particularly meaningful because it appears that convex-optimization-based algorithms tend to underperform on real data for which (1) the noise level is high, (2) there are many near-duplicated columns, and (3) the data set violates the model assumptions.

Finally, it is important to keep in mind that it is often possible to improve solutions by applying various heuristics that locally look for better solutions, as in the second phase of FAW (Algorithm 7.2), in the spirit of NFINDR.

7.8 ▪ Separable tri-symNMF

In Section 5.4.9.1, we presented the tri-symNMF model of Arora et al. [14] for topic modeling: Given the word-by-document matrix X, it considers the following decomposition:

$$A = XX^\top \approx WSW^\top.$$

The matrix A is the word co-occurrence matrix, the matrix W is the word-by-topic matrix such that $W(i, k)$ is the probability for word i to be picked under the topic k, and the matrix S is a topic-by-topic matrix that accounts for the interactions between the topics. The tri-symNMF problem is to recover (W, S) from A.

In topic modeling, the following separability assumption makes sense: for each topic, there exists an anchor word, that is, a word that is only used by that topic. This is equivalent to requiring that the matrix $W^\top$ is separable, that is, there exists $\mathcal{K}$ such that $W(\mathcal{K}, :)$ is a diagonal matrix. As explained in the next paragraph, (W, S) can be extracted from A under the separability assumption, in polynomial time and even in the presence of noise. Hence separable tri-symNMF is *identifiable* and *solvable in polynomial time* with *provably robust* algorithms, as for separable NMF. This is in contrast to most topic models, such as collapsed Gibbs sampling or latent Dirichlet allocation, that do not come with such guarantees. Moreover, separable tri-symNMF provides state-of-the-art results compared to these approaches [14].

Computation of W and S under the separability assumption   For simplicity, let us assume the noiseless case, that is, $A = WSW^\top$, and let us assume that $W^\top$ is separable. We refer to [14] for the analysis in noisy settings.

Since $W^\top$ is separable, applying a separable NMF algorithm on $A = (WS)W^\top$ allows us to identify an index set $\mathcal{K}$ such that

$$W(\mathcal{K}, :) = \operatorname{diag}(z) \text{ for some } z \in \mathbb{R}^r_{++}.$$

Note that, by symmetry of tri-symNMF, the matrix $W^\top$ of tri-symNMF plays the role of the matrix H in separable NMF. Therefore, given A, separable NMF algorithms will identify $\mathcal{K}$ such that

$$A(:, \mathcal{K}) = (WS)W(\mathcal{K}, :)^\top = (WS)\operatorname{diag}(z).$$

Hence $A(:, \mathcal{K})$ is equal to WS, up to scaling of the columns. However, as opposed to separable NMF, it is not necessarily the case that $A(:, \mathcal{K}) = W$, up to scaling, since S is usually not diagonal. (If S is diagonal, this is the symNMF model for which separability was also considered in [271] but is rather unrealistic as A needs to contain a diagonal matrix as a submatrix.)

By using the scaling degree of freedom in tri-symNMF, we may assume w.l.o.g. $W^\top e = e$. Note that this makes sense when interpreting W in terms of probabilities; see Section 5.4.9.1. Let us define $q = A(\mathcal{K}, :)e$; we have

$$q = A(\mathcal{K}, :)e = W(\mathcal{K}, :)SW^\top e = W(\mathcal{K}, :)Se = \operatorname{diag}(z)Se. \tag{7.17}$$

Moreover,

$$A(\mathcal{K}, \mathcal{K}) = W(\mathcal{K}, :)SW(\mathcal{K}, :)^\top = \operatorname{diag}(z)S\operatorname{diag}(z);$$

hence

$$S = \operatorname{diag}(z)^{-1}A(\mathcal{K}, \mathcal{K})\operatorname{diag}(z)^{-1}. \tag{7.18}$$

Putting (7.18) in (7.17), we obtain

$$q = A(\mathcal{K}, \mathcal{K})\operatorname{diag}(z)^{-1}e = A(\mathcal{K}, \mathcal{K})\left(\frac{[e]}{[z]}\right),$$

where $[\cdot]/[\cdot]$ is the componentwise division. Therefore, z can be computed by solving a linear system. Once z is recovered, we obtain S via (7.18), while W is obtained by solving the linear system

$$A(\mathcal{K}, :) = W(\mathcal{K}, :)SW^\top = \operatorname{diag}(z)SW^\top = A(\mathcal{K}, \mathcal{K})\operatorname{diag}(z)^{-1}W^\top.$$

Algorithm 7.6 Tri-symNMF under the separability assumption [14] [Matlab file: septrisymNMF.m]

Input: The matrix $A \approx WSW^\top \in \mathbb{R}^{m \times m}$ where $W^\top \in \mathbb{R}^{r \times m}$ is separable.
Output: The matrices W^* and S^*, up to permutation and scaling, so that $A \approx W^* S^* W^{*\top}$.

1: Using a separable NMF algorithm, extract $\mathcal{K}$ from A so that $A(:, \mathcal{K})$ contains the r extreme rays of cone(A).
2: Let
$$y^* = \operatorname{argmin}_{y \in \mathbb{R}_+^r} D\big(A(\mathcal{K}, :)e, A(\mathcal{K}, \mathcal{K})y\big),$$
where $D(\cdot, \cdot)$ is an error measure such as a β-divergence; see Section 5.1.3. (The quantity y^* plays the role of $\frac{[e]}{[z]}$ in the notation in the text, so that $\operatorname{diag}(y^*) = \operatorname{diag}(z)^{-1}$.)
3: Let
$$S^* = \operatorname{diag}(y^*)A(\mathcal{K}, \mathcal{K})\operatorname{diag}(y^*).$$

4: Obtain W^* by solving
$$\min_{W \in \mathbb{R}_+^{n \times r}} D\big(A(\mathcal{K}, :), A(\mathcal{K}, \mathcal{K})\operatorname{diag}(y^*)W^\top\big).$$

In the presence of noise, to recover z and W, one should instead solve optimization problems under nonnegativity constraints (such as NNLS problems); see Arora et al. [14] for the details where the KL divergence is used to estimate W as it is better suited for word count data sets; see Section 5.1.1. Algorithm 7.6 summarizes this procedure; for a numerical example, see [Matlab file: septrisymNMF_example.m].

Minimum-volume tri-symNMF Min-vol NMF relies on a relaxed condition of separability, namely the SSC; see Section 4.3.3. Similarly, the separability assumption in tri-symNMF can be relaxed to the SSC, which does not require the presence of anchor words. In this case, one should solve a minimum-volume tri-symNMF problem, that is, approximate $A = XX^\top$ with $WSW^\top$ where W has a small volume [171]. Numerical results showed that min-vol tri-symNMF outperforms separable tri-symNMF; see also [170] for a discussion and some numerical examples.

7.9 ▪ Further readings

Let us briefly mention a few works closely related to separable NMF. Liu and Tan [326] use the link between ONMF and separable NMF to obtain algorithms with identifiability guarantees for ONMF in the presence of noise. The authors in [123] generalized separability to tackle convolutive NMF (Section 5.4.6), providing provably correct algorithms in the presence of noise. Finally, separable NMF has recently been generalized as follows: Given X and r, find the index sets $\mathcal{K}$ and $\mathcal{L}$ such that $|\mathcal{K}| + |\mathcal{L}| = r$ and

$$X = X(:, \mathcal{K})P_1 + P_2X(\mathcal{L}, :) \quad \text{with } P_1 \geq 0 \text{ and } P_2 \geq 0.$$

This allows selecting both columns and rows of X to reconstruct itself. For example, in text mining, this means that for each topic, we need only a pure document *or* an anchor word. However, this model is still not well-understood, even in the noiseless case. An issue is that, as opposed to separable NMF, it is not always identifiable; see [373] for more details.

7.10 ▪ Take-home messages

Near-separable NMF is a powerful NMF model as it is identifiable and can be solved efficiently, even in the presence of noise. We have presented in this chapter a variety of algorithms allowing one to perform this task; see Table 7.3 (page 257) for a summary of provably robust near-separable NMF algorithms. However, it relies on a relatively strong assumption. Nevertheless even if separability is not approximately satisfied by the input matrix, near-separable NMF algorithms can be used as an initialization for iterative NMF algorithms; see Section 8.6. This is particularly meaningful for ONMF (see for example Figures 4.10 and 4.11) or min-vol NMF algorithms as these problems are closely related to separable NMF; see Section 4.3.5. If no better option is available, we recommend in general to initialize NMF algorithms with a fast greedy near-separable NMF algorithm, such as SPA or SNPA applied on X or its transpose depending on the application; see the discussion in Section 7.1.

Chapter 8

Iterative algorithms for NMF

In this chapter, we present several optimization strategies to tackle the standard NMF problem (Problem 1.1, page 4), that is, given $X \in \mathbb{R}_+^{m \times n}$ and r, solve

$$\min_{\substack{W \in \mathbb{R}_+^{m \times r}, \\ H \in \mathbb{R}_+^{r \times n}}} D(X, WH), \tag{8.1}$$

where $D(X, WH)$ is an error measure between X and WH, such as the Frobenius norm of the residual $X - WH$; see Section 5.1 for more examples. Recall that (8.1) is NP-hard in general (Chapter 6); hence the algorithms presented in this chapter are iterative local optimization schemes converging to local solutions (in general, they are guaranteed to converge to stationary points; see Section 8.1). Such algorithms come with no global optimality guarantee.

Most iterative NMF algorithms optimize alternatively over the variable W for H fixed and over the variable H for W fixed. This is a two-block coordinate descent (2-BCD) method where the subproblems can be solved exactly or approximately; see Algorithm 8.1. This approach is also referred to as alternating optimization.

Algorithm 8.1 Two-block coordinate descent framework of most NMF algorithms

---

Input: Input nonnegative matrix $X \in \mathbb{R}_+^{m \times n}$ and factorization rank r.
Output: $(W, H) \geq 0$: A rank-r NMF of $X \approx WH$.

1: Generate some initial matrices $W^{(0)} \geq 0$ and $H^{(0)} \geq 0$; see Chapter 8.6.
2: **for** $t = 1, 2, \ldots$ [†] **do**
3:      $W^{(t)} = \text{update}\big(X, H^{(t-1)}, W^{(t-1)}\big)$, typically such that

$$D(X, W^{(t)} H^{(t-1)}) \leq D(X, W^{(t-1)} H^{(t-1)}).$$

4:      $H^{(t)\top} = \text{update}\left(X^\top, W^{(t)\top}, H^{(t-1)\top}\right)$, typically such that

$$D(X, W^{(t)} H^{(t)}) \leq D(X, W^{(t)} H^{(t-1)}).$$

5: **end for**

[†] See Chapter 8.5 for stopping criteria.

There are at least two reasons for this. First, for most error measures, these subproblems are convex and hence have many nice properties and are typically efficiently solvable. This is the case for $D(X, WH) = \|X - WH\|$ for any norm $\|.\|$: norms are convex functions while $X - WH$ is a linear function in W for H fixed and vice versa. Another important class of measures that are convex are the β-divergences for $\beta \in [1, 2]$. Second, by symmetry of the problem, $X = WH$ if and only if $X^\top = H^\top W^\top$ so that $D(X, WH) = D(X^\top, H^\top W^\top)$ for most error measures. Therefore deriving an update for one of the factors, W or H, directly leads to an update for the other factor; see Algorithm 8.1. These two observations make the design of NMF algorithms based on 2-BCD much easier than trying to optimize the variables (W, H) simultaneously. As far as we know, there exist very few such approaches[58] and they have not been very successful so far. Therefore, we consider in this chapter only 2-BCD, so that, by the symmetry of the problem, the main focus will be on the following nonnegatively constrained optimization problem: Given $X \in \mathbb{R}_+^{m \times n}$ and $W \in \mathbb{R}_+^{m \times r}$, solve

$$\min_{H \geq 0} D(X, WH). \tag{8.2}$$

There are many algorithms designed to tackle (8.2), and they are not specific to NMF. In this chapter, we present such algorithms and see how to embed them in the 2-BCD strategy.

Moreover, the focus of this chapter is on error measures $D(X, WH)$ based on β-divergences, that is, $D_\beta(X, WH)$, and, in particular, $\beta = 2$, which corresponds to the Frobenius norm, that is, $D_2(X, WH) = 1/2\|W - WH\|_F^2$. Although we will discuss a few important NMF models (such as orthogonal, sparse, and min-vol NMF), we present mostly optimization strategies for the standard NMF formulation (8.1). However, the techniques described in this chapter apply to other objective functions and other NMF models. This is important to keep in mind because of the following two reasons:

- There exist many other measures including weighted norms,[59] componentwise ℓ_p norms, α-divergences, and Bregman divergences; see Chapter 5.1.

- As discussed in Section 4.3 and Chapter 5, using regularizers in the objective and/or additional constraints to the NMF model is often crucial in practice. The main motivation is to obtain solutions that better satisfy the structure of the sought solution and hence lead to identifiability.

Organization of the chapter We start by presenting notions that will be useful throughout this chapter, namely the structure of the NMF problem, the first-order optimality conditions, the majorization-minimization (MM) framework, and convergence of BCD methods (Section 8.1). MM includes in particular the MU that are presented for β-divergences in Section 8.2. In Section 8.3, we focus on the case where $D(X, WH)$ is the Frobenius norm of $X - WH$, which is arguably the most studied objective function for NMF. Moreover, it is a situation where the MU are outperformed by other strategies. In Section 8.4, we explain how to choose the number of inner iterations of NMF algorithms, as well as acceleration techniques. In Section 8.4.3, we provide a numerical comparison of the algorithms for the Frobenius norm on dense and sparse data sets and discuss their practical performances. In Sections 8.5 and 8.6, we discuss the stopping criteria and the initialization schemes of NMF algorithms, respectively. In Section 8.7, we present several other approaches to tackle NMF which are not solely based on a well-chosen

---

[58]We refer the interested reader to [357] for a recent gradient-based nonalternating approach for constrained low-rank matrix approximation problems. However, in their numerical experiments with NMF, their approach does not compete with a gradient-based BCD approach.

[59]An algorithm for weighted NMF will be presented in Section 9.5 in the context of recommender systems. It is a simple extension of an algorithm presented in this chapter, namely hierarchical alternating least squares (Section 8.3.3).

optimization strategy applied on (8.1). In Section 8.9 we provide the link to online resources containing NMF codes. We conclude with further readings and take-home messages in Sections 8.8 and 8.10, respectively.

8.1 ▪ Preliminaries

In this section, we first briefly discuss the structure of the NMF problem (Section 8.1.1). Then we present three important tools used throughout this chapter, namely the first-order optimality conditions of NMF (8.1) (Section 8.1.2), the MM framework which is used to design optimization algorithms (Section 8.1.3), and the convergence theory of various BCD schemes (Section 8.1.4).

8.1.1 ▪ Structure of the NMF problem

As mentioned in the introduction, the NMF problem (8.1) is symmetric in variables W and H, as long as $D(X, WH) = D(X^\top, H^\top W^\top)$, which holds true for most error measures. Let us look at the subproblem (8.2) in variable H, which is a linear regression problem. In most cases, it can be decoupled into n independent problems. More precisely, solving (8.2) is equivalent to solving

$$\min_{H(:,j) \geq 0} D\big(X(:,j), WH(:,j)\big), \tag{8.3}$$

for $j = 1, 2, \ldots, n$, as long as

$$D(X, WH) = \sum_{j=1}^{n} D\big(X(:,j), WH(:,j)\big),$$

which holds true for most error measures such as the β-divergences $D_\beta(\cdot, \cdot)$. In particular it holds true for error measures of the form $D(A, B) = \sum_{i,j} d(A_{i,j}, B_{i,j})$ for some scalar error function $d(\cdot, \cdot)$. However, in practice, these problems should not be solved independently as they share information, namely, the matrix W, which can be used to reduce the computational load significantly. For example, for the Frobenius norm, the gradient requires computing $W^\top W$, which can be done once for all subproblems.

8.1.2 ▪ First-order optimality conditions

Assume the objective function $D(X, WH)$ is differentiable, and let us denote $\nabla_H D(X, WH) \in \mathbb{R}^{r \times n}$ the gradient of $D(X, WH)$ with respect to H in matrix form, that is, for all k, j

$$[\nabla_H D(X, WH)]_{k,j} = \frac{\partial D(X, WH)}{\partial H(k, j)}.$$

For the β-divergences (see page 165 for their definition),

$$\nabla_H D_\beta(X, WH) = W^\top \left((WH)^{\circ(\beta-2)} \circ (WH - X) \right), \tag{8.4}$$

where $\circ$ is the componentwise multiplication, and $A^{\circ a}$ is the componentwise exponent of matrix A by the scalar a, that is, $A^{\circ a}(i, j) = A(i, j)^a$ for all i, j. For example, for the Frobenius norm ($\beta = 2$), we obtain

$$\nabla_H \frac{1}{2} \|X - WH\|_F^2 = W^\top (WH - X),$$

and, for the KL divergence, we obtain

$$\nabla_H D_1(X, WH) = W^\top \left(\frac{[WH - X]}{[WH]} \right),$$

where $\frac{[\cdot]}{[\cdot]}$ is the componentwise division between two matrices. At this point, it is important to recall that $\nabla_H D_\beta(X, WH)$ is not defined everywhere depending on the values of X and β; see Section 5.1.3. The domain of the gradient $\nabla_H D_\beta(X, WH)$ of β-divergences can be obtained directly from Table 5.3 (page 167):

- For $\beta \leq 0$: if X has an entry equal to zero, the domain is empty. Otherwise, for $X > 0$, it requires $(WH)_{i,j} > 0$ for all i, j.

- For $0 < \beta < 1$: it requires $(WH)_{i,j} > 0$ for all i, j.

- For $\beta \in [1, 2)$: it requires $(WH)_{i,j} > 0$ for all i, j such that $X_{i,j} > 0$.

- For $\beta \geq 2$: there is no condition beyond the nonnegativity of WH.

The point (W, H) satisfies the first-order optimality condition of (8.1), also known as the Karush–Kuhn–Tucker (KKT) conditions, if (W, H) satisfies

$$\begin{aligned} H \geq 0, \quad \nabla_H D(X, WH) \geq 0, \quad \langle H, \nabla_H D(X, WH) \rangle = 0, \\ W \geq 0, \quad \nabla_W D(X, WH) \geq 0, \quad \langle W, \nabla_W D(X, WH) \rangle = 0, \end{aligned} \tag{8.5}$$

and if D is differentiable at (W, H). Such a point is referred to as a (first-order) stationary point of (8.1). The equation $\langle H, \nabla_H D(X, WH) \rangle = 0$ implies that for all k, j

$$H(k, j) = 0 \quad \text{or} \quad [\nabla_H D(X, WH)]_{k,j} = 0,$$

which are the complementary slackness conditions. As we will see, most NMF algorithms are first-order methods, that is, methods that compute only the gradient at each iterate, and only convergence to stationary points can be achieved (such methods are stuck at stationary points). A stationary point is either a local minimum, a local maximum, or a saddle point (which is a point for which there exists a direction in which the objective function goes down and a direction in which the objective function goes up; hence the term "saddle").

The KKT conditions can be leveraged in several ways:

- To assess whether a given solution is a stationary point and, if not, quantify the violation of the conditions. This is often used as a stopping criterion for optimization algorithms; see Section 8.5.

- To design algorithms, for example based on fixed-point iterations. One of the most widely used class of NMF algorithms, namely the MU, can be interpreted in this way; see Section 8.2.

- To characterize and identify properties of stationary points. For example, for the KL divergence, we have seen that any stationary point (W, H) preserves the row and column sum of the input matrix, that is, $WHe = Xe$ and $e^\top WH = e^\top X$; see Theorem 6.9. For the Frobenius norm, it can be easily shown that for any stationary point (W, H), we have

$$\operatorname{argmin}_\alpha \|X - \alpha WH\|_F^2 = \frac{\langle X, WH \rangle}{\langle WH, WH \rangle} = 1,$$

which implies

$$\|X - WH\|_F^2 = \|X\|_F^2 - 2\langle X, WH\rangle + \|WH\|_F^2 = \|X\|_F^2 - \|WH\|_F^2,$$

since $\langle WH, WH\rangle = \|WH\|_F^2$; see [194, Theorem 11]. A consequence of this observation is that, at stationarity, $\|WH\|_F \leq \|X\|_F$; see also Example 5.1 for a discussion.

8.1.2.1 ▪ Trivial saddle points

It is interesting to note, and this will have implications when studying convergence of NMF algorithms, that if (W, H) is a stationary point of (8.1), then adding p zero columns to W and p zero rows to H generates another stationary point $\left([W, 0_{m \times p}], [H; 0_{p \times n}]\right)$ of (8.1) where r is increased by p. The reason is that the gradient of H is also added p zero rows when p zero columns are added to W and p zero rows to H, because when $W(:, k) = 0$, the variables $H(k, :)$ do not appear in the objective function; by symmetry, the same observation applies to W. In particular, $\left(0_{m \times r}, 0_{r \times n}\right)$ is a stationary point for any r.

Such stationary points are always saddle points unless $X = WH$. Let us show this by denoting the residual $R = X - WH$. If $R_{i,j} > 0$ for some (i, j), then the objective function strictly decreases in the direction defined by replacing a zero column of W by e_i and the corresponding zero row of H by $R_{(i,j)}e_j^\top$ so that the residual at position (i, j) goes to zero (while the other entries are untouched) as the solution moves toward this direction [239]. Moreover, $R = X - WH \leq 0$ is not possible for stationary points of NMF based on β-divergences (unless $R = 0$) since otherwise the gradient defined in (8.4) cannot be nonnegative. Another way to show this is to see that if $R \leq 0$ and $R \neq 0$, then $(-W, 0)$ and $(0, -H)$ are descent directions. Such situations can be easily identified and escaped.

Another type of trivial saddle points are rank-deficient solutions. For example, if (w, h) is a rank-one stationary point, then $([w, w], [h/2; h/2])$ is a rank-deficient stationary point of (8.1) for $r = 2$. In most data analysis applications, W and H are expected to have rank r, and hence such situations can also be easily identified and escaped, by monitoring the numerical ranks of W and H.

8.1.3 ▪ Majorization-minimization framework

The MM framework covers a large class of optimization methods. In this section, we give a brief overview of MM. For more details, we refer the reader to [441, 408] and the references therein.

Assume we want to minimize the function $f(h)$ over the set $\mathcal{H}$, that is, we want to solve

$$\min_{h \in \mathcal{H}} f(h).$$

MM algorithms are iterative algorithms. Let us denote $\tilde{h} \in \mathcal{H}$ the current iterate. At each iteration, MM uses a two-step strategy:

1. Majorization. The goal is find a function $g(h, \tilde{h})$ such that

   (i) $g(h, \tilde{h}) \geq f(h)$ for all $h \in \mathcal{H}$, and

   (ii) $g(\tilde{h}, \tilde{h}) = f(\tilde{h})$.

   The function $g(h, \tilde{h})$ is called an auxiliary function for *or* a majorizer of f at $\tilde{h}$: it is larger than f everywhere and is equal to f at the current iterate $\tilde{h}$.

2. Minimization. Instead of minimizing f, one minimizes the function g. The next iterate h^+ needs to be chosen such that

$$g(h^+, \tilde{h}) \; \leq \; g(\tilde{h}, \tilde{h}). \tag{iii}$$

In many cases, g is chosen simple enough so that one can compute a global minimizer of $g(h, \tilde{h})$ in closed form. A possible way to achieve this goal is to choose g as a separable[60] function, that is,

$$g(h, \tilde{h}) = \sum_i g_i(h_i, \tilde{h}), \tag{8.6}$$

for some functions g_i's, so that minimizing $g(h, \tilde{h})$ requires solving independent univariate subproblems. This is a typical choice in the NMF literature. However, this is not necessary. Finally, any update that decreases $g(h, \tilde{h})$ leads to an algorithm that monotonically decreases $f(h)$. This follows directly from the properties of g. We have

$$f(h^+) \; \underset{(i)}{\leq} \; g(h^+, \tilde{h}) \; \underset{(iii)}{\leq} \; g(\tilde{h}, \tilde{h}) \; \underset{(ii)}{=} \; f(\tilde{h}).$$

Figure 8.1 (page 276) illustrates one step of an MM algorithm applied on minimizing the scalar IS divergence $\min_{y \geq 0} d_0(1, y)$ where the current iterate is $\tilde{y} = 2$.

Example 8.1. Many standard algorithms belong to the MM framework. A notable example is gradient descent for smooth convex optimization which uses the update

$$h^+ = \tilde{h} - \frac{1}{L} \nabla f(\tilde{h}),$$

where L is the Lipschitz constant of the gradient of f, that is, for all x, y,

$$\|\nabla f(h) - \nabla f(y)\|_2 \leq L \|h - y\|_2. \tag{8.7}$$

Using the majorizer

$$g(h, \tilde{h}) = f(\tilde{h}) + \nabla f(\tilde{h})^\top (h - \tilde{h}) + \frac{L}{2} \|h - \tilde{h}\|_2^2 \; \geq \; f(h),$$

we obtain the above update by minimizing exactly the quadratic function $g(h, \tilde{h})$. ∎

The main difficulty in designing MM algorithms is the choice of the function g which should be a good approximation of f while being easy to optimize. In particular, having a closed-form formula for the minimizer of g is of particular interest as it prevents parameter tuning (no step size to tune) and is easy to implement.

8.1.4 ▪ Convergence to stationary points

In this section, we present a few convergence results for block coordinate descent (BCD) methods. It is beyond the scope of this book to dig deep into this technical and still very active area of research, in particular the recent results based on the Kurdyka–Łojasiewicz property; see for example [238] and the references therein. Hence we only focus on presenting relatively simple results that imply the convergence of some of the methods presented in this chapter.

---

[60]The term separable here has nothing to do with separable NMF.

8.1.4.1 ▪ Monotone algorithms and convergent subsequence

Most algorithms presented in this chapter are monotonically decreasing the objective function which is bounded below. This implies the convergence of the objective function values

$$f^{(t)} = D\big(X, W^{(t)} H^{(t)}\big) \quad \text{for } t = 1, 2, \ldots,$$

where $\big(W^{(t)}, H^{(t)}\big)$ is the tth iterate of the algorithm.

Moreover, if the feasible set is compact, then there exists at least one converging subsequence of the iterates (Bolzano–Weierstrass theorem). In NMF, because of the scaling degree of freedom of the rank-one factors, we might have that $W^{(t)} H^{(t)}$ converges but not $\big(W^{(t)}, H^{(t)}\big)$. Moreover, the feasible set of NMF, namely, $(W, H) \geq 0$, is not compact since it is not bounded. The level sets $\{(W, H) \geq 0 \mid D(X, WH) \leq c\}$ for some constant c are not compact either, because of the scaling degree of freedom. However, adding a constraint that fixes the scaling degree of freedom, for example imposing w.l.o.g. that

$$\|W(:,k)\|_2 = \|H(k,:)\|_2 \text{ for } k = 1, 2, \ldots, r,$$

can be used to guarantee compactness. Let us focus on the Frobenius norm for simplicity. Letting $\big(W^{(0)}, H^{(0)}\big)$ be the initial iterate, we have, by monotonicity of the algorithm, that all iterates (W, H) satisfy

$$\big\|X - WH\big\|_F \leq \big\|X - W^{(0)} H^{(0)}\big\|_F.$$

Moreover,

$$\|X - WH\|_F \geq \|WH\|_F - \|X\|_F,$$

by the triangle inequality, and

$$\|WH\|_F \geq \|W(:,k)H(k,:)\|_F = \|W(:,k)\|_2 \|H(k,:)\|_2 = \|W(:,k)\|_2^2 = \|H(k,:)\|_2^2,$$

for all j, because of the nonnegativity of (W, H). Finally, putting these inequalities together, we have for all k and any iterate (W, H) that

$$\|W(:,k)\|_2^2 = \|H(k,:)\|_2^2 \leq \|X\|_F + \|X - W^{(0)} H^{(0)}\|_F,$$

which is a compact set. Note that this is not a convex set, so one may prefer to use the constraints

$$\|H(k,:)\|_2^2 \leq \|X\|_F + \|X - W^{(0)} H^{(0)}\|_F \text{ for all } k,$$

and similarly for W. Another typical rescaling used in practice is $\|W(:,k)\|_1 = 1$ for all k (or similarly for the rows of H), for example when the columns of W can be interpreted as probabilities, which is the case, for example, in topic modeling (Section 5.4.9); see also the introduction of Chapter 9 for a discussion.

The above trick to make the feasible set compact can be used for most other error measures, as long as they are coercive, that is, as long as $D(X, WH)$ goes to infinity as $\|WH\|$ goes to infinity.

8.1.4.2 ▪ Exact two-block coordinate descent method

If the subproblems in steps 3 and 4 of Algorithm 8.1 are solved exactly, that is, an optimal solution is used for $W^{(t)}$ and $H^{(t)}$, Algorithm 8.1 is the so-called exact 2-BCD method for which we have the following convergence guarantee.

Theorem 8.2. *[222, Corollary 2] The limit points of the iterates of an exact 2-BCD algorithm are stationary points provided that the following two conditions hold:*

1. *the objective function is continuously differentiable, and*

2. *each block of variables is required to belong to a closed convex set.*

For NMF and Algorithm 8.1, the second condition is satisfied since the nonnegative orthant is a closed convex set. The first condition is met for many objective functions such as the Frobenius norm. However, this condition is not met by β-divergences for $\beta < 1$, that is, $d_\beta(z, \cdot)$ is not continuously differentiable at zero. Moreover, for $\beta < 2$, the derivative of $d_\beta(z, \cdot)$ is not defined at zero when $z > 0$; see Table 5.3 (page 167).

8.1.4.3 ▪ Exact block coordinate descent method

As we will see, it might be powerful to use updates in steps 3 and 4 of Algorithm 8.1 that are based on exact BCD. This amounts to solving NMF with an exact BCD method with more than two blocks of variables, in which case the following convergence results can be used.

Theorem 8.3. *[42, 41, Proposition 2.7.1] The limit points of the iterates of an exact BCD algorithm are stationary points provided that the following conditions hold:*

1. *the objective function is continuously differentiable,*

2. *each block of variables is required to belong to a closed convex set,*

3. *the minimum computed at each iteration for a given block of variables is uniquely attained, and*

4. *the objective function values in the interval between all iterates and the next (which is obtained by updating a single block of variables) is monotonically decreasing.*

Condition 4 can be dropped if each block of variables belongs to a convex and compact set.

The order in which the blocks are updated is arbitrary, as long as each block is updated at least once every K iterations, where K is a fixed constant; this is referred to as the essentially cyclic block update.

Compared to Theorem 8.2, Theorem 8.3 requires that the minimum is uniquely attained and that the objective function is monotonically decreasing between two iterates. This holds true, for example, if the subproblems in one block of variables are strongly convex.

There exist counterexamples of exact BCD not converging to stationary points when the assumptions of Theorem 8.3 are not met; see [385, 242] and [28, Chapter 14].

8.1.4.4 ▪ Block successive upper-bound minimization

Razaviyayn, Hong, and Luo [392] introduced the block successive upper-bound minimization (BSUM) framework. It is a BCD method where the subproblems in one block of variables are solved using an MM framework (Section 8.2.3). More precisely, for each block of variables, a majorizer is constructed which has more properties than in the MM framework. Recall that MM requires the majorizer to be a global upper bound of the objective function and be tight at the current iterate (see Section 8.2.3). On top of that, BSUM requires that the directional derivatives of the majorizers and of the objective function coincide at the current iterate for each block of variables (intuitively, their tangents need to exist and coincide), while the majorizers should be continuous functions in all the variables. Moreover, in the BSUM framework, the majorizers are minimized exactly, while an essentially cyclic block update is used. We refer the reader to [392] for more details.

Theorem 8.4. *[392, Theorem 2] Convergence of BSUM can be guaranteed in the following two scenarios:*

- *If the majorizers are quasi-convex[61] and their minimum is uniquely attained, then every limit point of the sequence of iterates generated by BSUM is a stationary point.*

- *If the level sets of the majorizers are compact and the subproblems have a unique solution for all blocks but one, then the sequence of iterates generated by BSUM converges to the set of stationary points.*

Let us mention two examples that will be encountered in this chapter and for which Theorem 8.4 can be used; we refer the interested reader to the survey [242] for more examples.

Example 8.5. Assume you are using an exact BCD scheme where the objective function is continuous and differentiable, and the subproblems in one block of variables are convex but not strongly convex. Let us denote $\min_{h \in \mathcal{H}} f(h)$ the convex subproblem in one block of variables. Because the subproblems are not strongly convex, unicity of the optimal solution of the subproblems is not guaranteed and Theorem 8.3 does not apply. An example that will be of particular interest to us is the NNLS problem $\min_{h \geq 0} \|Wh - x\|_2$ when W is not full column rank, in which case that problem is not strongly convex and does not admit a unique solution.

The update of exact BCD

$$h^+ \; = \; \mathrm{argmin}_{h \in \mathcal{H}} \, f(h)$$

can be replaced with

$$h^+ \; = \; \mathrm{argmin}_{h \in \mathcal{H}} \, f(h) + \mu \|h - \tilde{h}\|_2^2,$$

where $\tilde{h}$ is the current iterate. The majorizer $f(h) + \mu \|h - \tilde{h}\|_2^2$ of $f(h)$ at $h = \tilde{h}$ is strongly convex when f is convex and $\mu > 0$. Moreover, the directional derivatives of $\|h - \tilde{h}\|_2^2$ w.r.t. h at $h = \tilde{h}$ is zero, so that the directional derivatives of $f(h)$ and of the majorizer coincide. Therefore, the above regularized BCD scheme falls within the BSUM framework and convergence to stationary points is guaranteed by Theorem 8.4. The regularization term $\|h - \tilde{h}\|_2^2$ is known as a proximal term: it prevents the algorithm from going too far away from the current iterate $\tilde{h}$. In particular, it prevents the current iterate from moving if it is a globally optimal solution, while the original exact BCD scheme could move away from an optimal solution toward another optimal solution (when it is nonunique, which is always the case for NMF because of the scaling and permutation ambiguity). ∎

Example 8.6. In a BCD scheme, assume the subproblems in one block of variables cannot be solved easily in closed form but are convex with Lipschitz gradient with constant L (see Equation (8.7)) and have the form $\min_h f(h)$ where h is one block of variables. As in Example 8.1, we have

$$f(h) \; \leq \; f(\tilde{h}) + \nabla f(\tilde{h})^\top (h - \tilde{h}) + \frac{L}{2} \|h - \tilde{h}\|_2^2.$$

---

[61]A function u is quasi-convex if the level sets $\{x \mid u(x) \leq c\}$ are convex, for any constant c. A convex function is always quasi-convex. An example of a quasi-convex function which is not convex is $\sqrt{|x|}$.

Using the right-hand side as the majorizer of $f(h)$ at $\tilde{h}$ in the BSUM framework, assumptions of Theorem 8.4 are satisfied and convergence to stationary points is guaranteed. The majorizer above is minimized exactly at

$$h^+ = \tilde{h} - \frac{1}{L} \nabla f(\tilde{h}),$$

which is a gradient descent update with step size $1/L$. Hence BSUM applies to block coordinate gradient descent, given that the subproblems in each block are convex with Lipschitz continuous gradients.

The convergence results above generalizes to other cases such as the so-called block coordinate proximal gradient method (where only part of f is linearized as above; for example if the objective function contains a nondifferentiable term such as $\|x\|_1$, it is kept as is in the majorizer), as well as in the constrained case (see [242] for more details). This is closely related to the proximal alternating linearized minimization algorithm [52]; see also [28]. ∎

8.2 ▪ The multiplicative updates

The MU algorithms consist in simple updates for W and H that take the form of componentwise multiplications, that is,

$$H \quad \leftarrow \quad H \circ G(X, W, H),$$

where $G(X, W, H)$ is nonnegative when X, W, and H are. A first advantage of such updates is that they are easily implemented and do not require any parameter tuning. Moreover, for β-divergences corresponding to values of β smaller than two, such as the KL and IS divergences, the MU perform very well and are competitive with other approaches; see for example the recent survey [237] on NMF with the KL divergence.

Brief history   For the KL divergence, the MU were derived by Richardson [397] (1972) and Lucy [332] (1974), who used them for image restoration and for rectification and deconvolution in statistical astronomy, respectively. They considered the linear regression problem

$$\min_{h \geq 0} \mathrm{KL}(x, Wh)$$

for some given matrix W and vector x. The motivation to use the KL divergence (instead of least squares) comes from statistical considerations (see also Section 5.1.1). This algorithm was later referred to as the Richardson–Lucy algorithm.

For the Frobenius norm, the MU were proposed by Daube-Witherspoon and Muehllehner [116] (1986) and referred to as the iterative space reconstruction algorithm (ISRA). They used it for solving inverse problems in imaging.

The Richardson–Lucy algorithm and ISRA were shown to belong to the class of MM algorithms by De Pierro [121] (1993). Later, Lee and Seung rediscovered these updates in the context of NMF [303, 304] (1999, 2001). For the IS divergence, Cao, Eggermont, and Terebey [74] (1999) developed MU as well; see also [158]. These approaches were later generalized by Févotte and Idier [160] (2011) as MM algorithms (see Section 8.2.3). We refer the interested reader to their paper for more historical background on the MU.

The MU are the most popular in the NMF literature. In our opinion, the reason is at least fourfold:

1. They were proposed for the first time in the context of NMF in the seminal paper of Lee and Seung which can be considered as the "big bang" of NMF; see Section 1.4.9.

2. They can be easily derived and implemented (Section 8.2.1) and hence are easily adaptable to many NMF models (Section 8.2.7).

3. The MU are first-order methods (Section 8.2.2) so that they scale well when the dimensions of the input matrix increase (namely they scale linearly with the input dimensions and the factorization rank; see Section 8.2.6).

4. In some cases, the MU compete with state-of-the-art algorithms; this is the case, for example, for the KL divergence [237]. For some NMF models, they are the only available algorithm. However, for the Frobenius norm, they perform rather poorly compared to the state of the art (Section 8.3).

This section is organized as follows. In the next three sections, we present different ways to derive and interpret the MU, namely as

1. the ratio of the positive to the negative parts of the gradient which is an intuitive and heuristic approach often used in the literature (Section 8.2.1),

2. a rescaled gradient descent which is a quasi-Newton method where the Hessian is approximated by a diagonal matrix (Section 8.2.2), and

3. an MM algorithm which allows us to derive MU for any β-divergence with the guarantee that the objective function is nonincreasing (Section 8.2.3).

We then discuss the convergence of the MU (Section 8.2.4), the zero locking phenomenon that prevents the MU from modifying zero entries in the factors W and H and a modified variant that fixes this issue (Section 8.2.5), the computational cost of the MU (Section 8.2.6), and the flexibility of the MU (Section 8.2.7).

8.2.1 ▪ MU obtained via the gradient ratio heuristic

Let us focus on β-divergences for which the gradient is given in (8.4). Denoting

$$\nabla_H D_\beta^+(X, WH) = W^\top (WH)^{\circ(\beta-1)}$$

and

$$\nabla_H D_\beta^-(X, WH) = W^\top \left((WH)^{\circ(\beta-2)} \circ X \right),$$

we obtain

$$\nabla_H D_\beta(X, WH) = \nabla_H D_\beta^+(X, WH) - \nabla_H D_\beta^-(X, WH). \tag{8.8}$$

Moreover, the KKT conditions for H in (8.5) are $H(k, j) = 0$ or $[\nabla_H D_\beta(X, WH)]_{k,j} = 0$ for all k, j, while these two quantities are nonnegative. This implies that, at stationarity, when $H(k, j) > 0$, we must have

$$\left[\nabla_H D_\beta^+(X, WH) \right]_{k,j} = \left[\nabla_H D_\beta^-(X, WH) \right]_{k,j}. \tag{8.9}$$

If $[\nabla_H D_\beta(X, WH)]_{k,j} > 0$ (resp. < 0), a sufficiently small decrease (resp. increase) of $H(k, j)$ decreases the objective function. Therefore, it makes sense to decrease $H(k, j)$ if $\left[\nabla_H D_\beta^+(X, WH) \right]_{k,j} > \left[\nabla_H D_\beta^-(X, WH) \right]_{k,j}$, and vice versa. This is the intuitive idea behind the MU that update H as follows:

$$H \leftarrow H \circ \frac{\left[\nabla_H D_\beta^-(X, WH) \right]}{\left[\nabla_H D_\beta^+(X, WH) \right]}. \tag{8.10}$$

For example, for the Frobenius norm, we obtain

$$H \quad \leftarrow \quad H \circ \frac{[XH^\top]}{[W^\top WH]},$$

and, for the KL divergence,

$$H \quad \leftarrow \quad H \circ \frac{\left[W^\top \frac{[X]}{[WH]}\right]}{[W^\top 1_{m \times n}]},$$

where $1_{m \times n}$ is the m-by-n matrix of all ones.

The MU keep $H(k,j)$ unchanged if and only if $H(k,j) = 0$ or if (8.9) holds. With the MU, H should be initialized with positive entries because the MU cannot modify an entry of H equal to zero; this is the so-called zero locking phenomenon (see Section 8.2.5). If the entry $H(k,j)$ is initialized at zero, then the condition $[\nabla_W D_\beta(X, WH)]_{k,j} \geq 0$ is not taken into account in the MU and hence could be violated. The MU (8.10) can be seen as a fixed-point iteration of the KKT conditions. Assume that H is initialized with positive entries and that the MU converge to a fixed point H^f (see Section 8.2.4 for a discussion on convergence). There are two cases:

- If $H^f(k,j) > 0$, then we must have $\left[\nabla_H D_\beta(X, WH^f)\right]_{k,j} = 0$, otherwise H^f is not a fixed point.

- If $H^f(k,j) = 0$, this means that

$$\left[\nabla_H D_\beta^+(X, WH^f)\right]_{k,j} \geq \left[\nabla_H D_\beta^-(X, WH^f)\right]_{k,j},$$

  because the MU, by construction, cannot set an entry of H to zero, except at the limit. If $\left[\nabla_H D_\beta^+(X, WH^f)\right]_{k,j} < \left[\nabla_H D_\beta^-(X, WH^f)\right]_{k,j}$, then in a open neighborhood around $H^f(k,j)$ (by continuity of the gradient), the MU increase the value of $H^f(k,j)$, a contradiction. As we will see, this may cause some numerical issues when working with finite floating point precision (see Section 8.2.4).

At this point, nothing guarantees the MU (8.10) will monotonically decrease $D_\beta(X, WH)$. However, we will see in Section 8.2.3 that it holds true for $\beta \in [1,2]$, while, for $\beta \notin [1,2]$, similar MU can be obtained that ensure monotonicity of $D_\beta(X, WH)$; see Theorem 8.8.

8.2.2 • MU viewed as a rescaled gradient descent method

Let us consider the following general optimization problem with nonnegativity constraints:

$$\min_{h \geq 0} \ f(h). \tag{8.11}$$

The projected rescaled gradient descent method applied to (8.11) uses the update

$$h^+ = \mathcal{P}\left(\tilde{h} - B\nabla f(\tilde{h})\right),$$

where $\tilde{h}$ is the current iterate, h^+ is the next iterate, B is a diagonal matrix with positive diagonal elements, and $\mathcal{P}(h) = \max(h, 0)$ is the projection onto the feasible set. This method is related to quasi-Newton methods which use B as an approximation of the inverse of the Hessian of f.

Let $\nabla_+ f(h) > 0$ and $\nabla_- f(h) > 0$ be such that $\nabla f(h) = \nabla_+ f(h) - \nabla_- f(h)$. Taking $B = \mathrm{diag}\left(\frac{[h]}{[\nabla_+ f(h)]}\right)$ leads to the following update for the projected rescaled gradient descent method

$$h^+ = \tilde{h} - \frac{[\tilde{h}]}{[\nabla_+ f(\tilde{h})]} \circ \left(\nabla_+ f(\tilde{h}) - \nabla_- f(\tilde{h})\right) = \tilde{h} \circ \frac{[\nabla_- f(\tilde{h})]}{[\nabla_+ f(\tilde{h})]}, \tag{8.12}$$

which is multiplicative. For the NMF in variable H, one may use the decomposition given in (8.8), in which case we recover the MU (8.10) based on the gradient ratio. It is important to note that the strict positivity of $\nabla_+ f(h)$ and $\nabla_- f(h)$ is necessary; otherwise there is a division by zero or a variable is directly set to zero.

Interestingly, the interpretation of the MU in terms of a rescaled gradient method allows us to modify the MU to guarantee the monotonicity of the objective function. Since B has positive diagonal entries, $B\nabla f(h)$ is a descent direction. Therefore, for a sufficiently small $\alpha \in (0, 1]$, the iterate

$$h^+(\alpha) = \tilde{h} - \alpha B\nabla f(\tilde{h}) = \tilde{h} + \alpha(h^+ - \tilde{h}) = (1 - \alpha)\tilde{h} + \alpha h^+$$

decreases the objective function, while $h^+(\alpha)$ remains nonnegative for any $\alpha \in [0, 1]$ since $\tilde{h}$ and h^+ are nonnegative. This means that embedding the MU within a line-search procedure allows guaranteeing the nonincreasingness of the objective function; see for example [201]. However, implementing a line-search procedure requires additional computations, in particular the values of the objective functions for different values of α. In the next section, we show how to avoid line search.

8.2.3 ▪ MU as an MM algorithm

In this section, we follow the work of Févotte and Idier [160], who proposed a nice general framework to design MU based on the MM framework that are guaranteed to monotonically decrease the objective function for β-divergences. The key resides in the design of the majorizer.

As the NNLS subproblem in H (8.2) can be decomposed into n independent subproblems (see Section 8.1.1), we focus in the following on the problem for one column of H, which we denote $h \in \mathbb{R}^r$, and one column of X, which we denote $x \in \mathbb{R}^m$. We end up with the following regression problem: Given $x \in \mathbb{R}^m$ and $W \in \mathbb{R}^{m \times r}$, solve

$$\min_{h \in \mathbb{R}^r_+} D_\beta(x, Wh). \tag{8.13}$$

Let us apply the MM framework (Section 8.1.3) to (8.13), and let us denote the current iterate $\tilde{h}$. We need a majorizer $g(h, \tilde{h})$ for $D_\beta(x, Wh)$ with

$$g(h, \tilde{h}) \geq D_\beta(x, Wh) \text{ for all } h \geq 0, \text{ and } g(\tilde{h}, \tilde{h}) = D_\beta(x, W\tilde{h}).$$

Recall that on top of the above two properties, majorizers should be easy to optimize. The standard choice in the NMF literature is to have $g(h, \tilde{h})$ separable, as defined in (8.6), which will be the case for the majorizers described in this section. The design of the MU by Févotte and Idier [160] follows a three-step approach. The first two steps construct the majorizer for $f(h) = D_\beta(x, Wh)$, while the last step exactly minimizes the majorizer.

Step 1. The first step decomposes the objective function into the sum of simpler terms, motivated by the following observation.

Lemma 8.7. *Let $f(h) = \sum_{i=1}^{p} f_i(h)$, and let $g_i(h, \tilde{h})$ be majorizers at $\tilde{h}$ of f_i for $i = 1, 2, \ldots, p$. Then $g(h, \tilde{h}) = \sum_{i=1}^{p} g_i(h, \tilde{h})$ is a majorizer for f at $\tilde{h}$.*

Proof. This follows directly from the definition of a majorizer (Section 8.1.3).   □

The β-divergence $d_\beta(z, y)$ between the scalars z and y (see page 165 for the definition) can be decomposed as follows:

$$d_\beta(z, y) = \check{d}_\beta(z, y) + \hat{d}_\beta(z, y) + \bar{d}_\beta(z, y), \tag{8.14}$$

where $\check{d}_\beta(z, y)$ is convex in y, $\hat{d}_\beta(z, y)$ is concave in y, and $\bar{d}_\beta(z, y)$ is constant in y. This is referred to as a convex-concave-constant decomposition of $d_\beta(z, y)$ with respect to y. Table 8.1 reports such decompositions for all β-divergences.

Table 8.1. *Scalar convex-concave-constant decomposition* (8.14) *of $d_\beta(z, y)$ with respect to variable y.*

	$\check{d}_\beta(z, y)$	$\hat{d}_\beta(z, y)$	$\bar{d}_\beta(z, y)$
	convex	concave	constant
$\beta < 1, \beta \neq 0$	$-\frac{1}{\beta-1} z y^{\beta-1}$	$\frac{1}{\beta} y^\beta$	$\frac{1}{\beta(\beta-1)} z^\beta$
$\beta = 0$	$z y^{-1}$	$\log y$	$z(\log z - 1)$
$1 \leq \beta \leq 2$	$d_\beta(z, y)$	0	0
$\beta > 2$	$\frac{1}{\beta} y^\beta$	$-\frac{1}{\beta-1} z y^{\beta-1}$	$\frac{1}{\beta(\beta-1)} z^\beta$

Since Wh is a linear function of h, it preserves convexity and concavity, implying that

$$D_\beta(x, Wh) = \sum_{i=1}^{m} d_\beta(x_i, (Wh)_i)$$

$$= \sum_{i=1}^{m} \check{d}_\beta(x_i, (Wh)_i) + \hat{d}_\beta(x_i, (Wh)_i) + \bar{d}_\beta(x_i, (Wh)_i) \tag{8.15}$$

is a convex-concave-constant decomposition of $D_\beta(x, Wh)$ with respect to h.

Step 2. Using Lemma 8.7, the second step is to find majorizers for each term in the convex-concave-constant decomposition (8.15):

- *Constant part.* There is no need to find a majorizer for the constant term $\bar{d}(x_i, (Wh)_i)$ since it does not play any role when minimizing $d(x, Wh)$ with respect to h.

- *Concave part.* Let us first analyze the scalar function $\hat{d}_\beta(z, y)$: it is concave in y and hence can be upper bounded with its first-order Taylor expansion at any point $\tilde{y}$ (that is, its tangent at $\tilde{y}$):

$$\hat{d}_\beta(z, y) \leq \hat{d}_\beta(z, \tilde{y}) + (y - \tilde{y}) \hat{d}'_\beta(z, \tilde{y}),$$

where $\hat{d}'_\beta(z, y)$ is the derivative of $\hat{d}_\beta(z, y)$ with respect to y. Denoting $\tilde{x} = W\tilde{h}$, this implies that, for $i = 1, 2, \ldots, m$,

$$\hat{d}_\beta(x_i, \tilde{x}_i) + (Wh - \tilde{x})_i \, \hat{d}'_\beta(x_i, \tilde{x}_i)$$

is a majorizer for $\hat{d}_\beta(x_i, (Wh)_i)$ at $\tilde{x}_i$. Moreover, this majorizer is linear in h and hence separable in h.

- *Convex part.* For the convex part, let us introduce the matrix $\Lambda \in \mathbb{R}^{m \times r}$ defined as follows: for all i, k,

$$\Lambda(i, k) = \frac{W(i, k)\tilde{h}_k}{\sum_k W(i, k)\tilde{h}_k} = \frac{W(i, k)\tilde{h}_k}{W(i, :)\tilde{h}},$$

which does not depend on h. By construction,

$$\Lambda \geq 0 \quad \text{and} \quad \Lambda e = e.$$

In other words, each row of Λ belongs to the unit simplex. Note that Λ is well-defined only when $W\tilde{h} > 0$. This condition can be met by assuming that $W > 0$ and $\tilde{h} > 0$. This makes sense since, as explained above, the MU should be initialized with positive matrices and can set entries of W and H to zero only at the limit. In other words, if W and H are initialized with positive entries, these entries remain positive throughout the executions of the MU, although some entries may converge to zero (see Section 8.2.4 for a discussion). Let us denote S_i the support of the ith row of W and Λ, that is,

$$S_i = \{k \mid W(i, k) \neq 0\} = \{k \mid \Lambda(i, k) \neq 0\} \text{ for } i = 1, 2, \ldots, m,$$

which coincide by construction. We can now construct majorizers for $\check{d}_\beta(x_i, (Wh)_i)$ for $i = 1, 2, \ldots, m$ as follows:

$$\check{d}_\beta(x_i, (Wh)_i) = \check{d}_\beta\left(x_i, \sum_{k \in S_i} W(i, k)h_k\right)$$

$$= \check{d}_\beta\left(x_i, \sum_{k \in S_i} \Lambda(i, k)\frac{W(i, k)h_k}{\Lambda(i, k)}\right)$$

$$\leq \sum_{k \in S_i} \Lambda(i, k) \check{d}_\beta\left(x_i, \frac{W(i, k)h_k}{\Lambda(i, k)}\right),$$

where the inequality follows from convexity of $\check{d}_\beta$ with respect to its second argument since rows of Λ define convex combinations; this is sometimes referred to as the Jensen's inequality.

Figure 8.1 shows an example for the IS divergence in the scalar case, namely $d_0(1, y)$ with $\tilde{y} = 2$ for which

$$d_0(1, y) \leq \underbrace{\frac{1}{y}}_{\check{d}_0(1,y)} + \underbrace{\log 2 + \frac{y - 2}{2}}_{\hat{d}_0(1,2)+(y-2)\hat{d}'_0(1,2)} + \underbrace{-1}_{\bar{d}_0(1,y)},$$

and equality holds for $y = \tilde{y} = 2$.

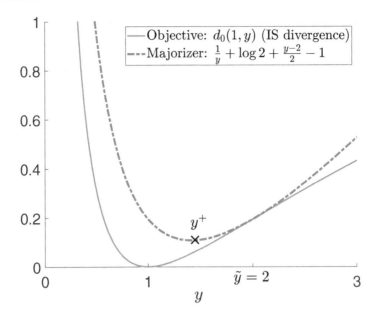

Figure 8.1. *Illustration of the majorizer at $\tilde{y} = 2$ of $d_\beta(z, y)$ for $\beta = 0$ (IS divergence) and $z = 1$. The minimizer of the majorizer is denoted y^+.*

Step 3. The last step is to find a closed form for the optimal solution of the minimization of the majorizer, that is,

$$\min_{h \geq 0} g(h, \tilde{h}).$$

Since $g(\cdot, \tilde{h})$ is constructed such that it is convex and admits a Slater point (any $h > 0$), the KKT conditions are necessary and sufficient so that the optimal solution satisfies $h_i = 0$ or $[\nabla_h g(h, \tilde{h})]_i = 0$ for all i (see page 5). Moreover, since $g(\cdot, \tilde{h})$ is separable, these equations boil down to finding the roots of r independent univariate functions. It turns out these have closed-form solutions, leading to the following theorem.

Theorem 8.8. *[160] Given $X \in \mathbb{R}_+^{m \times n}$, $W \in \mathbb{R}_{++}^{m \times n}$, and $H \in \mathbb{R}_{++}^{m \times n}$, the MU*

$$H \quad \leftarrow \quad H \circ \left(\frac{\left[W^\top \left((WH)^{\circ(\beta-2)} \circ X \right) \right]}{\left[W^\top (WH)^{\circ(\beta-1)} \right]} \right)^{\gamma(\beta)},$$

where

$$\gamma(\beta) = \begin{cases} \frac{1}{2-\beta} & \text{for } \beta < 1, \\ 1 & \text{for } 1 \leq \beta \leq 2, \\ \frac{1}{\beta-1} & \text{for } \beta > 1, \end{cases} \tag{8.16}$$

do not increase the objective function $D_\beta(X, WH)$.

For $\beta \in [1, 2]$, $\gamma(\beta) = 1$ so that the above MU coincide with (8.10) derived heuristically (Section 8.2.1). For $\beta \notin [1, 2]$, $\gamma(\beta) < 1$, which implies that the above MU are more conservative and modify H more slowly than the heuristic ones (8.10). However, they are guaranteed to decrease the objective function, as they follow an MM framework. We refer the interested reader to [160] and [491] for more details on the design of MU for NMF.

8.2.4 ▪ Convergence

For $\beta \geq 2$, the MU satisfy the BSUM framework and Theorem 8.4 applies, as long as the initial iterate $(W^{(0)}, H^{(0)})$ has only positive entries; see the discussion in [242], where the case $\beta = 2$ is described in detail.

For $\beta < 2$, convergence is more difficult to establish. In particular, β-divergences for $\beta < 2$ are not defined (nor differentiable) everywhere: the derivatives of β-divergences with respect to their second argument are not defined for $\beta < 2$ at $(WH)_{i,j} = 0$ when $X_{i,j} > 0$, while most convergence results require the objective function to be differentiable everywhere, such as BSUM (Theorem 8.4), and the analysis in [508]. A possible way to overcome this issue is to use a small lower bound for the entries of W and H and consider the following modified NMF problem:

$$\min_{\substack{W \in \mathbb{R}^{m \times r}, \\ H \in \mathbb{R}^{r \times n}}} D_\beta(X, WH) \quad \text{such that} \quad W \geq \epsilon \text{ and } H \geq \epsilon, \tag{8.17}$$

where $\epsilon \geq 0$ is a parameter. For $\epsilon = 0$, this is standard NMF. This modification was proposed in [194] to guarantee convergence of the MU to stationary points in the case $\beta = 2$, and the analysis was provided for $\beta = 1$ in [444]. The so-called modified MU simply apply the operator $\max(\epsilon, H)$ to the MU. Using this modification, we obtain the following result.

Theorem 8.9. *Let $\epsilon > 0$ and let us define the modified MU as*

$$H \quad \leftarrow \quad \max\left(\epsilon \, , \, H \circ \left(\frac{\left[W^\top \left((WH)^{\circ(\beta-2)} \circ X\right)\right]}{\left[W^\top (WH)^{\circ(\beta-1)}\right]} \right)^{\gamma(\beta)} \right),$$

where $\gamma(\beta)$ is given in (8.16), and the update of W is obtained by symmetry. Then,

- *the modified MU do not increase the objective function of (8.17), given that $W \geq \epsilon$ and $H \geq \epsilon$;*

- *for any initial matrices (W, H), every limit point of the modified MU that alternatively update H and W converge to a stationary point of (8.17);*

- *for $\beta \geq 2$, we may take $\epsilon = 0$ (that is, the standard MU) and the same convergence result as in the case $\epsilon > 0$ applies given that the initial iterate has only positive entries.*

Proof. Monotonicity follows from Theorem 8.8. In fact, all the derivations used in Section 8.2.3 apply directly to (8.17); the only difference is that the minimizer is obtained by taking the maximum with ϵ instead of zero.

For the convergence to stationary points,

- for $\epsilon > 0$, this result was proved in [194, Theorem 2] and [444, Theorem 1] for $\beta = 2$, and in [444, Theorem 3] for $\beta = 1$, and for the other values of β, the proof follows from BSUM theory since the objective function is differentiable everywhere;

- for $\epsilon = 0$ and $\beta \geq 2$, this follows from BSUM theory; see [242], which analyzes the particular case $\beta = 2$ in detail. □

The MU from Theorem 8.9 are available from [Matlab file: betaNMF.m].

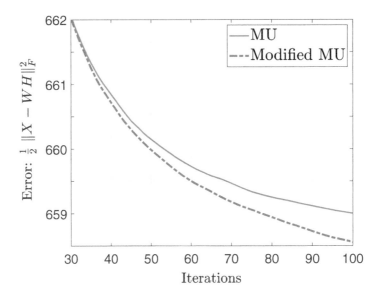

Figure 8.2. *Illustration of the zero locking phenomenon of the MU on a synthetic data set $X =$ sprand(500,1000,0.01), $r = 40$ and $\beta = 2$. The modified MU use a lower bound of machine epsilon ($\epsilon = 2^{-52}$) which guarantees convergence (Theorem 8.9) and prevents the MU from setting entries to zero, which allows both the convergence to stationary points and the decrease of the objective function.* [Matlab file: MU_vs_modifiedMU.m].

8.2.5 ▪ Zero locking phenomenon of the MU

In practice, even when $\beta \geq 2$, we recommend using the modified MU with $\epsilon > 0$ (more precisely, to take ϵ as the machine epsilon). The reason is because of numerical issues and the zero locking phenomenon of the MU (see Figure 8.2 for a numerical example). Some authors have argued that using $\epsilon > 0$ is not reasonable as one requires zeros in the factors W and H in many applications. Although it is true that zeros are expected and desired in many applications (see for example Sections 1.3, 4.3.4, and 5.4.4), we believe that having zero entries in W and H or having a value that matches the machine epsilon makes no difference. After the modified MU have run, the user is welcome to set the machine epsilon entries to zero, which does not affect the solution or the objective function much; see [185, pp. 66–68], where this statement is quantified. Let us discuss this interesting case where theory does not meet practice, because of finite machine precision.

As mentioned above, the initial iterates of the MU should be chosen positive. In theory, by the nature of the MU, all iterates remain positive, and some entries may tend to zero at convergence. However, numerically, some entries can be set to zero during the course of the MU (because of finite precision) while the corresponding entry of the gradient becomes negative later in the MU which prevents the MU to converge to a stationary point. This may also make the MU run into numerical problems as some entries of the denominator might become equal to zero. This numerical issue has been sometimes overlooked in the literature. Using a lower bound on the entries of W and H (such as the machine epsilon) fixes this numerical issue, and this can have an important impact on the behavior and performance of the MU, especially when W and H are expected to be very sparse, which is the case, for example, for sparse input matrices [197]. Figure 8.2 illustrates this behavior on a randomly generated sparse data set with $m = 500$, $n = 1000$, and 99% of the entries equal to zero (sprand(m,n,0.01) in MATLAB), and we used $\beta = 2$ and $r = 40$. We run the MU and the modified MU with $\epsilon = 2^{-52} = 2.2204 \, 10^{-16}$ (MATLAB machine epsilon) with the same positive initial matrices $(W^{(0)}, H^{(0)})$ where each entry is randomly generated using the uniform distribution $\mathcal{U}[0,1]$ ($\mathtt{rand(m,r)}$ and $\mathtt{rand(r,n)}$ in MATLAB). In this example, we observe that, after only 100 iterations, the MU have set 3369 entries of W to zero (about 17% of its entries) and 14551 entries of H to zero (about 36% of its entries). Since these entries are locked, some of the corresponding gradient entries may become negative and convergence to a stationary point is not possible. It also prevents the corresponding variables from reducing the objective function value by being increased. For example, out of the 14551 zeros in H, 81 have a negative gradient (and this is only after 100 iterations), the minimal entry of the gradient corresponding to a zero entry being -0.058 (which is far from machine epsilon). This behavior is observed for most sufficiently sparse input matrix; see [197, Figure 9] for an example on a real document data set. For dense input matrices X (for example $\mathtt{rand(m,n)}$), this behavior is typically less damaging as W and H are denser.

Another way around the zero locking phenomenon   Lin [322] proposed another way to avoid entries of W and H being stuck at zero while the corresponding entries of the gradient are negative, which prevents convergence to stationary points (see the discussion above). The idea was later slightly improved by Chi and Kolda [86]; let us describe it. For the entries of W and H equal to zero, one can monitor whether the corresponding entries in the gradient are negative (or, equivalently, monitor whether the multiplicative factors in the MU are larger than one). If this is the case (such zeros are referred to as inadmissible), then the corresponding entries are reinitialized at a small positive value (such as the machine epsilon). As for the lower bound used in the previous paragraph, this trick leads to better numerical performances, especially for sparse matrices; see [86, Figure 6.1] for an experiment on a randomly generated data set where the behavior is similar to that in Figure 8.2. However, as far as we know, this scheme has not been proved to converge to stationary points. Note that Lin [322] uses a lower bound for all the entries of W and H whose corresponding entries of the gradient is negative. Moreover, Lin does not treat this case in his convergence result (Lin considers W and H to remain positive throughout the iterations, as done for Theorem 8.9 in the case $\epsilon = 0$).

8.2.6 ▪ Computational cost

The MU are first-order methods: to apply the MU at a given iterate (W, H), one only needs to compute the gradient as the MU require to take the ratio of the two terms within the gradient; see (8.4). More precisely, for the update of H, we need to compute

$$W^\top (WH)^{\circ(\beta-1)} \quad \text{and} \quad W^\top \left((WH)^{\circ(\beta-2)} \circ X \right).$$

The first term $W^\top (WH)^{\circ(\beta-1)}$ requires $\mathcal{O}(mnr)$ operations: compute WH, and then multiply $(WH)^{\circ(\beta-1)}$ by $W^\top$. However, there are two notable exceptions:

- For $\beta = 1$ (KL divergence), $(WH)^{\circ 0} = 1_{m \times n}$ is the all-one matrix and the cost reduces to $\mathcal{O}(mr)$ operations, namely to compute

$$W^\top 1_{m \times n} = \left(W^\top 1_{m \times 1} \right) 1_{1 \times n}.$$

- For $\beta = 2$ (Frobenius norm), $W^\top (WH) = (W^\top W)H$ which can be computed in $\mathcal{O}((m+n)r^2)$ operations, by computing first $W^\top W$. One should not compute WH, which requires $\mathcal{O}(mnr)$ operations.

The second term $W^\top \left((WH)^{\circ(\beta-2)} \circ X \right)$ also requires $\mathcal{O}(mnr)$ operations as it also requires the computation of WH (which can be computed only once). For $\beta = 2$, this term reduces to $W^\top X$, which requires $\mathcal{O}(mnr)$ operations.

Therefore, in total, the MU require $\mathcal{O}(mnr)$ operations for dense input matrices, regardless of the β-divergence used, although the cost can be reduced for $\beta = 1, 2$ as the first and/or the second terms can be computed with fewer operations.

For sparse input matrices, let us denote $\mathrm{nnz}(X)$ the number of nonzero entries of X. The second term does not require $\mathcal{O}(mnr)$ operations but only $\mathcal{O}(r\,\mathrm{nnz}(X))$ operations since we only need to compute the entries of WH corresponding to nonzero entries of X, while premultiplying by $W^\top$ a matrix with $\mathrm{nnz}(X)$ nonzero entries requires $\mathcal{O}(r\,\mathrm{nnz}(X))$ operations. For $\beta \in \{1, 2\}$, the first term can be computed in fewer than $\mathcal{O}(r\,\mathrm{nnz}(X))$ operations, assuming $\mathrm{nnz}(X) \geq (m + n)r$, which is reasonable as it assumes that (W, H) has fewer entries than $\mathrm{nnz}(X)$ (otherwise no compression is achieved). Therefore, for sparse matrices, using $\beta \in \{1, 2\}$ leads to computationally much cheaper MU, namely from $\mathcal{O}(mnr)$ to $\mathcal{O}(r\,\mathrm{nnz}(X))$ operations, as mn can be much larger than $\mathrm{nnz}(X)$. For example, for document data sets, it is common to have more than 99% of the entries equal to zero in a word-count matrix, in which case the MU for $\beta \in \{1, 2\}$ are at least 100 times faster than for the other β's.

8.2.7 ▪ Flexibility of the MU

The MU are very flexible and can be adapted in most situations, in particular if using the gradient ratio heuristic (Section 8.2.1) to design them, which is straightforward (although it does not necessarily guarantee monotonicity). The MU are used for almost all NMF models: sparse NMF [96, 302], ONMF [132, 495], projective NMF [490], convolutive NMF [150], graph-regularized NMF [150], and min-vol β-NMF [310], to cite a few. We refer the reader to [98] for more examples and other error measures such as α-divergences.

8.3 ▪ Algorithms for the Frobenius norm

In this section, we focus on the β-divergence for $\beta = 2$, that is,

$$D_2(X, WH) \;=\; \frac{1}{2}\|X - WH\|_F^2 \;=\; \frac{1}{2}\sum_{i,j}(X - WH)_{i,j}^2,$$

as the error measure for NMF. This is arguably the most widely used norm for NMF, because it corresponds to Gaussian noise which is reasonable in many situations while allowing the design of particularly efficient schemes. However, it is important to keep in mind that, for nonnegative data, Gaussian noise does not necessarily make sense; this is especially true if the input data is sparse; see the discussion in Section 5.1. However, the Frobenius norm offers avenues for computationally much more efficient algorithms than the MU. We present in this section the most standard and widely used algorithms, namely alternating nonnegative least squares (ANLS), which is an exact 2-BCD method (Section 8.3.1), the alternating least squares (ALS) heuristic (Section 8.3.2), hierarchical alternating least squares (HALS), which is a BCD method (Section 8.3.3), a fast projected gradient method (FPGM) (Section 8.3.4), and the ADMM (Section 8.3.5).

Implementation   All algorithms presented in this section can be run using [Matlab file: FroNMF.m]. Since all algorithms follow the 2-BCD framework of Algorithm 8.1, they differ in the way the subproblems in W and H are solved. These subproblems are convex NNLS problems with multiple right-hand sides. For example, when W is fixed, the NNLS subproblem in variable H is

$$\min_{H \geq 0} \|X - WH\|_F^2.$$

The MATLAB code for NNLS using the different algorithms discussed in this section can be run using [Matlab file: NNLS.m], where options.algo allows one to choose the desired algorithm.

8.3.1 ▪ Alternating nonnegative least squares

Following the framework of 2-BCD (Algorithm 8.1), a first possibility is to solve the subproblems in W and H up to global optimality. This is an exact 2-BCD approach for NMF and was first proposed by Paatero and Tapper [371] (1994). By Theorem 8.2, this approach is guaranteed to converge to stationary points. This 2-BCD approach for NMF is ANLS.

In practice, an efficient method to solve NNLS problems with high accuracy is active-set methods. For simplicity, let us consider the case where X and H have a single column, denoted x and h, respectively:

$$\min_{h \geq 0} f(h), \quad \text{where } f(h) = \frac{1}{2}\|x - Wh\|_F^2. \tag{8.18}$$

The KKT conditions are given by

$$h \geq 0, \ \nabla_h f(h) = W^\top(Wh - x) \geq 0 \ \text{and} \ h^\top \nabla_h f(h) = 0;$$

see Section 8.1.2. Since (8.18) is convex and there exists a point in the relative interior of the feasible domain (a Slater point), namely any $h > 0$, the KKT conditions are necessary and sufficient for global optimality. Let us denote h^* an optimal solution of the NNLS. Assume we are given the set

$$\mathcal{I} = \{i \mid h_i^* > 0\}.$$

The complement of $\mathcal{I}$ is the so-called active set, where the corresponding constraints are active, that is, the active set contains the indices i such that $h_i^* = 0$. The nonzero entries of h^* can be computed by solving a linear system

$$[\nabla_h f(h)]_\mathcal{I} = 0 \iff \left[W^\top(Wh - x)\right]_\mathcal{I} = 0$$
$$\iff W(:,\mathcal{I})^\top W(:,\mathcal{I})h(\mathcal{I}) = W(:,\mathcal{I})^\top x.$$

These are the normal equations of the unconstrained least squares problem for $h(\mathcal{I})$, that is,

$$\min_{h(\mathcal{I})} \|W(:,\mathcal{I})h(\mathcal{I}) - x\|_2.$$

Akin to the simplex method for linear optimization, active-set methods iteratively update the active set via pivoting (that is, entering and removing variables from the active set) that guarantees the objective function to decrease; see [300] for more details. The MATLAB function lsqnonneg implements this method. More aggressive strategies are possible by entering and leaving more than one variable at a time. Moreover, in the context of NMF, one should use the structure, namely, the multiple right-hand sides, which allows solving the linear systems more efficiently, reusing some computations, such as the products $W(:,\mathcal{I})^\top W(:,\mathcal{I})$ and their inverses. Kim, Kim, and Park[62] have developed such dedicated active-set methods designed specifically for NMF [277, 278, 280].

The computational cost of the active-set method for NNLS resides in two main steps: compute $W^\top X$ which requires $\mathcal{O}(\text{nnz}(X)r)$ operations, and solve the n NNLS problems in r variables which requires $\mathcal{O}(nr^3 s(r))$ operations, where $s(r) \leq 2^r$ is the number of active sets explored. In practice, $s(r)$ is typically much smaller than 2^r (like in the simplex method for linear

---

[62]https://www.cc.gatech.edu/~hpark/nmfsoftware.html.

optimization where the worst-case complexity is exponential, while the average case complexity is polynomial [430]). Moreover, the coefficient n in front of $r^3 s(r)$ is pessimistic because it assumes the NNLS problems in the column of H are solved independently, not taking advantage of the fact that they share the matrix W. For example, at the first step, the active sets are typically initialized as the empty sets and the first ANLS iteration, if properly implemented as in [280], requires solving unconstrained least squares which require computing the inverse $W^\top W$ in $\mathcal{O}(r^3)$ operations, which is then multiplied by $W^\top X$ in $\mathcal{O}(mr^2)$ operations.

Finally, accounting for the update of W, ANLS requires, in the worst case, a total of $\mathcal{O}(mnr + (m+n)r^3 s(r))$ operations per iteration. For r rather small (more precisely, for $r^2 s(r) \le \frac{mn}{m+n} \le \min(m,n)$), NNLS requires $\mathcal{O}(\mathrm{nnz}(X)r)$ operations, like the MU, and should be preferred as it solves the subproblems exactly. However, as r gets larger, NNLS tends to be rather slow compared to more sophisticated iterative methods that do not attempt to solve the NNLS subproblems exactly (see Section 8.4.3).

Another option for solving NNLS problems with high accuracy would be to use second-order methods such as interior-point methods. However, these are typically more expensive in practice and have not been used much in the context of NMF. Moreover, interior-point methods cannot be easily warm started (this is a well-known drawback), as opposed to active-set methods. This is an important feature within the ANLS framework.

Flexibility of ANLS   ANLS based on active sets can be directly extended to any regularized NMF model where the regularizer is quadratic in W and H, that is, as long as the subproblems in W and H are NNLS problems. This is, for example, the case of sparse NMF using an ℓ_1 penalty since for a nonnegative matrix H, $\|H\|_1 = \sum_{k,j} H(k,j)$ is linear [277]. However, as soon as this structure is lost, and as soon as additional constraints are added, it is more difficult to solve the subproblems in W or H exactly using active-set methods. (Note that interior-point methods are more flexible in that respect.)

ANLS for ONMF   An interesting exception is ONMF, that is,

$$\min_{W \ge 0, H \ge 0} \|X - WH\|_F \text{ such that } HH^\top = I_r.$$

As shown in Lemma 5.2, for H fixed and $HH^\top = I_r$, the optimal solution for W has a closed form given by $XH^\top$. For W fixed, the problem in H amounts to assigning each data point to its closest centroid (the columns of W) where the distance is measured in terms of angles. Recall that the constraint $HH^\top = I_r$ together with nonnegativity implies that H has a single positive entry per column, and ONMF is similar to a spherical k-means problem; see Section 5.5.3. Using this ANLS approach was described in [384] and referred to as EM-ONMF; see [Matlab file: alternatingONMF.m].

8.3.2 ▪ Alternating least squares heuristic

Because the NNLS subproblem in ANLS is nontrivial to solve, some researchers have resorted to a rather radical idea: solve the least squares problems without the nonnegativity constraints and project the solution onto the nonnegative orthant. In other words, the NNLS solution is approximated by using the update

$$\max\left(0, \mathrm{argmin}_Y \|X - WY\|_F^2\right)$$

for H, and similarly for W. In MATLAB, this is written very easily as max(0,W\X). This is the ALS method.

If one uses the normal equations[63] $(W^\top W)H = W^\top X$ to solve the least squares problem in H, this requires $\mathcal{O}(r^3 + r^2 m)$ operations, which is negligible for $r \ll \min(m, n)$. In total, ALS requires $\mathcal{O}\left(r\,\text{nnz}(X) + (m+n)r^2\right)$ operations per iteration, to compute $W^\top X$ and $XH^\top$ and solve the m and n least squares problem with the same Hessian matrix. Hence, in total, ALS has the same cost as a first-order method such as the MU.

ALS is a easy to implement (given that a least squares solver is available) and sometimes provides reasonable solutions, typically for sparse matrices; see for example Figure 8.4 in Section 8.4.3. As for ANLS, it can be easily adapted to regularized NMF models where the regularizer is quadratic in W and H. However, it comes with no theoretical guarantee and in fact often diverges in practice, especially for dense input matrices; see Figure 8.3 in Section 8.4.3. Therefore, ALS can be recommended only as a warm-start stage for theoretically better grounded approaches [98].

Example 8.10 (Beware that MATLAB default NMF algorithm is ALS). As surprising as it may be, ALS is currently the default NMF algorithm in MATLAB (function nnmf, version R2019b). The other algorithm available is the MU for the Frobenius norm. Here is a numerical example in MATLAB showing why you should not use ALS (see Section 8.4.3 for more numerical experiments):

```
» rng(2020); X = rand(100); r = 50;
» [W,H] = nnmf(X,r); norm(X - W*H,'fro')/norm(X,'fro')*100
  ans = 70.7262
» [W,H] = FroNMF(X,r); norm(X - W*H,'fro')/norm(X,'fro')*100
  ans = 23.8642
```

FroNMF is our provided implementation with the default settings (which uses the extrapolated A-HALS (E-A-HALS) algorithm described in Section 8.4.3). On a randomly generated matrix, ALS provides a solution with relative error of 70.73%, while our default algorithm provides 23.86%. This is a reminder that one should be careful when using software blindly. ∎

8.3.3 ▪ Hierarchical alternating least squares

HALS solves the NNLS subproblem using an exact BCD method, where the blocks of variables are the rows of H. Two observations are key to the design of HALS:

1. The variables on a single row of H are independent in the NNLS problem: only entries in the same column interact in the objective function since

$$\|X - WH\|_F^2 \ = \ \sum_{i=1}^{n} \|X(:,j) - WH(:,j)\|_F^2;$$

see Section 8.1.1. Because of that, HALS can also be interpreted as an exact coordinate descent method which updates a single variable while keeping the others fixed at each iteration.

2. Solving a univariate least squares problem over the nonnegative orthant has a closed-form solution. More precisely, given $x, w \in \mathbb{R}_+^m$,

$$\text{argmin}_{h \in \mathbb{R}_+} \|x - wh\|_2^2 = \max\left(0, \frac{w^\top x}{\|w\|_2^2}\right).$$

---

[63]Using the normal equations leads to a lower numerical accuracy as it squares the condition number [456]. However, this is not crucial in our settings because (1) projecting the least squares solution is a very crude approximation of the NNLS problem, and (2) in practice, because of the presence of noise, NMF problems typically do not need to be solved to high accuracy.

In fact, $\|x - wh\|_2^2 = \|x\|_2^2 - 2hw^\top x + h^2\|w\|_2^2$ is a quadratic function of h and has a unique minimizer $\frac{w^\top x}{\|w\|_2^2}$, given that $w \neq 0$; otherwise any h is optimal. If this unconstrained minimum is negative, the minimum over the nonnegative orthant is attained at zero because of the convexity of the objective function.

Let us now put this in practice and focus on the ℓth row of H while all other variables are kept fixed. We want to solve

$$\min_{H(\ell,:) \geq 0} \left\| \left(X - \sum_{k \neq \ell} W(:,k)H(k,:)\right) - W(:,\ell)H(\ell,:) \right\|_F^2. \tag{8.19}$$

Denoting

$$R_\ell = X - \sum_{k \neq \ell} W(:,k)H(k,:) = X - WH + W(:,\ell)H(\ell,:)$$

the residual with respect to the ℓth rank-one factor $W(:,\ell)H(\ell,:)$, (8.19) can be written as

$$\min_{H(\ell,:) \geq 0} \|R_\ell - W(:,\ell)H(\ell,:)\|_F^2.$$

We have

$$\|R_\ell - W(:,\ell)H(\ell,:)\|_F^2 = \sum_{j=1}^n \|R_\ell(:,j) - W(:,\ell)H(\ell,j)\|_F^2$$

which is an n independent univariate least squares problem over the nonnegative orthant, with the closed-form solution

$$\operatorname{argmin}_{H(\ell,j) \geq 0} \|X - WH\|_F^2 = \max\left(0, \frac{W(:,\ell)^\top R_\ell(:,j)}{\|W(:,\ell)\|_2^2}\right) \quad \text{for all } \ell, j,$$

which follows from the second observation above. In vector form, this gives

$$\operatorname{argmin}_{H(\ell,:) \geq 0} \|X - WH\|_F^2 = \max\left(0, \frac{W(:,\ell)^\top R_\ell}{\|W(:,\ell)\|_2^2}\right) \quad \text{for all } \ell.$$

HALS cyclically updates each row of H as above, and similarly for the columns of W; see Algorithm 8.2.

The computational cost of HALS is the same as the MU, namely, $\mathcal{O}(mnr)$ operations in the dense case, and $\mathcal{O}(r \, \mathrm{nnz}(X))$ in the sparse case; the main computational cost is to compute $W^\top X$, $XH^\top$, $(W^\top W)H$, and $W(HH^\top)$. Note that the residuals R_ℓ are not computed explicitly in Algorithm 8.2 (they were just used to simplify the presentation); otherwise, for sparse input matrices, the algorithm would run in $\mathcal{O}(mnr)$ operations.

HALS was observed to converge significantly faster than the MU while having essentially the same computational cost; see Section 8.4.3 for some numerical examples, and see [98, 197, 279] and the references therein. This can be explained theoretically as follows: the MU can be interpreted as updating the rows of H and columns of W independently (since the majorizers are separable) but do not use an optimal solution of these subproblems, as opposed to HALS [185, p. 131]. Another reason why HALS works well is because W and H are expected to be sparse. Therefore, the number of nonzero variables in the NNLS subproblems in (8.3) is typically smaller than r.

In HALS (Algorithm 8.2), each row of H is updated only once before updating W. However, HALS can be accelerated significantly by updating H several times before updating W [197], as it allows us to reuse the computations of $W^\top W$ and $W^\top X$; see Section 8.4, where this trick is

---

Algorithm 8.2 Hierarchical alternating least squares (HALS)

Input: Input nonnegative matrix $X \in \mathbb{R}_+^{m \times n}$ and factorization rank r.
Output: $(W, H) \geq 0$: A rank-r NMF of $X \approx WH$.

1: Generate some initial matrices $W \geq 0$ and $H \geq 0$; see Chapter 8.6.
2: **for** $t = 1, 2, \ldots$ **do**
3:     % Update H
4:     **for** $\ell = 1, 2, \ldots, r$ **do**
5:         $H(\ell, :) \leftarrow \max\left(0, \dfrac{W(:,\ell)^\top X - \sum_{k \neq \ell}\left(W(:,\ell)^\top W(:,k)\right)H(k,:)}{\|W(:,\ell)\|_2^2}\right)$
6:     **end for**
7:     % Update W
8:     **for** $\ell = 1, 2, \ldots, r$ **do**
9:         $W(:, \ell) \leftarrow \max\left(0, \dfrac{XH(\ell,:)^\top - \sum_{k \neq \ell} W(:,\ell)\left(H(\ell,:)H(\ell,:)^\top\right)}{\|H(\ell,:)\|_2^2}\right)$
10:    **end for**
11: **end for**

---

described and can be applied to any first-order methods such as the MU. It can also be accelerated by selecting the entries of H that are updated by HALS that will decrease the objective function the most; this is referred to as a Gauss–Southwell-type rule [245]. The rationale is that many entries of W and H remain zero in the course of HALS, and it is more efficient to avoid updating these entries.

Remark 8.1 (HALS is a block projected gradient descent method). *Interestingly, HALS can be interpreted as a block projected gradient descent method. The problem*

$$\min_{H(\ell,:)\geq 0} \|R_\ell - W(:,\ell)H(\ell,:)\|_F^2$$

is a convex optimization problem with Lipschitz continuous gradient with constant $L = \|W(:,\ell)\|_2^2$ (see Example 8.1). Performing a gradient step with step size $1/L$, we obtain

$$H(\ell,:) - \frac{1}{L}\left(W(:,\ell)^\top \left(W(:,\ell)H(\ell,:) - R_\ell\right)\right) = \frac{W(:,\ell)^\top R_\ell}{\|W(:,\ell)\|_2^2},$$

which is the HALS update, before the projection.

The aficionados of numerical linear algebra will also recognize that the update of HALS is closely related to the power method, applied on the matrix R_ℓ with a projection onto the nonnegative orthant.

8.3.3.1 ▪ Convergence

HALS is an exact BCD method with $2r$ blocks of variables updated cyclically. To guarantee convergence, Theorem 8.3 can be invoked and relies on four conditions. The first two conditions are satisfied, as for ANLS. The third and fourth conditions (the objective is monotone between two iterates, the subproblems admit a unique solution) are satisfied as long as $H(\ell, :)$ and $W(:, \ell)$ are not set to zero in the course of the algorithm (in which case the HALS updates are not well-defined, with a division by zero). At this point, it is important to point out that HALS might be rather sensitive to initialization: for example, if $WH \gg X$, then the first update of $H(1, :)$ will most likely set it to zero, running into numerical problems. To avoid this issue, one should properly scale the initial matrices (W, H) such that

$$\text{argmin}_\alpha \|X - \alpha W H\|_F = \frac{\langle X, WH \rangle}{\langle WH, WH \rangle} = \frac{\langle W^\top X, H \rangle}{\langle W^\top W, HH^\top \rangle} = 1;$$

see the discussions in, for example, [239, 185]. However, this does not guarantee that no rank-one factor is set to zero in the course of HALS.

If a rank-one factor is set to zero, there are two options: it can be discarded, or it can be reinitialized; see for example the strategy described in Section 8.1.2.1.

In order to guarantee the convergence to stationary points without the requirement that HALS does not set any factor to zero, here are two options:

- One can resort to BSUM and in particular the trick described in Example 8.5, that is, solve the subproblems

$$\min_{H(\ell,:) \geq 0} \left\| R_\ell - W(:,\ell) H(\ell,:) \right\|_F^2 + \mu \left\| H(\ell,:) - \tilde{H}(\ell,:) \right\|_2^2$$

for some $\mu > 0$ and where $\tilde{H}(\ell,:)$ is the value of the current iterate. Because of the regularizer, these subproblems are strictly convex, regardless of the fact that $W(:,\ell) \neq 0$, and have the closed-form solution

$$H(\ell,:) \leftarrow \max \left(0, \frac{W(:,\ell)^\top R_\ell + \mu \tilde{H}(\ell,:)}{\|W(:,\ell)\|_2^2 + \mu} \right).$$

However, this does not really resolve our issue because this modified HALS may also set a factor to zero and could be stuck there (if $W(:,\ell) = 0$, then $H(\ell,:)$ remains unchanged, that is, it remains equal to $\tilde{H}(\ell,:)$). From a theoretical point of view, this is fine since this corresponds to a trivial saddle point (see Section 8.1.2.1). However, this is not very satisfactory from a practical point of view and is equivalent to discarding that rank-one factor. Hence we do not recommend this strategy.

- As for the MU, one can use a lower bound ϵ (such as the machine epsilon) for the entries of W and H. This guarantees that HALS converges to a stationary point by Theorem 8.3; see also [185, Theorem 4.3]. Moreover, this allows us to avoid trivial saddle points as, by construction, these are absent from the feasible set since $W \geq \epsilon$ and $H \geq \epsilon$.

An interesting observation is that if your current rank-r iterate has an error smaller than the best possible rank-$(r-1)$ solution, then HALS cannot set any factor to zero as it monotonically reduces the objective function. Therefore, as soon as you have an iterate that satisfies this condition, you are guaranteed that HALS converges to a nontrivial stationary point (more precisely, every limit point is a stationary point with nonnegative rank r). Of course, it is not possible to check that condition since NMF is NP-hard; however, this could be guaranteed by checking that your current rank-r solution has an error smaller than the rank-$(r-1)$ truncated SVD which is a lower bound for the rank-$(r-1)$ NMF.

8.3.3.2 ▪ History

HALS has been rediscovered several times, originally in [97] (2007) (see also [95]), then as the rank-one residue iteration in [239] (2008), as FastNMF in [313] (2009), and also in [324] (2012). HALS was actually first described in Rasmus Bro's thesis [63, pp. 161–170] in 1998, although it was not investigated thoroughly (ANLS was preferred):

to solve for a row of H it is only necessary to solve the unconstrained problem and subsequently set negative values to zero. Though the algorithm for imposing non-negativity is thus simple and may be advantageous in some situations, it is not pursued here. Since it optimizes a smaller subset of parameters than the other approaches it may be unstable in difficult situations.

8.3.3.3 ▪ Flexibility of HALS

HALS can be generalized to other NMF models as long as the entries in the columns of W and rows of H do not interact in the objective function, and as long as the subproblems are solvable in closed form. For example, for sparse NMF using an ℓ_1 penalty (see Section 5.3), the suproblems are

$$\min_{H(\ell,:)\geq 0} \frac{1}{2}\|R_\ell - W(:,\ell)H(\ell,:)\|_F^2 + \lambda_\ell\|H(\ell,:)\|_1$$

for some parameter $\lambda_\ell > 0$. Since $H \geq 0$, $\|H(\ell,:)\|_1 = H(\ell,:)e$ is simply a linear term, and hence the HALS update can be directly modified for sparse NMF as

$$H(\ell,:) \;\leftarrow\; \max\left(0, \frac{W(:,\ell)^\top R_\ell - \lambda_\ell}{\|W(:,\ell)\|_2^2}\right),$$

which is reminiscent of the soft-thresholding operator used to solve ℓ_1 penalized least squares problem. The parameters λ_ℓ can be tuned to achieve a desired level of sparsity [186] and to avoid setting rows of H to zero.

HALS can be easily adapted to many other NMF models, including weighted NMF [239] (see Section 9.5, where the update is given), symNMF via a penalty approach [290, 30], ONMF [315], and tri-NMF [106].

8.3.4 ▪ Fast projected gradient method

Among the first methods proposed for NMF, after ANLS and the MU, was the projected gradient method (PGM) by Lin [323]. Let us consider again the NNLS subproblem for one column of X and H:

$$\min_{h\geq 0} \frac{1}{2}\|x - Wh\|_F^2.$$

It is a quadratic optimization problem with nonnegativity constraints, whose Hessian is $W^\top W$. Hence the gradient is Lipschitz continuous with constant $L = \lambda_1(W^\top W) = \sigma_1(W)^2 = \|W\|_2^2$. As explained in Example 8.1, the gradient update is given by

$$h \;\leftarrow\; \max\left(0, h - \frac{1}{L}W^\top(Wh - x)\right),$$

which guarantees the decrease of the objective function. In matrix form, this gives

$$H \;\leftarrow\; \max\left(0, H - \frac{1}{L}\left((W^\top W)H - W^\top X\right)\right). \tag{8.20}$$

This update has the same computational cost as the MU and HALS, the main cost being to compute $(W^\top W)H$ and $W^\top X$. PGM tends to perform better than the MU but worse than HALS; see [197].

Fast PGM If H is updated several times before W, the above scheme can be accelerated by using an FPGM. Such methods introduce an additional sequence of iterates in order to add momentum: Letting $Y^{(0)} = H^{(0)}$, for $t = 1, 2, \ldots$, FPGM uses the following updates:

$$H^{(t)} = \max\left(0, Y^{(t-1)} - \frac{1}{L}\left((W^\top W)Y^{(t-1)} - W^\top X\right)\right),$$

$$Y^{(t)} = H^{(t)} + \beta_t\left(H^{(t)} - H^{(t-1)}\right), \tag{8.21}$$

where $\beta_t \in (0, 1]$ are properly chosen parameters. The update of $Y^{(t)}$ is also referred to as an extrapolation step. This scheme allows us to reduce the objective function from the rate $\mathcal{O}(1/t)$ for PGM to $\mathcal{O}(1/t^2)$ for FPGM when $\mathrm{rank}(W) < r$, and from linear convergence with rate $(1 - \mu/L)$ for PGM to rate $(1 - \sqrt{\mu/L})$ for FPGM where $\mu = \sigma_r(W)^2 > 0$ (strongly convex case) [364]. Note, however, that FPGM is not monotonically decreasing the objective function. This was used in [225] to design the so-called Nesterov-NMF algorithm, which we prefer to refer to as FPGM.

FPGM performs better than HALS because it updates H several times before updating W and hence can reuse the computation of $(W^\top W)H$ and $W^\top X$. FPGM typically performs slightly worse than accelerated HALS (A-HALS; where H is updated several times before the update of W, and vice versa); see Section 8.4.3 for some numerical experiments. However, FPGM is very flexible: to extend it to other NMF models, it is only required that the subproblems in W and H are smooth convex optimization problems. Let us describe an example on min-vol NMF.

Example 8.11 (Min-vol NMF via MM and FPGM). As we have seen in Chapter 4.3.3, min-vol NMF is a key NMF model as it leads to identifiability under the rather mild condition that H is sufficiently scattered. Let us describe at a high level an algorithm combining MM and FGM for min-vol NMF; see [172, 309] for more details.

Let us consider the model

$$\min_{W \geq 0, H \geq 0, H^\top e \leq e} \|X - WH\|_F^2 + \lambda\nu(W),$$

where

$$\nu(W) = \mathrm{logdet}\left(W^\top W + \delta I\right).$$

The problem in H is the same as in NMF, except that the entries in each column have to sum to at most one. Hence, in FPGM, we only need to replace the projection onto the nonnegative orthant, namely, $\max(0, .)$, with the projection of the columns of H onto the set $\mathcal{S}^r$ which can be performed efficiently; see for example [188, Appendix A].

The problem in W is nonconvex, because of the term $\nu(W)$. Let us derive a smooth convex majorizer for $\nu(W)$, so that we can then apply PFGM on the majorizer—we keep the quadratic term $\|X - WH\|_F^2$ unchanged. The function $\mathrm{logdet}(Q)$ is concave in $Q \succ 0$ so that it can be majorized locally by its first-order Taylor expansion, as was done for $\hat{d}_\beta(z, y)$ to derive the MU in Section 8.2.3. Denoting $\tilde{Q} \succ 0$ the point around which we perform the expansion, we have for all $Q \succ 0$ that

$$\mathrm{logdet}(Q) \leq \mathrm{logdet}(\tilde{Q}) + \mathrm{tr}\left(\tilde{Q}^{-1}(Q - \tilde{Q})\right)$$

$$= \mathrm{tr}\left(\tilde{Q}^{-1}Q\right) + \mathrm{logdet}(\tilde{Q}) - r.$$

Replacing Q by $W^\top W + \delta I$, and denoting $\tilde{Q} = (\tilde{W}^\top \tilde{W} + \delta I) \succ 0$, leads to the following majorizer for $\nu(W)$ at $\tilde{W}$: for all W,

$$\nu(W) \leq g(W, \tilde{W}) = \mathrm{tr}\left(\tilde{Q}^{-1}(W^\top W)\right) + c = \left\langle \tilde{Q}^{-1}, W^\top W \right\rangle + c,$$

where c is a constant independent of W. The majorizer $g(W, \tilde{W})$ is strongly convex because $\tilde{Q}^{-1} \succ 0$. Its gradient is given by $2W\tilde{Q}^{-1}$, and FPGM can easily be applied on the majorizer, that is, on

$$\min_{W \geq 0} \|X - WH\|_F^2 + \lambda g(W, \tilde{W}).$$

Interestingly, as in NMF, this objective function can be written as m independent problem in the columns of W, so that HALS could also be used. ∎

Scaling of the rank-one factors of WH As we have already discussed several times, the objective function of NMF solutions is invariant to the scaling of the rank-one factors, that is, replacing $W(:, k)$ by $\alpha_k W(:, k)$ and $H(k, :)$ by $H(k, :)/\alpha_k$ for any scalars $\alpha_k > 0$, $k = 1, 2, \ldots, r$. However, it is key to notice that some NMF algorithms may be very sensitive to this scaling. In particular, this is the case for PGM and FPGM as their theoretical convergence depends on the conditioning of $W^\top W$ when updating H (and of $HH^\top$ when updating W). Hence very badly scaled columns of W and rows of H might slow down convergence significantly, for example when some columns of W have much larger norms than others. Let us illustrate this on a toy example. Let

$$X = W = \begin{pmatrix} 1 & 0 \\ 0 & 1 \end{pmatrix} \text{ and } H = \begin{pmatrix} 0.1 & 0 \\ 0 & 0.1 \end{pmatrix}.$$

Then, one can check that the PGM update (8.20) (which coincides with the first FPGM update) gives $H \leftarrow I_2$ after one iteration, leading to a perfect (trivial) NMF decomposition of X with error zero. However, take now

$$W = \begin{pmatrix} 1 & 0 \\ 0 & 0.1 \end{pmatrix} \text{ and } H = \begin{pmatrix} 0.1 & 0 \\ 0 & 1 \end{pmatrix}.$$

This gives to the same product $WH = 0.1I_2$ as the initial factors above (we use the scalings $\alpha_1 = 1$ and $\alpha_2 = 0.1$). One can check that the PGM update (8.20) of H becomes

$$H \leftarrow \begin{pmatrix} 1 & 0 \\ 0 & 1.09 \end{pmatrix} \quad \text{so that} \quad WH = \begin{pmatrix} 1 & 0 \\ 0 & 0.109 \end{pmatrix}$$

with error $\|X - WH\|_F = 0.891 \gg 0$.

There are many ways to scale W and H. A good choice for PGM and FPGM is, for example, to scale (W, H) such that $\|W(:, k)\|_2 = 1$ for all k before the update of H, and such that $\|H(k, :)\|_2 = 1$ for all k before the update of W. In our MATLAB code [Matlab file: FroNMF.m], we scale (W, H) so that $\|W(:, k)\|_2 = \|H(k, :)\|_2$ for all k. The reason for this choice is that we will also add momentum between NMF updates; hence we cannot scale (W, H) between every update of W and H; see Section 8.4.2.

The other methods presented so far, namely ALS, ANLS, MU, and HALS, are not sensitive to the scaling of the rank-one factors. Of course, if W and/or H are highly ill-conditioned, these methods could run into numerical problems (for example, ALS and ANLS need to solve least squares problems that depend on this conditioning); hence using a scaling is also recommended.

8.3.5 ▪ Alternating direction method of multipliers

The NNLS problem

$$\min_{h \in \mathbb{R}_+^r} \frac{1}{2} \|Wh - x\|_2^2$$

can be reformulated as

$$\min_{h\in\mathbb{R}^r,y\in\mathbb{R}^r_+} \frac{1}{2}\|Wh - x\|_2^2 \quad \text{such that} \quad y = h,$$

where $y \in \mathbb{R}^r_+$ is an auxiliary variable. This reformulation allows us to decouple the least squares term and the nonnegativity constraint in the NNLS problem. The augmented Lagrangian is given by

$$\mathcal{L}(h, y, z) = \frac{1}{2}\|Wh - x\|_2^2 + z^\top (h - y) + \frac{\rho}{2}\|h - y\|_2^2,$$

where $z \in \mathbb{R}^r$ contains the Lagrange multipliers of the constraint $y = h$. ADMM uses such reformulations to decouple difficult constraints and/or terms in the objective, and then solves the reformulation by alternatively minimizing the augmented Lagrangian with respect to its variables. For NNLS,

- the problem in h is an unconstrained convex quadratic optimization problem whose solution is obtained by solving a linear system, namely,

$$\left(W^\top W + \rho I\right) h = W^\top x - z - \rho y,$$

which follows from making the gradient of the Lagrangian $\mathcal{L}$ with respect to h equal to zero;

- the problem in y is a quadratic problem over the nonnegative orthant but where the Hessian is the identity matrix, and the optimal solution is given by

$$\text{argmin}_{y\geq 0}\, \mathcal{L}(h, y, z) = \max\left(0, h + \frac{1}{\rho}z\right).$$

Finally, ADMM for NNLS consists in the following updates:

$$h \leftarrow \text{argmin}_h\, \mathcal{L}(h, y, z) = \left(W^\top W + \rho I\right)^{-1} \left(W^\top x + \rho y - z\right),$$

$$y \leftarrow \text{argmin}_{y\geq 0}\, \mathcal{L}(h, y, z) = \max\left(0, h + \frac{1}{\rho}z\right),$$

$$z \leftarrow z + \rho(h - y).$$

The update of z is also sometimes modified to $z + \alpha(h - y)$ where α is a step-size parameter. If many ADMM iterations are performed, it is worth computing the inverse of the r-by-r matrix $\left(W^\top W + \rho I\right)$ which is well-conditioned for ρ sufficiently large, instead of solving a linear system at each iteration. This is especially true in the context of NMF since we have to solve n such least squares problems with the same Hessian matrix. Moreover, the term $\left(W^\top W + \rho I\right)^{-1} W^\top x$ can also be precomputed. Computing the inverse of $W^\top W + \rho I$ requires $\mathcal{O}(r^3)$ operations, which is negligible for $r \ll \min(m, n)$. In total, ADMM requires $\mathcal{O}\left(r\, \text{nnz}(X) + (m + n)r^2\right)$, as first-order methods. Typically, ADMM would initialize $y = 0$ and $z = 0$, so h is initially the solution of the unconstrained least squares problem with a Tikhonov regularization. Interestingly, ADMM can be interpreted as an ALS variant where the right-hand sides are adapted iteratively to achieve nonnegativity.

For smooth convex problems, ADMM is guaranteed to converge to an optimal solution, regardless of the values of $\rho > 0$; see for example [28, Chapter 15] for more details.

ADMM is relatively easy to implement for NNLS. The only issue is to properly choose the parameter $\rho > 0$, which is nontrivial. In [250], the authors use $\rho = \text{tr}(W^\top W)/r$, and $\alpha = 1$

in the update of z. Although any positive value guarantees convergence, this choice can have an important impact in practice, especially in the context of NMF since only a few iterations of ADMM are performed on the NNLS subproblems (see Section 8.4). In our experience, NMF algorithms relying on ADMM are very sensitive to this parameter. Moreover, because the iterates of ADMM are not feasible (feasibility of h is attained only at convergence), terminating ADMM early typically results in bad performances within the alternating framework of Algorithm 8.1, for example, with the objective function which is not monotonically decreasing (this is a similar issue as for ALS). Hence ADMM should be used to solve the NNLS subproblems with relatively high precision; for example, in [250], ADMM is stopped when $\|h - y\|_2 \leq \delta \|h\|_2$ for δ sufficiently small.

As for PGM and FPGM, ADMM is sensitive to the scaling of the columns of W and rows of H since the update of h requires solving a linear system with matrix $W^\top W + \rho I$; see the discussion in the previous section.

The 2-BCD strategy (that is, Algorithm 8.1) using ADMM, referred to as AO-ADMM (where AO stands for alternating optimization), is rather flexible and can be generalized to most NMF models and other matrix factorization problems [250].

Remark 8.2 (ADMM directly on NMF). *ADMM could be directly applied on the (nonconvex) NMF problem, using the reformulation*

$$\min_{W,H,U,V} \|X - WH\|_F^2 \quad \text{such that} \quad U \geq 0, V \geq 0, U = W, \text{ and } V = H;$$

see [502, 438, 230, 485]. However, in our experience, this appears to be less effective because it does not allow us to reuse the computations of $W^\top W$ and $W^\top X$ when optimizing over H (and similarly for W), which are the most expensive computations. When applying ADMM on the NNLS subproblems in W and H separately, this is done automatically; see the discussion in Section 8.4 and the numerical experiments in [250]. Note, however, that one could design an ADMM scheme where (H, V) are optimized several times before (W, U) and vice versa.

8.4 ▪ Number of inner iterations and acceleration

Beyond the choice of the optimization method used to solve the NNLS subproblems in W and H, a key tuning aspect in the design of 2-BCD for NMF is the number of inner iterations used to solve these NNLS subproblems. In all the first-order methods, which include MU, HALS, and (F)PGM, the main computational cost resides in computing the gradient. More precisely, to update H, we need to compute $W^\top X$ in $\mathcal{O}(\text{nnz}(X)r)$ operations and $(W^\top W)H$ in $\mathcal{O}((m+n)r^2)$ operations. For ADMM, the same terms also dominate the computational cost. For ANLS, where the NNLS subproblems are solved exactly, there is a priori no need to choose the number of inner iterations. However, when using active-set methods, the number of iterations can potentially be large, and hence it is possible to use an upper bound on the number of active sets explored; this is, for example, a strategy built into the code of Park and coauthors (they use the default value of $5r$).

Key observation As explained above, the first gradient computation when updating H requires $\mathcal{O}(\text{nnz}(X)r + (m+n)r^2)$ operations. This includes the computation of $W^\top X$ in $\mathcal{O}(\text{nnz}(X)r)$ operations, and $W^\top W$ in $\mathcal{O}(mr^2)$. As long as W is not updated, these (small) matrices can be kept in memory, and the next gradient computations for H require only computing $(W^\top W)H$ in $\mathcal{O}(nr^2)$ operations. Therefore,

> The first gradient computation is $\left(1 + \frac{\text{nnz}(X)+mr}{nr}\right)$ times more expensive than the next ones.

Therefore, if $\operatorname{nnz}(X)+mr \gg nr$, the first gradient computation is much more expensive than the next ones, and this should be leveraged. One should, in most cases, not update H only once; this would be a waste of computation. For example, for dense input matrices with $\operatorname{nnz}(X) = mn$, the cost of the first gradient computation is more than $\frac{m}{r}$ times more expensive than the next ones, which is typically much larger than one since $r \ll \min(m,n)$ in most applications.

Of course, it is not necessary to update H too many times because W will be modified at the next iteration; hence a high accuracy solution of the NNLS problem is not really useful. It was observed that the following heuristic works well in practice [197]:

- Perform at most $1 + \alpha \frac{\operatorname{nnz}(X)+mr}{nr}$ updates of H, where $\alpha \in [0.5, 1]$ works well.

- Stop the updates when H is not modified much compared to the first update, more precisely when

$$\|H^{(t)} - H^{(t-1)}\|_2 \le \delta \|H^{(0)} - H^{(1)}\|_2,$$

where $H^{(t)}$ is the tth update of H in the current inner iteration of the NNLS problem, and $\delta = 0.1$ works well in practice.

Remark 8.3 (Number of inner iterations for other β-divergences). *For other β-divergences, the computation of the gradient (see Equation (8.4)) does not allow for much reuse of computations. The case $\beta = 2$ is very special because the term $(WH)^{\beta-2}$ simplifies to the matrix of all ones. Hence, for other β-divergence, updating W and H alternatively once is a good strategy.*

8.4.1 ▪ Low-rank approximations and sketching

For dense matrices, the main cost to update H is to compute $W^\top X$ in $\mathcal{O}(mnr)$ operations. To update W, it is to compute $XH^\top$ on $\mathcal{O}(mnr)$ operations as well. In order to reduce this cost, a popular approach is to compute a low-rank approximation of $X \approx UV$, for example using the truncated SVD of size $p \ll \min(m,n)$ (typically, p is chosen equal to r). Although this initial setup cost might be relatively high, it will reduce the cost of all NMF iterations since computing $(W^\top U)V$ requires only $\mathcal{O}(pr(m+n))$ operations. This is particularly useful if both m and n are large. In order to further reduce the computational cost, and since X is assumed to be of low rank, sketching techniques can be used to compute a low-rank approximation of X at a lower computational cost than the SVD; see [511, 146] and the references therein for more details on such approaches.

In the same spirit, a multilevel approach can be used to compress structured data sets and obtain good solutions much faster. For example, on images, one would apply NMF on compressed versions of the images (for example, lower resolutions images) to obtain good solutions for the original NMF problem [198].

8.4.2 ▪ Adding momentum

It is possible to add momentum between the updates of W and H in the 2-BCD algorithm for NMF (Algorithm 8.1). More precisely, this can be done as follows: Given $Y^{(0)} = H^{(0)}$ and $Z^{(0)} = W^{(0)}$, compute, for $t = 1, 2, \ldots,$

$$W^{(t)} = \operatorname{update}(X, Y^{(t-1)}, Z^{(t-1)}),$$
$$Z^{(t)} = W^{(t)} + \beta_t(W^{(t)} - W^{(t-1)}),$$
$$H^{(t)} = \operatorname{update}\left(X^\top, Z^{(t)\top}, Y^{(t-1)\top}\right)^\top,$$
$$Y^{(t)} = H^{(t)} + \beta_t(H^{(t)} - H^{(t-1)}),$$

where "update" is any iterative or exact method for NNLS, as in Algorithm 8.1. Because NMF is nonconvex, tuning the parameter β_t is nontrivial and various heuristic approaches are possible. Moreover, convergence to stationary points is not yet understood for such schemes. However, the nonincreasingness of the objective can be guaranteed by restarting the method when the objective function increases. Other extrapolation strategies for NMF with strong convergence guarantees were proposed in [486, 238]. However, these methods empirically converge in general more slowly than the simple strategy described above. We refer the interested reader to [11, 10] for more details.

8.4.3 ▪ Numerical comparison of NMF algorithms for the Frobenius norm

Table 8.2 summarizes the NNLS algorithms covered in this chapter.

Table 8.2. *Comparison of NMF algorithms for the Frobenius norm. Flexibility means that the algorithm can be easily adapted to other NMF models, on a scale from ✓ to ✓✓✓ (see the discussion in the section corresponding to each algorithm). Monotonicity means that the algorithm monotonically reduces the objective function. Speed is the empirical convergence speed of the algorithm, on a scale from ✗ to ✓✓✓ ; see Figures 8.3 and 8.4 for some examples. All algorithms run in $\mathcal{O}\left(r\,\mathrm{nnz}(X) + (m+n)r^2\right)$ operations corresponding to computing the gradient, that is, computing the terms $W^T W$, $W^T X$, $X H^T$, and $X H^T$, except ANLS, for which $\mathcal{O}\left((m+n)r^3 s(r)\right)$ operations need to be added.*

Algorithm	Flexibility	Monotonicity	Speed
MU	✓✓✓	✓	✗
A-MU	✓✓✓	✓	✓
ANLS (active-set)	✓	✓	✓
ALS	✓	✗	✗
HALS	✓✓	✓	✓
A-HALS	✓✓	✓	✓✓
PGM	✓✓✓	✓	✓
FPGM	✓✓✓	✗	✓✓
AO-ADMM	✓✓✓	✗	✓✓
E-A-HALS	✓✓	✗	✓✓✓

We now compare various algorithms presented in this chapter:

- The MU of Lee and Seung [303] that update W (resp. H) once between two updates of H (resp. W).

- The accelerated MU (A-MU) that update W (resp. H) several times between two updates of H (resp. W).

- ANLS with active set [280] described in Section 8.3.1.

- ALS described in Section 8.3.2.

- Accelerated HALS (A-HALS) that updates the rows of H several times before updating the columns of W using the HALS update described in Section 8.3.3.

- E-A-HALS that adds momentum to A-HALS; see Section 8.3.3. Note that momentum can be added to any method [10]; however, we only apply it to A-HALS for simplicity.

- FPGM described in Section 8.3.4.

- AO-ADMM described in Section 8.3.2, for which we use the parameters from [250], that is, requiring $\|h - y\|_2 \le \delta\|h\|_2$ with $\delta = 0.01$ to stop the inner iterations. However, with this strategy, we observed in practice that, after sufficiently many iterations, this precision might not be enough to guarantee the objective function to decrease. Hence, when the objective function increases, we divide δ by 10.

For the methods that require choosing a number of inner iterations (namely A-MU, A-HALS, E-A-HALS, and FPGM), we use the same strategy, namely the one described in Section 8.4 with $\alpha = 0.5$ and $\delta = 0.1$. To limit the number of tested algorithms, we do not report the results of HALS, which is outperformed by A-HALS [197], or PGM, which is outperformed by FPGM [225].

The goal of this section is to provide some general observations. Comparing such algorithms and providing strong statements (such as "this algorithm outperforms this other algorithm") is typically not possible because the behavior of these algorithms depends on many parameters including tuning parameters (such as α and δ for the number of inner iterations), the size of the input matrix, the sparsity of the input matrix and of the factors, the factorization rank, the machine used (for example, number of processors), the implementation used, and the software used (for example, matrix multiplication is faster in MATLAB than in C, but using loops is much slower).

8.4.3.1 ▪ Dense data sets

We first compare the algorithms on the most famous NMF data set from the Lee and Seung paper [303], namely the CBCL data set with $m = 361$, $n = 2429$, and $r \in \{10, 49\}$ (Figure 1.2). We use $r = 49$ as in [303], while we use $r = 10$ to illustrate that ANLS performs better when r is small; see the discussion in Section 8.3.1.

Figure 8.3 reports the evolution of the average of

$$\frac{\|X - WH\|_F}{\|X\|_F} - e_{best}, \tag{8.22}$$

where e_{best} is the smallest relative error $\|X-WH\|_F / \|X\|_F$ obtained by any algorithm and any initialization. We run the algorithms for 30 seconds with the same 30 random initializations using the uniform distribution $\mathcal{U}(0, 1)$ to generate all entries of W and H.

Before commenting on the results, it is important to properly interpret these figures based on the measure (8.22). The use of this measure allows us to better visualize the differences between the algorithms.[64] However, one has to be careful about how such figures are interpreted:

---

[64]Plotting the average relative error $\frac{\|X-WH\|_F}{\|X\|_F}$ does not allow us to see the differences as well since the relative errors of most algorithms converge around 0.1% away from one another.

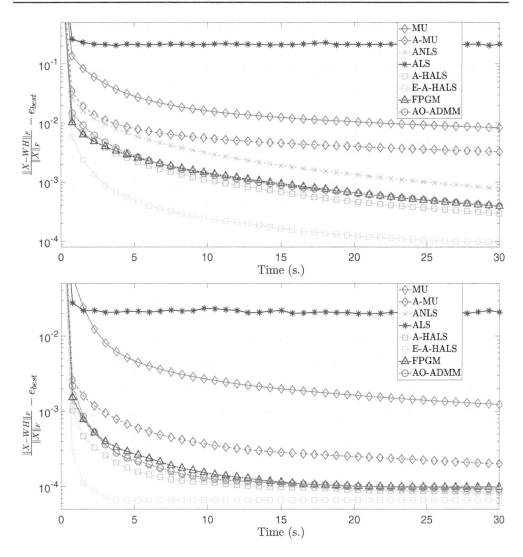

Figure 8.3. *Comparison of NMF algorithms on the CBCL dense data set with* $m = 361$, $n = 2429$, $r = 49$ *(top), and* $r = 10$ *(bottom).* [Matlab file: FroNMF_algo_comparison.m].

- Looking at the right-hand side tells us which algorithm obtains on average the best solution. Since NMF algorithms are rather sensitive to initialization (see Section 8.6), this might change from one experiment to another. This is particularly true for sparse data sets which are more sensitive to initialization; see the discussion in the next subsection.

- Looking at the left-hand side tells us which algorithm initially converges the fastest.

In Figure 8.3, we observe the following:

- Since the input matrices are dense, ALS performs very poorly; see Section 8.3.2.

- MU is rather slow, being the worse algorithm after ALS. This is consistent with all the observations made in the literature: MU with the Frobenius norm on dense data sets performs poorly.

- Adapting the number of iterations accelerates the MU, that is, A-MU outperforms MU. Moreover, it converges initially almost as fast as ANLS, although it produces worse solutions within the 30 seconds. A-MU performs worse than A-HALS, AO-ADMM, and FPGM.

- ANLS is not very effective for CBCL with $r = 49$, while it performs among the best for $r = 10$. The reason is that when r becomes large, the term $\mathcal{O}\left((m + n)r^3 s(r)\right)$ kicks in (Table 8.2). For r larger, ANLS is less competitive.

- A-HALS performs better than FPGM, which performs similarly as AO-ADMM; however, the difference in performance is not significant. (In the paper [250], it is reported that AO-ADMM works better than A-HALS). In general, we have observed that, in the dense case, there is no clear winner between these three methods depending on the many factors influencing the performances of these methods (see the discussion above).

- E-A-HALS improves upon A-HALS and performs significantly better than the other algorithms. Note that extrapolated variants of the other algorithms can also be implemented (this can be done easily with our MATLAB code) and allows us to accelerate their convergence as well; see the extensive numerical experiments in [10].

8.4.3.2 ▪ Sparse data sets

In our experience, sparse matrices are more difficult to factorize in the sense that the corresponding optimization problem tends to have stationary points where the objective function has rather different values. In other words, the objective function value gaps between stationary points tend to be larger for sparse matrices. The intuitive reason is that NMF is looking for dense rank-one submatrices corresponding to the nonzero entries of the rank-one factors $W(:, k)H(k, :)$ for $k = 1, 2, \ldots, r$: when X is sparse, this amounts to finding clusters, which is known to be a hard combinatorial problem, where the gap between local minima can be arbitrarily large. For example, for a block diagonal matrix, any rank-one factor that identifies one of the blocks (which may have different sizes) is a local minimum. Another difficulty is that such data sets are typically not low-rank; see the discussion in the introduction of Chapter 9.

Hence, this makes the comparison of algorithms (even more) difficult because they converge to different stationary points with different objective function values. Let us compare NMF algorithms on two widely used data sets: the TDT2 data set with $m = 9394$, $n = 19528$, and $r = 30$ with 99.37% of the entries equal to zero (TDT stands for topic detection task), and the Classic data set [509] with $m = 7094$, $n = 41681$, and $r = 30$ with 99.92% of the entries equal to zero. We use the same experimental settings as in the previous paragraph: algorithms are run with the same 30 random initializations during 30 seconds, and Figure 8.4 reports the average value of (8.22).

In our experience, most of the observations from the dense case are still valid for sparse data sets since all algorithms adapt well to sparse matrices. However, a few notable differences can be observed for sparse matrices:

- MU does not perform as bad; for example, it converges initially faster than ANLS.

- In general, the differences between the algorithms are less significant than in the dense case.

- ALS performs much better: although it stagnates at a higher error than the other algorithms, it is able to quickly decrease the objective function values. It is initially among the fastest algorithms, and hence it especially makes sense to use it as a warm-start strategy for sparse matrices.

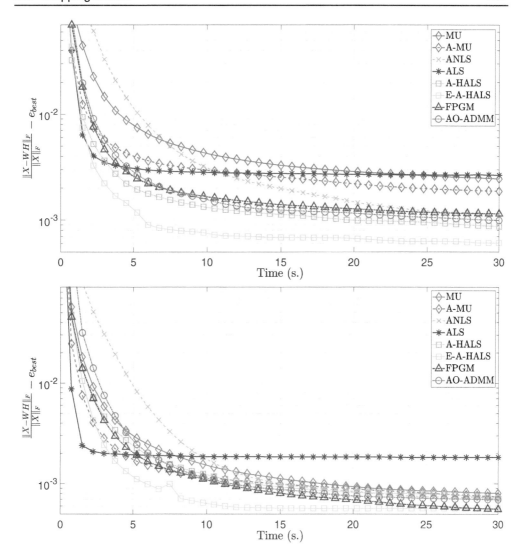

Figure 8.4. *Comparison of NMF algorithms on sparse data sets: the TDT2 data set with $m = 9394$, $n = 19528$, and $r = 30$ with 99.37% of the entries equal to zero, and on the Classic data set with $m = 7094$, $n = 41681$, and $r = 30$ with 99.92% of the entries equal to zero. [`Matlab file: FroNMF_algo_comparison.m`].*

E-A-HALS still outperforms the other algorithms, and it is the default algorithm of [`Matlab file: FroNMF.m`].

8.5 ▪ Stopping criteria

As for many aspects in this book (such as the choice of the objective function or of the factorization rank), the choice of the stopping criterion is not specific to NMF and is a choice encountered by any iterative (optimization) algorithm. Let us simply mention the most standard strategies:

- *Iterations.* An upper bound on the number of iterations of Algorithm 8.1 is chosen, typically between 200 and 5000.

- *Time.* An upper bound on the time allotted to the algorithm is given. We recommend to always use such an upper bound to guarantee the algorithm will stop in a reasonable amount of time.

- *Error.* One can monitor the evolution of the error and stop the algorithm when no significant changes are observed for T iterations. For example, one can use

$$|e(t - T) - e(t)| \leq \delta\, e(t), \tag{8.23}$$

where $e(t) = D(X, W^{(t)} H^{(t)})$ is the error at iteration t, $T \geq 1$, and $0 < \delta \ll 1$. Using values around $T = 10$ and $\delta = 10^{-4}$ is a good choice in practice; it means that the relative error has not changed more than 0.01% within the last 10 iterations.

One has to be careful when using this stopping criterion as it can incur additional computational cost. However, most algorithms involve computations that can be reused to compute $e(t)$ cheaply. For example, for β-divergence with $\beta \neq 1, 2$, the term WH is computed to update W or H even when X is sparse (see Section 8.2.6), and hence computing $D(X, WH)$ only requires $\mathcal{O}(mn)$ additional operations. For $\beta = 2$,

$$\|X - WH\|_F^2 = \|X\|_F^2 - 2\left\langle W^\top X, H\right\rangle + \left\langle W^\top W, HH^\top\right\rangle,$$

where the term $\|X\|_F$ can be computed once in $\mathrm{nnz}(X)$ operations, the terms $W^\top X$ and $W^\top W$ are computed by any first-order method to update H. In the next update of W, $HH^\top$ is computed. Therefore, computing the error requires once $\mathrm{nnz}(X)$ operations, and $\mathcal{O}(rn + r^2)$ additional operations at each iteration, which is negligible for first-order methods. For $\beta = 1$, to update W and H, we need to compute the entries of WH where X is positive (see Section 8.2.6). This allows us to compute $\sum_{(i,j), X_{i,j} \neq 0} d_1(X_{i,j}, (WH)_{i,j})$ in $\mathcal{O}(\mathrm{nnz}(X))$ operations. For the other terms in $D_1(X, WH)$, we obtain

$$\sum_{(i,j), X_{i,j}=0} d_1(X_{i,j}, (WH)_{i,j}) = \sum_{(i,j), X_{i,j}=0} (WH)_{i,j}$$

$$= \sum_{i,j} (WH)_{i,j} - \sum_{(i,j), X_{i,j} \neq 0} (WH)_{i,j},$$

where $\sum_{i,j} (WH)_{i,j} = \|WH\|_1 = \|W\|_1 \|H\|_1$ since $(W, H) \geq 0$ which can be computed in $\mathcal{O}(mr + nr)$ operations, while the second term requires $\mathcal{O}(\mathrm{nnz}(X))$ operations (since these entries of WH corresponding to positive entries of X are already computed).

- *Iterates.* A simple yet effective stopping criterion is to monitor the evolution of the iterates, that is,

$$\|H^{(t)} - H^{(t-T)}\|_F \leq \delta \|H^{(t-T)}\|_F,$$

and similarly for W. As for the error, using $T = 10$ and $\delta = 10^{-4}$ is a reasonable choice in practice.

- *KKT conditions.* It is possible to use the KKT conditions. However, as pointed out in [189], one has to be careful when using such criteria because they are not invariant to the scaling of the rank-one factors $W(:, k) H(k, :)$. Hence one has to be particularly careful when using a criterion based on the KKT conditions. Let us consider the criterion $C(W, H) = C_W(W) + C_H(H)$ where

$$C_W(W) = \underbrace{\|\min(W, 0)\|_F}_{W \geq 0} + \underbrace{\|\min(\nabla_W D, 0)\|_F}_{\nabla_W D \geq 0} + \underbrace{\|W \circ \nabla_W D\|_F}_{W \circ \nabla_W D = 0}, \tag{8.24}$$

where $\nabla_W D$ denotes the gradient of the objective function with respect to W, and $C_H(H)$ is defined similarly for H. We have $C(W, H) = 0$ if and only if (W, H) is a stationary point of (8.1). However, consider the following:

- $C(W, H)$ is sensitive to scaling. For $\alpha > 0$ and $\alpha \neq 1$, we have in general that

$$C(W, H) \neq C(\alpha W, \alpha^{-1} H),$$

  since the terms in (8.24) are sensitive to scaling. For example, with the Frobenius norm, multiplying W by α and dividing H by α multiplies $\nabla_H D$ by α and divides $\nabla_W D$ by α. If one solves the subproblem in H exactly (for example, with an active-set method), $C_H(H) = 0$ and this holds true for any scaling. However, the second term in (8.24) can be made arbitrarily small by dividing H by a large constant which divides $\nabla_W D$ by the same constant. This issue can be handled with proper normalization; for example, imposing $\|W(:, k)\|_2 = \|H(k, :)\|_2$ for all k [239].

- The value of $C(W, H)$ after the update of W can be very different from the value after an update of H, in particular, if the scaling is bad or if $|m-n|$ is large. Therefore, one should be very careful when using this type of criteria to compare ANLS with other algorithms such as the MU or HALS as the evolution of $C(W, H)$ can be misleading. A potential fix would be to scale the columns of W and the rows of H so that $C_W(W)$ after the update of H and $C_H(H)$ after the update of W have the same order of magnitude, that is, use a scaling α such that $C_W(\alpha W) = C_H(H/\alpha)$ (actually, such a scaling should preferably be performed on each rank-one factor). Another possibility would be to use a proper normalization of (W, H) and apply one step of ANLS before evaluating the criterion (hence implicitly focusing on one of the two factors since after an update of ANLS, one of the two terms, $C_W(W)$ or $C_H(H)$, is equal to zero).

- One has to be careful about trivial stationary points (see the discussion in Section 8.1.2): convergence to such stationary points should be avoided, and one should check within the algorithm if this happens (although, for reasonable initializations, this usually does not happen). If your only goal is to find a stationary point quickly, here is one: $W = 0$ and $H = 0$.

In [Matlab file: betaNMF.m] and [Matlab file: FroNMF.m], we have implemented the first three possibilities. To stop the inner iterations of NNLS algorithms within NMF, we use the stopping criterion based on the iterates.

8.6 ▪ Initialization

Since NMF is an NP-hard nonconvex optimization problem, iterative algorithms such as the ones presented in this chapter are sensitive to the initial iterate $(W^{(0)}, H^{(0)})$.

Random initialization   The standard strategy in the literature is to use the uniform distribution $\mathcal{U}(0, 1)$ in the interval $[0, 1]$ to generate the entries of $(W^{(0)}, H^{(0)})$. This is typically combined with a multistart heuristic that keeps the best solution obtained out of a certain number of random initializations; see the discussions in [98, 460]. Note that using distributions that generate sparse factors as initial matrices (such as sprand in MATLAB) is sometimes a good strategy as it allows us to explore the search space better: the uniform distributions generates dense factors and hence does not attain local minima located close to the faces of the nonnegative orthant; see [460] for a discussion and numerical experiments.

In the following, we describe more sophisticated initialization strategies, which typically have two advantages compared to random initializations:

1. They often allow one to identify better local minima as they use the structure of the input data X to compute $(W^{(0)}, H^{(0)})$.

2. They allow the iterative scheme to converge faster since they are initialized close to a reasonable solution. However, the cost of the initialization scheme should be taken into account for a fair comparison.

Figure 8.5 illustrates a typical behavior of a good initialization scheme (in this case, SNPA) compared to a random initialization. However, without further assumptions (such as separability), such strategies do not come with theoretical guarantees because of the NP-hardness of NMF.

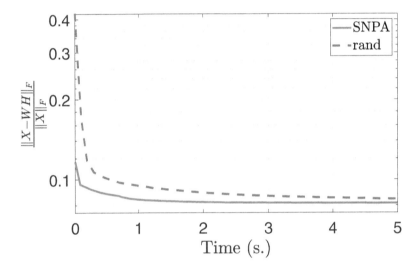

Figure 8.5. *Comparison of the relative error of the iterates generated by A-HALS initialized with SNPA and with a random initialization (using the uniform distribution $\mathcal{U}(0,1)$ for all entries of W and H) on the CBCL data set with $m = 361$, $n = 2429$, and $r = 49$.* [Matlab file: FroNMF_init_randvsSNPA.m].

SVD   Let $UV = \sum_{k=1}^{r} U(:,k)V(k,:)$ be an optimal unconstrained rank-r approximation of X which can be computed via the truncated SVD; see Section 6.1. Each rank-one factor $U(:,k)V(k,:)$ might contain positive and negative entries, except for the first one, by the Perron–Frobenius theorem given that the input matrix is irreducible; see Section 6.1.2. Denoting $[x]_+ = \max(x,0)$, we have for all $1 \leq k \leq r$ that

$$U(:,k)V(k,:) = [U(:,k)]_+[V(k,:)]_+ + [-U(:,k)]_+[-V(k,:)]_+$$
$$- [-U(:,k)]_+[V(k,:)]_+ - [U(:,k)]_+[-V(k,:)]_+,$$

where the first two rank-one factors in this decomposition are nonnegative. Boutsidis and Gallopoulos [54] replaced each rank-one factor in $\sum_{k=1}^{r} U(:,k)V(k,:)$ with either $[U(:,k)]_+[V(k,:)]_+$ or $[-U(:,k)]_+[-V(k,:)]_+$, selecting the one with larger norm and scaling it properly. Observe that half of the information is lost when using this selection step. Motivated by this observation, two alternative approaches were proposed:

- In [388], the author simply replaces each rank-one factor by its absolute values, that is, the proposed initialization is $(|U|, |V|))$. However, this creates an approximation $|U||V|$ which is much larger than X as all the negative terms in (U, V) are ignored.

- In order the reduce the computational cost of the initialization, and keep all of the information of the SVD computation, the authors in [18] propose an initialization based on a rank-r' approximation of X where $r' = 1 + \lfloor \frac{r}{2} \rfloor$. The first factor $U(:, 1)V(1, :)$ is nonnegative and is kept as is, while the next rank-one factors $U(:, k)V(k, :)$ for $k \geq 2$ are split into two parts, namely

$$([U(:,k)]_+, [V(k,:)]_+) \quad \text{and} \quad ([-U(:,k)]_+, [-V(k,:)]_+),$$

to initialize two columns of W and two rows of H. This initialization produces sparser factors as the average sparsity of the generated factor is 50% for $k \geq 2$, by construction.

Clustering As NMF is closely related to clustering techniques such as k-means and spherical k-means (Section 5.5.3), it makes sense to use such techniques to initialize NMF. For example, applying k-means on the columns of X, W can be initialized as the computed centroids, while H as the membership indicator matrix (with a single nonzero entry per column). This approach based on k-means and spherical k-means was proposed by Wild, Curry, and Dougherty [481]; see also [75] and the references therein for more sophisticated strategies based on other clustering techniques.

Near-separable NMF As discussed at length in Chapter 7, near-separable NMF approximates the matrix X as $X(:, \mathcal{K})H$ where $\mathcal{K} \subseteq \{1, 2, \ldots, n\}$ contains r indices and $H \geq 0$. This model not only makes sense in many applications, it also allows us to compute such decompositions in polynomial time, given that the input matrix admits a separable factorization. Moreover, greedy algorithms such as SPA and SNPA can quickly produce an index set $\mathcal{K}$ such that $X(:, \mathcal{K})H \approx X$, even when X does not admit a separable factorization. Therefore, using near-separable NMF algorithms to initialize NMF algorithms especially makes sense, and, if you do not know better, we recommend such an initialization scheme: if your input matrix satisfies the separability assumption approximately, it provides a very good initialization. You need to decide whether you apply such an algorithm on X or $X^\top$ depending on which of these two matrices is more likely to admit a separable factorization; see the discussion in Section 7.1 for some examples. Most NMF solutions shown in this book were obtained with SPA or SNPA initialization; see for example Figures 4.10, 4.11, 5.2, and 9.2.

Figure 8.5 compares a random initialization with SNPA on the CBCL data sets applied on $X^\top$ (see Section 7.1) using A-HALS. We observe that SNPA allows faster convergence. SNPA took 0.6 seconds to be computed, and hence the random initialization actually achieves a lower error within that time limit (below 10% relative error, while SNPA produces an initial solution with relative error about 12%). However, in the long run, SNPA is beneficial: using the stopping criterion (8.23) based on the evolution of the error with $T = 10$ and $\delta = 10^{-3}$, A-HALS with SNPA converges in 4.4 seconds while with the random initialization it converges in 8.6 seconds.

8.7 · Alternative algorithmic approaches

In this section, we briefly review other approaches to solve NMF (8.1), beyond near-separable NMF (Chapter 7), and nonlinear optimization schemes which we have focused on so far in this chapter.

8.7.1 ▪ Sequential approaches using deflation

Instead of trying to compute all rank-one factors simultaneously when solving (8.1), one could be tempted to compute one rank-one factor at a time as this approach is optimal in the unconstrained case; see Section 6.1. However, in the context of NMF, these rank-one factors should satisfy some additional constraints. If $W(:,1)H(1,:)$ minimizes $\|X - W(:,1)H(1,:)\|_F$ under only the nonnegativity constraints of $W(:,1)$ and $H(1,:)$, the residual $R_1 = X - W(:,1)H(1,:)$ obtained after the first deflation will contain roughly half positive and half negative entries. These negative entries cannot be approximated by any of the following rank-one factors.

The most natural approach is therefore to impose that $W(:,1)H(1,:)$ is an underapproximation of X, that is,

$$W(:,1)H(1,:) \leq X.$$

With this underapproximation condition, the residual $R_1 = X - W(:,1)H(1,:)$ is nonnegative and the next factor $W(:,2)H(2,:)$ can be computed as an underapproximation of R_1, and then $W(:,3)H(3,:)$ as an underapproximation of $R_2 = R_1 - W(:,2)H(2,:)$, etc.

This sequential strategy requires solving a rank-one nonnegative matrix underapproximation (NMU) problem: given $A \geq 0$, solve

$$\min_{w \in \mathbb{R}^m_+, h \in \mathbb{R}^n_+} \|A - wh^\top\|_F \quad \text{such that} \quad wh^\top \leq A. \tag{8.25}$$

This idea dates back to Levin [311] (1985), although he considers only one rank-one factor, and does not use it within a deflation scheme. It was later used in the context of NMF in [184] (2007) and in [195] (2010), where (8.25) is solved via a Lagrangian approach [Matlab file: recursiveNMU.m]. We refer the interested reader to [447] for a more recent and efficient algorithm for NMU based on ADMM. Figure 5.6 illustrates an NMU factorization result on the CBCL data set (page 177) and Figure 1.4 on the Swimmer data set (page 7).

NMU has several nice properties:

- The factors do not need to be recomputed when r is modified.

- NMU provably generates sparser solutions than NMF [195]; for example, it is easy to show that the residual R_1 has a least one zero per row and per column, implying that $W(:,2)H(2,:)$ will have zeros in the same locations. The rank-one factors tend to become sparser as NMU unfolds because the residuals become sparser; see Figure 5.6 for an illustration (page 177).

- It has a unique solution (up to scaling) under mild conditions [208]. For example, for a binary matrix X, its optimal rank-one NMU is unique if and only if the largest submatrix of all ones is unique. Considering X as the biadjacency matrix of a bipartite graph, this requires its maximum biclique to be unique (see the proof of Theorem 6.5). Moreover, algorithms can be meaningfully initialized with the best rank-one nonnegative approximation of the residual using the SVD (this follows from the Perron–Frobenius theorem as the residuals are nonnegative).

However, the problem (8.25) is NP-hard in general [184] (see the proof of Theorem 6.5 that provides a reduction from the maximum-edge biclique problem). Moreover, because of its sequential nature, NMU produces solutions with larger error than NMF.

Other similar strategies were later developed. For example, Biggs, Ghodsi, and Vavasis [44] (2008) proposed to compute the rank-one factors using a power-like method enforcing sparsity. Intuitively, their method tries to identify sequentially large and dense submatrices in X. Dong, Lin, and Chu [137] (2014) also use a deflation idea but in the context of Exact NMF, and their deflation is based on the Wedderburn rank reduction formula to compute the rank-one factors.

8.7.2 ▪ Hierarchical/divide-and-conquer approaches

The rank-two NMF problem, that is, (8.1) with $r = 2$, is easier to solve; see the discussion in Section 6.1.3. Hence a possible way to construct an NMF solution is via a hierarchical/divide-and-conquer strategy. First, a rank-two NMF of X is computed, that is,

$$WH = W(:,1)H(1,:) + W(:,2)H(2,:) \approx X.$$

Then the columns of X are divided into two clusters depending on the values of the entries of $H(1,:)$ and $H(2,:)$. In its simplest form, this approach generates the clusters as

$$C_1 = \{j \mid H(1,j) > H(2,j)\} \quad \text{and} \quad C_2 = \{j \mid H(1,j) \le H(2,j)\},$$

after a proper scaling of WH (for example, $\|H(1,:)\|_\infty = \|H(2,:)\|_\infty = 1$). Then, the scheme is applied recursively on $X(:,C_1)$ and/or $X(:,C_2)$. This constructs a binary tree structure of the data where the clusters are split in two at each level. Figure 8.6 illustrate such a binary tree structure for the Urban hyperspectral image using [`Matlab file: hierclust2nmf.m`] [202].

This idea has been used successfully for classification of pixels in medical [318, 406] and hyperspectral images [202], as well as in document data sets [291, 139]. The algorithms proposed in these papers differ in how the data points are split in the two clusters and in the criterion used to select the cluster(s) to split.

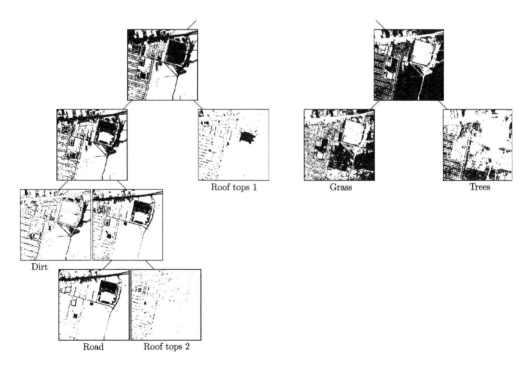

Figure 8.6. *Example of a hierarchical rank-two NMF solution on the Urban data set (see Figure 1.6). At the first level, rank-two NMF splits pixels between the vegetation (on the right) and the other materials (on the left). The vegetation is then split into grass and trees, while the other materials, namely the two types of roof tops, the roads, and dirt, are split at lower levels in the tree. Figure modified from [202]. [`Matlab file: Urban.m`].*

8.7.3 ▪ Transformations of unconstrained solutions

For Exact NMF, Theorem 2.21 showed that if $X = WH$ where $\text{rank}(X) = \text{rank}_+(X) = r$, then any NMF of X has the form

$$X = \underbrace{UQ}_{W \geq 0} \underbrace{Q^{-1}V}_{H \geq 0},$$

where Q is an r-by-r invertible matrix, and UV is an unconstrained decomposition of X obtained, for example, via the SVD. Hence, given $X = UV$, it makes sense to try to find an invertible r-by-r matrix Q such that $UQ \geq 0$ and $Q^{-1}V \geq 0$. This problem has much fewer variables than the original NMF problem, namely r^2 instead of $mr + nr$. In noisy cases, UV can be replaced by the optimal rank-r approximation of X obtained via the truncated SVD (Theorem 6.3).

This idea was already discussed by Paatero and Tapper [371] and has been investigated particularly in the SMCR literature where the goal is to compute all possible factorizations (see Section 1.4.1 and [366]). In his seminal paper, Vavasis [465] discusses this approach and proposes a local search heuristic.

However, as far as we know, outside the SMCR literature, this approach has not been much investigated nor been used successfully in practice. The reason is that finding Q such that $UQ \geq 0$ and $Q^{-1}V \geq 0$ is difficult and is impossible when $\text{rank}_+(UV) > \text{rank}(UV)$. Finding a Q such that UQ and $Q^{-1}V$ have mostly nonnegative entries, and then projecting this solution onto the nonnegative orthant, that is, using $\max(0, UQ)$ and $\max(0, Q^{-1}V)$, may lead to reasonable solutions. However, the projection step deteriorates the solution in a way that is not controlled and hence typically leads to larger values of the objective function than standard NMF algorithms. Also, computing an optimal unconstrained rank-r approximation requires some computational effort (the computational cost is of the same order as that of first-order methods for NMF).

8.8 ▪ Further readings

We have covered in this chapter the most fundamental algorithms used for NMF, focusing on first-order methods. The design of NMF algorithms is still a very active area of research; see for example [238] for a recent efficient NMF algorithm using two extrapolation points with strong convergence guarantees (including convergence rates), [446] for an algorithm specifically designed for sparse NMF with similar convergence guarantees, and [247] for a recent second-order method.

Moreover, we have not covered in this chapter stochastic algorithms, such as stochastic gradient descent (SGD). Such techniques are important when dealing with large-scale problems. The core idea is to use randomly chosen subsamples of the data set at each iteration. For example, to update H, one could select only a subset $\mathcal{I}$ of the rows of X and W and use a gradient direction of the objective function $\|X(\mathcal{I}, :) - W(\mathcal{I}, :)H\|_F$. In the dense case, this reduces the cost of an iteration from $\mathcal{O}(mnr)$ to $\mathcal{O}(|\mathcal{I}|nr)$; classic SGD uses $|\mathcal{I}| = 1$. (Note that this idea is related to sketching, briefly discussed in Section 8.4.1.) SGD and related methods have a very rich theory and many strategies exist depending on the problem at hand. However, such methods are typically difficult to tune and should be designed carefully. An example of such algorithms is a BCD method combined with proximal SGD in the context of low-rank tensor decompositions under various constraints including nonnegativity [173]. We refer the interested reader to the survey [177] for details on recent advances on optimization techniques for structured low-rank matrix and tensor decompositions, including BCD, SGD, and second-order methods.

Finally, we refer the interested reader to the paper [248], where the accuracy of the nonnegative latent factor estimates of NMF algorithms are put to the test against the Cramér–Rao lower bound.

8.9 ▪ Online resources

Let us mention a few online resources containing codes for NMF algorithms in different languages:

- NIMFA [518]. This toolbox in Python is available from

    `http://nimfa.biolab.si/index.html`.

    It contains algorithms for several NMF models, including standard, separable, sparse, and binary NMF. It also contains several initialization strategies, and numerical examples for several applications, namely in bioinformatics, functional genomics, text analysis, image processing, and recommendation systems.

- NMFLibrary by Hiroyuki Kasai. This toolbox in MATLAB is available from

    `https://github.com/hiroyuki-kasai/NMFLibrary`.

    It contains many algorithms (including ANLS, ALS, MU, A-MU, HALS, PGD, FPGD) for many NMF models including standard, symmetric, sparse, robust, online, semi-, and orthogonal NMF.

    Another MATLAB toolbox focused on biological data mining is available from `https://sites.google.com/site/nmftool/`; see [317]. It contains not many algorithms (namely MU and ANLS) but quite a few models, including standard, semi-, orthogonal, sparse, convex, weighted, and kernel NMF.

- NMF Toolbox [328]. This toolbox in Python and MATLAB is available from

    `https://www.audiolabs-erlangen.de/resources/MIR/NMFtoolbox/`.

    It focuses on audio source separation, including algorithms for standard and convolutive NMF.

- NMF package [181]. This toolbox in R is available from

    `http://renozao.github.io/NMF/`.

    It contains a few NMF algorithms, including several initialization strategies.

- libNMF [258]. This toolbox in C is available from

    `https://www.univie.ac.at/rlcta/software/`.

    It contains a few NMF algorithms for the standard NMF model and includes several initialization strategies.

8.10 ▪ Take-home messages

Over the years, numerous algorithms for various NMF models have been developed, in particular for the Frobenius norm. In this particular scenario, various highly efficient methods exist, including ANLS, A-HALS, FPGM, ADMM, and E-A-HALS. For β-divergences with $\beta \neq 2$, not many algorithms have been developed beyond the MU, which are easy to implement and flexible. Moreover, the MU are an effective optimization strategy for small β; see [237] for a discussion on the case $\beta = 1$.

Chapter 9

Applications

In Chapter 1, we presented four applications of NMF, namely feature extraction from a set of images, topic extraction from a set of documents, blind unmixing of hyperspectral images into their constitutive materials, and blind audio source separation. Moreover, we have discussed applications in other sections, for example community detection in Section 5.4.7 and topic modeling in Sections 5.4.9.1, 5.5.4, and 7.8.

Organization of the chapter   We present three other applications of NMF: SMCR (Section 9.3), gene expression analysis (Section 9.4), and recommender systems (Section 9.5). We do not dig deep into these applications, that is, we do not discuss the specifics and which NMF models and algorithms are the most appropriate. We rather focus on explaining why the NMF model makes sense. Then, we provide a list of applications with useful references (Section 9.6). Before doing so, we discuss two important aspects of the use of NMF in practice, namely the scaling of the rank-one factors (Section 9.1) and whether your input matrix should be low-rank for NMF to make sense in practice (Section 9.2).

9.1 ▪ Beware of scaling ambiguity

In most NMF applications, the matrix H corresponds to activation of the basis elements within the data points. For example,

- in blind HU, $H(k, j)$ is the abundance of endmember k in pixel j,

- in text mining, $H(k, j)$ is the importance of topic k in document j,

- in audio source separation, $H(k, j)$ is the intensity of source k during the time window j.

Therefore, it is important to properly postprocess the result of your factorization algorithm to interpret it meaningfully. This (minor) issue is sometimes overlooked in the literature. Let us discuss it briefly.

In some NMF models, there is no such ambiguity because the normalization is incorporated within the model, for example NMF with the sum-to-one constraint $H^\top e = e$ often used in blind HU. Other NMF models do not have the scaling ambiguity such as projective NMF, ONMF, convex NMF, separable NMF, binary NMF, and symNMF.

Normalizing the rows of H to have unit ℓ_1 or ℓ_2 norm is typically not a good idea: there is in general no reason to believe that all the sources activate with the same total energy. Moreover, such normalization could lead to ill-conditioned W; see the discussion on page 142. What could make sense is to normalize the rows of H to have unit ℓ_∞ norm, that is, given $\alpha_k = \|H(k,:)\|_\infty$ for all k, use the normalization

$$H(k,:) \leftarrow \frac{H(k,:)}{\alpha_k} \quad \text{and} \quad W(:,k) \leftarrow \alpha_k W(:,k).$$

Doing so, each row of H has an entry equal to one which corresponds to the data point that has the largest activation for the corresponding basis element. This allows us to interpret the entries of H easily: they are proportions of this highest possible activation.

However, in most cases, it makes more sense to normalize the columns of W to have unit ℓ_1 or ℓ_2 norm. Hence the entries of H can be directly interpreted as the importance of each column of W to reconstruct each data point since the columns of W have the same norm.

9.2 ▪ Should your data set be approximately of low rank?

In some applications, the input matrix is not close to a low-rank matrix. Three typical examples are word count data sets used in text mining (Sections 1.3.3, 5.5.4, and 5.4.9.1), evaluation matrices used in recommender systems (Section 9.5), and adjacency matrices used for community detection (Section 5.4.7). However, in all these applications, low-rank models have had tremendous success [287, 14, 488]. The reason is that although the input matrix might have high rank, it still makes sense to try to find a low-rank structure within it. For example, in recommender systems, it makes sense to look for subsets of items that were appreciated by several users. In text mining, although all documents might not be well-approximated by a small subset of topics, it makes sense to look for topics, that is, subsets of words found simultaneously in different documents. This amounts to finding a global behavior within the data set. As we have seen in Chapter 7.8, a more appropriate model to achieve this goal might be to consider the tri-symNMF of $XX^\top$. Similarly, in community detection, low-rank models are applied on other transformations of the data to obtain more powerful models [246]. In audio source separation, the short-time Fourier transform is typically used to construct the input matrix X (see Section 1.3.4). However, the transformation can be optimized to best fit a low-rank model [149]. In this section, we stick to applications of the standard NMF model. In many applications, such as hyperspectral unmixing or image analysis where the input matrices are in most cases close to being low rank, NMF is a suitable and powerful model.

9.3 ▪ Self-modeling curve resolution

SMCR is one of the first areas of research where the NMF model has been used (Section 1.4.1) and is closely related to blind HU (Section 1.3.2). As for blind HU, rows of the input matrix X correspond to different wavelengths where light intensities are recorded. However, each column sample corresponds not to a spatial location (that is, a pixel in the image) like in blind HU but to a time point when the measure is taken. In SMCR, the evolution of the spectra of a chemical reaction is measured over time, and the input matrix X is a wavelength-by-time matrix. Given X, the goal is to find the spectra of the pure chemical components as the columns of W, and their concentrations over time in the matrix H that evolve as the chemical reaction takes place.

Let us illustrate this application using the synthetic example of Luce et al. [331] inspired from Raman spectroscopy. The reaction is made up of five components (A, B, C, D, E). Let us denote $h(t) \in \mathbb{R}_+^5$ the vector of concentrations over time of these five components. The kinetics of the reaction in this example is given by

$$h(t) = e^{Kt} h_0,$$

where $h_0 \in \mathbb{R}_+^5$ contains the initial concentrations, and e^{Kt} is the matrix exponential of Kt with

$$
K = \begin{pmatrix}
-0.53 & 0.02 & 0 & 0 & 0 \\
0.53 & -0.66 & 0.25 & 0 & 0 \\
0 & 0.43 & -0.36 & 0 & 0.1 \\
0 & 0.21 & 0 & 0 & 0 \\
0 & 0 & 0.11 & 0 & -0.1
\end{pmatrix}.
$$

Graphically, the matrix K translates the following reaction coefficients:

$$
A \underset{0.02}{\overset{0.53}{\rightleftharpoons}} B \underset{0.25}{\overset{0.43}{\rightleftharpoons}} C \underset{0.1}{\overset{0.11}{\rightleftharpoons}} E
$$
$$
\downarrow 0.21
$$
$$
D
$$

Assuming the spectrum of this reaction is measured every time step Δt for a time period of $(n-1)\Delta t$ starting at $t = 0$, the observed data matrix X satisfies, for all $j = 1, 2, \ldots, n$,

$$
X(:,j) = WH(:,j), \quad \text{where} \quad H(:,j) = h\big((j-1)\Delta t\big),
$$

and the columns of W are the five component spectra constructed as the spectra of various organic compounds. Figure 9.1 displays the matrix W and H of this example, with initial concentration $h_0 = (1, 0, 0, 0, 0)^\top$.

Although in spirit SMCR is the same problem as blind HU (except that spatial location is replaced by time), the structure of the data sets is usually rather different, which plays a key role when designing models and algorithms. For example,

- the concentrations in SMCR are typically not sparse because most components are present at all times in the chemical reaction (in the example above, this is the case except when $t = 0$); moreover, the concentrations are typically smooth functions of time (as in Figure 9.1) which can be leveraged using parametrized NMF models (see for example [234]);

- the spectra in SMCR can be rather sparse and peaky (as in Figure 9.1, but this is not necessarily the case), which is typically not the case for airborne hyperspectral images (see for example Figure 1.6).

A consequence of the two observations above is that H in SMCR typically does not satisfy the separability condition or the SSC (see Chapter 4). However, the component spectra $W^\top$ may satisfy the separability assumption or the SSC. In Figure 9.1, the spectra are separable as every component contains a peak where the other spectra are equal to zero, and SNPA can be used to decompose this particular data set; see [331].

Note also that typically in the SMCR literature, the goal is to compute all possible solutions (which is of course highly nontrivial and becomes quickly impractical as r increases) so that the chemist can then identify the one that makes more sense; see [366] and the references therein for more details.

9.4 ▪ Gene expression analysis

Microarray data analysis aims to analyze gene expression data obtained using microarray experiments to extract information among genes and across different conditions and different samples; see [66, 277, 125, 147] and the references therein for more details. Standard microarrays are gene-by-sample matrices (which we denote X, as usual) recording gene expression levels on

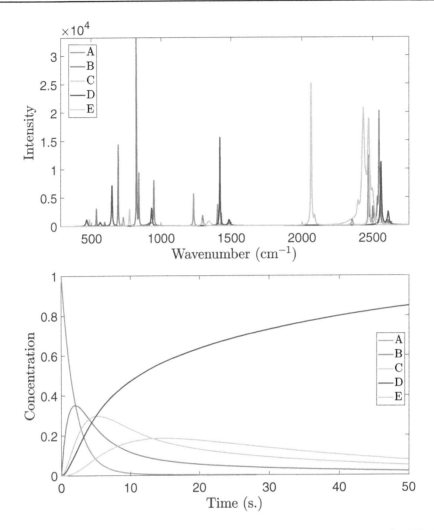

Figure 9.1. *On the top, spectral signatures of the components contained in W. On the bottom, the concentrations over time contained in H.* [`Matlab file: Raman.m`].

different samples. Decomposing such a data matrix X using NMF allows us to extract common behavior among the genes: the columns of the matrix W are so-called metagenes and gather subsets of genes displaying a similar behavior on a subset of the samples; the matrix H indicates which metagene is active in which sample.

Let us illustrate this on a simple numerical example taken from[65] Baranzini et al. [22]. The data set collects gene expression levels about patients affected by multiple sclerosis and treated during different time steps with the protein interferon β. Analyzing such data sets allows us to understand the medical responses of the treatment. After some preprocessing [147], we obtain a data set consisting of 52 genes measured for 27 patients over 7 time periods. For simplicity, we concatenate these measures as a gene-by-sample matrix $X \in \mathbb{R}^{52 \times 189}$. Performing an NMF with $r = 3$ using A-HALS initialized with SNPA identifies three metagenes represented in Figure 9.2, where we normalize the columns of W so that $\|W(:,k)\|_\infty = 1$ for all k. Using a threshold of 0.2

---

[65]The data set is available online as supplementary material; see `https://doi.org/10.1371/journal.pbio.0030002.sd001`.

Figure 9.2. *Heatmap of the matrix* $W \in \mathbb{R}_+^{52 \times 3}$ *(white corresponds to 1, black to 0) obtained by factorizing the gene-by-sample matrix* $X \in \mathbb{R}_+^{52 \times 189}$. *[Matlab file:* `Microarray.m`*].*

to decide whether a gene belongs to a metagene, the first metagene contains one gene (MIP1a gene), the second metagene contains two genes (CD6 and RANTES), and the third metagene contains three genes (RANTES, Tbet and IRF5). These genes are well-known in the literature to play a key role in multiple sclerosis disease studies; see [147] and the references therein.

9.5 ▪ Recommender systems and collaborative filtering

The goal of collaborative filtering is to predict the preferences of users for some items (filtering), based on the preferences or taste information from many users (collaborative). It has been used extensively by electronic commercial sites as good recommendations increase the propensity of a purchase.

The most famous example is the prediction of how much someone is going to like a movie based on her/his movie ratings and the ratings of others. The Netflix prize competition was launched in 2006 and aimed at improving by 10% the predictions of the Netflix collaborative filtering algorithm with a reward of US$1,000,000. This prize launched a considerable research effort on this problem. The winners of the competition combined several models, one of the central ones being low-rank models; see [287]. The rationale behind such models is that the behavior of most users can be well-modeled as a linear combination of a few feature users, since the low-rank model gives $X(:, j) \approx \sum_{k=1}^{r} W(:, k)H(k, j)$ for all j where X is the movie-by-user matrix. Equivalently, looking at how the rows are reconstructed, that is, $X(i, :) \approx \sum_{k=1}^{r} W(i, k)H(k, :)$ for all i, movie preferences can be explained via linear combinations of feature movies related to genres (such as child oriented, thriller, and romantic). For example, performing a rank-two factorization on the Netflix data set, Koren, Bell, and Volinsky [287, Figure 2] were able to identify two main genres: male-oriented versus female-oriented movies (for example, *Lethal Weapon* versus *The Princess Diaries*), and serious versus escapist movies (for example, *Braveheart* versus *The Lion King*).

Although nonnegativity is not often used in this context, it would make sense as it provides easily interpretable feature movies and users; see [124, 85, 226] for some recent works. In unconstrained factorizations, both factors are typically dense and contain both positive and negative entries; hence they cannot be easily interpreted. With NMF, it is easy to interpret the feature users: $W(i, k)$ is the preference of the kth feature user for the ith movie. It is also easy to interpret the behavior of users as linear combinations of the feature users: $H(k, j)$ quantifies how much the jth user behaves as the kth feature user. Let us illustrate this through a simple example.

Example 9.1. Let us consider the matrix

$$X = \begin{pmatrix} 2 & 3 & 2 & ? & ? \\ ? & 1 & ? & 3 & 2 \\ 1 & ? & 4 & 1 & ? \\ 5 & 4 & ? & 3 & 2 \\ ? & 1 & 2 & ? & 4 \\ 1 & ? & 3 & 4 & 3 \end{pmatrix},$$

where each row corresponds to a movie, each column corresponds to a user, and each entry is the score in $\{1, 2, 3, 4, 5\}$ of a given user for a given movie. Let us define $P \in \{0, 1\}^{m \times n}$ as

$$P(i, j) = \begin{cases} 1 & \text{if } X(i, j) \text{ is observed,} \\ 0 & \text{if } X(i, j) = ?. \end{cases}$$

Most algorithms for NMF can be easily adapted to the NMF problem with missing data and, more generally, to the weighted NMF problem, that is, to the problem

$$\min_{W \geq 0, H \geq 0} \sum_{i,j} P_{i,j}(X - WH)_{i,j}^2.$$

For example, for HALS, the optimal update for the ℓth column of W is given by

$$W(:, \ell) \leftarrow \frac{[(R_\ell \circ P)H(\ell, :)^\top]}{[P(H(\ell, :)^{\circ 2})^\top]},$$

where $R_\ell = X - \sum_{k \neq \ell} W(:, k)H(k, :)$, and similarly for the rows of H. The derivations of these updates are obtained in the same way as HALS, and we leave it as an exercise for the interested reader; see [Matlab file: WLRA.m].

Using $r = 3$ and a random initialization,[66] weighted NMF provides the solution (displayed with one digit of accuracy)

$$X = \begin{pmatrix} 2 & 3 & 2 & ? & ? \\ ? & 1 & ? & 3 & 2 \\ 1 & ? & 4 & 1 & ? \\ 5 & 4 & ? & 3 & 2 \\ ? & 1 & 2 & ? & 4 \\ 1 & ? & 3 & 4 & 3 \end{pmatrix} \approx \begin{pmatrix} 1.5 & 1.7 & 1.8 \\ 2.3 & 0.4 & 0.3 \\ 0.0 & 5.0 & 1.2 \\ 2.4 & 0.1 & 5.0 \\ 5.0 & 0.0 & 0.0 \\ 2.8 & 2.4 & 0.0 \end{pmatrix} \begin{pmatrix} 0.4 & 0.2 & 0.4 & 1.2 & 0.8 \\ 0.0 & 0.9 & 0.8 & 0.2 & 0.3 \\ 0.8 & 0.7 & 0.1 & 0.0 & 0.0 \end{pmatrix}$$

$$= \begin{pmatrix} 2.0 & 3.0 & 2.0 & 2.1 & 1.7 \\ 1.0 & 1.0 & 1.3 & 3.0 & 2.0 \\ 1.0 & 5.3 & 4.0 & 1.0 & 1.6 \\ 5.0 & 4.0 & 1.4 & 3.0 & 2.0 \\ 1.8 & 1.0 & 2.0 & 6.2 & 4.0 \\ 1.0 & 2.7 & 3.0 & 4.0 & 3.0 \end{pmatrix},$$

with small root-mean-square error, namely

$$\frac{1}{\|P\|_1} \sqrt{\sum_{i,j} P_{i,j}(X - WH)_{i,j}^2} = 1.9 \, 10^{-6};$$

---

[66]Weighted NMF is even more sensitive to initialization than NMF because already the rank-one problem is NP-hard [196]; see Section 6.4.

see [Matlab file: RecomSys.m]. Based on this model, we should, for example, recommend the fifth movie to the fourth user (with a prediction[67] of 6.2), while we should not recommend the fourth movie to the third user (with a prediction of 0.6).

We have scaled the factorization above so that $\|W(:,k)\|_\infty = 5$ for all k which allows us to interpret the feature users (that is, the columns of W) meaningfully. There are three feature users (the columns of W) with rather different behaviors, as they all have a different favorite movie (that is, W has a single entry equal to 5 in each column). The users behave as different linear combinations of these feature users. For example, the fourth user behaves mostly as the first feature user with $H(:,4) = (1.2, 0.2, 0)^\top$, and the first user behaves midway between the first and third feature users with $H(:,1) = (0.4, 0, 0.8)^\top$. ∎

Low-rank matrix completion based on weighted NMF can be used in other contexts such as image completion [404, 233].

9.6 ▪ Other applications

Other applications of NMF include the following:

- Identification of hidden Markov models [296].

- Community detection [386, 488, 246]: NMF finds subsets of rows and columns of X via the rank-one factors $W(:,k)H(k,:)$ that are highly connected; see also Sections 5.4.7 and 5.4.9, and in particular Figure 5.9 (page 180).

- Air quality control by identifying the compounds present in the atmosphere [371, 382].

- Explanation of contingency tables using as few independent variables as possible [120]; see also Section 1.4.6.

- Decomposition of various types of spectra into their constitutive materials such as in blind HU (Section 1.3.2) and SMCR (Section 9.3). Other examples include gas chromatography-mass spectrometry [371, 65], the analysis of time-resolved Raman spectra [331], and the decomposition of medical images such as magnetic resonance (spectroscopic) imaging [318, 406].

- Decomposition of temperature time series via the extraction of physical meaningful sources [480].

- Decompositions of global spatio-temporal atmospheric chemistry data in order to extract major features of atmospheric chemistry, such as summertime surface pollution and biomass burning activities [466].

- Discovering the signatures characterizing geothermal resources and favorable geothermal systems sites from geothermal data sets [467].

- The problem of nonintrusive appliance load monitoring, which is the energy disaggregation of fixed and shiftable loads [346, 499]. For instance, daily energy consumption of a household unit can be disaggregated into various loads such as refrigerators, air conditioner, and lighting.

---

[67]Nothing prevents entries of WH from being larger than five nor smaller than one. Kannan, Ishteva, and Park [272] have developed a bounded low-rank approximation model with the constraints $l \leq WH \leq u$ for some parameters $l < u$. One should use $l = 1$ and $u = 5$ for the Netflix prize, preventing predictions outside the admissible range.

- Prediction of epileptic seizures using electroencephalographic signals [436].

- Identification of low-dimensional features within large-scale neural recordings [336].

- Computation of the temporal psychovisual modulation which is a paradigm of information display [484].

We refer the reader to the book [98, Chapter 8] and the surveys [433, 40, 125, 382, 478, 425, 512, 189, 190, 170] and the references therein for other applications of NMF.

9.7 ▪ Take-home messages

Since the paper of Lee and Seung [303] in 1999, NMF and its variants have been used in many different contexts. As soon as your data is nonnegative, it is worth considering such models. They are easily interpretable and provide sparse and part-based representations. The basis matrix W is nonnegative and hence can be interpreted in the same way as the data, while the activation/weight matrix H is nonnegative (and typically sparse) and hence tells us the importance of each basis vector (that is, each column of W) in each data sample (that is, each column of X). Moreover, in several applications, NMF is motivated by physical models where the entries of W and H represent nonnegative physical quantities, for example in blind HU (Sections 1.4.2 and 1.3.2), SMCR (Sections 1.4.1 and 9.3), and audio source separation (Section 1.3.4).

Bibliography

[1] Abbe, E.: Community detection and stochastic block models: recent developments. Journal of Machine Learning Research **18**(177), 1–86 (2018) (Cited on p. 184)

[2] Abdolali, M., Gillis, N.: Simplex-structured matrix factorization: sparsity-based identifiability and provably correct algorithm. arXiv preprint arXiv:2007.11446 (2020) (Cited on pp. 155, 213)

[3] Aggarwal, A., Booth, H., O'Rourke, J., Suri, S.: Finding minimal convex nested polygons. Information and Computation **83**(1), 98–110 (1989) (Cited on pp. xxi, 13, 44, 45, 46, 49)

[4] Aharon, M., Elad, M., Bruckstein, A., et al.: K-SVD: An algorithm for designing overcomplete dictionaries for sparse representation. IEEE Transactions on Signal Processing **54**(11), 4311 (2006) (Cited on pp. 3, 150)

[5] Airoldi, E.M., Blei, D.M., Fienberg, S.E., Xing, E.P.: Mixed membership stochastic blockmodels. Journal of Machine Learning Research **9**(Sep), 1981–2014 (2008) (Cited on p. 184)

[6] Alexe, G., Alexe, S., Crama, Y., Foldes, S., Hammer, P.L., Simeone, B.: Consensus algorithms for the generation of all maximal bicliques. Discrete Applied Mathematics **145**(1), 11–21 (2004) (Cited on p. 72)

[7] Ambikapathi, A., Chan, T.H., Chi, C.Y., Keizer, K.: Hyperspectral data geometry-based estimation of number of endmembers using p-norm-based pure pixel identification algorithm. IEEE Transactions on Geoscience and Remote Sensing **51**(5), 2753–2769 (2012) (Cited on p. 229)

[8] Ambikapathi, A., Chan, T.H., Ma, W.K., Chi, C.Y.: Chance-constrained robust minimum-volume enclosing simplex algorithm for hyperspectral unmixing. IEEE Transactions on Geoscience and Remote Sensing **49**(11), 4194–4209 (2011) (Cited on p. 138)

[9] Ang, A.M.S.: Nonnegative matrix and tensor factorizations: Models, algorithms and applications. Ph.D. thesis, University of Mons (2020), `https://angms.science/doc/PhDThesis.pdf` (Cited on p. 169)

[10] Ang, A.M.S., Cohen, J.E., Gillis, N., Hien, L.T.K.: Accelerating block coordinate descent for nonnegative tensor factorization. arXiv preprint arXiv:2001.04321 (2020) (Cited on pp. 293, 294, 296)

[11] Ang, A.M.S., Gillis, N.: Accelerating nonnegative matrix factorization algorithms using extrapolation. Neural Computation **31**(2), 417–439 (2019) (Cited on p. 293)

[12] Ang, A.M.S., Gillis, N.: Algorithms and comparisons of nonnegative matrix factorizations with volume regularization for hyperspectral unmixing. IEEE Journal of Selected Topics in Applied Earth Observations and Remote Sensing **12**(12), 4843–4853 (2019) (Cited on pp. 144, 148)

[13] Araújo, U., Saldanha, B., Galvão, R., Yoneyama, T., Chame, H., Visani, V.: The successive projections algorithm for variable selection in spectroscopic multicomponent analysis. Chemometrics and Intelligent Laboratory Systems **57**(2), 65–73 (2001) (Cited on pp. 224, 229)

[14] Arora, S., Ge, R., Halpern, Y., Mimno, D., Moitra, A., Sontag, D., Wu, Y., Zhu, M.: A practical algorithm for topic modeling with provable guarantees. In: Proceedings of the 30th International Conference on Machine Learning, pp. 280–288 (2013) (Cited on pp. 183, 193, 208, 210, 226, 230, 231, 257, 258, 259, 308)

[15] Arora, S., Ge, R., Kannan, R., Moitra, A.: Computing a nonnegative matrix factorization–provably. In: Proceedings of STOC '12, pp. 145–162 (2012) (Cited on pp. 52, 53, 200, 208, 232, 249, 257)

[16] Arora, S., Ge, R., Kannan, R., Moitra, A.: Computing a nonnegative matrix factorization—provably. SIAM Journal on Computing 45(4), 1582–1611 (2016) (Cited on pp. 52, 53, 200)

[17] Asteris, M., Papailiopoulos, D., Dimakis, A.G.: Orthogonal NMF through subspace exploration. In: Advances in Neural Information Processing Systems (NIPS), pp. 343–351 (2015) (Cited on p. 136)

[18] Atif, S.M., Qazi, S., Gillis, N.: Improved SVD-based initialization for nonnegative matrix factorization using low-rank correction. Pattern Recognition Letters 122, 53–59 (2019) (Cited on p. 301)

[19] Ban, F., Bhattiprolu, V., Bringmann, K., Kolev, P., Lee, E., Woodruff, D.P.: A PTAS for ℓ_p-low rank approximation. In: Proceedings of the Thirtieth Annual ACM-SIAM Symposium on Discrete Algorithms, pp. 747–766. SIAM (2019) (Cited on pp. 204, 205)

[20] Ban, F., Woodruff, D., Zhang, R.: Regularized weighted low rank approximation. In: Advances in Neural Information Processing Systems (NIPS), pp. 4061–4071 (2019) (Cited on p. 203)

[21] Bancilhon, F.: A geometric model for stochastic automata. IEEE Transactions on Computers 100(12), 1290–1299 (1974) (Cited on p. 13)

[22] Baranzini, S.E., Mousavi, P., Rio, J., Caillier, S.J., Stillman, A., Villoslada, P., Wyatt, M.M., Comabella, M., Greller, L.D., Somogyi, R., et al.: Transcription-based prediction of response to IFNβ using supervised computational methods. PLoS Biology 3(1), e2 (2004) (Cited on p. 310)

[23] Barefoot, C., Hefner, K.A., Jones, K.F., Lundgren, J.R.: Biclique covers of the complements of cycles and paths in a digraph. Congressus Numerantium 53, 133–146 (1986) (Cited on p. 72)

[24] Barioli, F., Berman, A.: The maximal cp-rank of rank k completely positive matrices. Linear Algebra and Its Applications 363, 17–33 (2003) (Cited on p. 79)

[25] Bärmann, A.: Polyhedral Approximation of Second-Order Cone Robust Counterparts, pp. 163–177. Springer Fachmedien Wiesbaden (2016) (Cited on p. 90)

[26] Basu, S., Pollack, R., Roy, M.F.: On the combinatorial and algebraic complexity of quantifier elimination. Journal of the ACM 43(6), 1002–1045 (1996) (Cited on p. 52)

[27] Beasley, L.B., Laffey, T.J.: Real rank versus nonnegative rank. Linear Algebra and Its Applications 431(12), 2330–2335 (2009) (Cited on pp. 65, 70)

[28] Beck, A.: First-Order Methods in Optimization. MOS-SIAM Series on Optimization vol. 25. SIAM (2017) (Cited on pp. 268, 270, 290)

[29] Beer, A.: Bestimmung der Absorption des rothen Lichts in farbigen Flussigkeiten. Annalen der Physik 162, 78–88 (1852) (Cited on p. 12)

[30] Belachew, M.T.: Efficient algorithm for sparse symmetric nonnegative matrix factorization. Pattern Recognition Letters 125, 735–741 (2019) (Cited on pp. 185, 287)

[31] Bell, R.M., Koren, Y.: Lessons from the Netflix prize challenge. SiGKDD Explorations 9(2), 75–79 (2007) (Cited on p. 3)

[32] Ben-Tal, A., Nemirovski, A.: On polyhedral approximations of the second-order cone. Mathematics of Operations Research 26(2), 193–205 (2001) (Cited on p. 90)

[33] Benetos, E., Dixon, S., Duan, Z., Ewert, S.: Automatic music transcription: An overview. IEEE Signal Processing Magazine 36(1), 20–30 (2019) (Cited on p. 11)

[34] Bennett, J., Lanning, S., et al.: The Netflix prize. In: Proceedings of KDD Cup and Workshop, p. 35. New York (2007) (Cited on p. 3)

[35] Benson, A.R., Lee, J.D., Rajwa, B., Gleich, D.F.: Scalable methods for nonnegative matrix factorizations of near-separable tall-and-skinny matrices. In: Advances in Neural Information Processing Systems (NIPS), pp. 945–953 (2014) (Cited on p. 216)

[36] Berman, A., Dur, M., Shaked-Monderer, N.: Open problems in the theory of completely positive and copositive matrices. Electronic Journal of Linear Algebra 29(1), 46–58 (2015) (Cited on p. 179)

[37] Berman, A., Plemmons, R.J.: Rank factorization of nonnegative matrices. SIAM Review **15**(3), 655 (1973) (Cited on p. 13)

[38] Berman, A., Plemmons, R.J.: Nonnegative Matrices in the Mathematical Sciences, Classics in Applied Mathematics vol. 9. SIAM (1994) (Cited on pp. 105, 174, 197)

[39] Berman, A., Shaked-Monderer, N.: Completely Positive Matrices. World Scientific (2003) (Cited on pp. 79, 179)

[40] Berry, M.W., Browne, M., Langville, A.N., Pauca, V.P., Plemmons, R.J.: Algorithms and applications for approximate nonnegative matrix factorization. Computational Statistics & Data Analysis **52**(1), 155–173 (2007) (Cited on p. 314)

[41] Bertsekas, D.: Corrections for the Book Nonlinear Programming: Second Edition (1999). `http://www.athenasc.com/nlperrata.pdf` (Cited on p. 268)

[42] Bertsekas, D.: Nonlinear Programming: Second Edition. Athena Scientific (1999) (Cited on p. 268)

[43] Bhadury, J., Chandrasekaran, R.: Finding the set of all minimal nested convex polygons. In: Proceedings of the 8th Canadian Conference on Computational Geometry, pp. 26–31 (1996) (Cited on p. 44)

[44] Biggs, M., Ghodsi, A., Vavasis, S.: Nonnegative matrix factorization via rank-one downdate. In: Proceedings of the 25th International Conference on Machine Learning, pp. 64–71 (2008) (Cited on p. 302)

[45] Bioucas-Dias, J.M., Plaza, A., Dobigeon, N., Parente, M., Du, Q., Gader, P., Chanussot, J.: Hyperspectral unmixing overview: Geometrical, statistical, and sparse regression-based approaches. IEEE Journal of Selected Topics in Applied Earth Observations and Remote Sensing **5**(2), 354–379 (2012) (Cited on pp. 13, 209, 247, 248)

[46] Blei, D.M.: Probabilistic topic models. Communicatoins of the ACM **55**(4), 77–84 (2012) (Cited on p. 192)

[47] Blei, D.M., Ng, A.Y., Jordan, M.I.: Latent Dirichlet allocation. Journal of Machine Learning Research **3**(Jan), 993–1022 (2003) (Cited on pp. 183, 192, 193)

[48] Boardman, J.W.: Automating spectral unmixing of AVIRIS data using convex geometry concepts. In: Summaries of the 4th Annual JPL Airborne Geoscience Workshop, vol. 1, pp. 11–14 (1993) (Cited on p. 13)

[49] Boardman, J.W., Kruse, F.A., Green, R.O.: Mapping target signatures via partial unmixing of AVIRIS data. In: Summaries of the JPL Airborne Earth Science Workshop, Pasadena, CA, pp. 23–26 (1995) (Cited on pp. 209, 247)

[50] Bobin, J., Starck, J.L., Fadili, J., Moudden, Y.: Sparsity and morphological diversity in blind source separation. IEEE Transactions on Image Processing **16**(11), 2662–2674 (2007) (Cited on p. 170)

[51] Bocci, C., Carlini, E., Rapallo, F.: Perturbation of matrices and nonnegative rank with a view toward statistical models. SIAM Journal on Matrix Analysis and Applications **32**(4), 1500–1512 (2011) (Cited on pp. 47, 57)

[52] Bolte, J., Sabach, S., Teboulle, M.: Proximal alternating linearized minimization for nonconvex and nonsmooth problems. Mathematical Programming **146**(1–2), 459–494 (2014) (Cited on p. 270)

[53] Borgen, O.S., Davidsen, N., Mingyang, Z., Øyen, Ø.: The multivariate n-component resolution problem with minimum assumptions. Microchimica Acta **89**(1–6), 63–73 (1986) (Cited on p. 12)

[54] Boutsidis, C., Gallopoulos, E.: SVD based initialization: A head start for nonnegative matrix factorization. Pattern Recognition **41**(4), 1350–1362 (2008) (Cited on p. 300)

[55] Boutsidis, C., Mahoney, M.W., Drineas, P.: An improved approximation algorithm for the column subset selection problem. In: Proceedings of the 20th Annual ACM-SIAM Symposium on Discrete Algorithms, pp. 968–977. SIAM (2009) (Cited on p. 211)

[56] Boyd, S., Vandenberghe, L.: Convex Optimization. Cambridge University Press (2004) (Cited on p. 238)

[57] Braun, G., Fiorini, S., Pokutta, S., Steurer, D.: Approximation limits of linear programs (beyond hierarchies). In: Proceedings of the Annual IEEE Symposium on Foundations of Computer Science, pp. 480–489 (2012) (Cited on pp. 76, 92)

[58] Braun, G., Fiorini, S., Pokutta, S., Steurer, D.: Approximation limits of linear programs (beyond hierarchies). Mathematics of Operations Research **40**(3), 756–772 (2015) (Cited on pp. 76, 92)

[59] Braun, G., Pokutta, S.: Common information and unique disjointness. In: 2013 IEEE 54th Annual Symposium on Foundations of Computer Science, pp. 688–697. IEEE (2013) (Cited on p. 92)

[60] Braun, G., Pokutta, S.: The matching polytope does not admit fully-polynomial size relaxation schemes. In: Proceedings of the 26th Annual ACM-SIAM Symposium on Discrete Algorithms, pp. 837–846. SIAM (2015) (Cited on pp. 15, 92)

[61] Braun, G., Pokutta, S.: Common information and unique disjointness. Algorithmica **76**(3), 597–629 (2016) (Cited on pp. 82, 83, 92, 95)

[62] Braverman, M., Moitra, A.: An information complexity approach to extended formulations. In: Proceedings of the 45th Annual ACM Symposium on Theory of Computing, pp. 161–170. ACM (2013) (Cited on p. 92)

[63] Bro, R.: Multi-way analysis in the food industry: Models, algorithms, and applications. Ph.D. thesis, University of Copenhagen (1998). http://curis.ku.dk/ws/files/13035961/Rasmus_Bro.pdf (Cited on p. 286)

[64] Bro, R., Acar, E., Kolda, T.G.: Resolving the sign ambiguity in the singular value decomposition. Journal of Chemometrics **22**(2), 135–140 (2008) (Cited on p. 197)

[65] Bro, R., Sidiropoulos, N.D.: Least squares algorithms under unimodality and non-negativity constraints. Journal of Chemometrics **12**(4), 223–247 (1998) (Cited on pp. 169, 313)

[66] Brunet, J.P., Tamayo, P., Golub, T.R., Mesirov, J.P.: Metagenes and molecular pattern discovery using matrix factorization. Proceedings of the National Academy of Sciences **101**(12), 4164–4169 (2004) (Cited on p. 309)

[67] Buciu, I., Nikolaidis, N., Pitas, I.: Nonnegative matrix factorization in polynomial feature space. IEEE Transactions on Neural Networks **19**(6), 1090–1100 (2008) (Cited on p. 185)

[68] Burer, S., Monteiro, R.D.: A nonlinear programming algorithm for solving semidefinite programs via low-rank factorization. Mathematical Programming **95**(2), 329–357 (2003) (Cited on p. 3)

[69] Businger, P., Golub, G.: Linear least squares solutions by Householder transformations. Numerische Mathematik **7**, 269–276 (1965) (Cited on p. 229)

[70] de Caen, D., Gregory, D., Pullman, N.: The Boolean rank of zero-one matrices. In: Proceedings of the 3rd Caribbean Conference on Combinatorics and Computing, pp. 169–173 (1981) (Cited on pp. 69, 70)

[71] Cai, D., He, X., Han, J., Huang, T.S.: Graph regularized nonnegative matrix factorization for data representation. IEEE Transactions on Pattern Analysis and Machine Intelligence **33**(8), 1548–1560 (2011) (Cited on p. 170)

[72] Campbell, S., Poole, G.: Computing nonnegative rank factorizations. Linear Algebra and Its Applications **35**, 175–182 (1981) (Cited on p. 14)

[73] Candès, E.J., Li, X., Ma, Y., Wright, J.: Robust principal component analysis? Journal of the ACM **58**(3), 11 (2011) (Cited on p. 3)

[74] Cao, Y., Eggermont, P.P., Terebey, S.: Cross burg entropy maximization and its application to ringing suppression in image reconstruction. IEEE Transactions on Image Processing **8**(2), 286–292 (1999) (Cited on p. 270)

[75] Casalino, G., Del Buono, N., Mencar, C.: Subtractive clustering for seeding non-negative matrix factorizations. Information Sciences **257**, 369–387 (2014) (Cited on p. 301)

[76] Casalino, G., Gillis, N.: Sequential dimensionality reduction for extracting localized features. Pattern Recognition **63**, 15–29 (2017) (Cited on p. 177)

[77] Çivril, A., Magdon-Ismail, M.: On selecting a maximum volume sub-matrix of a matrix and related problems. Theoretical Computer Science **410**(47–49), 4801–4811 (2009) (Cited on pp. 204, 227, 229)

[78] Çivril, A., Magdon-Ismail, M.: Exponential inapproximability of selecting a maximum volume sub-matrix. Algorithmica **65**(1), 159–176 (2013) (Cited on pp. 204, 229)

[79] Chan, T.H., Chi, C.Y., Huang, Y.M., Ma, W.K.: A convex analysis-based minimum-volume enclosing simplex algorithm for hyperspectral unmixing. IEEE Transactions on Signal Processing **57**(11), 4418–4432 (2009) (Cited on p. 138)

[80] Chan, T.H., Ma, W.K., Ambikapathi, A., Chi, C.Y.: A simplex volume maximization framework for hyperspectral endmember extraction. IEEE Transactions on Geoscience and Remote Sensing **49**(11), 4177–4193 (2011) (Cited on pp. 227, 229)

[81] Chan, T.H., Ma, W.K., Chi, C.Y., Wang, Y.: A convex analysis framework for blind separation of non-negative sources. IEEE Transactions on Signal Processing **56**(10), 5120–5134 (2008) (Cited on p. 209)

[82] Chen, G., Wang, F., Zhang, C.: Collaborative filtering using orthogonal nonnegative matrix tri-factorization. Information Processing & Management **45**(3), 368–379 (2009) (Cited on p. 181)

[83] Chen, J., Huo, X.: Theoretical results on sparse representations of multiple-measurement vectors. IEEE Transactions on Signal processing **54**(12), 4634–4643 (2006) (Cited on p. 252)

[84] Chen, J.C.: The nonnegative rank factorizations of nonnegative matrices. Linear Algebra and Its Applications **62**, 207–217 (1984) (Cited on p. 14)

[85] Chen, R., Varshney, L.R.: Optimal recovery of missing values for non-negative matrix factorization. bioRxiv 647560 (2019) (Cited on p. 311)

[86] Chi, E.C., Kolda, T.G.: On tensors, sparsity, and nonnegative factorizations. SIAM Journal on Matrix Analysis and Applications **33**(4), 1272–1299 (2012) (Cited on pp. 163, 164, 279)

[87] Chi, Y., Lu, Y.M., Chen, Y.: Nonconvex optimization meets low-rank matrix factorization: An overview. IEEE Transactions on Signal Processing **67**(20), 5239–5269 (2019) (Cited on p. 3)

[88] Chistikov, D., Kiefer, S., Marusic, I., Shirmohammadi, M., Worrell, J.: On restricted nonnegative matrix factorization. In: I. Chatzigiannakis, M. Mitzenmacher, Y. Rabani, D. Sangiorgi (eds.) 43rd International Colloquium on Automata, Languages, and Programming (ICALP 2016), Leibniz International Proceedings in Informatics (LIPIcs), vol. 55, pp. 103:1–103:14. Schloss Dagstuhl–Leibniz-Zentrum fuer Informatik, Dagstuhl, Germany (2016). DOI:10.4230/LIPIcs.ICALP.2016.103 (Cited on pp. 13, 35, 36, 40, 54)

[89] Chistikov, D., Kiefer, S., Marušić, I., Shirmohammadi, M., Worrell, J.: Nonnegative matrix factorization requires irrationality. SIAM Journal on Applied Algebra and Geometry **1**(1), 285–307 (2017) (Cited on p. 54)

[90] Choi, H., Choi, S., Katake, A., Choe, Y.: Learning α-integration with partially-labeled data. In: IEEE International Conference on Acoustics, Speech and Signal Processing (ICASSP), pp. 2058–2061. IEEE (2010) (Cited on p. 164)

[91] Choi, S.: Algorithms for orthogonal nonnegative matrix factorization. In: Proceedings of the International Joint Conference on Neural Networks, pp. 1828–1832 (2008) (Cited on p. 136)

[92] Chouh, M., Hanafi, M., Boukhetala, K.: Semi-nonnegative rank for real matrices and its connection to the usual rank. Linear Algebra and Its Applications **466**, 27–37 (2015) (Cited on p. 173)

[93] Chu, M.T., Lin, M.M.: Low-dimensional polytope approximation and its applications to nonnegative matrix factorization. SIAM Journal on Scientific Computing **30**(3), 1131–1155 (2008) (Cited on p. 22)

[94] Chung, F.: Spectral Graph Theory. CBMS Regional Conference Series in Mathematics 92. AMS (1997) (Cited on p. 3)

[95]   Cichocki, A., Phan, A.: Fast local algorithms for large scale nonnegative matrix and tensor factor-
       izations. IEICE Transactions on Fundamentals of Electronics **E92-A**(3), 708–721 (2009) (Cited on
       p. 286)

[96]   Cichocki, A., Zdunek, R., Amari, S.i.: New algorithms for non-negative matrix factorization in
       applications to blind source separation. In: 2006 IEEE International Conference on Acoustics Speech
       and Signal Processing Proceedings, vol. 5, pp. 621–624. IEEE (2006) (Cited on p. 280)

[97]   Cichocki, A., Zdunek, R., Amari, S.i.: Hierarchical ALS Algorithms for Nonnegative Matrix and
       3D Tensor Factorization. In: Lecture Notes in Computer Science, vol. 4666, Springer, pp. 169–176
       (2007) (Cited on p. 286)

[98]   Cichocki, A., Zdunek, R., Phan, A.H., Amari, S.i.: Nonnegative Matrix and Tensor Factorizations:
       Applications to Exploratory Multi-Way Data Analysis and Blind Source Separation. John Wiley &
       Sons (2009) (Cited on pp. xii, xiv, 5, 163, 168, 193, 280, 283, 284, 299, 314)

[99]   Clarkson, K.: Algorithms for polytope covering and approximation. In: Proceedings of the Third
       Workshop on Algorithms and Data Structures, pp. 246–252 (1993) (Cited on p. 51)

[100] Cohen, J., Rothblum, U.: Nonnegative ranks, decompositions and factorization of nonnegative ma-
       trices. Linear Algebra and Its Applications **190**, 149–168 (1993) (Cited on pp. 14, 23, 53, 55,
       63)

[101] Cohen, J.E., Gillis, N.: Dictionary-based tensor canonical polyadic decomposition. IEEE Transac-
       tions on Signal Processing **66**(7), 1876–1889 (2017) (Cited on p. 172)

[102] Cohen, J.E., Gillis, N.: Identifiability of complete dictionary learning. SIAM Journal on Mathematics
       of Data Science **1**(3), 518–536 (2019) (Cited on pp. 152, 153)

[103] Cohen, J.E., Gillis, N.: Nonnegative low-rank sparse component analysis. In: IEEE International
       Conference on Acoustics, Speech and Signal Processing (ICASSP), pp. 8226–8230 (2019) (Cited
       on p. 153)

[104] Comon, P., Jutten, C.: Handbook of Blind Source Separation: Independent Component Analysis and
       Applications. Academic Press (2010) (Cited on p. 3)

[105] Conforti, M., Cornuéjols, G., Zambelli, G.: Extended formulations in combinatorial optimization.
       4OR: A Quarterly Journal of Operations Research **10**(1), 1–48 (2010) (Cited on p. 84)

[106] Čopar, A., Zupan, B., Zitnik, M.: Fast optimization of non-negative matrix tri-factorization. PloS
       One **14**(6) (2019) (Cited on p. 287)

[107] Cotter, S.F., Rao, B.D., Engan, K., Kreutz-Delgado, K.: Sparse solutions to linear inverse problems
       with multiple measurement vectors. IEEE Transactions on Signal Processing **53**(7), 2477–2488
       (2005) (Cited on p. 252)

[108] Craig, M.D.: Unsupervised unmixing of remotely sensed images. In: Proceedings of the Fifth
       Australasian Remote Sensing Conference, pp. 324–330 (1990) (Cited on p. 12)

[109] Craig, M.D.: Minimum-volume transforms for remotely sensed data. IEEE Transactions on Geo-
       science and Remote Sensing **32**(3), 542–552 (1994) (Cited on pp. 12, 138)

[110] Cutler, A., Breiman, L.: Archetypal analysis. Technometrics **36**(4), 338–347 (1994) (Cited on
       p. 172)

[111] Damle, A., Sun, Y.: A geometric approach to archetypal analysis and nonnegative matrix factoriza-
       tion. Technometrics **59**(3), 361–370 (2017) (Cited on p. 247)

[112] Das, G.: Approximation schemes in computational geometry. Ph.D. thesis, University of Wisconsin–
       Madison (1990) (Cited on pp. 20, 36, 50, 51)

[113] Das, G., Goodrich, M.: On the complexity of optimization problems for three-dimensional convex
       polyhedra and decision trees. Computational Geometry: Theory and Applications **8**, 123–137 (1997)
       (Cited on p. 51)

[114] Das, G., Joseph, D.: The complexity of minimum convex nested polyhedra. In: Proceedings of the
       2nd Canadian Conference on Computational Geometry, pp. 296–301 (1990) (Cited on pp. 13, 20,
       36, 50, 51)

[115] d'Aspremont, A., El Ghaoui, L., Jordan, M.I., Lanckriet, G.R.G.: A direct formulation for sparse PCA using semidefinite programming. SIAM Review **49**(3), 434–448 (2007) (Cited on pp. 3, 150)

[116] Daube-Witherspoon, M., Muehllehner, G.: An iterative image space reconstruction algorithm suitable for volume ECT. IEEE Transactions on Medical Imaging **5**, 61–66 (1986) (Cited on p. 270)

[117] De Handschutter, P., Gillis, N., Siebert, X.: Deep matrix factorizations. arXiv:2010.00380 (2020) (Cited on p. 184)

[118] De Handschutter, P., Gillis, N., Vandaele, A., Siebert, X.: Near-convex archetypal analysis. IEEE Signal Processing Letters **27**(1), 81–85 (2020) (Cited on pp. 172, 193, 211)

[119] De Juan, A., Tauler, R.: Multivariate curve resolution (MCR) from 2000: progress in concepts and applications. Critical Reviews in Analytical Chemistry **36**(3-4), 163–176 (2006) (Cited on p. 12)

[120] De Leeuw, J., Van der Heijden, P.G.: Reduced rank models for contingency tables. Biometrika **78**(1), 229–232 (1991) (Cited on pp. 14, 313)

[121] De Pierro, A.R.: On the relation between the ISRA and the EM algorithm for positron emission tomography. IEEE Transactions on Medical Imaging **12**(2), 328–333 (1993) (Cited on p. 270)

[122] Debals, O., Van Barel, M., De Lathauwer, L.: Nonnegative matrix factorization using nonnegative polynomial approximations. IEEE Signal Processing Letters **24**(7), 948–952 (2017) (Cited on p. 169)

[123] Degleris, A., Gillis, N.: A provably correct and robust algorithm for convolutive nonnegative matrix factorization. IEEE Transactions on Signal Processing **68** (2020). DOI:10.1109/TSP.2020.2984163 (Cited on pp. 185, 259)

[124] Del Corso, G.M., Romani, F.: Adaptive nonnegative matrix factorization and measure comparisons for recommender systems. Applied Mathematics and Computation **354**, 164–179 (2019) (Cited on p. 311)

[125] Devarajan, K.: Nonnegative matrix factorization: an analytical and interpretive tool in computational biology. PLoS Computational Biology **4**(7), e1000,029 (2008) (Cited on pp. 309, 314)

[126] Deville, Y.: From separability/identifiability properties of bilinear and linear-quadratic mixture matrix factorization to factorization algorithms. Digital Signal Processing **87**, 21–33 (2019) (Cited on p. 185)

[127] Dewez, J., Gillis, N., Glineur, F.: A geometric lower bound on the extension complexity of polytopes based on the f-vector. Discrete Applied Mathematics, to appear (Cited on p. 75)

[128] Dewez, J., Glineur, F.: Lower bounds on the nonnegative rank using a nested polytopes formulation. In: 28th European Symposium on Artificial Neural Networks-Computational Intelligence and Machine Learning (ESANN) (2020) (Cited on p. 47)

[129] Dikmen, O., Yang, Z., Oja, E.: Learning the information divergence. IEEE Transactions on Pattern Analysis and Machine Intelligence **37**(7), 1442–1454 (2015) (Cited on p. 164)

[130] Ding, C., He, X., Simon, H.D.: On the equivalence of nonnegative matrix factorization and spectral clustering. In: Proceedings of the 5th SIAM Conference on Data Mining, pp. 606–610 (2005) (Cited on p. 136)

[131] Ding, C., Li, T., Peng, W.: On the equivalence between non-negative matrix factorization and probabilistic latent semantic indexing. Computational Statistics & Data Analysis **52**(8), 3913–3927 (2008) (Cited on pp. 191, 192)

[132] Ding, C., Li, T., Peng, W., Park, H.: Orthogonal nonnegative matrix t-factorizations for clustering. In: Proceedings of the 12th ACM SIGKDD International Conference on Knowledge Discovery and Data Mining, pp. 126–135. ACM (2006) (Cited on pp. 136, 181, 280)

[133] Ding, C.H., Li, T., Jordan, M.I.: Convex and semi-nonnegative matrix factorizations. IEEE Transactions on Pattern Analysis and Machine Intelligence **32**(1), 45–55 (2008) (Cited on pp. 172, 173)

[134] Doan, X.V., Vavasis, S.: Finding the largest low-rank clusters with Ky Fan 2-k-norm and ℓ_1-norm. SIAM Journal on Optimization **26**(1), 274–312 (2016) (Cited on p. 197)

[135] Dobbins, M.G., Holmsen, A.F., Miltzow, T.: On the complexity of nesting polytopes. In: 35th European Workshop on Computational Geometry (EuroCG 2019) (2019) (Cited on pp. 20, 36, 54)

[136] Dobigeon, N., Tourneret, J.Y., Richard, C., Bermudez, J.C.M., McLaughlin, S., Hero, A.O.: Nonlinear unmixing of hyperspectral images: Models and algorithms. IEEE Signal Processing Magazine **31**(1), 82–94 (2013) (Cited on pp. 185, 209)

[137] Dong, B., Lin, M.M., Chu, M.T.: Nonnegative rank factorization—a heuristic approach via rank reduction. Numerical Algorithms **65**(2), 251–274 (2014) (Cited on pp. 53, 110, 128, 302)

[138] Donoho, D., Stodden, V.: When does non-negative matrix factorization give a correct decomposition into parts? In: Advances in Neural Information Processing Systems (NIPS), pp. 1141–1148 (2004) (Cited on pp. 6, 104, 108, 208)

[139] Du, R., Kuang, D., Drake, B., Park, H.: DC-NMF: nonnegative matrix factorization based on divide-and-conquer for fast clustering and topic modeling. Journal of Global Optimization **68**(4), 777–798 (2017) (Cited on p. 303)

[140] Dür, M.: Copositive programming—a survey. In: Recent Advances in Optimization and Its Applications in Engineering, pp. 3–20. Springer (2010) (Cited on p. 79)

[141] Eckart, C., Young, G.: The approximation of one matrix by another of lower rank. Psychometrika **1**(3), 211–218 (1936) (Cited on p. 196)

[142] Edmonds, J.: Maximum matching and a polyhedron with 0, 1-vertices. Journal of Research of the National Bureau of Standards B **69**(125–130), 55–56 (1965) (Cited on p. 15)

[143] Eftekhari, A., Hauser, R.A.: Principal component analysis by optimization of symmetric functions has no spurious local optima. SIAM Journal on Optimization **30**(1), 439–463 (2020) (Cited on p. 196)

[144] Eldén, L.: Matrix Methods in Data Mining and Pattern Recognition, 2nd ed. SIAM (2019) (Cited on p. 2)

[145] Elhamifar, E., Sapiro, G., Vidal, R.: See all by looking at a few: Sparse modeling for finding representative objects. In: 2012 IEEE Conference on Computer Vision and Pattern Recognition, pp. 1600–1607. IEEE (2012) (Cited on pp. 211, 251, 252)

[146] Erichson, N.B., Mendible, A., Wihlborn, S., Kutz, J.N.: Randomized nonnegative matrix factorization. Pattern Recognition Letters **104**, 1–7 (2018) (Cited on p. 292)

[147] Esposito, F., Gillis, N., Del Buono, N.: Orthogonal joint sparse NMF for microarray data analysis. Journal of Mathematical Biology **79**(1), 223–247 (2019) (Cited on pp. 309, 310, 311)

[148] Esser, E., Moller, M., Osher, S., Sapiro, G., Xin, J.: A convex model for nonnegative matrix factorization and dimensionality reduction on physical space. IEEE Transactions on Image Processing **21**(7), 3239–3252 (2012) (Cited on pp. 211, 251, 252)

[149] Fagot, D., Wendt, H., Févotte, C.: Nonnegative matrix factorization with transform learning. In: 2018 IEEE International Conference on Acoustics, Speech and Signal Processing (ICASSP), pp. 2431–2435. IEEE (2018) (Cited on p. 308)

[150] Fagot, D., Wendt, H., Févotte, C., Smaragdis, P.: Majorization-minimization algorithms for convolutive NMF with the beta-divergence. In: IEEE International Conference on Acoustics, Speech and Signal Processing (ICASSP), pp. 8202–8206. IEEE (2019) (Cited on p. 280)

[151] Fawzi, H.: On representing the positive semidefinite cone using the second-order cone. Mathematical Programming **175**(1), 109–118 (2019) (Cited on p. 93)

[152] Fawzi, H., Gouveia, J., Parrilo, P., Robinson, R., Thomas, R.: Positive semidefinite rank. Mathematical Programming **153**(1), 133–177 (2015) (Cited on p. 93)

[153] Fawzi, H., Gouveia, J., Parrilo, P.A., Saunderson, J., Thomas, R.R.: Lifting for simplicity: concise descriptions of convex sets. arXiv preprint arXiv:2002.09788 (2020) (Cited on pp. 84, 93)

[154] Fawzi, H., Parrilo, P.A.: Lower bounds on nonnegative rank via nonnegative nuclear norms. Mathematical Programming **153**(1), 41–66 (2015) (Cited on pp. 77, 78, 79, 80)

[155] Fawzi, H., Parrilo, P.A.: Self-scaled bounds for atomic cone ranks: applications to nonnegative rank and cp-rank. Mathematical Programming **158**(1-2), 417–465 (2016) (Cited on pp. 47, 80, 81, 82)

[156] Fazel, M., Hindi, H., Boyd, S.P.: Log-det heuristic for matrix rank minimization with applications to Hankel and Euclidean distance matrices. In: Proceedings of the 2003 American Control Conference, vol. 3, pp. 2156–2162. IEEE (2003) (Cited on p. 148)

[157] Févotte, C.: Majorization-minimization algorithm for smooth Itakura-Saito nonnegative matrix factorization. In: IEEE International Conference on Acoustics, Speech and Signal Processing (ICASSP), pp. 1980–1983. IEEE (2011) (Cited on p. 169)

[158] Févotte, C., Bertin, N., Durrieu, J.L.: Nonnegative matrix factorization with the Itakura-Saito divergence: With application to music analysis. Neural Computation **21**(3), 793–830 (2009) (Cited on pp. 10, 162, 163, 165, 270)

[159] Févotte, C., Dobigeon, N.: Nonlinear hyperspectral unmixing with robust nonnegative matrix factorization. arXiv preprint arXiv:1401.5649 (2014) (Cited on p. 164)

[160] Févotte, C., Idier, J.: Algorithms for nonnegative matrix factorization with the β-divergence. Neural Computation **23**(9), 2421–2456 (2011) (Cited on pp. 163, 270, 273, 276)

[161] Fiorini, S., Guo, K., Macchia, M., Walter, M.: Lower bound computations for the nonnegative rank. In: 17th Cologne-Twente Workshop on Graphs and Combinatorial Optimization (2019) (Cited on p. 82)

[162] Fiorini, S., Kaibel, V., Pashkovich, K., Theis, D.O.: Combinatorial bounds on nonnegative rank and extended formulations. Discrete Mathematics **313**(1), 67–83 (2013) (Cited on pp. 63, 64, 71, 72, 73, 83, 85)

[163] Fiorini, S., Massar, S., Pokutta, S., Tiwary, H.R., De Wolf, R.: Linear vs. semidefinite extended formulations: exponential separation and strong lower bounds. In: Proceedings of the 44th Annual ACM Symposium on Theory of Computing, pp. 95–106. ACM (2012) (Cited on pp. 15, 83, 85, 86, 95)

[164] Fiorini, S., Massar, S., Pokutta, S., Tiwary, H.R., Wolf, R.D.: Exponential lower bounds for polytopes in combinatorial optimization. Journal of the ACM **62**(2), 17 (2015) (Cited on pp. 15, 86, 95)

[165] Fiorini, S., Rothvoss, T., Tiwary, H.R.: Extended formulations for polygons. Discrete & Computational Geometry **48**(3), 658–668 (2012) (Cited on pp. 82, 88, 90, 92)

[166] Flamant, J., Miron, S., Brie, D.: Quaternion non-negative matrix factorization: Definition, uniqueness, and algorithm. IEEE Transactions on Signal Processing, **68** 1870–1883 (2020) (Cited on p. 185)

[167] Fortunato, S.: Community detection in graphs. Physics Reports **486**(3–5), 75–174 (2010) (Cited on p. 178)

[168] Freund, R.M., Orlin, J.B.: On the complexity of four polyhedral set containment problems. Mathematical Programming **33**(2), 139–145 (1985) (Cited on p. 118)

[169] Fu, X., Huang, K., Sidiropoulos, N.D.: On identifiability of nonnegative matrix factorization. IEEE Signal Processing Letters **25**(3), 328–332 (2018) (Cited on pp. 142, 144, 145)

[170] Fu, X., Huang, K., Sidiropoulos, N.D., Ma, W.K.: Nonnegative matrix factorization for signal and data analytics: Identifiability, algorithms, and applications. IEEE Signal Processing Magazine **36**(2), 59–80 (2019) (Cited on pp. 5, 108, 122, 124, 125, 148, 149, 184, 259, 314)

[171] Fu, X., Huang, K., Sidiropoulos, N.D., Shi, Q., Hong, M.: Anchor-free correlated topic modeling. IEEE Transactions on Pattern Analysis and Machine Intelligence **41**(5), 1056–1071 (2018) (Cited on pp. 139, 193, 259)

[172] Fu, X., Huang, K., Yang, B., Ma, W.K., Sidiropoulos, N.D.: Robust volume minimization-based matrix factorization for remote sensing and document clustering. IEEE Transactions on Signal Processing **64**(23), 6254–6268 (2016) (Cited on pp. 144, 149, 288)

[173] Fu, X., Ibrahim, S., Wai, H., Gao, C., Huang, K.: Block-randomized stochastic proximal gradient for low-rank tensor factorization. IEEE Transactions on Signal Processing **68**, 2170–2185 (2020) (Cited on p. 304)

[174] Fu, X., Ma, W.K.: Robustness analysis of structured matrix factorization via self-dictionary mixed-norm optimization. IEEE Signal Processing Letters **23**(1), 60–64 (2015) (Cited on p. 252)

[175] Fu, X., Ma, W.K., Chan, T.H., Bioucas-Dias, J.M.: Self-dictionary sparse regression for hyperspectral unmixing: Greedy pursuit and pure pixel search are related. IEEE Journal of Selected Topics in Signal Processing **9**(6), 1128–1141 (2015) (Cited on pp. 229, 252)

[176] Fu, X., Ma, W.K., Huang, K., Sidiropoulos, N.D.: Blind separation of quasi-stationary sources: exploiting convex geometry in covariance domain. IEEE Transactions on Signal Processing **63**(9), 2306–2320 (2015) (Cited on pp. 139, 140, 141)

[177] Fu, X., Vervliet, N., De Lathauwer, L., Huang, K., Gillis, N: Computing large-scale matrix and tensor decomposition with structured factors: A unified nonconvex optimization perspective. IEEE Signal Processing Magazine, **37**(5), 78–94 (2020) (Cited on p. 304)

[178] Full, W.E., Ehrlich, R., Klovan, J.: EXTENDED QMODEL–objective definition of external end members in the analysis of mixtures. Journal of the International Association for Mathematical Geology **13**(4), 331–344 (1981) (Cited on p. 138)

[179] Gabriel, K.R., Zamir, S.: Lower rank approximation of matrices by least squares with any choice of weights. Technometrics **21**(4), 489–498 (1979) (Cited on pp. 163, 203)

[180] Garey, M.R., Johnson, D.S.: Computers and Intractability. Freeman (2002) (Cited on pp. 51, 64)

[181] Gaujoux, R., Seoighe, C.: A flexible r package for nonnegative matrix factorization. BMC Bioinformatics **11**(1), 367 (2010) (Cited on p. 305)

[182] Gaussier, E., Goutte, C.: Relation between PLSA and NMF and implications. In: Proceedings of the 28th Annual International ACM SIGIR Conference on Research and Development in Information Retrieval, pp. 601–602. ACM (2005) (Cited on p. 191)

[183] Ge, R., Zou, J.: Intersecting faces: Non-negative matrix factorization with new guarantees. In: Proceedings of the 32nd International Conference on Machine Learning (ICML), pp. 2295–2303 (2015) (Cited on p. 155)

[184] Gillis, N.: Approximation et sous-approximation de matrices par factorisation positive: algorithmes, complexité et applications. Master's thesis, Université catholique de Louvain (2007) (Cited on p. 302)

[185] Gillis, N.: Nonnegative matrix factorization: Complexity, algorithms and applications. Ph.D. thesis, Université catholique de Louvain (2011) (Cited on pp. 75, 278, 284, 286)

[186] Gillis, N.: Sparse and unique nonnegative matrix factorization through data preprocessing. Journal of Machine Learning Research **13**(Nov), 3349–3386 (2012) (Cited on pp. 127, 129, 169, 176, 287)

[187] Gillis, N.: Robustness analysis of Hottopixx, a linear programming model for factoring nonnegative matrices. SIAM Journal on Matrix Analysis and Applications **34**(3), 1189–1212 (2013) (Cited on pp. 222, 253, 254, 255, 257)

[188] Gillis, N.: Successive nonnegative projection algorithm for robust nonnegative blind source separation. SIAM Journal on Imaging Sciences **7**(2), 1420–1450 (2014) (Cited on pp. 230, 232, 233, 234, 235, 236, 243, 257, 288)

[189] Gillis, N.: The why and how of nonnegative matrix factorization. In: J. Suykens, M. Signoretto, A. Argyriou (eds.) Regularization, Optimization, Kernels, and Support Vector Machines, Machine Learning and Pattern Recognition, chap. 12, pp. 257–291. Chapman & Hall/CRC (2014) (Cited on pp. 1, 5, 298, 314)

[190] Gillis, N.: Introduction to nonnegative matrix factorization. SIAG/OPT Views and News **25**(1), 7–16 (2017) (Cited on pp. 1, 29, 35, 314)

[191] Gillis, N.: Learning with nonnegative matrix factorizations. SIAM News **52**(5), 1–3 (2019), `https://sinews.siam.org/Portals/Sinews2/Issue%20Pdfs/sn_June2019.pdf` (Cited on pp. 1, 8, 11)

[192] Gillis, N.: Separable simplex-structured matrix factorization: Robustness of combinatorial approaches. In: IEEE International Conference on Acoustics, Speech and Signal Processing (ICASSP), pp. 5521–5525. IEEE (2019) (Cited on pp. 213, 217, 218, 219, 220, 257)

[193] Gillis, N.: Successive projection algorithm robust to outliers. In: IEEE International Workshop on Computational Advances in Multi-sensor Adaptive Processing (CAMSAP 2019), pp. 331–335. IEEE (2019) (Cited on p. 228)

[194] Gillis, N., Glineur, F.: Nonnegative factorization and the maximum edge biclique problem. arXiv preprint arXiv:0810.4225 (2008) (Cited on pp. 166, 198, 265, 277)

[195] Gillis, N., Glineur, F.: Using underapproximations for sparse nonnegative matrix factorization. Pattern Recognition **43**(4), 1676–1687 (2010) (Cited on pp. 176, 302)

[196] Gillis, N., Glineur, F.: Low-rank matrix approximation with weights or missing data is NP-hard. SIAM Journal on Matrix Analysis and Applications **32**(4), 1149–1165 (2011) (Cited on pp. 203, 312)

[197] Gillis, N., Glineur, F.: Accelerated multiplicative updates and hierarchical ALS algorithms for nonnegative matrix factorization. Neural Computation **24**(4), 1085–1105 (2012) (Cited on pp. 278, 279, 284, 287, 292, 294)

[198] Gillis, N., Glineur, F.: A multilevel approach for nonnegative matrix factorization. Journal of Computational and Applied Mathematics **236**(7), 1708–1723 (2012) (Cited on p. 292)

[199] Gillis, N., Glineur, F.: On the geometric interpretation of the nonnegative rank. Linear Algebra and Its Applications **437**(11), 2685–2712 (2012) (Cited on pp. 35, 36, 37, 42, 65, 73, 74, 75, 82, 85, 89, 91)

[200] Gillis, N., Glineur, F.: A continuous characterization of the maximum-edge biclique problem. Journal of Global Optimization **58**(3), 439–464 (2014) (Cited on p. 198)

[201] Gillis, N., Hien, L.T.K., Leplat, V., Tan, V.Y.: Distributionally robust and multi-objective nonnegative matrix factorization. arXiv preprint arXiv:1901.10757 (2019) (Cited on pp. 163, 164, 273)

[202] Gillis, N., Kuang, D., Park, H.: Hierarchical clustering of hyperspectral images using rank-two nonnegative matrix factorization. IEEE Transactions on Geoscience and Remote Sensing **53**(4), 2066–2078 (2014) (Cited on pp. 199, 303)

[203] Gillis, N., Kumar, A.: Exact and heuristic algorithms for semi-nonnegative matrix factorization. SIAM Journal on Matrix Analysis and Applications **36**(4), 1404–1424 (2015) (Cited on pp. 135, 173, 174)

[204] Gillis, N., Luce, R.: Robust near-separable nonnegative matrix factorization using linear optimization. Journal of Machine Learning Research **15**(1), 1249–1280 (2014) (Cited on pp. 243, 254, 255, 256, 257)

[205] Gillis, N., Luce, R.: A fast gradient method for nonnegative sparse regression with self-dictionary. IEEE Transactions on Image Processing **27**(1), 24–37 (2017) (Cited on pp. 243, 255, 256, 257)

[206] Gillis, N., Ma, W.-K.: Enhancing pure-pixel identification performance via preconditioning. SIAM Journal on Imaging Sciences **8**(2), 1161–1186 (2015) (Cited on pp. 237, 241, 257)

[207] Gillis, N., Ohib, R., Plis, S., Potluru, V.: Grouped sparse projection. arXiv preprint arXiv:1912.03896 (2019) (Cited on p. 176)

[208] Gillis, N., Plemmons, R.J.: Dimensionality reduction, classification, and spectral mixture analysis using non-negative underapproximation. Optical Engineering **50**(2), 027,001 (2011) (Cited on p. 302)

[209] Gillis, N., Shitov, Y.: Low-rank matrix approximation in the infinity norm. Linear Algebra and Its Applications **581**, 367–382 (2019) (Cited on pp. 162, 163, 202)

[210] Gillis, N., Vavasis, S.A.: Fast and robust recursive algorithms for separable nonnegative matrix factorization. IEEE Transactions on Pattern Analysis and Machine Intelligence 36(4), 698–714 (2013) (Cited on pp. 223, 226, 229, 231, 248, 257)

[211] Gillis, N., Vavasis, S.A.: Semidefinite programming based preconditioning for more robust near-separable nonnegative matrix factorization. SIAM Journal on Optimization 25(1), 677–698 (2015) (Cited on pp. 231, 238, 239, 240, 257)

[212] Gillis, N., Vavasis, S.A.: On the complexity of robust PCA and ℓ_1-norm low-rank matrix approximation. Mathematics of Operations Research 43(4), 1072–1084 (2018) (Cited on pp. 3, 162, 204)

[213] Girvan, M., Newman, M.E.: Community structure in social and biological networks. Proceedings of the National Academy of Sciences 99(12), 7821–7826 (2002) (Cited on p. 179)

[214] Glineur, F.: Computational experiments with a linear approximation of second order cone optimization (2000). Image Technical Report 0001, Service de Mathématique et de Recherche Opérationnelle, Faculté Polytechnique de Mons (Cited on p. 90)

[215] Goemans, M.X.: Smallest compact formulation for the permutahedron. Mathematical Programming 153(1), 5–11 (2015) (Cited on pp. 65, 66, 84)

[216] Golub, G.H., Van Loan, C.: Matrix Computations, 4th ed. Johns Hopkins University Press (2013) (Cited on pp. 2, 196)

[217] Gouveia, J., Parrilo, P.A., Thomas, R.R.: Lifts of convex sets and cone factorizations. Mathematics of Operations Research 38(2), 248–264 (2013) (Cited on p. 93)

[218] Gouvert, O., Oberlin, T., Févotte, C.: Ordinal non-negative matrix factorization for recommendation. In: Proceedings of the 37th International Conference on Machine learning (ICML) (2020) (Cited on p. 185)

[219] Gribling, S., de Laat, D., Laurent, M.: Lower bounds on matrix factorization ranks via noncommutative polynomial optimization. Foundations of Computational Mathematics 19, 1013–1070 (2019) (Cited on p. 82)

[220] Gribonval, R., Jenatton, R., Bach, F.: Sparse and spurious: dictionary learning with noise and outliers. IEEE Transactions on Information Theory 61(11), 6298–6319 (2015) (Cited on p. 150)

[221] Gribonval, R., Zibulevsky, M.: Sparse component analysis. In: Handbook of Blind Source Separation, pp. 367–420. Elsevier (2010) (Cited on p. 3)

[222] Grippo, L., Sciandrone, M.: On the convergence of the block nonlinear Gauss–Seidel method under convex constraints. Operations Research Letters 26(3), 127–136 (2000) (Cited on p. 268)

[223] Groemer, H.: On some mean values associated with a randomly selected simplex in a convex set. Pacific Journal of Mathematics 45(2), 525–533 (1973) (Cited on p. 62)

[224] Grussler, C., Rantzer, A.: On optimal low-rank approximation of non-negative matrices. In: 2015 54th IEEE Conference on Decision and Control (CDC), pp. 5278–5283 (2015) (Cited on p. 100)

[225] Guan, N., Tao, D., Luo, Z., Yuan, B.: NeNMF: An optimal gradient method for nonnegative matrix factorization. IEEE Transactions on Signal Processing 60(6), 2882–2898 (2012) (Cited on pp. 288, 294)

[226] Guglielmi, N., Scalone, C.: An efficient method for non-negative low-rank completion. Advances in Computational Mathematics 46 (2020). Article number: 31 (Cited on p. 311)

[227] Guillamet, D., Vitrià, J.: Non-negative matrix factorization for face recognition. In: Catalonian Conference on Artificial Intelligence, pp. 336–344. Springer (2002) (Cited on p. 101)

[228] Guo, Z., Zhang, S.: Sparse deep nonnegative matrix factorization. Big Data Mining and Analytics, 3(1), 13–28 (2019) (Cited on p. 184)

[229] Haeffele, B.D., Vidal, R.: Structured low-rank matrix factorization: Global optimality, algorithms, and applications. IEEE Transactions on Pattern Analysis and Machine Intelligence 42(6) (2019) (Cited on p. 197)

[230] Hajinezhad, D., Chang, T.H., Wang, X., Shi, Q., Hong, M.: Nonnegative matrix factorization using ADMM: Algorithm and convergence analysis. In: IEEE International Conference on Acoustics, Speech and Signal Processing (ICASSP), pp. 4742–4746. IEEE (2016) (Cited on p. 291)

[231] Hannah, J., Laffey, T.J.: Nonnegative factorization of completely positive matrices. Linear Algebra and Its Applications **55**, 1–9 (1983) (Cited on p. 79)

[232] Hasinoff, S.W.: Photon, Poisson noise. (2014). `http://people.csail.mit.edu/hasinoff/pubs/hasinoff-photon-2011-preprint.pdf` (Cited on p. 161)

[233] Hautecoeur, C., Glineur, F.: Image completion via nonnegative matrix factorization using HALS and B-splines. In: 28th European Symposium on Artificial Neural Networks-Computational Intelligence and Machine Learning (ESANN) (2020) (Cited on p. 313)

[234] Hautecoeur, C., Glineur, F.: Nonnegative matrix factorization over continuous signals using parametrizable functions. Neurocomputing (2020) (Cited on pp. 169, 309)

[235] He, Z., Xie, S., Zdunek, R., Zhou, G., Cichocki, A.: Symmetric nonnegative matrix factorization: Algorithms and applications to probabilistic clustering. IEEE Transactions on Neural Networks **22**(12), 2117–2131 (2011) (Cited on p. 179)

[236] Hendrix, E.M., Garcia, I., Plaza, J., Martin, G., Plaza, A.: A new minimum-volume enclosing algorithm for endmember identification and abundance estimation in hyperspectral data. IEEE Transactions on Geoscience and Remote Sensing **50**(7), 2744–2757 (2011) (Cited on p. 138)

[237] Hien, L.T.K., Gillis, N.: Algorithms for nonnegative matrix factorization with the Kullback-Leibler divergence. arXiv:2010.01935 (2020) (Cited on pp. 270, 271, 305)

[238] Hien, L.T.K., Gillis, N., Patrinos, P.: Inertial block proximal methods for non-convex non-smooth optimization. In: Proceedings of the 37th International Conference on Machine learning (ICML) (2020) (Cited on pp. 266, 293, 304)

[239] Ho, N.D.: Nonnegative matrix factorization algorithms and applications. Ph.D. thesis, Université catholique de Louvain (2008) (Cited on pp. 196, 203, 265, 286, 287, 299)

[240] Ho, N.D., Van Dooren, P.: Non-negative matrix factorization with fixed row and column sums. Linear Algebra and Its Applications **429**(5–6), 1020–1025 (2008) (Cited on pp. 200, 201)

[241] Hong, D., Kolda, T.G., Duersch, J.A.: Generalized canonical polyadic tensor decomposition. SIAM Review **62**(1), 133–163 (2020) (Cited on p. 163)

[242] Hong, M., Razaviyayn, M., Luo, Z.Q., Pang, J.S.: A unified algorithmic framework for block-structured optimization involving big data: With applications in machine learning and signal processing. IEEE Signal Processing Magazine **33**(1), 57–77 (2015) (Cited on pp. 268, 269, 270, 277)

[243] Hoyer, P.O.: Non-negative matrix factorization with sparseness constraints. Journal of Machine Learning Research **5**(Nov), 1457–1469 (2004) (Cited on pp. 150, 154, 175)

[244] Hrubeš, P.: On the nonnegative rank of distance matrices. Information Processing Letters **112**(11), 457–461 (2012) (Cited on p. 82)

[245] Hsieh, C.J., Dhillon, I.S.: Fast coordinate descent methods with variable selection for non-negative matrix factorization. In: Proceedings of the 17th ACM SIGKDD International Conference on Knowledge Discovery and Data Mining, pp. 1064–1072 (2011) (Cited on p. 285)

[246] Huang, K., Fu, X.: Detecting overlapping and correlated communities without pure nodes: Identifiability and algorithm. In: Proceedings of the 36th International Conference on Machine Learning, pp. 2859–2868 (2019) (Cited on pp. 139, 184, 210, 308, 313)

[247] Huang, K., Fu, X.: Low-complexity proximal Gauss-Newton algorithm for nonnegative matrix factorization. In: 2019 IEEE Global Conference on Signal and Information Processing (GlobalSIP) (2019) (Cited on p. 304)

[248] Huang, K., Sidiropoulos, N.D.: Putting nonnegative matrix factorization to the test: a tutorial derivation of pertinent Cramer–Rao bounds and performance benchmarking. IEEE Signal Processing Magazine **31**(3), 76–86 (2014) (Cited on p. 304)

[249] Huang, K., Sidiropoulos, N.D.: Kullback-Leibler principal component for tensors is not NP-hard. In: 2017 51st Asilomar Conference on Signals, Systems, and Computers, pp. 693–697. IEEE (2017) (Cited on p. 201)

[250] Huang, K., Sidiropoulos, N.D., Liavas, A.P.: A flexible and efficient algorithmic framework for constrained matrix and tensor factorization. IEEE Transactions on Signal Processing **64**(19), 5052–5065 (2016) (Cited on pp. 290, 291, 294, 296)

[251] Huang, K., Sidiropoulos, N.D., Swami, A.: Non-negative matrix factorization revisited: Uniqueness and algorithm for symmetric decomposition. IEEE Transactions on Signal Processing **62**(1), 211–224 (2013) (Cited on pp. 53, 108, 111, 112, 116, 118, 122, 125, 127, 133)

[252] Ibrahim, S., Fu, X.: Recovering joint probability of discrete random variables from pairwise marginals. arXiv preprint arXiv:2006.16912 (2020) (Cited on p. 211)

[253] Ibrahim, S., Fu, X., Kargas, N., Huang, K.: Crowdsourcing via pairwise co-occurrences: Identifiability and algorithms. In: Advances in Neural Information Processing Systems (NIPS), pp. 7845–7855 (2019) (Cited on p. 210)

[254] Imbrie, J., Van Andel, T.H.: Vector analysis of heavy-mineral data. Geological Society of America Bulletin **75**(11), 1131–1156 (1964) (Cited on p. 12)

[255] Iordache, M.D., Bioucas-Dias, J.M., Plaza, A.: Total variation spatial regularization for sparse hyperspectral unmixing. IEEE Transactions on Geoscience and Remote Sensing **50**(11), 4484–4502 (2012) (Cited on p. 170)

[256] Iordache, M.D., Bioucas-Dias, J.M., Plaza, A.: Collaborative sparse regression for hyperspectral unmixing. IEEE Transactions on Geoscience and Remote Sensing **52**(1), 341–354 (2013) (Cited on p. 172)

[257] Jameson, G.J.O.: Summing and Nuclear Norms in Banach Space Theory, vol. 8. Cambridge University Press (1987) (Cited on p. 79)

[258] Janecek, A.G.K., Grotthoff, S.S., Gansterer, W.N.: libNMF—a library for nonnegative matrix factorization. Computing and Informatics **30**(2), 205–224 (2011) (Cited on p. 305)

[259] Jang, B., Hero, A.: Minimum volume topic modeling. In: 22nd International Conference on Artificial Intelligence and Statistics (AISTATS), pp. 3013–3021 (2019) (Cited on p. 139)

[260] Janzamin, M., Ge, R., Kossaifi, J., Anandkumar, A., et al.: Spectral learning on matrices and tensors. Foundations and Trends in Machine Learning **12**(5–6), 393–536 (2019) (Cited on p. 183)

[261] Jeter, M., Pye, W.: A note on nonnegative rank factorizations. Linear Algebra and Its Applications **38**, 171–173 (1981) (Cited on p. 14)

[262] Jia, S., Qian, Y.: Constrained nonnegative matrix factorization for hyperspectral unmixing. IEEE Transactions on Geoscience and Remote Sensing **47**(1), 161–173 (2008) (Cited on p. 169)

[263] Jiang, J.H., Liang, Y., Ozaki, Y.: Principles and methodologies in self-modeling curve resolution. Chemometrics and Intelligent Laboratory Systems **71**(1), 1–12 (2004) (Cited on pp. 5, 12)

[264] Jiang, J.H., Liang, Y.Z., Ozaki, Y.: On simplex-based method for self-modeling curve resolution of two-way data. Chemometrics and Intelligent Laboratory Systems **65**(1), 51–65 (2003) (Cited on p. 229)

[265] John, F.: Extremum problems with inequalities as subsidiary conditions. In: J. Moser (ed.) Fritz John, Collected Papers, pp. 543–560. Birkhaüser (1985). First published in 1948 (Cited on p. 240)

[266] Jolliffe, I.: Principal Component Analysis. Springer (2011) (Cited on p. 2)

[267] Kaibel, V.: Extended formulations in combinatorial optimization. Optima **85**, 2–7 (2011) (Cited on p. 84)

[268] Kaibel, V., Pashkovich, K.: Constructing extended formulations from reflection relations. In: Integer Programming and Combinatorial Optimization, pp. 287–300. Springer (2011) (Cited on p. 90)

[269] Kaibel, V., Pashkovich, K.: Constructing extended formulations from reflection relations. In: M. Jünger, G. Reinelt (eds.) Facets of Combinatorial Optimization, pp. 77–100. Springer (2013) (Cited on p. 90)

[270] Kaibel, V., Pashkovich, K., Theis, D.O.: Symmetry matters for sizes of extended formulations. SIAM Journal on Discrete Mathematics **26**(3), 1361–1382 (2012) (Cited on p. 15)

[271] Kalofolias, V., Gallopoulos, E.: Computing symmetric nonnegative rank factorizations. Linear Algebra and Its Applications **436**(2), 421–435 (2012) (Cited on pp. 179, 258)

[272] Kannan, R., Ishteva, M., Park, H.: Bounded matrix factorization for recommender system. Knowledge and Information Systems **39**(3), 491–511 (2014) (Cited on p. 313)

[273] Ke, Q., Kanade, T.: Robust l_1 norm factorization in the presence of outliers and missing data by alternative convex programming. In: 2005 IEEE Computer Society Conference on Computer Vision and Pattern Recognition (CVPR'05), vol. 1, pp. 739–746. IEEE (2005) (Cited on pp. 162, 163, 203)

[274] Kervazo, C., Bobin, J., Chenot, C., Sureau, F.: Use of palm for ℓ_1 sparse matrix factorization: Difficulty and rationalization of a two-step approach. Digital Signal Processing, 102611 (2019) (Cited on p. 170)

[275] Kervazo, C., Gillis, N., Dobigeon, N.: Provably robust blind source separation of linear-quadratic near-separable mixtures. In preparation. (Cited on p. 185)

[276] Keshava, N., Mustard, J.F.: Spectral unmixing. IEEE Signal Processing Magazine **19**(1), 44–57 (2002) (Cited on p. 13)

[277] Kim, H., Park, H.: Sparse non-negative matrix factorizations via alternating non-negativity-constrained least squares for microarray data analysis. Bioinformatics **23**(12), 1495–1502 (2007) (Cited on pp. 150, 154, 168, 281, 282, 309)

[278] Kim, H., Park, H.: Nonnegative matrix factorization based on alternating nonnegativity constrained least squares and active set method. SIAM Journal on Matrix Analysis and Applications **30**(2), 713–730 (2008) (Cited on p. 281)

[279] Kim, J., He, Y., Park, H.: Algorithms for nonnegative matrix and tensor factorizations: A unified view based on block coordinate descent framework. Journal of Global Optimization **58**(2), 285–319 (2014) (Cited on p. 284)

[280] Kim, J., Park, H.: Fast nonnegative matrix factorization: An active-set-like method and comparisons. SIAM Journal on Scientific Computing **33**(6), 3261–3281 (2011) (Cited on pp. 281, 282, 294)

[281] Kingman, J.F.: Random secants of a convex body. Journal of Applied Probability **6**(3), 660–672 (1969) (Cited on p. 62)

[282] Kohjima, M., Matsubayashi, T., Toda, H.: Generalized interval valued nonnegative matrix factorization. In: IEEE International Conference on Acoustics, Speech and Signal Processing (ICASSP), pp. 3412–3416. IEEE (2019) (Cited on p. 185)

[283] Kolda, T.G., Bader, B.W.: Tensor decompositions and applications. SIAM Review **51**(3), 455–500 (2009) (Cited on p. xiv)

[284] Kopriva, I., Hadžija, M., Hadžija, M.P., Korolija, M., Cichocki, A.: Rational variety mapping for contrast-enhanced nonlinear unsupervised segmentation of multispectral images of unstained specimen. American Journal of Pathology **179**(2), 547–554 (2011) (Cited on p. 177)

[285] Kopriva, I., Jerić, I., Brkljačić, L.: Nonlinear mixture-wise expansion approach to underdetermined blind separation of nonnegative dependent sources. Journal of Chemometrics **27**(7–8), 189–197 (2013) (Cited on p. 185)

[286] Kopriva, I., Ju, W., Zhang, B., Shi, F., Xiang, D., Yu, K., Wang, X., Bagci, U., Chen, X.: Single-channel sparse non-negative blind source separation method for automatic 3-D delineation of lung tumor in PET images. IEEE Journal of Biomedical and Health Informatics **21**(6), 1656–1666 (2016) (Cited on p. 177)

[287] Koren, Y., Bell, R., Volinsky, C.: Matrix factorization techniques for recommender systems. Computer **42**(8), 30–37 (2009) (Cited on pp. 3, 308, 311)

[288] Krone, R., Kubjas, K.: Uniqueness of nonnegative matrix factorizations by rigidity theory. SIAM Journal on Matrix Analysis and Applications. To appear (Cited on pp. 127, 128, 130, 132, 133)

[289] Kruskal, J.B.: Three-way arrays: rank and uniqueness of trilinear decompositions, with application to arithmetic complexity and statistics. Linear Algebra and Its Applications **18**(2), 95–138 (1977) (Cited on p. 151)

[290] Kuang, D., Ding, C., Park, H.: Symmetric nonnegative matrix factorization for graph clustering. In: Proceedings of the 12th SIAM International Conference on Data Mining, pp. 106–117. SIAM (2012) (Cited on pp. 179, 185, 287)

[291] Kuang, D., Park, H.: Fast rank-2 nonnegative matrix factorization for hierarchical document clustering. In: Proceedings of the 19th ACM SIGKDD international conference on knowledge discovery and data mining, pp. 739–747 (2013) (Cited on p. 303)

[292] Kubjas, K., Robeva, E., Sturmfels, B.: Fixed points EM algorithm and nonnegative rank boundaries. Annals of Statistics pp. 422–461 (2015) (Cited on pp. 14, 47)

[293] Kumar, A., Sindhwani, V.: Near-separable non-negative matrix factorization with ℓ_1 and Bregman loss functions. In: Proceedings of the 2015 SIAM International Conference on Data Mining, pp. 343–351. SIAM (2015) (Cited on p. 211)

[294] Kumar, A., Sindhwani, V., Kambadur, P.: Fast conical hull algorithms for near-separable non-negative matrix factorization. In: Proceedings of the 30th International Conference on Machine Learning, pp. 231–239 (2013) (Cited on pp. 142, 189, 236)

[295] Kwan, M., Sauermann, L., Zhao, Y.: Extension complexity of low-dimensional polytopes. arXiv preprint arXiv:2006.08836 (2020) (Cited on p. 60)

[296] Lakshminarayanan, B., Raich, R.: Non-negative matrix factorization for parameter estimation in hidden Markov models. In: 2010 IEEE International Workshop on Machine Learning for Signal Processing, pp. 89–94. IEEE (2010) (Cited on p. 313)

[297] Lambert, J.H.: Photometria sive de mensura et gradibus luminis, colorum et umbrae. Klett (1760) (Cited on p. 12)

[298] Laurberg, H., Christensen, M.G., Plumbley, M.D., Hansen, L.K., Jensen, S.H.: Theorems on positive data: On the uniqueness of NMF. Computational Intelligence and Neuroscience **2008** (2008) (Cited on pp. 100, 104, 105, 107, 108, 111, 126, 127, 208)

[299] Laurberg, H., Hansen, L.K.: On affine non-negative matrix factorization. In: IEEE International Conference on Acoustics, Speech and Signal Processing (ICASSP), vol. 2, pp. II–653. IEEE (2007) (Cited on p. 176)

[300] Lawson, C.L., Hanson, R.J.: Solving Least Squares Problems. Classics in Applied Mathematics, vol. 15. SIAM (1995) (Cited on p. 281)

[301] Lawton, W.H., Sylvestre, E.A.: Self modeling curve resolution. Technometrics **13**(3), 617–633 (1971) (Cited on pp. 12, 103)

[302] Le Roux, J., Weninger, F.J., Hershey, J.R.: Sparse NMF—half-baked or well done? Mitsubishi Electric Research Labs (MERL), Tech. Rep. TR2015-023 **11**, 13–15 (2015) (Cited on p. 280)

[303] Lee, D.D., Seung, H.S.: Learning the parts of objects by non-negative matrix factorization. Nature **401**, 788–791 (1999) (Cited on pp. 4, 5, 7, 16, 163, 270, 293, 294, 314)

[304] Lee, D.D., Seung, H.S.: Algorithms for non-negative matrix factorization. In: Advances in Neural Information Processing Systems (NIPS), pp. 556–562 (2001) (Cited on pp. 16, 270)

[305] Lee, H., Yoo, J., Choi, S.: Semi-supervised nonnegative matrix factorization. IEEE Signal Processing Letters **17**(1), 4–7 (2009) (Cited on p. 170)

[306] Lee, J.R., Raghavendra, P., Steurer, D.: Lower bounds on the size of semidefinite programming relaxations. In: Proceedings of the 47th Annual ACM Symposium on Theory of Computing, pp. 567–576. ACM (2015) (Cited on p. 93)

[307] Lee, T., Friedman, L.: Nondeterministic communication complexity (2010). Lecture notes, `http://research.cs.rutgers.edu/~troyjlee/lec2.pdf` (Cited on p. 94)

[308] Lee, T., Shraibman, A.: Lower Bounds in Communication Complexity. Now Publishers (2009) (Cited on p. 95)

[309] Leplat, V., Ang, A.M., Gillis, N.: Minimum-volume rank-deficient nonnegative matrix factorizations. In: IEEE International Conference on Acoustics, Speech and Signal Processing (ICASSP), pp. 3402–3406. IEEE (2019) (Cited on pp. 148, 149, 288)

[310] Leplat, V., Gillis, N., Ang, M.S.: Blind audio source separation with minimum-volume beta-divergence NMF. IEEE Transactions on Signal Processing **68**, 3400–3410 (2020) (Cited on pp. 143, 145, 280)

[311] Levin, B.: On calculating maximum rank one underapproximations for positive arrays (1985). http://biostats.bepress.com/cgi/viewcontent.cgi?article=1009&context=columbiabiostat (Cited on p. 302)

[312] Li, B., Zhou, G., Cichocki, A.: Two efficient algorithms for approximately orthogonal nonnegative matrix factorization. IEEE Signal Processing Letters **22**(7), 843–846 (2014) (Cited on p. 136)

[313] Li, L., Zhang, Y.J.: FastNMF: highly efficient monotonic fixed-point nonnegative matrix factorization algorithm with good applicability. J. Electron. Imaging **18**, 033004 (2009) (Cited on p. 286)

[314] Li, M.L., Di Mauro, F., Candan, K.S., Sapino, M.L.: Matrix factorization with interval-valued data. IEEE Transactions on Knowledge and Data Engineering (2019) (Cited on p. 185)

[315] Li, W., Li, J., Liu, X., Dong, L.: Two fast vector-wise update algorithms for orthogonal nonnegative matrix factorization with sparsity constraint. Journal of Computational and Applied Mathematics **375**, 112,785 (2020) (Cited on p. 287)

[316] Li, Y., Liang, Y., Risteski, A.: Recovery guarantee of non-negative matrix factorization via alternating updates. In: Advances in Neural Information Processing Systems (NIPS), pp. 4987–4995 (2016) (Cited on p. 156)

[317] Li, Y., Ngom, A.: The non-negative matrix factorization toolbox for biological data mining. Source Code for Biology and Medicine **8**(1), 10 (2013) (Cited on p. 305)

[318] Li, Y., Sima, D.M., Cauter, S.V., Croitor Sava, A.R., Himmelreich, U., Pi, Y., Van Huffel, S.: Hierarchical non-negative matrix factorization (hNMF): a tissue pattern differentiation method for glioblastoma multiforme diagnosis using MRSI. NMR in Biomedicine **26**(3), 307–319 (2013) (Cited on pp. 303, 313)

[319] Lin, C.H., Bioucas-Dias, J.M.: Nonnegative blind source separation for ill-conditioned mixtures via John ellipsoid. IEEE Transactions on Neural Networks and Learning Systems (2020) (Cited on p. 149)

[320] Lin, C.H., Ma, W.K., Li, W.C., Chi, C.Y., Ambikapathi, A.: Identifiability of the simplex volume minimization criterion for blind hyperspectral unmixing: The no-pure-pixel case. IEEE Transactions on Geoscience and Remote Sensing **53**(10), 5530–5546 (2015) (Cited on pp. 111, 118, 119, 139, 141)

[321] Lin, C.-H., Wu, R., Ma, W.-K., Chi, C.-Y., Wang, Y.: Maximum volume inscribed ellipsoid: A new simplex-structured matrix factorization framework via facet enumeration and convex optimization. SIAM Journal on Imaging Sciences **11**(2), 1651–1679 (2018) (Cited on pp. 149, 213)

[322] Lin, C.J.: On the convergence of multiplicative update algorithms for nonnegative matrix factorization. IEEE Transactions on Neural Networks **18**(6), 1589–1596 (2007) (Cited on p. 279)

[323] Lin, C.J.: Projected gradient methods for nonnegative matrix factorization. Neural Computation **19**(10), 2756–2779 (2007) (Cited on p. 287)

[324] Liu, J., Liu, J., Wonka, P., Ye, J.: Sparse non-negative tensor factorization using columnwise coordinate descent. Pattern Recognition **45**(1), 649–656 (2012) (Cited on p. 286)

[325] Liu, Z.: Model selection for nonnegative matrix factorization by support union recovery. In: IEEE International Conference on Acoustics, Speech and Signal Processing (ICASSP), pp. 3407–3411. IEEE (2019) (Cited on p. 167)

[326] Liu, Z., Tan, V.Y.F.: Rank-one NMF-based initialization for NMF and relative error bounds under a geometric assumption. IEEE Transactions on Signal Processing **65**(18), 4717–4731 (2017) (Cited on p. 259)

[327] Lopes, M.B., Wolff, J.C., Bioucas-Dias, J.M., Figueiredo, M.A.: Near-infrared hyperspectral unmixing based on a minimum volume criterion for fast and accurate chemometric characterization of counterfeit tablets. Analytical Chemistry **82**(4), 1462–1469 (2010) (Cited on p. 139)

[328] López-Serrano, P., Dittmar, C., Özer, Y., Müller, M.: NMF Toolbox: Music processing applications of nonnegative matrix factorization. In: Proceedings of the 22nd International Conference on Digital Audio Effects (DAFx-19), Birmingham, UK (2019) (Cited on p. 305)

[329] Lu, Z., Yang, Z., Oja, E.: Selecting β-divergence for nonnegative matrix factorization by score matching. In: International Conference on Artificial Neural Networks, pp. 419–426. Springer (2012) (Cited on pp. 163, 164)

[330] Lubell, D.: A short proof of Sperner's lemma. Journal of Combinatorial Theory **1**(2), 299 (1966) (Cited on p. 69)

[331] Luce, R., Hildebrandt, P., Kuhlmann, U., Liesen, J.: Using separable nonnegative matrix factorization techniques for the analysis of time-resolved Raman spectra. Applied Spectroscopy **70**(9), 1464–1475 (2016) (Cited on pp. 210, 308, 309, 313)

[332] Lucy, L.B.: An iterative technique for the rectification of observed distributions. Astronomical Journal **79**, 745 (1974) (Cited on p. 270)

[333] Luo, W., Gao, L., Plaza, A., Marinoni, A., Yang, B., Zhong, L., Gamba, P., Zhang, B.: A new algorithm for bilinear spectral unmixing of hyperspectral images using particle swarm optimization. IEEE Journal of Selected Topics in Applied Earth Observations and Remote Sensing **9**(12), 5776–5790 (2016) (Cited on p. 247)

[334] Ma, W.K., Bioucas-Dias, J.M., Chan, T.H., Gillis, N., Gader, P., Plaza, A.J., Ambikapathi, A., Chi, C.Y.: A signal processing perspective on hyperspectral unmixing: Insights from remote sensing. IEEE Signal Processing Magazine **31**(1), 67–81 (2014) (Cited on pp. 13, 209, 216, 247, 248)

[335] Machida, K., Takenouchi, T.: Statistical modeling of robust non-negative matrix factorization based on γ-divergence and its applications. Japanese Journal of Statistics and Data Science **2**(2), 441–464 (2019) (Cited on p. 163)

[336] Mackevicius, E.L., Bahle, A.H., Williams, A.H., Gu, S., Denisenko, N.I., Goldman, M.S., Fee, M.S.: Unsupervised discovery of temporal sequences in high-dimensional datasets, with applications to neuroscience. eLife **8**, e38,471 (2019) (Cited on pp. 178, 314)

[337] Magron, P., Badeau, R., David, B.: Phase reconstruction of spectrograms with linear unwrapping: application to audio signal restoration. In: 23rd European Signal Processing Conference (EUSIPCO). IEEE (2015) (Cited on p. 11)

[338] Magron, P., Badeau, R., David, B.: Phase recovery in NMF for audio source separation: an insightful benchmark. In: IEEE International Conference on Acoustics, Speech and Signal Processing (ICASSP), pp. 81–85. IEEE (2015) (Cited on p. 11)

[339] Majmudar, J., Vavasis, S.A.: Provable overlapping community detection in weighted graphs. arXiv preprint arXiv:2004.07150 (2020) (Cited on p. 210)

[340] Mao, X., Sarkar, P., Chakrabarti, D.: On mixed memberships and symmetric nonnegative matrix factorizations. In: Proceedings of the 34th International Conference on Machine Learning, pp. 2324–2333 (2017) (Cited on p. 184)

[341] Markovsky, I.: Low Rank Approximation: Algorithms, Implementation, Applications. Springer (2011) (Cited on p. 3)

[342] Megiddo, N.: Towards a genuinely polynomial algorithm for linear programming. SIAM Journal on Computing **12**(2), 347–353 (1983) (Cited on p. 202)

[343] Miao, L., Qi, H.: Endmember extraction from highly mixed data using minimum volume constrained nonnegative matrix factorization. IEEE Transactions on Geoscience and Remote Sensing **45**(3), 765–777 (2007) (Cited on pp. 138, 139)

[344] Miettinen, P., Mielikäinen, T., Gionis, A., Das, G., Mannila, H.: The discrete basis problem. IEEE Transactions on Knowledge and Data Engineering **20**(10), 1348–1362 (2008) (Cited on p. 185)

[345] Mitchell, J., Suri, S.: Separation and approximation of polyhedral surfaces. Operations Research Letters **11**, 255–259 (1992) (Cited on p. 51)

[346] Miyasawa, A., Fujimoto, Y., Hayashi, Y.: Energy disaggregation based on smart metering data via semi-binary nonnegative matrix factorization. Energy and Buildings **183**, 547–558 (2019) (Cited on p. 313)

[347] Mizutani, T.: Ellipsoidal rounding for nonnegative matrix factorization under noisy separability. Journal of Machine Learning Research **15**(1), 1011–1039 (2014) (Cited on pp. 237, 238, 239)

[348] Mizutani, T.: Robustness analysis of preconditioned successive projection algorithm for general form of separable NMF problem. Linear Algebra and Its Applications **497**, 1–22 (2016) (Cited on p. 239)

[349] Mizutani, T., Tanaka, M.: Efficient preconditioning for noisy separable nonnegative matrix factorization problems by successive projection based low-rank approximations. Machine Learning **107**(4), 643–673 (2018) (Cited on pp. 239, 241)

[350] Mohan, K., Fazel, M.: Iterative reweighted algorithms for matrix rank minimization. Journal of Machine Learning Research **13**(Nov), 3441–3473 (2012) (Cited on p. 148)

[351] Moitra, A.: An almost optimal algorithm for computing nonnegative rank. In: Proceedings of the 24th Annual ACM-SIAM Symposium on Discrete Algorithms (SODA'13), pp. 1454–1464 (2013) (Cited on p. 53)

[352] Mollah, M.N.H., Eguchi, S., Minami, M.: Robust prewhitening for ICA by minimizing β-divergence and its application to fastica. Neural Processing Letters **25**(2), 91–110 (2007) (Cited on p. 164)

[353] Mond, D., Smith, J., Van Straten, D.: Stochastic factorizations, sandwiched simplices and the topology of the space of explanations. Proceedings of the Royal Society of London. Series A: Mathematical, Physical and Engineering Sciences **459**(2039), 2821–2845 (2003) (Cited on pp. 14, 29)

[354] Montanari, A., Richard, E.: Non-negative principal component analysis: Message passing algorithms and sharp asymptotics. IEEE Transactions on Information Theory **62**(3), 1458–1484 (2015) (Cited on p. 188)

[355] Moussaoui, S., Brie, D., Idier, J.: Non-negative source separation: range of admissible solutions and conditions for the uniqueness of the solution. In: IEEE International Conference on Acoustics, Speech and Signal Processing (ICASSP), vol. 5, pp. v–289. IEEE (2005) (Cited on pp. 103, 127)

[356] Moutier, F., Vandaele, A., Gillis, N.: Off-diagonal symmetric nonnegative matrix factorization. arXiv preprint arXiv:2003.04775 (2020) (Cited on p. 179)

[357] Mukkamala, M.C., Ochs, P.: Beyond alternating updates for matrix factorization with inertial Bregman proximal gradient algorithms. In: Advances in Neural Information Processing Systems, pp. 4266–4276 (2019) (Cited on p. 262)

[358] Naanaa, W., Nuzillard, J.M.: Blind source separation of positive and partially correlated data. Signal Processing **85**(9), 1711–1722 (2005) (Cited on p. 249)

[359] Nadisic, N., Vandaele, A., Cohen, J.E., Gillis, N.: Sparse separable nonnegative matrix factorization. In: Proceedings of the European Conference on Machine Learning and Principles and Practice of Knowledge Discovery in Databases (ECML-PKDD) (2020) (Cited on p. 185)

[360] Nascimento, J.M., Bioucas-Dias, J.M.: Hyperspectral unmixing based on mixtures of dirichlet components. IEEE Transactions on Geoscience and Remote Sensing **50**(3), 863–878 (2011) (Cited on pp. 144, 242)

[361] Nascimento, J.M., Dias, J.M.: Vertex component analysis: A fast algorithm to unmix hyperspectral data. IEEE Transactions on Geoscience and Remote Sensing **43**(4), 898–910 (2005) (Cited on pp. 216, 232)

[362] Nascimento, J.M.P., Bioucas-Dias, J.M.: Nonlinear mixture model for hyperspectral unmixing. In: L. Bruzzone, C. Notarnicola, F. Posa (eds.) Image and Signal Processing for Remote Sensing XV, vol. 7477, pp. 157–164. International Society for Optics and Photonics, SPIE (2009) (Cited on p. 185)

[363] Natarajan, B.K.: Sparse approximate solutions to linear systems. SIAM Journal on Computing **24**(2), 227–234 (1995) (Cited on p. 205)

[364] Nesterov, Y.: Introductory Lectures on Convex Optimization: A basic course. Applied Optimization 137, 2nd ed. Springer (2018) (Cited on p. 288)

[365] Newman, D.J.: The double dixie cup problem. American Mathematical Monthly **67**(1), 58–61 (1960) (Cited on p. 122)

[366] Neymeyr, K., Sawall, M.: On the set of solutions of the nonnegative matrix factorization problem. SIAM Journal on Matrix Analysis and Applications **39**(2), 1049–1069 (2018) (Cited on pp. 12, 41, 53, 304, 309)

[367] Nus, L., Miron, S., Brie, D.: An ADMM-based algorithm with minimum dispersion regularization for on-line blind unmixing of hyperspectral images. Chemometrics and Intelligent Laboratory Systems **204**, 104,090 (2020) (Cited on p. 168)

[368] Oelze, M., Vandaele, A., Weltge, S.: Computing the extension complexities of all 4-dimensional 0/1-polytopes. arXiv preprint arXiv:1406.4895 (2014) (Cited on pp. 71, 72, 73)

[369] Ott, G.H.: Reconsider the state minimization problem for stochastic finite state systems. In: 7th Annual Symposium on Switching and Automata Theory (SWAT 1966), pp. 267–273. IEEE (1966) (Cited on p. 13)

[370] Ouedraogo, W.S.B., Souloumiac, A., Jaidane, M., Jutten, C.: Non-negative blind source separation algorithm based on minimum aperture simplicial cone. IEEE Transactions on Signal Processing **62**(2), 376–389 (2013) (Cited on p. 139)

[371] Paatero, P., Tapper, U.: Positive matrix factorization: A non-negative factor model with optimal utilization of error estimates of data values. Environmetrics **5**(2), 111–126 (1994) (Cited on pp. 4, 15, 128, 159, 281, 304, 313)

[372] Packer, A.: NP-hardness of largest contained and smallest containing simplices for V-and H-polytopes. Discrete and Computational Geometry **28**(3), 349–377 (2002) (Cited on p. 148)

[373] Pan, J., Gillis, N.: Generalized separable nonnegative matrix factorization. IEEE Transactions on Pattern Analysis and Machine Intelligence (2019). DOI:10.1109/TPAMI.2019.2956046 (Cited on p. 259)

[374] Panov, M., Slavnov, K., Ushakov, R.: Consistent estimation of mixed memberships with successive projections. In: International Conference on Complex Networks and their Applications, pp. 53–64. Springer (2017) (Cited on pp. 184, 210)

[375] Pashkovich, K.: Extended formulations for combinatorial polytopes. Ph.D. thesis, Otto-von-Guericke-Universität Magdeburg (2012) (Cited on p. 84)

[376] Pashkovich, K.: Tight lower bounds on the sizes of symmetric extensions of permutahedra and similar results. Mathematics of Operations Research **39**(4), 1330–1339 (2014) (Cited on p. 84)

[377] Pauca, V.P., Piper, J., Plemmons, R.J.: Nonnegative matrix factorization for spectral data analysis. Linear Algebra and Its Applications **416**(1), 29–47 (2006) (Cited on p. 13)

[378] Paz, A.: Homomorphisms between stochastic sequential machines and related problems. Mathematical Systems Theory **2**(3), 223–245 (1968) (Cited on p. 13)

[379] Paz, A.: Introduction to Probabilistic Automata. Academic Press (1971) (Cited on p. 13)

[380] Peeters, R.: The maximum edge biclique problem is NP-complete. Discrete Applied Mathematics **131**(3), 651–654 (2003) (Cited on pp. 72, 198, 203)

[381] Peyerimhoff, N.: Areas and intersections in convex domains. American Mathematical Monthly **104**(8), 697–704 (1997) (Cited on p. 62)

[382] Plumbley, M., Cichocki, A., Bro, R.: Non-negative mixtures. In: Handbook of Blind Source Separation, pp. 515–547. Elsevier (2010) (Cited on pp. 193, 313, 314)

[383] Plumbley, M.D.: Algorithms for nonnegative independent component analysis. IEEE Transactions on Neural Networks **14**(3), 534–543 (2003) (Cited on p. 188)

[384] Pompili, F., Gillis, N., Absil, P.A., Glineur, F.: Two algorithms for orthogonal nonnegative matrix factorization with application to clustering. Neurocomputing **141**, 15–25 (2014) (Cited on pp. 137, 189, 282)

[385] Powell, M.J.: On search directions for minimization algorithms. Mathematical Programming **4**(1), 193–201 (1973) (Cited on p. 268)

[386] Psorakis, I., Roberts, S., Ebden, M., Sheldon, B.: Overlapping community detection using Bayesian non-negative matrix factorization. Physical Review E **83**(6), 066,114 (2011) (Cited on p. 313)

[387] Qian, Y., Jia, S., Zhou, J., Robles-Kelly, A.: Hyperspectral unmixing via $l_{1/2}$ sparsity-constrained nonnegative matrix factorization. IEEE Transactions on Geoscience and Remote Sensing **49**(11), 4282–4297 (2011) (Cited on p. 150)

[388] Qiao, H.: New SVD based initialization strategy for non-negative matrix factorization. Pattern Recognition Letters **63**, 71–77 (2015) (Cited on p. 301)

[389] Rajaraman, A., Ullman, J.D.: Mining of Massive Datasets. Cambridge University Press (2011) (Cited on p. 9)

[390] Rajkó, R., István, K.: Analytical solution for determining feasible regions of self-modeling curve resolution (SMCR) method based on computational geometry. Journal of Chemometrics **19**(8), 448–463 (2005) (Cited on p. 12)

[391] Rapin, J., Bobin, J., Larue, A., Starck, J.L.: Sparse and non-negative BSS for noisy data. IEEE Transactions on Signal Processing **61**(22), 5620–5632 (2013) (Cited on p. 170)

[392] Razaviyayn, M., Hong, M., Luo, Z.-Q.: A unified convergence analysis of block successive minimization methods for nonsmooth optimization. SIAM Journal on Optimization **23**(2), 1126–1153 (2013) (Cited on pp. 268, 269)

[393] Razenshteyn, I., Song, Z., Woodruff, D.P.: Weighted low rank approximations with provable guarantees. In: Proceedings of the 48th Symposium on Theory of Computing (STOC '16), pp. 250–263 (2016) (Cited on p. 203)

[394] Recht, B., Fazel, M., Parrilo, P.A.: Guaranteed minimum-rank solutions of linear matrix equations via nuclear norm minimization. SIAM Review **52**(3), 471–501 (2010) (Cited on pp. 77, 148)

[395] Recht, B., Re, C., Tropp, J., Bittorf, V.: Factoring nonnegative matrices with linear programs. In: Advances in Neural Information Processing Systems (NIPS), pp. 1214–1222 (2012) (Cited on pp. 253, 254, 256, 257)

[396] Ren, H., Chang, C.I.: Automatic spectral target recognition in hyperspectral imagery. IEEE Transactions on Aerospace and Electronic Systems **39**(4), 1232–1249 (2003) (Cited on p. 229)

[397] Richardson, W.H.: Bayesian-based iterative method of image restoration. JoSA **62**(1), 55–59 (1972) (Cited on p. 270)

[398] Ritov, Y., Gilula, Z.: Analysis of contingency tables by correspondence models subject to order constraints. Journal of the American Statistical Association **88**(424), 1380–1387 (1993) (Cited on p. 14)

[399] Rockafellar, R.T.: Convex Analysis. Princeton Landmarks in Mathematics 28. Princeton University Press (1970) (Cited on p. 106)

[400] Rothvoss, T.: The matching polytope has exponential extension complexity. In: Proceedings of the 46th Annual ACM Symposium on Theory of Computing, pp. 263–272. ACM (2014) (Cited on pp. 15, 76, 77, 83, 85, 97)

[401] Rothvoss, T.: The matching polytope has exponential extension complexity. Journal of the ACM **64**(6), 41 (2017) (Cited on pp. 15, 77, 86, 97)

[402] Roughgarden, T., et al.: Communication complexity (for algorithm designers). Foundations and Trends in Theoretical Computer Science **11**(3–4), 217–404 (2016) (Cited on p. 95)

[403] Ruckebusch, C., Blanchet, L.: Multivariate curve resolution: a review of advanced and tailored applications and challenges. Analytica Chimica Acta **765**, 28–36 (2013) (Cited on p. 12)

[404] Sadowski, T., Zdunek, R.: Image completion with smooth nonnegative matrix factorization. In: International Conference on Artificial Intelligence and Soft Computing, pp. 62–72. Springer (2018) (Cited on p. 313)

[405] Sandler, R., Lindenbaum, M.: Nonnegative matrix factorization with earth mover's distance metric. In: 2009 IEEE Conference on Computer Vision and Pattern Recognition, pp. 1873–1880. IEEE (2009) (Cited on p. 163)

[406] Sauwen, N., Sima, D.M., Van Cauter, S., Veraart, J., Leemans, A., Maes, F., Himmelreich, U., Van Huffel, S.: Hierarchical non-negative matrix factorization to characterize brain tumor heterogeneity using multi-parametric MRI. NMR in Biomedicine **28**(12), 1599–1624 (2015) (Cited on pp. 303, 313)

[407] Schachtner, R., Pöppel, G., Tomé, A.M., Lang, E.W.: Minimum determinant constraint for nonnegative matrix factorization. In: T. Adali, C. Jutten, J.M.T. Romano, A.K. Barros (eds.) Independent Component Analysis and Signal Separation, pp. 106–113. Springer (2009) (Cited on p. 139)

[408] Scutari, G., Sun, Y.: Parallel and distributed successive convex approximation methods for big-data optimization. In: Multi-agent Optimization, pp. 141–308. Springer (2018) (Cited on p. 265)

[409] Shaked-Monderer, N.: On the number of CP factorizations of a completely positive matrix. arXiv: 2009.12290 (2020) (Cited on p. 133)

[410] Shitov, Y.: Sublinear extensions of polygons. arXiv preprint arXiv:1412.0728v1 (2014) (Cited on p. 82)

[411] Shitov, Y.: An upper bound for nonnegative rank. Journal of Combinatorial Theory, Series A **122**, 126–132 (2014) (Cited on pp. 82, 90)

[412] Shitov, Y.: Nonnegative rank depends on the field II. arXiv:1605.07173 (2016) (Cited on p. 54)

[413] Shitov, Y.: A universality theorem for nonnegative matrix factorizations. arXiv preprint arXiv: 1606.09068 (2016) (Cited on p. 54)

[414] Shitov, Y.: The nonnegative rank of a matrix: Hard problems, easy solutions. SIAM Review **59**(4), 794–800 (2017) (Cited on pp. 52, 54)

[415] Shitov, Y.: Euclidean distance matrices and separations in communication complexity theory. Discrete & Computational Geometry **61**(3), 653–660 (2019) (Cited on p. 93)

[416] Shitov, Y.: Nonnegative rank depends on the field. Mathematical Programming (2019), `doi.org/ 10.1007/s10107-019-01448-2` (Cited on p. 54)

[417] Shitov, Y.: Sublinear extensions of polygons. arXiv preprint arXiv:1412.0728v2 (2020) (Cited on pp. 82, 92)

[418] Sidiropoulos, N.D., De Lathauwer, L., Fu, X., Huang, K., Papalexakis, E.E., Faloutsos, C.: Tensor decomposition for signal processing and machine learning. IEEE Transactions on Signal Processing **65**(13), 3551–3582 (2017) (Cited on p. xiv)

[419] Siewert, D.J.: Biclique covers and partitions of bipartite graphs and digraphs and related matrix ranks of {0, 1}-matrices. Ph.D. thesis, University of Colorado at Denver (2000) (Cited on p. 71)

[420] Silio, C.B.: An efficient simplex coverability algorithm in E2 with application to stochastic sequential machines. IEEE Transactions on Computers **28**(2), 109–120 (1979) (Cited on pp. xxi, 13, 44, 45, 46, 49)

[421] Simsekli, U., Cemgil, A.T., Yilmaz, Y.K.: Learning the beta-divergence in Tweedie compound Poisson matrix factorization models. In: Proceedings of the 30th International Conference on Machine Learning, pp. 1409–1417 (2013) (Cited on p. 163)

[422] Singh, A.P., Gordon, G.J.: A unified view of matrix factorization models. In: Joint European Conference on Machine Learning and Knowledge Discovery in Databases, pp. 358–373. Springer (2008) (Cited on p. 3)

[423] Smaragdis, P.: Non-negative matrix factor deconvolution; extraction of multiple sound sources from monophonic inputs. In: International Conference on Independent Component Analysis and Signal Separation, pp. 494–499. Springer (2004) (Cited on p. 177)

[424] Smaragdis, P.: Convolutive speech bases and their application to supervised speech separation. IEEE Transactions on Audio, Speech, and Language Processing **15**(1), 1–12 (2007) (Cited on pp. 11, 177)

[425] Smaragdis, P., Févotte, C., Mysore, G.J., Mohammadiha, N., Hoffman, M.: Static and dynamic source separation using nonnegative factorizations: A unified view. IEEE Signal Processing Magazine **31**(3), 66–75 (2014) (Cited on pp. 126, 314)

[426] Smaragdis, P., Venkataramani, S.: A neural network alternative to non-negative audio models. In: IEEE International Conference on Acoustics, Speech and Signal Processing (ICASSP), pp. 86–90. IEEE (2017) (Cited on p. 193)

[427] Smith, R.B.: Introduction to hyperspectral imaging with TNTmips by micro images (1999), `https://www.microimages.com/documentation/Tutorials/hyprspec.pdf` (Cited on p. 7)

[428] Song, Z., Woodruff, D.P., Zhong, P.: Low rank approximation with entrywise ℓ_1-norm error. In: Proceedings of the 49th Annual ACM SIGACT Symposium on Theory of Computing, pp. 688–701. ACM (2017) (Cited on p. 204)

[429] Sperner, E.: Ein Satz über Untermengen einer endlichen Menge. Mathematische Zeitschrift **27**(1), 544–548 (1928) (Cited on p. 68)

[430] Spielman, D.A., Teng, S.H.: Smoothed analysis of algorithms: Why the simplex algorithm usually takes polynomial time. Journal of the ACM **51**(3), 385–463 (2004) (Cited on p. 282)

[431] Spielman, D.A., Wang, H., Wright, J.: Exact recovery of sparsely-used dictionaries. In: Conference on Learning Theory, pp. 37.1–37.18 (2012) (Cited on p. 122)

[432] Squires, S., Prügel-Bennett, A., Niranjan, M.: Rank selection in nonnegative matrix factorization using minimum description length. Neural Computation **29**(8), 2164–2176 (2017) (Cited on p. 167)

[433] Sra, S., Dhillon, I.S.: Nonnegative matrix approximation: algorithms and applications. Computer Science Department, University of Texas at Austin (2006) (Cited on pp. 4, 5, 163, 314)

[434] Srebro, N., Jaakkola, T.: Weighted low-rank approximations. In: Proceedings of the 20th International Conference on Machine learning (ICML), vol. 3, pp. 720–727 (2003) (Cited on pp. 3, 203)

[435] Stoica, P., Selen, Y.: Model-order selection: a review of information criterion rules. IEEE Signal Processing Magazine **21**(4), 36–47 (2004) (Cited on pp. 167, 168)

[436] Stojanović, O., Kuhlmann, L., Pipa, G.: Predicting epileptic seizures using nonnegative matrix factorization. PloS One **15**(2), e0228,025 (2020) (Cited on p. 314)

[437] Strang, G.: Linear Algebra and Its Application, 3rd ed. MIT Press (1988) (Cited on p. 139)

[438] Sun, D.L., Fevotte, C.: Alternating direction method of multipliers for non-negative matrix factorization with the beta-divergence. In: IEEE International Conference on Acoustics, Speech and Signal Processing (ICASSP), pp. 6201–6205. IEEE (2014) (Cited on p. 291)

[439] Sun, K., Geng, X., Wang, P., Zhao, Y.: A fast endmember extraction algorithm based on gram determinant. IEEE Geoscience and Remote Sensing Letters **11**(6), 1124–1128 (2013) (Cited on pp. 227, 229)

[440] Sun, P., Freund, R.: Computation of minimum-volume covering ellipsoids. Operations Research **52**(5), 690–706 (2004) (Cited on p. 240)

[441] Sun, Y., Babu, P., Palomar, D.P.: Majorization-minimization algorithms in signal processing, communications, and machine learning. IEEE Transactions on Signal Processing **65**(3), 794–816 (2016) (Cited on p. 265)

[442] Sun, Y., Xin, J.: Underdetermined sparse blind source separation of nonnegative and partially overlapped data. SIAM Journal on Scientific Computing **33**(4), 2063–2094 (2011) (Cited on p. 135)

[443] Suppes, P., Zanotti, M.: When are probabilistic explanations possible? Synthese **48**(2), 191–199 (1981) (Cited on p. 14)

[444] Takahashi, N., Hibi, R.: Global convergence of modified multiplicative updates for nonnegative matrix factorization. Computational Optimization and Applications **57**(2), 417–440 (2014) (Cited on p. 277)

[445] Tan, V.Y.F., Févotte, C.: Automatic relevance determination in nonnegative matrix factorization with the β-divergence. IEEE Transactions on Pattern Analysis and Machine Intelligence **35**(7), 1592–1605 (2013) (Cited on p. 167)

[446] Teboulle, M., Vaisbourd, Y.: Novel proximal gradient methods for nonnegative matrix factorization with sparsity constraints. SIAM Journal on Imaging Sciences **13**(1), 381–421 (2020) (Cited on p. 304)

[447] Tepper, M., Sapiro, G.: Nonnegative matrix underapproximation for robust multiple model fitting. In: Proceedings of the IEEE Conference on Computer Vision and Pattern Recognition, pp. 2059–2067 (2017) (Cited on pp. 177, 302)

[448] Theis, F.J., Stadlthanner, K., Tanaka, T.: First results on uniqueness of sparse non-negative matrix factorization. In: 13th European Signal Processing Conference (EUSIPCO). IEEE (2005) (Cited on p. 150)

[449] Thom, M., Rapp, M., Palm, G.: Efficient dictionary learning with sparseness-enforcing projections. International Journal of Computer Vision **114**(2–3), 168–194 (2015) (Cited on pp. 175, 176)

[450] Thomas, L.B.: Solution to Problem 73-14*: Rank factorization of nonnegative matrices (A. Berman). SIAM Review **16**(3), 393–394 (1974) (Cited on pp. 14, 24, 26, 27, 28, 43, 56, 63, 64, 77, 105)

[451] Thomas, R.R.: Spectrahedral lifts of convex sets. In: Proceedings of the International Congress of Mathematicians, pp. 3819–3842 (2018) (Cited on p. 93)

[452] Thurau, C., Kersting, K., Wahabzada, M., Bauckhage, C.: Descriptive matrix factorization for sustainability adopting the principle of opposites. Data Mining and Knowledge Discovery **24**(2), 325–354 (2012) (Cited on p. 168)

[453] Tikhonov, A.N.: Regularization of incorrectly posed problems. In: Soviet Mathematics Doklady, vol. 4, pp. 1624–1627 (1963) (Cited on p. 168)

[454] Tiwary, H.R., Weltge, S., Zenklusen, R.: Extension complexities of Cartesian products involving a pyramid, Information Processing Letters, **128**, 11–13 (2017) (Cited on p. 94)

[455] Tosic, I., Frossard, P.: Dictionary learning. IEEE Signal Processing Magazine **28**(2), 27–38 (2011) (Cited on p. 3)

[456] Trefethen, L.N., Bau III, D.: Numerical Linear Algebra. SIAM (1997) (Cited on pp. 196, 283)

[457] Udell, M., Horn, C., Zadeh, R., Boyd, S.: Generalized low rank models. Foundations and Trends in Machine Learning **9**(1), 1–118 (2016) (Cited on p. 3)

[458] Udell, M., Townsend, A.: Why are big data matrices approximately low rank? SIAM Journal on Mathematics of Data Science **1**(1), 144–160 (2019) (Cited on p. 3)

[459] Vandaele, A., Gillis, N., Glineur, F.: On the linear extension complexity of regular n-gons. Linear Algebra and Its Applications **521**, 217–239 (2017) (Cited on pp. xxi, 70, 72, 82, 87, 88, 90, 91)

[460] Vandaele, A., Gillis, N., Glineur, F., Tuyttens, D.: Heuristics for exact nonnegative matrix factorization. Journal of Global Optimization **65**(2), 369–400 (2016) (Cited on pp. 53, 88, 96, 132, 299)

[461] Vandaele, A., Gillis, N., Lei, Q., Zhong, K., Dhillon, I.: Efficient and non-convex coordinate descent for symmetric nonnegative matrix factorization. IEEE Transactions on Signal Processing **64**(21), 5571–5584 (2016) (Cited on p. 180)

[462] Vanderbeck, F., Wolsey, L.A.: Reformulation and decomposition of integer programs. In: 50 Years of Integer Programming 1958-2008, pp. 431–502. Springer (2010) (Cited on p. 84)

[463] Varol, E., Nejatbakhsh, A., McGrory, C.: Temporal Wasserstein non-negative matrix factorization for non-rigid motion segmentation and spatiotemporal deconvolution. arXiv preprint arXiv:1912.03463 (2019) (Cited on p. 185)

[464] Vavasis, S.: Nonlinear Optimization: Complexity Issues. Oxford University Press (1991) (Cited on p. 196)

[465] Vavasis, S.A.: On the complexity of nonnegative matrix factorization. SIAM Journal on Optimization **20**(3), 1364–1377 (2009) (Cited on pp. 14, 36, 51, 52, 104, 195, 304)

[466] Velegar, M., Erichson, N.B., Keller, C.A., Kutz, J.N.: Scalable diagnostics for global atmospheric chemistry using Ristretto library (version 1.0). Geoscientific Model Development **12**(4), 1525–1539 (2019) (Cited on p. 313)

[467] Vesselinov, V.V., Mudunuru, M.K., Ahmmed, B., Karra, S., Middleton, R.S.: Discovering signatures of hidden geothermal resources based on unsupervised learning. In: 45th Workshop on Geothermal Reservoir Engineering, Stanford University (2020) (Cited on p. 313)

[468] Vial, P.-H., Magron, P., Oberlin, T., Févotte, C.: Phase retrieval with Bregman divergences and application to audio signal recovery. arXiv:2010.00392 (2020) (Cited on p. 11)

[469] Vincent, E., Bertin, N., Badeau, R.: Adaptive harmonic spectral decomposition for multiple pitch estimation. IEEE Transactions on Audio, Speech, and Language Processing **18**(3), 528–537 (2010) (Cited on p. 164)

[470] Vincent, E., Bertin, N., Gribonval, R., Bimbot, F.: From blind to guided audio source separation: how models and side information can improve the separation of sound. IEEE Signal Processing Magazine **31**(3), 107–115 (2014) (Cited on p. 159)

[471] Virtanen, T.: Monaural sound source separation by nonnegative matrix factorization with temporal continuity and sparseness criteria. IEEE Transactions on Audio, Speech, and Language Processing **15**(3), 1066–1074 (2007) (Cited on pp. 11, 150)

[472] Wall, J.: Rank factorizations of positive operators. Linear and Multilinear Algebra **8**(2), 137–144 (1979) (Cited on p. 14)

[473] Wallace, R.M.: Analysis of absorption spectra of multicomponent systems. Journal of Physical Chemistry **64**(7), 899–901 (1960) (Cited on p. 12)

[474] Wang, C.: Finding minimal nested polygons. BIT Numerical Mathematics **31**, 230–236 (1991) (Cited on p. 44)

[475] Wang, H., Huang, H., Ding, C.: Simultaneous clustering of multi-type relational data via symmetric nonnegative matrix tri-factorization. In: Proceedings of the 20th ACM International Conference on Information and Knowledge Management, pp. 279–284. ACM (2011) (Cited on p. 182)

[476] Wang, S., Chang, T.H., Cui, Y., Pang, J.S.: Clustering by orthogonal non-negative matrix factorization: a sequential non-convex penalty approach. In: IEEE International Conference on Acoustics, Speech and Signal Processing (ICASSP), pp. 5576–5580. IEEE (2019) (Cited on p. 136)

[477] Wang, Y., Fu, T., Gao, M., Ding, S.: Performance of orthogonal matching pursuit for multiple measurement vectors with noise. In: 2013 IEEE China Summit and International Conference on Signal and Information Processing, pp. 67–71. IEEE (2013) (Cited on p. 252)

[478] Wang, Y.X., Zhang, Y.J.: Nonnegative matrix factorization: A comprehensive review. IEEE Transactions on Knowledge and Data Engineering **25**(6), 1336–1353 (2013) (Cited on pp. 5, 168, 193, 314)

[479] Wedderburn, J.H.M.: Lectures on Matrices. American Mathematical Society Colloquium Publications 17. American Mathematical Society (1934) (Cited on p. 53)

[480] Weiderer, P., Tomé, A.M., Lang, E.W.: Decomposing temperature time series with non-negative matrix factorization. arXiv preprint arXiv:1904.02217 (2019) (Cited on p. 313)

[481] Wild, S., Curry, J., Dougherty, A.: Improving non-negative matrix factorizations through structured initialization. Pattern Recognition **37**(11), 2217–2232 (2004) (Cited on p. 301)

[482] Winter, M.E.: N-FINDR: An algorithm for fast autonomous spectral end-member determination in hyperspectral data. In: Proceedings of the SPIE Conference on Imaging Spectrometry V, vol. 3753, pp. 266–276. International Society for Optics and Photonics (1999) (Cited on pp. 13, 248)

[483] Wolsey, L.: Using extended formulations in practice. Optima **85**, 7–9 (2011) (Cited on p. 84)

[484] Wu, X., Zhai, G.: Temporal psychovisual modulation: a new paradigm of information display [exploratory DSP]. IEEE Signal Processing Magazine **30**(1), 136–141 (2012) (Cited on p. 314)

[485] Xu, L., Yu, B., Zhang, Y.: An alternating direction and projection algorithm for structure-enforced matrix factorization. Computational Optimization and Applications **68**(2), 333–362 (2017) (Cited on p. 291)

[486] Xu, Y., Yin, W.: A block coordinate descent method for regularized multiconvex optimization with applications to nonnegative tensor factorization and completion. SIAM Journal on Imaging Sciences **6**(3), 1758–1789 (2013) (Cited on p. 293)

[487] Yang, H., Wen, W., Li, H.: Deephoyer: Learning sparser neural network with differentiable scale-invariant sparsity measures. arXiv preprint arXiv:1908.09979 (2019) (Cited on p. 176)

[488] Yang, J., Leskovec, J.: Overlapping community detection at scale: a nonnegative matrix factorization approach. In: Proceedings of the 6th ACM International Conference on Web Search and Data Mining, pp. 587–596. ACM (2013) (Cited on pp. 308, 313)

[489] Yang, Z., Oja, E.: Linear and nonlinear projective nonnegative matrix factorization. IEEE Transactions on Neural Networks **21**, 734–749 (2010) (Cited on p. 136)

[490] Yang, Z., Oja, E.: Linear and nonlinear projective nonnegative matrix factorization. IEEE Transactions on Neural Networks **21**(5), 734–749 (2010) (Cited on pp. 171, 172, 173, 280)

[491] Yang, Z., Oja, E.: Unified development of multiplicative algorithms for linear and quadratic nonnegative matrix factorization. IEEE Transactions on Neural Networks **22**(12), 1878–1891 (2011) (Cited on p. 276)

[492] Yannakakis, M.: Expressing combinatorial optimization problems by linear programs. In: Proceedings of the 20th Symposium on Theory of Computing (STOC '88), pp. 223–228 (1988) (Cited on pp. 15, 85)

[493] Yannakakis, M.: Expressing combinatorial optimization problems by linear programs. Journal of Computer and System Sciences **43**(3), 441–466 (1991) (Cited on pp. 15, 68, 85, 95)

[494] Yannakakis, M.: On extended LP formulations. Optima **85**, 9–10 (2011) (Cited on pp. 15, 66)

[495] Yoo, J., Choi, S.: Orthogonal nonnegative matrix factorization: multiplicative updates on Stiefel manifolds. In: Proceedings of the 9th International Conference on Intelligent Data Engineering and Automated Learning, pp. 140–147. Springer (2008) (Cited on pp. 136, 280)

[496] Yu, J., Zhou, G., Cichocki, A., Xie, S.: Learning the hierarchical parts of objects by deep non-smooth nonnegative matrix factorization. IEEE Access **6**, 58,096–58,105 (2018) (Cited on p. 184)

[497] Yuan, Z., Oja, E.: Projective nonnegative matrix factorization for image compression and feature extraction. In: Scandinavian Conference on Image Analysis, pp. 333–342. Springer (2005) (Cited on p. 171)

[498] Zachary, W.W.: An information flow model for conflict and fission in small groups. Journal of Anthropological Research **33**(4), 452–473 (1977) (Cited on p. 179)

[499] Zarabie, A.K., Das, S.: An l0-norm constrained non-negative matrix factorization algorithm for the simultaneous disaggregation of fixed and shiftable loads. arXiv preprint arXiv:1908.00142 (2019) (Cited on p. 313)

[500] Zdunek, R.: Approximation of feature vectors in nonnegative matrix factorization with gaussian radial basis functions. In: International Conference on Neural Information Processing, pp. 616–623. Springer (2012) (Cited on p. 172)

[501] Zhang, D., Zhou, Z.H., Chen, S.: Non-negative matrix factorization on kernels. In: Pacific Rim International Conference on Artificial Intelligence, pp. 404–412. Springer (2006) (Cited on p. 185)

[502] Zhang, Y.: An alternating direction algorithm for nonnegative matrix factorization. Rice University, Tech. Rep. (2010) (Cited on p. 291)

[503] Zhang, Y., Yeung, D.Y.: Overlapping community detection via bounded nonnegative matrix tri-factorization. In: Proceedings of the 18th ACM SIGKDD International Conference on Knowledge Discovery and Data Mining, pp. 606–614. ACM (2012) (Cited on p. 182)

[504] Zhang, Z., Li, T., Ding, C., Zhang, X.: Binary matrix factorization with applications. In: Seventh IEEE International Conference on Data Mining (ICDM 2007), pp. 391–400. IEEE (2007) (Cited on p. 185)

[505] Zhang, Z., Zha, H., Simon, H.: Low-rank approximations with sparse factors I: Basic algorithms and error analysis. SIAM Journal on Matrix Analysis and Applications **23**(3), 706–727 (2002) (Cited on p. 150)

[506] Zhang, Z., Zhang, Y, Zhang, L., Yan, S.: A Survey on concept factorization: From shallow to deep representation learning. arXiv:2007.15840 (2020) (Cited on p. 172)

[507] Zhao, R., Tan, V.Y.F.: Online nonnegative matrix factorization with outliers. IEEE Transactions on Signal Processing **65**(3), 555–570 (2017) (Cited on p. 185)

[508] Zhao, R., Tan, V.Y.F.: A unified convergence analysis of the multiplicative update algorithm for regularized nonnegative matrix factorization. IEEE Transactions on Signal Processing **66**(1), 129–138 (2018) (Cited on p. 277)

[509] Zhong, S., Ghosh, J.: Generative model-based document clustering: a comparative study. Knowledge and Information Systems **8**(3), 374–384 (2005) (Cited on p. 296)

[510] Zhong, Y., Zhao, L., Zhang, L.: An adaptive differential evolution endmember extraction algorithm for hyperspectral remote sensing imagery. IEEE Geoscience and Remote Sensing Letters **11**(6), 1061–1065 (2013) (Cited on p. 247)

[511] Zhou, G., Cichocki, A., Xie, S.: Fast nonnegative matrix/tensor factorization based on low-rank approximation. IEEE Transactions on Signal Processing **60**(6), 2928–2940 (2012) (Cited on p. 292)

[512] Zhou, G., Cichocki, A., Zhao, Q., Xie, S.: Nonnegative matrix and tensor factorizations: An algorithmic perspective. IEEE Signal Processing Magazine **31**(3), 54–65 (2014) (Cited on p. 314)

[513] Zhou, G., Xie, S., Yang, Z., Yang, J.M., He, Z.: Minimum-volume-constrained nonnegative matrix factorization: Enhanced ability of learning parts. IEEE Transactions on Neural Networks **22**(10), 1626–1637 (2011) (Cited on pp. 139, 143, 145)

[514] Zhu, F.: Hyperspectral unmixing: ground truth labeling, datasets, benchmark performances and survey. arXiv preprint arXiv:1708.05125 (2017) (Cited on p. 148)

[515] Zhu, Z., Li, Q., Tang, G., Wakin, M.B.: Global optimality in low-rank matrix optimization. IEEE Transactions on Signal Processing **66**(13), 3614–3628 (2018) (Cited on p. 197)

[516] Zhuang, L., Lin, C.H., Figueiredo, M.A., Bioucas-Dias, J.M.: Regularization parameter selection in minimum volume hyperspectral unmixing. IEEE Transactions on Geoscience and Remote Sensing **57**(12), 9858–9877 (2019) (Cited on pp. 138, 170)

[517] Ziegler, G.: Lectures on Polytopes. Springer (1995) (Cited on pp. 21, 74)

[518] Žitnik, M., Zupan, B.: NIMFA: A python library for nonnegative matrix factorization. Journal of Machine Learning Research **13**(1), 849–853 (2012) (Cited on p. 305)

Index

The letters "f" and "t" following page numbers indicate figures and tables, respectively.